D0151778

# WORLDS APART:
## A Textbook
## in Planetary Sciences

# WORLDS APART:
## A Textbook
## in Planetary Sciences

**GUY J. CONSOLMAGNO, S.J.**
*Vatican Observatory*
*Tucson, Arizona*

**MARTHA W. SCHAEFER**
*University of Maryland*
*at College Park*

Prentice Hall, Englewood Cliffs, New Jersey 07632

*Library of Congress Cataloging-in-Publication Data*

CONSOLMAGNO, GUY, (date)
Worlds apart: a textbook in planetary sciences / Guy J.
Consolmagno, Martha W. Schaefer.
p.    cm.
Includes bibliographical references and index.
ISBN 0-13-964131-9
1. Planetology.   2. Astronomy.   I. Schaefer, Martha W., (date)
II. Title.
QB601. C66     1994                                        93-44585
523.4—dc20                                                    CIP

Acquisitions editor: *Ray Henderson*
Editorial/production supervision
    and interior design: *Kathleen M. Lafferty*
Proofreader: *Bruce D. Colegrove*
Cover design: *Tweet Graphics*
Manufacturing buyer: *Trudy Pisciotti*

Copyright © 1994 by Guy Consolmagno and Martha W. Schaefer

Prentice-Hall, Inc.
A Paramount Communications Company
Englewood Cliffs, New Jersey 07632

*All rights reserved. No part of this book*
*may be reproduced, in any form or by any means,*
*without permission in writing from the publisher.*

Printed in the United States of America

10   9   8   7   6   5   4   3   2   1

ISBN 0-13-964131-9

Prentice-Hall International (UK) Limited, *London*
Prentice-Hall of Australia Pty. Limited, *Sydney*
Prentice-Hall Canada Inc., *Toronto*
Prentice-Hall Hispanoamericana, S.A., *Mexico*
Prentice-Hall of India Private Limited, *New Delhi*
Prentice-Hall of Japan, Inc., *Tokyo*
Simon & Schuster Asia Pte. Ltd., *Singapore*
Editora Prentice-Hall do Brasil, Ltda., *Rio de Janeiro*

# Contents

# List of
# For the Record Boxes

# List of
# Further Information Boxes

# Preface

What's out there? And how did it get to be the way it is today?

The ancient Greeks and Romans charted the motions of the planets, and the inventions of the Renaissance age—the telescope, the Copernican system, the mathematics of Newton—aided the development of the theory of planetary motions. But only with the first large telescopes of the nineteenth century did *planetary science*, the study of planets as worlds like our own Earth, first become possible, a study that has been carried forward in a sudden rush since the advent of the space age.

But the space age is also the television age. Planetary science has progressed in the full sight of millions of viewers, live, as the first rocks arrived from the Moon and the first pictures came back to reveal the black geysers of Triton. Planetary science has developed a universal appeal. It is the newest realm of exploration and carries with it the romance and excitement that explorers have always shared.

Planetary science is accessible to everyone. You don't need a Ph.D. to be astonished by the rings of Saturn. In astronomy, quasars and black holes may have a mystical attraction, but planets—places you can stand on and walk around on and breathe (or not breathe) the air—have real gut appeal. Planets are places where human beings can have adventures.

*Worlds Apart* is a textbook of planetary sciences. But more than a compendium of facts, it is designed to be taught. It was born in a classroom at the Massachusetts Institute of Technology more than 10 years ago where we were the instructor and the teaching assistant of a course in planetary science for undergraduate nonmajors. The book has gone through four major rewrites, based on our experience using earlier versions in classrooms at Lafayette College (Easton, Pennsylvania) and Loyola College (Baltimore). Others have used earlier versions in the classroom as well, including Bill Hubbard at the University of Arizona and Dave Hogenboom at Lafayette College; we thank them for their wise suggestions.

Designed to be used by undergraduate science majors, *Worlds Apart* can be covered in one semester, with judicious choices based on the backgrounds and interests of the students. We assume that the students are not afraid of equations, but the use of integral calculus and most derivations are confined to separate boxes marked Further Information, usually found toward the end of each chapter. We assume that the students are familiar with freshman physics and chemistry, but other boxes, labelled For the Record, help remind them of details they may have forgotten. Many students may be geology or astronomy majors; however, we assume no knowledge of geology or astronomy.

It is fashionable for planetary scientists to divide their field into topic areas such as tectonics or atmospheres, but for most beginners it makes more sense to look at the solar system planet by planet. Thus most chapters begin with a purely descriptive section about a planet. However, along with each planet are two sections on topics that are related to the planet in question. For example, atmospheric chemistry is discussed with Venus and the generation of magnetic fields is discussed with Jupiter. More difficult topics are covered later in the book, and within each chapter the easier section comes first. Individual sections can be skipped without a serious loss of continuity.

It is our experience that students are more interested in planets than equations; thus we wait until the beginning of the outer solar system to discuss the details of celestial mechanics. We forget neither the largest nor the most interesting members of our solar system: both the Sun and Earth are given equal treatment with the other planets. Nor are particles and fields slighted, but they are placed in the later chapters.

In addition, there is a deeper agenda to this book. Very few students, even science majors, become planetary scientists. The point of a planetary science course, then, should not be to fill someone's head with many facts about the moons of Saturn. Instead, this course is an opportunity to use an interesting subject to illustrate by example how science is done.

Constantly in this book we go beyond the issue of merely what do we know. Instead, we ask: What motivates our questions? How do we go about finding the answers? The problems at the end of each chapter include those that ask the questions behind the questions, probing the hidden assumptions and, dare we mention, the philosophical implications of the science that asked those questions. These are issues about which all science students need to learn. Indeed, one effective way to teach this course is to assign the chapters for reading and then, instead of lecturing, spend class time discussing some of the problems from the end of the chapter.

We have tried always to remember that planetary science is an activity of human beings, by human beings, for human beings that started long before the space age. We've tried to keep history alive both with separate historical sections and by identifying when possible the names of notable past practitioners. On the other hand, to avoid showing favoritism in matters of current priority, we mention by name only two living scientists, both for work they did before 1960.

In keeping with the gender of the authors, it was originally decided to scrupulously mix equal numbers of male and female pronouns in the writing of this text. However, we changed our minds. Thus all gender-specific pronouns in this text are female.

In addition to the many people mentioned in the figure credits at the back of this book, we would like to acknowledge the assistance of Robert Clayton, Daniel Davis, William Higgins, Randy Jones, Walter Kiefer, David Slavsky, and Clifford Stoll. We also wish to thank the following reviewers for their useful comments: Joseph Cain, Florida State University; Bruce Hapke, University of Pittsburgh; Anne Hofmeister, University of California at Davis; Robert Nowak, Purdue University; and Joe Veverka, Cornell University.

Finally, we would especially like to acknowledge five people here. First, our thanks to Ray Henderson, our editor at Prentice Hall, for seeing what this book could be and for supporting it and its authors. We also thank Kathleen Lafferty, our production editor, who made our manuscript into a real book. We thank Pat Earl, S.J., and the Maryland Province of the Society of Jesus, for their generous support during the latter part of this book's production. Our single most valuable advisor and merciless critic has been Brad Schaefer, who also kept us sane with movies, chocolate, and tales from the world of gamma-ray astronomy. Finally, we thank John Lewis, who taught us equilibrium chemistry, advanced Frisbee, and fearless irreverence along with the fun of planetary science. We hope that this love and sense of fun has come through in this book.

*Guy J. Consolmagno, S.J.*
*Martha W. Schaefer*

# WORLDS APART:
## A Textbook
## in Planetary Sciences

**FIGURE 1.1** The solar eclipse of July 11, 1991.

CHAPTER **1**

# A Place Near the Sun

## 1.1 THE PUZZLE OF THE PLANETS

*In the novel* A Study in Scarlet *by Sir Arthur Conan Doyle, Dr. Watson says of Sherlock Holmes, soon after their first meeting,*

> ... I found incidentally that he was ignorant of the Copernican theory and of the composition of the Solar System. That any civilized human being in this nineteenth century should not be aware that the Earth travelled round the Sun appeared to me to be such an extraordinary fact that I could hardly realize it.
>
> "You appear to be astonished," Holmes said, smiling at my expression of surprise. "Now that I do know it I shall do my best to forget it..."
>
> "But the Solar System!"
>
> "What the deuce is it to me?" he interrupted impatiently. "You say that we go round the Sun. If we went round the Moon it would not make a pennyworth of difference to me or to my work."

*Now really, could Holmes not be aware of the motion of Earth around the Sun? In fact, he was clearly being less than forthcoming with Watson here; once the two friends got to know each other better (in the case of "The Greek In-terpreter"), Holmes could casually chat with Watson about "the causes of the change in the obliquity of the ecliptic," which is a rather detailed question of celestial mechanics.*

*The fact is, the mystery of the solar system and our place in it is perhaps the single greatest puzzle that has ever tantalized the reason of humankind. And it's just the sort of logic problem that would have completely delighted Sherlock Holmes. Unfortunately for Holmes, however, by the time he was born this most wonderful and elaborate puzzle had already been solved. He'd missed out on the great game. That must have been frustrating for him. The only challenge left to him was the mundane world of crime.*

Where do we start this game?

We start in the night. The game begins up above us, night after night, with those small bright spots in the utter black dome, unreachably far away; those brilliant gems we call stars. With time and with repeated viewings, various groups of stars become familiar. You may eventually notice that the same pattern of stars rises a little earlier every night, so that as the seasons change you begin to see different stars at dusk. But after a year has gone by the same stars repeat their same pattern of risings and settings.

In a world young with humans, a world without air pollution or city street lights, the stars are brilliant. They can also be your only source of entertainment. You can make games with the patterns you see, playing connect-the-dots to make the pictures that we now call **constellations**. (The Germans are much more direct with their name; they say *Sternbild*, or "star-picture.") And, year after year, the stars keep these same patterns and these same pictures; some are prominent in the winter and others visible in the summer. The stars never move among themselves.

Except there's the Sun and the Moon, although they are too big to be called stars. They seem to move along a certain restricted path among the fixed stars, always passing through the same 12 constellations, groupings that, to the ancient Greeks, sort of looked like animals or people. This *zoo* they called the **zodiac**. But after a while you might also see an occasional starlike light moving through the zodiac constellations.

These lights slowly wander among the stars in a way that seems to defy any simple explanation. You can actually see five of these wandering lights altogether. Two are found always near the horizon: a beautiful bright one, which the Romans named for their goddess of beauty, and a quickly moving dimmer one, which they named for their fleet-footed messenger god. A stately moving bright wanderer was named for the king of their gods. A slowly moving pale yellow one represented that king's ancient father. And a red one, prone to brighten and dim in an ominous fashion, they named for their god of war.

Venus, Mercury, Jupiter, Saturn, and Mars are all fine names. But what in the world were these *wanderers*, in Latin, *planetes*? To answer this question, we need a tool, and the first tool we use, invented by the Greek geometers, is the triangle.

The shape of a triangle is fixed by the three angles that make up the triangle. Once you measure each angle in a triangle, you can draw another triangle that looks just like the first triangle. Indeed, because in plane geometry the angles inside a triangle always add up to 180°, all you need to know are any two of the angles to determine the shape completely. You can always find the third angle by adding up the measures of the other two angles and subtracting that sum from 180°.

The *shape* of the triangle does not determine the *size* of the triangle, however. Consider as an example the classic 3-4-5 triangle. This is a right triangle whose sides have the length of 3 units, 4 units, and 5 units. It will always be a right triangle; it will always have the same shape. The angles at the corners will always be 90°, 53.13°, and 36.87°. But the sides can be any measure of length; the triangle could have sides that were 3, 4, and 5 centimeters (cm) long or 3, 4, and 5 kilometers (km) long.

Two identical triangles have both identically sized angles and sides. However, triangles where only the angles are the same, but where the lengths of their sides are different, are no longer identical; at best they can be called **similar**

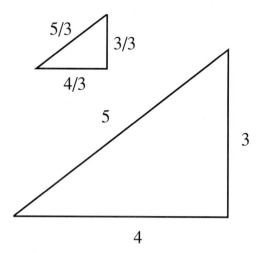

**FIGURE 1.2** Similar triangles have the same shapes because the angles inside are the same, even though the sides may have different lengths.

**triangles** (Figure 1.2). They'll have the same shape, but different sizes.

But similar triangles are just the tool we need. The important point about similar triangles is that, even though the sides may vary from inches to light years, the proportions always stay the same. The short side of a 3-4-5 triangle is always 60% of the length of the longest side, regardless of whether we're measuring in miles or millimeters. It is this property of similar triangles that makes them so useful in measuring long distances.

Take a simple example. Say that you were standing between two trees and wanted to find the distance to a third, far-off tree (see Figure 1.3). The three trees mark the points of a triangle. If you stand at the first tree, you can measure the angle when you look from the second tree to the third. Then you can walk over to the second tree and measure another angle inside the triangle, from the first to the third tree. With these two angles, you have completely determined the shape of the triangle. (Remember, you can calculate the third angle simply by adding your two angles together and subtracting this sum from 180°.) You can draw a triangle with those angles on a piece of paper that is the same shape as the triangle between the trees.

Now you can look at the triangle you drew, and with a ruler you can measure the lengths of each of the sides. Maybe you find that the short side of the triangle—the side equivalent to the line between tree 1 and tree 2—is 1 cm, and one long side—going from the point representing the first tree off to that representing the distant third tree—is 25 cm. Then you know that the distant tree must be 25 times farther away from tree 1 than the distance from tree 1 to tree 2. Pace off the distance between the two nearby trees; say it is 5 paces. If you multiply that by 25, you will have

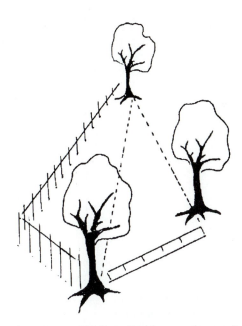

**FIGURE 1.3** Similar triangles can be used to measure distances. Just measure the angles between the trees and the distance between any two of the trees. Then draw a similar triangle on a piece of paper. You can measure on the map how far away the third tree lies from the other two and scale it up.

the distance to the far tree: 125 paces. There is no need to actually walk to the distant tree.

The laws of trigonometry, with their sines and cosines, are just a mathematical way of bypassing the map-and-ruler stage of this process. The basic idea stays the same. To measure how far away a distant object is, you must determine its position from two different vantage points, physically measure the distance between those points, and then draw up the appropriate similar triangle to find how many times farther away the distant object is, compared to the distance between the two vantage points.

Now, what about the planets? How can we apply the power of triangles to figuring out what they are and how far away they lie?

Careful observations of the positions of two planets, Mercury and Venus, show an interesting pattern. Recall that these two planets are never very far away from the Sun in the sky. They appear close to the horizon just after sunset or just before dawn. Venus is often called the **evening star** or the **morning star** for this reason. By measuring its position over a period of time, in fact, one can see that Venus never gets more than 47° away from the Sun.

To make sense out of this observation, we make use of a **model**; not a physical model, like a model airplane, but a model in the sense of a conceptual framework. We start with the idea that Venus and Mercury are two bodies travelling in circles around the Sun. (As we'll see later, their paths

are not exactly circles, but it's not a bad approximation for now.) Then the distance of each orbit, compared to the Earth–Sun distance, can be found with similar triangles.

Look at Figure 1.4. The planets are farthest from the Sun when there's a right angle between the planet–Earth line and the planet–Sun line; this gives one angle for our triangle, 90° in each case. The observations show that there's a 47° angle for Venus and a 28° angle for Mercury when they are at their highest on the horizon at sunset. So we have our two angles. From this, we can work out the relative positions of these planets. Just measure off the lines in the picture or use high-school trigonometry. The sine of the maximum angular distance for Venus, 47°, is 0.7; for Mercury, it's sin 28° = 0.4. Call the distance from Earth to the Sun—which is the hypotenuse of our two triangles—a distance of 1 **astronomical unit** (**AU**). Then Venus is 0.7 AU away from the Sun, or seven tenths the distance from Sun to Earth, and Mercury is 0.4 AU away.

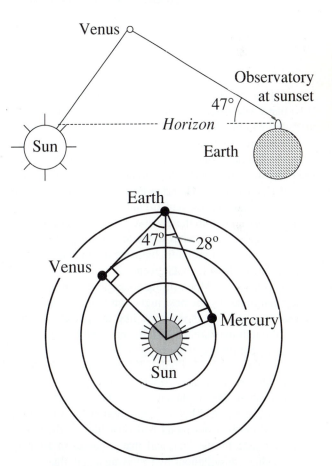

**FIGURE 1.4** Above, an observer on Earth sees Venus at an angle of 47° away from the Sun, at the time when the Sun has just dipped below the horizon. Below, this observation can be tied in with a model of how the planets move to give the locations of Venus (and Mercury) relative to the Earth–Sun distance.

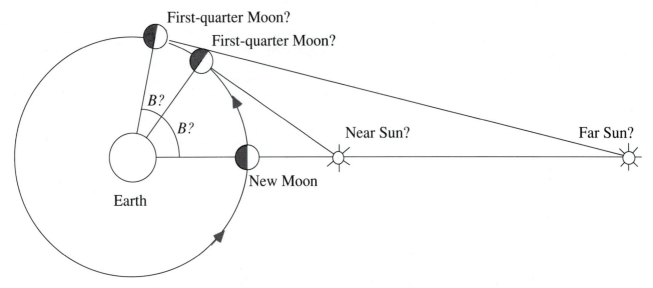

**FIGURE 1.5**   Aristarchus' way of comparing the distances to the Sun and Moon. If the Sun were at the "far Sun" position, it would take longer for the Moon to move from new moon to first quarter than it would if the Sun were closer, at the "near Sun" position.

Our next task is to try to find this Sun–Earth distance. How many kilometers are there in an AU? As it turns out, that's the hardest nut of the puzzle to crack.

Around the year 275 B.C. there lived an astronomer from the Greek island of Samos named Aristarchus, who tried to solve this puzzle. Aristarchus followed the teachings of Heracleides of Heraclea Pontica, who had realized that Mercury and Venus went around the Sun, following arguments very similar to those we've just gone through. This concept inspired Aristarchus to try to find out just how far away the Sun was from Earth, by using a Sun–Earth–Moon triangle. Here's how his logic went.

If you see the Moon sitting among one group of stars on a given night, by the next night you will see that it has moved in relation to those stars; like the planets, it only moves through the **zodiac** constellations, and it takes about 27 days to pass through all 12 of these constellations and return back to its starting point. While it moves, it shows various **phases**: from new Moon, to first quarter, to full, and back again. The bright side of the Moon is always the side facing the Sun. Thus it seems reasonable to assume that the Moon is a big round ball (just like Earth) that is lit on one side by the Sun, and that it passes in front of the 12 zodiac constellations by following a path that moves around Earth.

When is the Moon exactly at first quarter, exactly half full? If the Sun were infinitely far away, then this would occur when the Moon was exactly one quarter of its orbit away from new moon. However, the closer the Sun is, the less time it takes for the Moon to appear half full, as is

illustrated in Figure 1.5. Aristarchus measured the angle between the Sun, Earth, and an exactly half-full Moon to be about 87°. Using trigonometric functions or drawing similar triangles again, this implies that the Sun is 19 times farther from Earth than the Moon is. As you can imagine, this angle—so close to a right angle of 90°—is extremely hard to measure. In fact, Aristarchus didn't get it quite right. The real value for angle $B$ shown in the picture should be 89.85°, or almost exactly 90°. The Sun is in fact 390 times as far away as the Moon.

Still, the point is made that the Sun is at least about 20 times farther from Earth than the Moon is. And because the Moon and the Sun appear to be the same size in the sky (the Moon is just barely big enough to eclipse the Sun), if the Sun is actually about 20 times farther away, then it must be about 20 times as big. Thus we're beginning to get a general feel of how big the things in the sky are and where they're located.

In fact, this general scale made Aristarchus wonder. If both Venus and Mercury orbited the Sun, and the Sun was so far away, wasn't it possible that the other planets—and Earth!—also orbited the Sun? Thus he proposed the heliocentric view of the solar system, more than 1700 years before Copernicus.

But this still didn't give a measure in terms of meters, or kilometers, of just how far off that Sun is. We know the Sun is about 20 times as far away as the Moon, but we don't know how far away the Moon is. To solve that problem, it turns out that we need to solve two more puzzles, discussed Section 1.2.

**FOR THE RECORD…**

# Space, Matter, and Time

The universe can be described in terms of three basic concepts. The first is **length**. How much space does an object take up? Where is it located? The second is **mass**. How many atoms are in that space? How much material is in each atom? The third is **time**. When did that object arrive? How long did it take to get here? How long will it stay here?

In medieval England, length was measured in terms of feet, mass in pounds, and time in hours. These units are still in common use in the United States (but virtually nowhere else) today. In science, along with the rest of the world, the metric system has been adopted because it is easier to work with powers of 10 than to remember "5280 feet to a mile" and other such arbitrary conversion factors.

Strangely enough, though, there are three different metric systems in use. Many engineers, dealing with practical work on a human-sized scale, use units of meters, kilograms, and seconds—the **mks** system. Scientists, dealing with much smaller objects in a laboratory, have tended to refer to centimeters, grams, and seconds—the **cgs** system. In a noble attempt to unite these conflicting systems (which not only involve powers of 10, but other conventions in electricity and magnetism that we needn't worry about here), an international system called **SI** (after the initials of "international system" in French) has been established. It is similar to the mks system, with certain additional definitions and conventions.

Scientists and engineers, however, are creatures of habit, and the other units are still used in most labs and found in many older textbooks. Most modern textbooks, including this one, use the SI system now so that in a generation's time we'll all be speaking the same units.

## SUMMARY

The ancient Greeks and Romans knew that there were planets in the sky, and they named them after their gods and goddesses.

Based on the law of similar triangles, one can figure out that Mercury and Venus must be closer to the Sun than Earth is (0.4 and 0.7 AU).

Aristarchus was able to figure out that the Sun must be much farther from Earth than the Moon is and, consequently, much larger. He proposed that all the planets orbited that Sun.

## STUDY QUESTIONS

1.  Is it possible to see planets in the sky without a telescope?
2.  What geometrical figure do we use to measure the distances to the Sun, the Moon, and the planets?
3.  Venus is never more than 47° above the horizon at sunset. What does this tell us about the orbit of Venus?
4.  Which period of time, on average, is *slightly* longer: the time from new moon to exactly first quarter, or from first quarter to full moon?

5.  When Aristarchus measured the angle from the Sun to the half-full Moon, what quantity was he attempting to determine?

## 1.2 MEASURING 1 AU

We've seen how measuring the positions of Venus and Mercury relative to the Sun allows us to calculate their distance from the Sun in terms of the astronomical unit (AU), the distance from Earth to the Sun. By the end of the sixteenth century, the German astronomer Johannes Kepler was able to deduce the distances to all the other planets, in terms of an AU, using the careful observations of the planets' positions provided by his mentor, Tycho Brahe.

But even the astronomers of the sixteenth century were still left with the problem of determining just how large 1 AU is. How is it possible to measure the distance from Earth to the Sun accurately, without ever leaving Earth?

Some of the best ancient Greek astronomers attempted several tricks to come up with a result. But a satisfactory approach to the answer was only finally worked out 2000 years after Aristarchus and over 100 years after Kepler.

## The Size of Earth

To make any measurement we need a yardstick, something whose size is already known. The largest item in the solar system that we can measure directly on Earth, is Earth itself. Thus the first step in solving the puzzle of planetary distances is to take on a different sort of puzzle: How big is this world we are living on right now?

With the power of the triangle, the philosophers of ancient Greece could answer this puzzle. About 225 B.C., roughly a generation after Aristarchus of Samos, a philosopher named Eratosthenes of Cyrene ran the great library of Alexandria in Egypt, a part of the Greek empire conquered by Alexander the Great. Eratosthenes' contemporaries thought highly of him; their nickname for him was *Beta*, or Number Two, because they felt he ranked only behind the great Aristotle in wisdom and learning.

His interests were certainly as broad as Aristotle's. He made major advances in the study of history; he published mathematics, philosophy, and poetry; he even wrote a history of comedy. Eratosthenes lived in the century following Alexander, when Egypt was ruled as a highly successful trading and commercial state, and so his most practical work was in map making. He developed the ideas of latitude and longitude, and his map of the known world set the standard for the classical era.

One point that puzzled Eratosthenes was that the length of the Sun's shadow changed as one moved from place to place in Egypt. On the longest day of the year a spire in Alexandria cast a shadow even when the Sun was at its highest. But in Syene, 5000 *stadia* (about 800 km, or a little over 500 miles) to the south, he noticed that at noon on this day, no shadows were cast: The Sun stood directly overhead, sunlight penetrating even down a deep well (Figure 1.6).

Eratosthenes went one step farther. He noticed that the length of the shadow in Alexandria was one eighth the height of the tower. Now, by the principle of similar triangles, he might have immediately jumped to a conclusion that was simple, obvious, and wrong. The tower is eight times as long as the shadow? Then obviously, by similar triangles, the distance to the Sun above Syene must be eight times the distance from Syene to Alexandria, or about 6400 km (4000 miles) up in the sky!

Fortunately, Eratosthenes was saved from making this error, because he had another important tool: a model. Just as Aristarchus could use a model of planet orbits to make sense of his observations, Eratosthenes needed a model of Earth to measure how big it really was. The little calculation placing the Sun 6400 km up in the air would make perfect sense, if it really were a straight line from Syene to Alexandria. But what if it were a curved line, curved because we're living on a round Earth instead of a flat one?

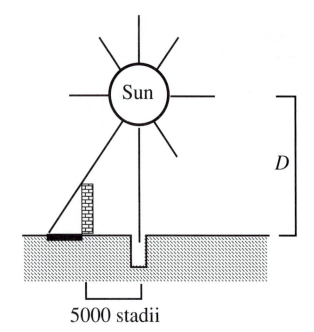

**FIGURE 1.6** A flat Earth view of Eratosthenes' measurement. The differences in the shadows might be used to conclude that the Sun was a relatively short distance, *D*, away from Earth.

Philosophers had long speculated that Earth was a round ball. For one thing, they could see that the Moon and the Sun were round. They could see that the Sun was the source of all the light in the sky, whereas it was only the side of the Moon facing the Sun that was lit up. If the Moon shines at night, it must be because the Sun, which has disappeared below the ground, can still shine its light past Earth and onto the Moon. And every couple of years, a Greek skygazer would be able to see the shadow of Earth itself move across the face of the Moon. We call this a **lunar eclipse**. During these eclipses Earth's shadow is revealed, even to the naked eye, as being visibly round.

Furthermore, from the similarity of the tides in the Atlantic and the Indian oceans, Eratosthenes deduced that these oceans were connected and thus that it should be possible to sail around Africa, a voyage not accomplished until Vasco de Gama did it some 1500 years later. In fact, Eratosthenes even suggested that it ought to be possible to sail westward from Spain and reach India. It was this work, quoted by later authors, that inspired Columbus.

So Eratosthenes knew that Earth was a round ball. How did this affect his calculations of the different shadows? The shadow of the tower at Alexandria signified, he reasoned, merely that Alexandria was situated on a part of Earth curved away from the Sun, while Syene was on a spot directly underneath the Sun. So instead of using this shadow to find the distance to the Sun, he made a different assumption. He assumed that the Sun was very far away, so far that, in effect, it was infinitely far away. This would mean that rays of light from the Sun down to Syene would be running parallel to rays of light from the Sun to Alexandria (Figure 1.7). Then he could use Euclid's geometry and the principle of parallel rays.

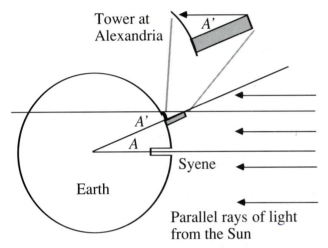

**FIGURE 1.7** Eratosthenes' actual way of determining the circumference of Earth, assuming a round Earth and a very distant Sun. The size of the tower and well are greatly exaggerated.

If you take two parallel lines and cut them with a third line, the third line lies at the same angle to both parallel lines. Look closely and you'll see that you've really made two similar triangles, the familiar triangle rule coming into play again.

The rays from the Sun at Syene and Alexandria make the two parallel lines. You can imagine the line at Syene continuing straight down toward the center of Earth. The tower at Alexandria also lies on a line that starts at the center of Earth and continues straight up through the surface at Alexandria. So the shadow and tower make one little triangle, and the triangle from Syene to Alexandria to the center of Earth makes another triangle. But by the parallel line rule they have the same angles, so they must be similar triangles.

Thus Eratosthenes could reason that the distance between Syene and Alexandria must be one eighth the radius of Earth. Thus the whole Earth was 8 × 5000 or 40,000 stadia from the center to the surface, or about 6400 km. Remembering that the circumference is $2\pi$ times this radius, he inferred that Earth must be a globe about 40,200 km around (in modern units). This is within a few percent of the modern measurement of 40,030 km.

So there we have our first numerical result. By making a key observation of shadows, measuring off a distance of 800 km, and adopting a certain theory about the shape of Earth, we can conclude just how big our planet is. Of course, as we also saw, if you plug even the best of numbers into the wrong theory, you'll come up with a totally wrong picture of the universe, with a small Sun hovering just a few thousand kilometers over a flat Earth.

## Hipparchus' Attempt

Recall that Aristarchus only determined a relative scale for the positions of the Sun and Moon. But how far away are the Sun and Moon really? Are they relatively small balls a few kilometers up in the air, or are they huge worlds much farther away?

To find out, we can take advantage of the way the Sun and the Moon appear during eclipses, as was done by yet another Greek, Hipparchus, around 150 B.C. It's a somewhat elaborate geometrical argument. Here's how it works.

We know that an eclipse of the Moon occurs whenever the Moon, orbiting around Earth, moves into the space exactly opposite Earth from the Sun and thus passes through Earth's shadow. During such an eclipse you can actually see the circular shadow of Earth passing over the Moon (a good indication, as you recall, that Earth is indeed a round ball).

On the other hand, when the Moon passes between Earth and the Sun and an eclipse of the Sun occurs, then the Moon almost exactly covers the Sun. The Moon just barely blocks out the Sun; sometimes you can actually see a thin rim of Sun peeking all around the Moon. This is what tells us that the Moon and the Sun are almost exactly the same size in the sky, a fact we've already used.

The shapes of the shadows are drawn in Figure 1.8. We can use them to start to make some deductions about the sizes of the various bodies, using the laws of similar triangles. For example, the small triangle made by the Moon's shadow on Earth during a solar eclipse is similar to the large triangle made by drawing the edge of this shadow back to the disk of the Sun. Because Aristarchus concluded that the Sun is 19 times farther away than the Moon is, we can conclude that the Sun is 19 Moon-diameters wide. The triangle Earth to bottom of the Moon to top of the Moon is similar in shape to the triangle Earth to bottom of the Sun to top of the Sun. So if one side of the Sun triangle is 19 times bigger, all the other sides must be 19 times bigger, too. (Actually, using the real Earth–Sun distance rather than Aristarchus' figure, the Sun is really 390 times bigger than the Moon, but to see how Hipparchus could arrive at some reasonable figures, we will continue to use the first number in this exercise.)

Now, consider an eclipse of the Moon, when the Moon passes into the shadow of Earth. The shadow is shown in the bottom part of Figure 1.8. Hipparchus was able

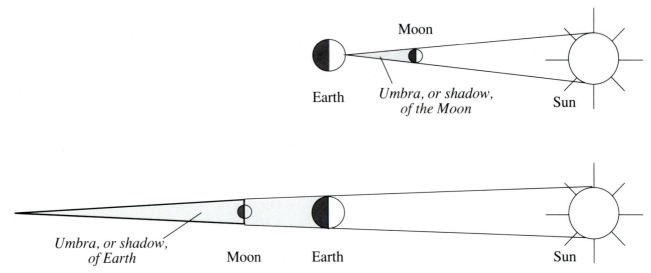

**FIGURE 1.8** Two similar triangles help determine the relative distances to Earth and the Sun. Because the Moon almost exactly covers the Sun during a solar eclipse (upper), the cone of the Moon's shadow makes a small similar triangle inside the larger triangle of the Sun's rays. During a lunar eclipse (lower), the shadow of Earth (shaded), the part of the shadow that reaches past the Moon, and the rays of the Sun (thick shaded lines) are all similar. Knowing the width of Earth and the relative sizes of the Moon and the Sun, we can determine how far apart these three celestial bodies lie.

to look at the curvature of Earth's shadow at the Moon and estimate its size; you could do this today just by measuring the length of time it takes for the Moon to pass through the shadow. He concluded, correctly, that this shadow was about $2\frac{1}{2}$ times as big as the Moon itself.

Note also that this shadow must extend a distance beyond the Moon. Using similar triangles again we can see that we have three embedded similar triangles. One is the shadow of Earth itself, from the endpoint of that shadow up to the diameter of Earth. A second triangle, inside this triangle, goes from the endpoint up to the Moon; here, we know, the shadow is $2\frac{1}{2}$ times as big as the Moon. Finally, we can trace the shadow all the way back to the Sun itself. The edge of the shadow marks the points where the first rays from each edge of the Sun can reach past Earth; thus, the base of this triangle must be the diameter of the Sun itself, which we figured earlier must be 19 times the diameter of the Moon.

This gives us enough information to draw out our triangles and measure up their sides with a ruler. And now that we have a number for the diameter of Earth, we can even use that as a scale to see how big—in kilometers—each of those triangles must be.

Or on the other hand, if you remember your high-school algebra, you could set up these three equations relating the relative sizes of each of our three triangles with three unknown distances: from the Moon to the endpoint

of the shadow (which is actually not particularly interesting by itself), from Earth to that endpoint, and from the Sun to that endpoint. The rules of algebra assure us that three equations can be solved to find three unknowns. Once you know those distances, then the diameters of the Moon and Sun, in terms of Earth's diameter, can also be calculated in a straightforward manner. Unfortunately, neither algebra nor high schools had been invented yet in the time of Hipparchus. He had to work at it, trying different numbers until the answers made consistent sense with themselves.

What is the result of all this calculation? First, you can calculate that the Moon must be about 3800 km in diameter. (The real value is 3450 km.) Here is our first important result: The Moon, at least, is a whole world, about one third the diameter of Earth, and the Sun is at least six times larger than Earth! And how far away from us is the Moon? The result from these calculations comes out to be about 450,000 km. Modern measurements indicate the distance is 400,000 km.

Finally, if the Sun were 19 times farther away than the Moon, its distance would be about 7,600,000 km, but this, like the Sun's diameter, is a factor of 20 smaller than the true value. (Recall, this error came from the difficult and imprecise observation of an "exactly half-full" Moon.) Still, even though this number is not too accurate, it begins to give a flavor of the size of the solar system. Earth and the Moon are roughly the same size, while the Sun is much bigger; and Earth and the Moon are about 30 Earth-diameters apart, while the Sun is 20 times farther away, at least!

Notice a remarkable thing about this whole exercise. We've deduced the sizes of Earth, the Moon, and (with less accuracy) the Sun, and the distances between these bodies and the planets Venus and Mercury, with nothing more elaborate than geometry and some simple observations of positions and angles. We didn't need any instrument more complicated than a protractor. We didn't need a telescope; we didn't need calculus or higher mathematics; we didn't need a computer, radar, spacecraft, Kepler's laws of the planets, or Newton's laws of motion. We wouldn't need anything that hadn't been invented by the time of the ancient Greeks. A high-school student could repeat these observations and calculations today.

## Transits

Although Hipparchus was able to show that the solar system was a big place, full of Earth-sized bodies, his measurement of the astronomical unit left much to be desired. It all hinged on Aristarchus' virtually impossible observation of a half-full Moon. There had to be a better way.

Can we use another set of similar triangles? Recall that by measuring the apparent angular size of the Sun, we can draw a similar triangle that tells us the distance to the Sun in terms of the Sun's diameter. The observed angle is about half a degree. The sine of $0.5°$ is roughly 0.01; this means that the diameter of the Sun is roughly one hundredth the distance between the Sun and Earth. If we had a way of measuring the size of the Sun, we'd be able to use this information to calculate the size of 1 astronomical unit.

That's where transits come in. A **transit** occurs when a planet passes between the Sun and Earth; thus only Venus and Mercury can have transits. However, because the orbits of both planets are slightly inclined to the plane of Earth's orbit, transits do not occur very often; usually when Venus and Earth are in conjunction (that is, lined up on the same side of the Sun), Venus's orbit is such that it passes a little bit above or below the Sun (Figure 1.9). But on rare occasions the orbits are lined up just right so that Venus can be seen to cross the face of the Sun, looking like a small black dot against the Sun's large bright disk. Because of the contrast in size and brightness, this transit could really only be observed carefully after the telescope had been developed, with thick dark filters to cut down the Sun's glare. And you had to know exactly when to look.

By 1629 Kepler had worked out tables for the motions of the planets, based on his rules of planetary motion. From these he could compute exactly when Venus would pass across the disk of the Sun, and he noticed that transits of Venus tend to occur in pairs, 8 years between them, about once a century. There was to be a transit in 1631 and again in 1639.

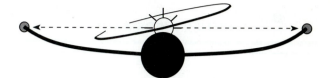

**FIGURE 1.9**   The orbits of Venus and Mercury are slightly inclined to Earth's orbit; thus, only a very few orientations have Venus actually cross in front of the Sun.

Unfortunately, the 1631 transit occurred when it was nighttime in Europe, so the sight of Venus crossing the disk of the Sun could not be observed there. However, in 1639 a 20-year-old Englishman, the Reverend Jeremiah Horrox, in the company of a local draper named William Crabtree, succeeded in observing the second transit. Apparently they were the only ones to observe that particular transit.

Horrox produced several other important astronomical works in his lifetime, however. He corrected, expanded, and defended the work of Kepler; he produced the first careful observations of the ellipticity of the Moon's orbit; and he made important advances in the study of the orbits of Jupiter, Saturn, and comets. His last, incomplete work was on tides. He died when he was 22 years old.

Horrox's transit report suggested that such observations could be useful for finding the solar distance, but he did not elaborate on this notion at the time. In 1663, as a side comment to another problem, the Scottish mathematician James Gregory stated more explicitly than Horrox how transits could be useful in determining the Sun's distance from Earth. However, not until 1716 did the English astronomer Edmund Halley (of comet fame; he was also the person responsible for getting Isaac Newton to publish the *Principia*) finally follow up on this suggestion with a detailed plan of how such transits could be used.

Look at Figure 1.10. If the transits are observed from two different spots on Earth, a distance $B$ apart, each spot sees Venus take a slightly different path across the Sun's disk. Some paths are shorter than others, because the Sun is round. Each observer can measure how long it takes Venus to cross the Sun's disk; the times are proportional to the distance across the disk that each observer sees Venus travel. Assuming that the Sun is round, these distances can be related to the length of a chord across the circle of the Sun. By measuring the difference in these paths from two different observation points, one can use similar triangles (again) to determine the radius of the Sun in kilometers.

The distance between two of these chord lines (the distance $D$ in the Figure 1.10) divided by the distance between the two observers (the distance $B$ in the figure) is the same as the ratio of $D_V$ to $D_E$. But that ratio is known; in the previous section, we saw that $D_V$ is equal to 0.7 AU,

# The Transits of the Eighteenth Century

In 1716, when he calculated how to use transits to find the value of the astronomical unit, Halley knew that the next transit of Venus would occur in 1761 with its mate occurring in 1769. Unfortunately, in 1716 Halley was already an old man. He died in 1742. It was left to others to do the observing.

In the first half of the eighteenth century France was the dominant country of European culture and science. Thus it was a Frenchman, Joseph-Nicholas Delisle, whose correspondence provided him the position of unofficial coordinator for transit observations.

In 1750 Delisle realized that an upcoming transit of Mercury might work much the same way as the Venus transit and, in any case, would be a good "dry run" for astronomers to perfect their transit-observing techniques. His first task was to assemble a worldwide network of observers. One possible observing site was in French Quebec. Accordingly, in late 1752 Delisle asked the Governor-general, the Marquis de la Galissonière, to send instructions on how to observe the transit to a Jesuit missionary/scientist at Cayenne, Father Bonnécamp.

Unfortunately, time was short. Recognizing that mail travelled faster to the British colonies than the French colonies the Governor-general sent the directions with a letter to the governor of New York, asking him to forward them on to Quebec. The letter was dutifully sent on to Quebec, but not before James Alexander had made a translation of the directions. Alexander realized that the Americas could play an important role not only in this matter, but also in the upcoming transit of Venus. He sent a copy of the letter and instructions to Benjamin Franklin, who announced the transit in his *Poor Richard's Almanac* for 1753.

Meanwhile, Delisle corresponded with astronomers in Lisbon, Stockholm, and Naples and prepared to observe the transit himself from Paris. On the day of the Mercury transit, he was joined in Paris by other astronomers, including Chappe d'Auteroche and Guillaume Le Gentil de la Galaisière.

Le Gentil read the report of the Mercury transit observers' results to the French Académie in 1753. The transit was successfully observed, but because Mercury was so small and moved so quickly, the precise time at which it entered and exited the disk of the Sun could not be determined with enough precision to give a very accurate measurement. Astronomers would have to wait for Venus.

For the Venus transit of 1761, the French consulted with British and other European astronomers. Observations were set up for the observatory in Vienna (to be witnessed by the Archduke Joseph himself), throughout Germany and Italy, at Uppsala and other Swedish observatories, and in Spain and Portugal. Delisle sent Chappe d'Auteroche to Siberia and Le Gentil to Pondicherry, a French colony in eastern India. Letters of instruction on how to observe the transit were sent to French Jesuits in Peking. In all, more than 120 observers in 62 different locations observed the transit.

The British, rushing to catch up with the French, decided to send expeditions to St. Helena Island in the south Atlantic and to Bencoolen, an outpost of the East India Company in present-day Indonesia. The Astronomer Royal proposed that his assistant, Charles Mason, should head one of the observing teams; because the astronomers needed to know the precise location of the telescope, a surveyor and amateur astronomer named Jeremiah Dixon was also hired to join him.

The Americas were not to be left out. Encouraged by

so $D_E$ must be 0.3 AU and the ratio is 7/3. Thus if our observers are 1000 km apart on Earth, the distance $D$ on the Sun is 2333 km.

Furthermore, by knowing the lengths of several such chords across the Sun (three chords give you an unambiguous picture) one can work out just where on the circle of the Sun's disk these chords lie. From this, one can determine just what fraction of a diameter of the Sun the distance $D$ represents. This gives us a measure of the Sun's diameter in kilometers.

Halley realized that the next transits of Venus would occur in 1761 and 1769, long after his death, but he left careful instructions for the next generation of astronomers. Difficulties with the observations made the final result less precise than the two-tenths of 1% accuracy that Halley had predicted. But the distance to the Sun was nevertheless measured to greater precision than had ever been contemplated before Halley's technique.

According to the results of the eighteenth-century transit observers, the Sun clearly lay some 100 million miles

Franklin, John Winthrop of Harvard travelled to St. John's, Newfoundland, and observed the transit from a hill above the town.

Amid all these preparations, the French and British faced an additional obstacle: Their two nations were engaged in the Seven Years' War. In an extraordinary gesture of cooperation, the governments of the two nations officially gave passports for free passage through naval lines to the astronomers on expedition.

Unfortunately, such passports were generally ignored by the forces in the war zone. Mason and Dixon's ship to Bencoolen was intercepted by the French Navy before they even got out of the English Channel, and by the time they reembarked, Bencoolen had been captured by the French. They travelled to South Africa instead. (Two years later, in 1763, they would survey the line between the British colonies of Maryland and Pennsylvania, known ever since as the Mason–Dixon line.)

But the worst luck of all befell Le Gentil. By the time he reached India, the French colony had fallen to the British. The captain was afraid to land, for fear of losing his ship, and so Le Gentil spent the day of the transit at sea, unable to observe.

With all these difficulties, the first set of observations were less than satisfactory. Many lessons had been learned by the hardships of the first observers, however. Difficulties in making the observations led to new techniques in timing the duration of the transit. Furthermore, the manufacture of accurate telescopes was spurred on by the excitement of the transit observations. By the time of the second transit in 1769, better and cheaper telescopes were available, and by then, the war was finally over.

The difficulty of travel was clear; Le Gentil, for one, opted to stay in India (once he had landed) and wait the 8 years for the next transit to occur. He travelled to the Philippines and set up an observatory there; but, fearing that the site might be clouded out, he left it to an assistant and returned to India. Chappe arranged with the Spanish to travel to Veracruz on the Gulf of Mexico. The British sent expeditions to Hudson Bay as well as to India and the South Seas. In all, 151 observers from 77 stations saw the second transit.

Chappe's observations in 1769 were superb. Unfortunately, after the transit, he fell ill to a local disease and died in Mexico. Le Gentil's assistant in Manila also reported good results.

Le Gentil's own site, however, was covered with clouds during the transit. After 10 years away from home and having failed to observe either transit, Le Gentil returned to France only to discover that his letters from India had been lost at sea, that he had been declared legally dead, and that his property had been distributed among his heirs!

One interesting development came from the British South Seas mission. The British astronomers, irritated with their difficulties as passengers on a naval ship, arranged to lease an entire ship from the Royal Navy for the next transit. But after leasing them the ship, the Navy balked at having a civilian command it. Instead, they promoted a young lieutenant, James Cook, to become captain of the *Endeavour*. Thus began Cook's legendary voyages to the South Seas, opening up the Pacific from Australia to Alaska. It also began the tradition, both historical and literary, of combining naval vessels and scientific exploration.

from Earth. Another pair of transits in 1874 and 1882 improved that figure to the well-accepted 150,000,000 km (93 million miles) used today. Other more precise techniques for making this measurement are available nowadays; however, if only for the rarity and beauty of the event, the next set of transits of Venus will certainly be observed by interested amateur astronomers around the world. They'll occur on June 8, 2004, and June 6, 2012.

## SUMMARY

The first step in measuring the distance between Earth and the Sun (the AU) is to measure the size of Earth. Eratosthenes did this about 2000 years ago by measuring the different heights of shadows in Alexandria and Syene.

Nearly 100 years later, another Greek, named Hipparchus, used eclipses of the Sun and Moon to deduce the

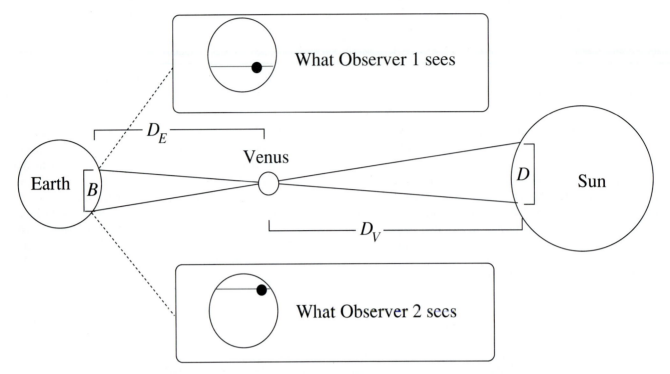

**FIGURE 1.10**  A transit of Venus. Observer 2 sees Venus cross the Sun in a path that is different in length from the path Observer 1 sees.

relative sizes of and distances to the Sun, the Moon, and Earth. The data he was working with were not very accurate, however.

The first accurate measurements of the size and distance to the Sun were made in the eighteenth century, based on the work of several astronomers and mathematicians: Horrox, Gregory, and Halley.

## STUDY QUESTIONS

1. Eratosthenes measured the shadows of a tower in Alexandria and a well in Syene. What did he deduce from his measurements?

2. Aristarchus determined that the Sun was 19 times farther from Earth than the Moon was. How far off was he, according to modern measurements: 3 times, 20 times, or 800 times short of the true distance?

3. What distances are you ultimately trying to measure when you observe a transit of Venus?

4. What was the weak link in Hipparchus' chain of calculations to find the Earth–Sun distance?

5. Where does Venus move during a transit?

## 1.3  THE MYSTERY OF EARTH

The observations of the transits of Venus finally established the immense size of the solar system. The size of the planets themselves could be deduced from this result. Venus lies about a third of an astronomical unit from Earth, and during a transit it subtends 1 **arc-minute**, or minute of arc ($\frac{1}{60}$ of a degree). From this, trigonometry (similar triangles again!) tells us that the diameter of Venus must be $\frac{1}{3}$ of an AU times the sine of 1 minute; that comes to 0.01% of an AU. But if an AU is 150 million km, then the diameter of Venus must be nearly 15,000 km, as big as Earth!

Knowing the value of an astronomical unit, similar calculations could be done for all the planets whose angular size can be measured in a telescope. We find that Mercury and Mars are bigger than the Moon, Venus is about the size of Earth, and the outer planets are all giants compared to our own world. Earth is but one planet among many.

So the other planets may be places like Earth. But what is Earth? What is our own world really like? What is it made of? How did it form?

When the British explorer Captain James Cook carried astronomers to the South Seas to observe the transit of Venus, he was travelling across a planet that was itself still widely unknown. Cook himself would play an important role in mapping out for Europeans the half of the world farthest from Europe. (Of course, that part of the world was well known to the island peoples, who had mapped and sailed the South Pacific for generations. On the other hand, they had never heard of Europe.) But it was only during the latter part of Cook's century, the eighteenth century, that Western science began to address the fundamental question: What is Earth itself?

## Ancient Speculations

As with astronomy, the roots of geology lie with the classical Greeks. Their questions had a very practical bent; living in a mountainous peninsula, much of Athens's wealth came from mining. Knowing how ores or gemstones were made clearly could be useful in determining where to look for more minerals.

But the mountains themselves were mysterious. Why was Greece mountainous, but Egypt flat? What controlled the source of springs and rivers? The Greek colony of Syracuse was near the volcano Mount Etna, on the island of Sicily, and in A.D. 79 the wealthy Roman cities of Pompeii and Herculaneum were destroyed by the eruptions of Mount Vesuvius. What caused volcanoes and earthquakes?

The evidence from astronomy had made it clear to all educated Greeks that the world was round, and we've seen that they estimated well its size. However, the origin of the world was an unsettled question. Philosophers disagreed as to whether Earth was formed out of original chaos (a concept found both in Plato and in ancient Hindu philosophy) or whether the world was eternal and essentially unchanging, as implied by Aristotle's physics.

The followers of Pythagoras, who were mathematicians and philosophers, postulated that in the center of the world was a central fire, the source of volcanoes and earthquakes. The role of this internal fire was important to the theories of Theophrastus (c. 300 B.C.). Arguing against those who said that the continued existence of mountains, in spite of erosion, proved that the world must be relatively young, Theophrastus maintained that the world was eternal. He postulated a balance between mountain building and erosion: The fire inside Earth, attempting to rise to its "natural place" above the air, would tend to push up new mountains.

Perhaps the most mysterious aspect of the mountains around the Mediterranean was the presence of **fossils**. High above sea level, one could find hard rocky shapes that looked in remarkable detail like oysters and the skeletons of fish. How did sea creatures get trapped inside rocks at the tops of mountains, thousands of feet above the ocean?

Among our first record of fossils are reports dating to 540 B.C. by Xenophanes of Colophon. Many explanations were proposed; some suggested that astrological influences were at work, just as "virtues" or other influences of the Sun and stars were held to be responsible for the formation of ores and minerals. Other writers thought that fos-

sils were deposits from underground gases. These theories continued throughout the Middle Ages. As late as the seventeenth century, Ulysses Aldrovandi suggested that fossils were the remains of partly generated creatures, as predicted by the biological theory of spontaneous generation. Perhaps the most far-fetched was the theory, attributed to Aristotle, that fossils were the remains of creatures that swam through Earth just like fish swam through the sea! This odd theory at least could explain how fossilized sea creatures could be found in rocks at the tops of mountains.

However, most ancient writers did recognize that fossils were once-living creatures that had somehow been turned into stone. The writings of the Dominican friar Albert the Great, sometimes called the father of geology, were perhaps the most influential in promoting this view during the Middle Ages.

One powerful theory to explain how these fossils could occur on mountain tops was proposed by Alexander of Aphrodisias about A.D. 200. Commenting on Aristotle's *Meteorologica,* he proposed that the entire Earth was once covered with an ocean. This water was eventually evaporated by the Sun's heat and destroyed. As the water ran off, he proposed, erosion cut valleys into the surface, making the mountains. Many philosophers before him had noted that parts of what was now dry land had once been under water; the historian Herodotus in the fifth century B.C. had described how lower Egypt was once the gulf of a sea. But Alexander's theory of a complete Earth-covering ocean proved to be extremely influential.

The eleventh-century Arab philosopher Avicenna suggested a variation on this theory, proposing that winds inside Earth caused earthquakes, which served to raise up mountains and continents above the level of the all-encompassing ocean. Albert of Saxony (c. 1350) carried this theory further. He had developed a theory of gravity suggesting that all heavy materials seek to locate themselves at the center of the universe, which he believed was the center of Earth. However, he proposed that the heat of the Sun caused part of Earth to expand, shifting it away from its natural position. Water rushing away from this uprising land filled the other side of the world with ocean, eroding away mountains and valleys in the process.

The concept that the far side of Earth was all ocean later influenced Christopher Columbus and other early explorers. With the embarrassing exception of the Americas, it proved to be almost correct, although not for the reasons proposed!

## Neptunism and Plutonism

Abraham Gottlob Werner (1749–1817), a Prussian instructor at the Freiberg Mining Academy at the turn of the nineteenth century, launched the modern approach to geology. He invented what he called **geognosy**, a science that attempted to study and understand the solid Earth as a whole. His theory had two key ideas, one of them brilliant and the other completely wrong.

His brilliant insight was realizing that the **stratigraphic sequence**, the layers of rocks that could be seen where river valleys cut through the sides of mountains, could be generalized into a global scheme. Rock layers of a certain type, which were found in a certain order in Germany, could also be found in in the same sequence in France or England. Subsequent workers, such as George Cuvier in France and Charles Lyell in Scotland, confirmed that similar fossils could be found in similar layers of rock throughout Europe. Different layers of rock contained different fossils, and the sequence of fossil types from layer to layer was indeed maintained in many locations.

Werner's second idea, however, was equally influential but not nearly as successful. To explain the origin of these layers and to explain why fossils could be found in strata on mountaintops, Werner returned to Alexander of Aphrodisias' ancient idea of a primordial ocean that covered all Earth. He proposed that all the rock layers originated either as sediments from this ocean or else formed from chemical reactions that caused minerals to crystallize along the side of mountains, like precipitates form on the sides of a beaker in a lab.

This idea, dubbed **Neptunism**, was bitterly attacked by geologists in Italy. The key issue was the origin of a rock type known as **basalt**. Werner used basalt as a prime example of a chemical precipitate. The Italians, who lived closer than the Germans to active volcanoes, realized that basalt was really frozen lava. The idea that rocks were formed by volcanoes was called **Volcanism**. Werner and his followers countered this idea, insisting that volcanoes were only relatively recent phenomena, caused by the heat of burning underground coal deposits!

The Italian geologists, such as Scipio Breislak, asked the Neptunists about the ultimate fate of their worldwide ocean. If so much water once covered Earth, where did the water go? Werner's answer was that, in some way not understood, it escaped to space. The Neptunists asked the Italians how fossils could occur on mountaintops. The Volcanists replied that, in some way not understood, volcanoes caused the uplift of the land.

The substance that both Neptunists and Volcanists assumed was primordial, the bedrock upon which sediments and precipitates (or volcanic flows of lava) was overlaid, was the large-crystalled rock type known as **granite**. The key advance that broke the Neptunist–Volcanist debate came when James Hutton realized that granite was formed by a subterraneous lava that intruded into rocks, so-called plutonic intrusions, leading to the term **Plutonism**. Not only did Plutonism provide a mechanism for the uplift of mountains, but more fundamentally, it changed the way both sides

approached the problem by pointing out that there was no such thing as a "primordial rock."

Hutton, a Scotsman, was part of the flowering of science in eighteenth-century Edinburgh. Writing in the *Transactions of the Royal Society of Edinburgh* in 1788, Hutton published his landmark paper, "Theory of the Earth; or an investigation of the laws observable in the composition, dissolution, and restoration of land upon the globe." The title of this paper gives a clue to its basic breakthrough. All Earth was subject to dissolution or restoration, he stated, and the forces that caused the formation or destruction of landforms obeyed laws, as basic and unshakable as Newton's laws of physics.

Rather than looking to a past very different from the present, where some primordial granite was covered with a long-lost ocean, Hutton maintained that whatever occurred in the past must be processes that continue and can be seen in operation, even at the present. "The operations of nature are equable and steady," he wrote. Earth did not behave any differently in the past than it behaves today. He concluded, "The result, therefore, of our present enquiry is, that we find no vestige of a beginning—no prospect of an end." This principle became known as **Uniformitarianism**.

These ideas were bitterly attacked at the time. Richard Kirwan, president of the Royal Irish Academy, accused Hutton of atheism and insisted that geology actually supported a version of history consistent with the description of the Flood in the Bible. Jean Andre de Luc, a Swiss emigré to England, insisted that the continents and mountains must have been formed by huge forces in the past. Following this dramatic episode of mountain building, he said, a catastrophic flood 4000 years ago deposited the fossils on these mountains. (He maintained that his evidence for the dating of this flood was geologic, not biblical.) These views became known as **Catastrophism**.

Hutton himself denied that he was an atheist, merely insisting that the authenticity of Genesis could not be supported, or denied, by the scientific data. The Reverend John Fleming, speaking in Hutton's defense, pointed out that the Catastrophists' version of a violent flood leading to mass extinctions was, in any event, inconsistent with many details in the story of Noah's ark. Cuvier's work with fossils indicated that there were many episodes of extinctions, not just one flood, and that some of these extinctions were local, not worldwide.

The final support for Hutton came from another Scotsman, Charles Lyell. Trained as a barrister, he was an excellent writer and debater. In 1830, based on his studies of fossils across Europe, he published his *Principles of Geology*. He argued that mountains were built slowly, like the pyramids were built, a brick at a time. Erosion was balanced by deposition, he argued; the world was in a steady state.

## Comparative Planetology

The Uniformitarian view dominated geology for nearly 200 years. It has slowly, gradually altered over the years; for instance, we know now that Earth has a finite, if very old, age. Plate tectonics has shown, in a way unimaginable to Hutton or Werner, how mountains can be built. But the key idea, that the processes that shaped Earth in the past are processes that must still occur today, has held firm.

Yet, as we will see, our study of the planets has shaken many of these foundations of classical geology. Catastrophes *do* occur. Giant impacts have completely broken apart icy worlds orbiting Saturn; they may have caused the mass extinctions once attributed to the biblical Flood. Primordial oceans of water have been proposed for both Venus and Mars, based on solid evidence; such oceans may be the best way to explain the isotopic composition of their atmospheres or the landforms seen on their surfaces. Where did their water go? Echoing Werner, one answer proposed is that it escaped to space.

A lesson to be learned from all this is that in planetary science, to quote the wizard Merlin (in T. H. White's classic fairytale, *The Once and Future King*), "Everything not forbidden is compulsory." If a process is possible, then at sometime or somewhere in the solar system, it probably happened.

Another lesson is that when we attempt to understand new geologic forms on other planets, we should take a warning from the mistakes of our past. Germans living near the ocean assumed that all the world was once covered with ocean; Italians living near volcanoes postulated that all rocks came from Volcanism. Even Hutton's Plutonism was inspired by an intrusive **sill** (a long, narrow intrusion of lava) he explored in Scotland. But just because we come from Earth, we must be careful not to think that the geology of our local part of the solar system can explain everything we see on all the other planets. There is no reason to doubt that each of these worlds will have as many surprises, and as much variety, as our own Earth has shown us.

But perhaps the most important lesson to be learned is the principle of **comparative planetology**. We only really began to understand Earth when we were able to travel and compare all the different types of geology available, to see what fossil strata were common everywhere and which were unique, to find both chains of mid-ocean volcanoes and deep ocean trenches, to find both plains of sediments and rough hills of ancient rock. Likewise, it is only through comparing one planet with another that we will be able to understand a whole system of planets and how any one planet (especially our home, Earth) has evolved in the past and is likely to behave in the future.

## SUMMARY

The past 400 years, the time when telescopes and physics, chemistry and geology have progressed enough for us to understand the nature of these worlds, has also been an age of exploration when the nature of planet Earth has finally begun to be grasped. Not until the twentieth century did explorers make it to the extremities of this planet and the final "unexploreds" could be wiped off the maps of Earth. And only now can this exploration be continued to the other worlds around us.

An important lesson we've learned about the planets in this chapter is the enormous distances that separate them from us. We've only sent a handful of spacecraft out to visit them, and with few exceptions these spacecraft have only been able to give us a couple days' look at each planet. How do we interpret the results of these spacecraft? What can we learn about the planets without leaving the surface of Earth? How do the Earth-based and the spacecraft observations complement each other? What do we want to find out about these worlds? And how do we do it?

The following chapters of this book start by describing some of the features we've discovered on each member of the solar system, the things we've learned in order to wipe away the "unexplored" parts on our maps. We describe in these chapters some of the scientific tools we use to break open the history of each planet, and we describe the ways we've learned how each planet has evolved—and continues to evolve—with time.

These are the things that make each planet unique, special, and from all other planets, a world apart.

## STUDY QUESTIONS

1.  The section lists two key ideas of Abraham Werner. Name them.
2.  What was observed in mountain rocks that led people to think that the world was once completely covered by an ocean?
3.  What do we call the series of rock layers bearing fossils from a sequence of narrowly defined time periods?
4.  Name one argument in favor of Neptunism. Name one argument against it.
5.  Name one way in which the Uniformitarian principle outlined by Hutton has been changed in the light of new knowledge.

## 1.4  PROBLEMS

1. Choose a local street, or a long hallway on campus, whose length can be measured and which goes past familiar landmarks. Say that the Sun was positioned at one end of this street and that Pluto is at the other. Figure out where, on the street or the hallway, the other planets would lie. How big would the Sun and Earth be, to scale? (You can find the size and position data in the appendix.)

2. Was Sherlock Holmes right? Has the idea that the Sun goes around Earth, and not vice versa, made any real difference in the daily lives of ordinary people? How do people's views of the universe and their culture reflect each other?

3. As we saw in the text, the argument used by Eratosthenes to measure the radius of Earth assumes that the Sun is very far away. But maybe Earth really is flat, and the Sun is close enough to cast different shadows on different parts of Earth (see Figure 1.6). What extra measurements could you make to test this idea, to differentiate between a round Earth with a distant Sun versus a flat Earth with a nearby Sun?

4. In 1766 Johann Titius noted a mathematical regularity in the positions of the planets; in an astronomy book published in 1772 Johann Bode restated this regularity, known ever since as **Bode's law**. According to this rule, the positions of the planets in AU are found in the series $4/10, 7/10, 10/10, \ldots, (4 + 3 \times 2^n)/10$, where $n$ is an integer. Compare this law with the positions of the planets given in the appendix. Most planetary scientists today believe that this correlation is merely coincidence; state your opinion, and give your reasons.

5. *Naked Eye Astronomy.* As the ancients demonstrated, an enormous amount of insight into the universe can be derived with good mathematics and the naked eye. Where calculations are called for in the following problems, work them out only to the nearest round number. Follow the ancient Babylonians who assumed a round 360 days in a year, the origin of our standard $360°$ in a circle. Feel free, however, to go one up on the Babylonians by using a calculator if you wish.
   (a) Which is larger, the Sun (or the full Moon) or a dime held at arm's length?
   (b) The famous harvest moon illusion: The Sun and Moon appear to be much larger than normal when they are near the horizon, because our eye judges sizes differently when it has familiar objects near them for comparison. Or is it an illusion after all? Devise an experiment to convince yourself that the sizes of the Sun and Moon do not really change when they are near the horizon.
   (c) How long does it take for the Sun to set, from the moment one edge of the Sun touches the horizon until it is completely gone? Make a rough estimate, from knowing the size of the Sun and the rate at which Earth turns. Does the actual length of sunset vary with latitude? What other effects might your simple calculation neglect?
   (d) The Sun goes around Earth, noted the Babylonians, once every 24 hours. But it takes only 23 hours and 56 minutes for the stars to go completely around Earth and return exactly to their original starting positions. Of course, we now know that Earth goes around the Sun, not vice versa. How does this explain the difference between the **solar day** (24 hr) and the **sidereal day** (23 hr 56 min)? (Hint: This 4-min difference represents what fraction of a solar day?)

6. Research the life of Captain James Cook. What parallels can you see between his adventures and that of the fictional Captain James Kirk of the starship *Enterprise*? What other literary figures may have been based on Captain Cook?

7. What is the philosophical difference between Plutonism and Volcanism?

8. Compare the theories of the eighteenth- and nineteenth-century geologists mentioned in the text with those of the ancient Greeks. How were they similar? In what important way were they different?

9. Astrophysicists assume that the laws of physics in an Earth-based laboratory are the same as the laws of physics in a distant galaxy. Is this a reasonable assumption? What principle of geology does this assumption echo?

## 1.5  FOR FURTHER READING

A classic tome on ancient astronomy, an exhaustive study of Greek sources, is *Aristarchus of Samos* by Sir Thomas Heath (Oxford: Clarendon Press, 1913). A shorter and more modern summary of the techniques of Greek astronomy can be found in Michael J. Crowe, *Theories of the World from Antiquity to the Copernican Revolution* (New York: Dover, 1990).

A fascinating and very readable history of the transits of Venus is by Harry Woolf, *The Transits of Venus: A Study of Eighteenth Century Science* (Princeton, N.J.: Princeton University Press, 1959).

Another classic work, this one on the history of geology, is Frank Dawson Adams, *The Birth and Development of the Geological Sciences*. It first appeared in 1938, but it is available in a Dover edition (New York: Dover, 1954). A lively modern treatment of the people and issues throughout the history of geology, including some current controversies, is A. Hallam, *Great Geological Controversies* (Oxford: Oxford University Press, 1989).

A standard work on medieval science is A. C. Crombie, *Augustine to Galileo (Volume I: Science in the Middle Ages; Volume II: Science in the Later Middle Ages and Early Modern Times)* (London: Heinemann Educational Books, 1970).

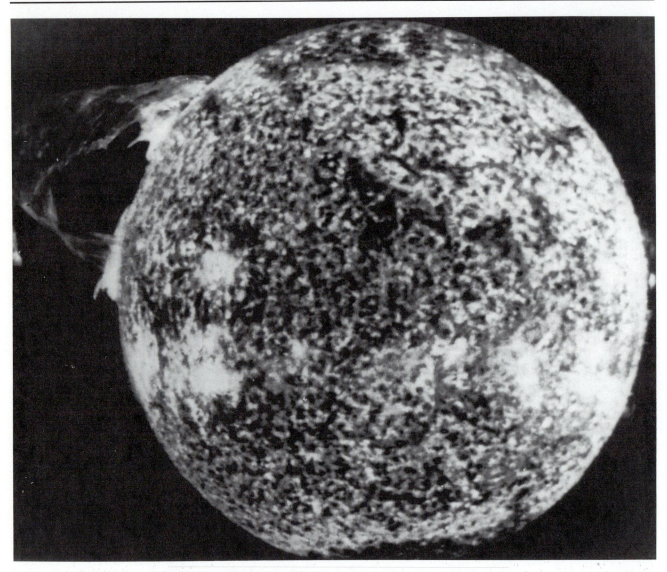

**FIGURE 2.1** This photograph of the Sun, showing a large prominence, was taken from Skylab on December 19, 1973, using the 304-angstrom (Å) radiation from He$^+$.

# CHAPTER 2

# The Sun

## 2.1 IN THE SUN

The center of our solar system, the source of light and heat for all the planets, the central mass that holds us in our orbits, is a star we call the Sun.

Just how big is the Sun? It is actually surprisingly easy to determine how much mass is concentrated in the center of the solar system. According to the laws of gravity that govern a body's orbit, the mass of the central body is proportional to the average distance of an orbiting body, cubed, divided by the period of time it takes for that body to complete an orbit, squared. With the Sun, we have nine orbiting bodies—the planets—whose periods can be measured directly. And once we know the scale of the solar system, we know how far from the Sun they are. We can thus use the laws of orbits nine times, once for each planet, to find the mass holding each planet in its orbit. The result in every case is that the mass of material at the center of the solar system, the mass we call the Sun, must be $1.99 \times 10^{30}$ kg. This represents 99.9% of the mass of the solar system, or 332,800 times the mass of Earth (Figure 2.2).

Likewise, we can observe transits of Venus across the Sun from various points on Earth and then use similar

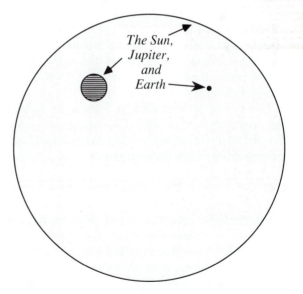

**FIGURE 2.2** The Sun contains 99.9% of the mass of the solar system. This diagram compares the relative sizes of the Sun, Jupiter (the biggest planet), and Earth.

triangles to find that the ball of the Sun is $6.96 \times 10^5$ km in radius. That's about 100 times bigger than Earth's radius.

From these two facts, we can begin to make some deductions about what the Sun is made of. To pack $1.99 \times 10^{30}$ kg of material into a ball of radius $6.96 \times 10^5$ km, every cubic centimeter of the Sun must have, on average, about 1.4 grams (g) of material. That's a rather ordinary density for material; water has a density of about 1 g/cm$^3$, and rocks are generally around 3 g/cm$^3$. Does this mean that the Sun is made up of, say, water and rock?

On further reflection, you can realize that that's not reasonable. The Sun is so big that one would expect the pressure inside the Sun to be enormous. A simple calculation (similar to what we'll eventually be doing for the planets) suggests that the pressure at the center of the Sun may be a billion times greater than the pressure of the air on the surface of Earth. Such a great pressure should certainly compress anything that exists inside the Sun. Because what we've calculated is only an average density, whatever the Sun is made of must have a much, much lower density under ordinary laboratory conditions. In fact, a careful calculation shows that the Sun must be made almost entirely of hydrogen and helium gases, the two lightest gases in the universe.

Besides being massive and large, we can see that the Sun is tremendously bright. Here at Earth, we can measure that the Sun pours two calories of energy onto a square centimeter every minute. In SI units, that's $1.4 \times 10^3$ joules per meter squared per second (J/m$^2$s). Adding up all the possible square centimeters that the Sun can be shining on—in other words, imagine a giant shell 1 AU in radius that completely enclosed the Sun, and find the surface area of that spherical shell—one calculates that, in the standard SI units of power, the Sun's energy output is $4 \times 10^{26}$ watts. We call that the **luminosity** of the Sun.

This energy is equivalent to trillions of hydrogen bombs exploding every second. The comparison between the Sun and bombs is apt; both the Sun and hydrogen bombs derive their energy from the same source.

## Fusion

Fusion is the basic energy source of all stars, including our Sun. And the clearest hint of how fusion works comes from noting that, even though atoms seem to be built up of protons and neutrons, when we try to keep track of the masses inside atoms something goes wrong. The masses don't add up properly.

Following common practice, we (arbitrarily) look at an isotope of carbon called carbon-12 and define a **unified atomic mass unit** (with the symbol $u$) as one-twelfth the mass of a $^{12}$C nucleus. One would expect that each proton or neutron should have a mass of 1 $u$ because we know that

there are exactly 12 of them in a $^{12}$C nucleus. However, by measuring an individual proton, we find that it has a mass of 1.00727 $u$. The mass of a single neutron is slightly larger, 1.00866 $u$. A $^4$He nucleus (often called an **alpha particle**) made up of two protons and two neutrons, has a mass of 4.0026033 $u$. Why isn't a proton exactly 1 $u$? Why is a $^4$He nucleus more massive than 4 $u$, but less massive than two protons and two neutrons?

The difference comes from the energy released when the protons and neutrons bind together into a single nucleus. A $^4$He nucleus is inherently more stable than a group of protons and neutrons, so when you take two protons and two neutrons and fuse them into one $^4$He nucleus, you release a certain amount of energy. You can even calculate how much energy is released, by taking the difference between the mass of two protons plus two neutrons, and one $^4$He nucleus, and multiplying by the speed of light squared. (That's the meaning of Einstein's famous equation, energy $E = mc^2$.) This energy is called the **binding energy** of the nucleus.

Exactly how does this fusion take place? It starts when two protons run into each other. Because both protons have a positive charge, they repel each other strongly, so it takes a lot of energy to get them to collide; that's why it takes the high pressure and temperature of the interior of a star. You need a cloud of gas at least 6% as massive as our Sun to get such a high pressure. When the two protons collide, they form a $^2$H nucleus, one proton and one neutron fused together, and they get rid of the extra charge from one of the protons by emitting a **positron** (or **beta particle**), which acts just like an electron except that it is positively charged. When the positron encounters an electron, the two annihilate each other, turning all their mass into energetic but massless bits of light, called **photons**. These energetic photons are called **gamma rays**. (The alpha/beta/gamma terminology was invented in the late 1800s, before physicists understood what each of these "rays" was made of.) This reaction is usually written as $^1$H $+ ^1$H $\rightarrow ^2$H $+ \beta^+$ (plus $\nu$, a **neutrino**, discussed below).

The $^2$H atom is called **deuterium** and is sometimes written as $^2$D. This notation comes from a time when scientists did not understand that deuterium was just an isotope of hydrogen, and not an element by itself. Once enough $^2$H is formed, it can start reacting with more $^1$H protons, $^2$H $+ ^1$H $\rightarrow ^3$He (plus a gamma ray), which can in turn react, most commonly as $^3$He $+ ^3$He $\rightarrow ^4$He $+ ^1$H $+ ^1$H (plus a gamma ray). And the cycle continues. Each reaction releases energy, which in turn builds up the temperatures inside the stars. This promotes even more reactions, because the higher the temperature, the faster the particles move and so the more likely they are to collide and fuse. This energy eventually gets radiated away from the surface of the star as light, thus making the star "shine."

# Elements, Isotopes, and Nuclides

## Elements

Elements are made up of atoms. An atom consists of a nucleus made up of protons and neutrons, surrounded by a cloud of electrons. Protons have a positive charge, electrons have a negative charge, and neutrons are neutral.

Most of the volume of an atom is taken up by the electron cloud, which is roughly $10^{-10}$ m in diameter, but almost all the mass is in the nucleus, which is four orders of magnitude smaller. If the atom were as large as a large lecture hall, the nucleus would be the size of a small grain of sand.

Protons and neutrons both have roughly the same mass, while electrons are 1/1800 as massive. Electrons can come and go as the atoms become chemically combined with other atoms to form compounds, while it takes a much more drastic nuclear reaction to change the number of protons or neutrons. So it is the nucleus, and more specifically the charge on the nucleus, determined by the number of protons, that controls the basic properties of the atom.

The term **element** is used to describe any set of atoms that has the same number of protons. The number of protons is called **Z**, the **atomic number**. Thus, hydrogen has the atomic number 1 and any atom that has only one proton is a hydrogen atom, carbon has an atomic number 6 and has 6 protons, and so on, through the periodic table of the elements.

## Isotopes

The number of neutrons that an atom of any element has can vary. Some carbon atoms have six neutrons, while other carbon atoms have seven or eight neutrons. The sum of the number of protons and neutrons is used to identify the isotope of the element. For example, carbon atoms with six protons and six neutrons are called carbon-12 (usually written $^{12}C$), while those with seven neutrons are carbon-13, or $^{13}C$, and so on.

It's theoretically possible to talk about an isotope of an element with as many neutrons as you want. For example, $^{153}C$ would have 6 protons and 147 neutrons.

But only a limited number of isotopes are stable, and the rest fall apart, or decay. Quantum physics says that it is impossible to tell with certainty when any given atom will decay, but given a large enough number of atoms, you can determine how long it will take before half of them have decayed. This time is called the **half-life** of the isotope. It always takes one half-life for half of the remaining atoms to decay. This behavior is called **exponential decay**.

The half-lives of different isotopes vary. An unlikely isotope such as $^{153}C$ would probably fall apart faster than we could measure, although nobody has ever tried to make any $^{153}C$ so we don't actually know. On the other hand, half a collection of $^{16}C$ atoms (which has been made in nuclear reactors) survives for 0.75 seconds. The isotope $^{14}C$, which is found in nature, has a half-life of 5730 years.

## Nuclides

The term nuclide is used as a generic term to describe any isotope of any element. Figures 2.14 to 2.17 are parts of the **chart of the nuclides**. The entire chart would show every isotope of every element, arranged by number of neutrons (the horizontal axis) and number of protons (the vertical axis).

## Getting the Energy Out

The fusion that gives energy to the Sun goes on in the center of the Sun; current theories estimate that only the innermost 10% of the Sun is actually hot and dense enough to allow fusion to take place. How does the energy get from the core to the surface?

We saw above that the energy released in the fusion process appears in the form of gamma-rays, which are very energetic bits of light. However, like any sort of light waves, gamma-rays can be stopped and absorbed by dust, gas, and even individual protons and electrons. For about 80% of the way out from the center of the Sun, the temperature is too high for anything that could completely absorb the gamma rays to exist, but individual electrons in the Sun's hot interior do scatter gamma rays back and forth; gamma rays tend to bounce around inside the Sun like balls in a pinball machine. As a result, it takes millions of years for gamma rays to reach the upper layers of the Sun.

For the last 20% of the way out of the Sun, the gamma rays have a hard time travelling very far because the temperature of the gas there is low enough for atoms to exist in a state where they can absorb the rays and turn their energy into heat.

"Heat" is simply a word to describe the energy that molecules have as they vibrate and jostle about. A particle such as a proton or electron that is moving contains **kinetic energy**; the amount of energy depends on how much mass the particle has and how fast it is moving. If a fast-moving particle bounces off a slower one, then it can transfer some of its energy to the particle it hit. This sort of energy transport is called conduction; energy is conducted when individual protons or electrons pass off their kinetic energy when they bounce into each other.

While heat conduction always moves heat from hotter areas to cooler areas, it is a relatively slow process. Particles travel far slower than gamma rays (which, after all are moving at the speed of light; they *are* light). The part that really makes it slow is waiting for the faster particles to bounce their way up to the surface of the Sun. This process is mathematically similar to watching a drunk stagger away from a street light. Eventually she may find herself up the street, but the odds are only slightly better than 50/50 that any given stagger will find her farther away from the light pole. (Such a process in mathematics is called a **random walk**.) Here, the light pole is the center of the Sun, the drunk represents an individual hot particle, and the end of the street is the surface of the Sun.

However, imagine if you could organize a whole army of drunks and get them to march together up the street; they could arrive much faster than one drunk wandering randomly. The equivalent happens in the upper reaches of the Sun when a whole parcel of hot gas molecules is physically carried, en masse, from the interior up to the surface of the Sun. This process is called **convection**.

Think of what happens when a parcel of gas is heated by absorbing gamma rays from the Sun's interior. As the gas heats up, it expands; as it expands, it becomes less dense than the surrounding gases; it becomes buoyant. It starts to float upward. The higher it goes, the cooler the surrounding gas is, and so the more buoyant our parcel appears; it moves upwards ever faster until it reaches the surface. As it pushes the cooler surrounding gas out of its way, it can make way for more hot gas from below to follow it up to the surface. The cooler gas gets pushed first sideways, then back down to replace the hot gas at the bottom of the convecting layer. Down there, it can absorb the gamma rays and become hot; meanwhile, when the hot gas reaches the surface it cools down by radiating the light we see as sunshine. Because it emits photons of light, the glowing surface of the Sun is called the **photosphere** (Figure 2.3).

"Cool" and "hot" are relative terms, of course. The gas that is "cooling down" on the surface of the Sun is,

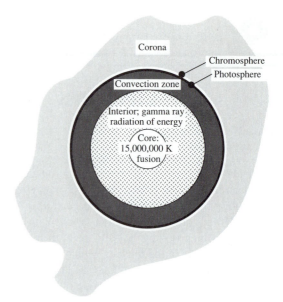

**FIGURE 2.3**   The Sun is divided into different regions. Fusion occurs in the core; the part we see from Earth is the photosphere.

in fact, at a temperature of 5770 K. We see the evidence for this convection when we look closely at the surface of the Sun with a telescope. A distinct polygonal pattern called **granulation** covers the surface. These are the tops of individual cells of convection. Hot material comes up at the center of these cells; as it cools, and gets darker, it goes back down along the dark rims of the cells. Individual cells can be thousands of kilometers across. Convection cells like this can be reproduced (on a much smaller scale!) in the lab; you can get a feeling for what it is like by watching how a pot of oatmeal bubbles on a stovetop.

But the Sun doesn't stop at the visible surface. In the nineteenth century when astronomers first began studying the colors in the solar spectra, they discovered that they could observe a very similar spectrum of color from the region just above the Sun's surface, even when the Sun itself was completely covered by the Moon during an eclipse. They named this region the **chromosphere**. It is a thin atmosphere of gas, continuing about 2000 km out into space beyond the top of the photosphere.

Above the chromosphere is an even hotter region (the **solar corona**) where atoms of the different elements exist, stripped of all their electrons. This state is called **plasma**. Violent activity such as **solar flares** can extend from the photosphere out into the corona. They play a part in producing a tenuous but energetic outflow of gas from the Sun called the **solar wind** (see Chapter 13 for more about the solar wind).

**FOR THE RECORD...**

# Units and Temperature

All physics is concerned with the description of matter in the universe, where it positions itself, and how it changes with time. Thus the basic units of mass, space, and time are of paramount importance. Velocity has the units of distance divided by time: miles per hour, centimeters per second, meters per second. Energy has units of mass times velocity squared; a kilogram–meter squared–per second squared, the SI unit for energy, is called a joule. In the cgs system, with grams and centimeters, the unit of energy is an erg. All the other quantities of physics, such as acceleration (length per time squared), force (mass times length per time squared), and so on, are defined in this way.

However, the temperature of an object is also a direct measure of the energy it has stored up in the motions of its molecules. A hotter object has more energy than a cooler one. Thus we have joules, ergs, and units of temperature all trying to describe the same concept, energy. The confusion doesn't end there. There are several different scales of temperature, such as Fahrenheit and Celsius. When Gabriel Fahrenheit, an eighteenth-century German physicist, was inventing his temperature scale, he picked "0" to be the freezing point of salt water and "100" to be roughly the body temperature of a human being. The Celsius scale, with 0 at the freezing point of pure water and 100 at its boiling point, may seem more rational, but in fact it's just as arbitrary as Fahrenheit's scale.

Instead, what we need to do is use a scale that registers 0 degrees when a body is completely devoid of thermal energy and every atom is frozen still. In reality, this never happens, but we can extrapolate from how molecules behave at low temperatures to define a point of lowest possible energy, which we can call absolute zero. This temperature is $-273°$ C, or $-460°$ F. Thus we can define two new, energy-related "absolute" temperature scales. The Kelvin scale starts at absolute zero and uses Celsius-sized degrees (so that water freezes at 273 K and room temperature is around 300 K). The Rankine scale does the same with Fahrenheit-sized degrees. From now on, we'll only refer to Kelvin temperature and follow the SI standard by referring to a temperature as, say, 273 K instead of 273° K. Finally, how do we relate a degree of temperature to a joule of energy? The conversion factor is called **Boltzmann's constant**, usually signified by the letter $k$. Thus $k$ times $T$, or $kT$, is the same as a unit of energy. The value of $k$ is approximately $1.38 \times 10^{-23}$ J/K (joules per kelvin).

## Sunspots and the Solar Cycle

Much more noticeable to the casual observer than granulation, on the surface of the Sun we see large dark spots where the temperature is much cooler than the surrounding surface (Figure 2.4). Instead of 5770 K these spots are at a mere 4500 K. These sunspots seem to come in pairs, forming between 5° and 40° away from the Sun's equator. Over a period of a few weeks, the positions of the spots move relative to the rotation of the Sun. It is clear that these spots are not features on a solid surface.

Because the Sun is a ball of gas, we can assume that it is in **hydrostatic equilibrium**, that the pressure depends only on the weight of the material above it and changes only as you go deeper into the Sun. That's the same law used to calculate the pressure inside of planets. It means that the pressure of the solar gas at the sunspots should be exactly the same as the pressure in the bright regions around the spots. But, because the spots are much cooler,

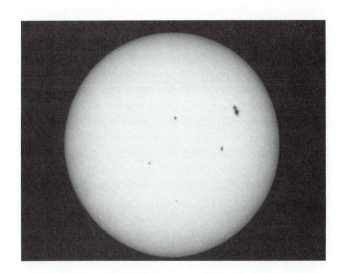

**FIGURE 2.4**  Sunspots.

Sunspot number

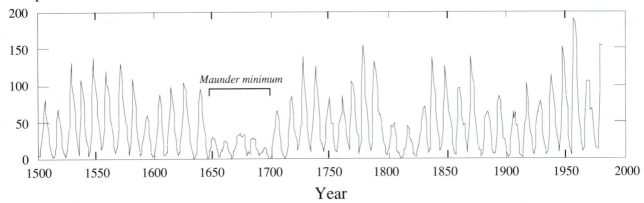

**FIGURE 2.5** The sunspot cycle, showing the number of sunspots seen every year over the past several hundred years. Numbers before 1610 are estimates based in part on reports of auroras.

they should be much denser. Why don't they sink into the hotter regions of the Sun? Something must be holding them up.

That something is the Sun's magnetic field. The Sun, much like Earth, has a strong dipole magnetic field that (we think) is created by fluid motions deep in the core of the Sun. Strong magnetic fields affect certain spectral lines in light emitted by various atoms in the Sun; by observing these spectral lines from Earth we can see that on the surface of the Sun the average magnetic field is about $2 \times 10^{-4}$ tesla (about seven times the field on Earth). The magnetic field in a sunspot is much higher, however, typically ranging from 0.025 to 0.5 tesla.

This strong magnetic field has enough energy to support the colder material in the sunspots from falling into the Sun's interior. Apparently, the sunspots are regions where the magnetic field lines of the Sun are "bundled together" and break through the surface of the Sun. The spots usually come in pairs; the first spot is where the field breaks out of the Sun, and the second spot is where it goes back inside.

These sunspots are not constant in number. New spots are continually forming as others are going away. There appears to be an 11-year cycle in the number of spots seen in the Sun (Figure 2.5). Every 11 years on average (the range is 7 years to 18 years), there is a year or two of low sunspot number, called a quiet Sun period, followed by increased sunspot activity that reaches its greatest extent during the solar maximum. This then drops down into another quiet Sun period.

The direction of the magnetic field stays constant throughout this cycle. But during the quiet Sun periods, the direction changes so that the polarity of the field during the next 11-year cycle is opposite to the polarity during the first 11-year cycle. So, all told, it takes 22 years for the Sun to go through one complete magnetic cycle.

## The Neutrino Problem

So far, we have described the source of the energy that makes the Sun shine, we have described how that energy gets out from the interior of the Sun to its surface, and we have given explanations for the major features seen on the surface of the Sun itself, the granulation and the sunspots. In the manner of most good textbooks, we have tried to make our explanations clear and lucid, firm in the conviction that these explanations can provide a complete and satisfying description of the events in the Sun. There is only one problem. Large chunks of what we have described here might be completely wrong.

Of course, all scientific descriptions are merely the "best theory we have at the moment," and it is always necessary for a scientist to hold onto a healthy skepticism. But our worries about the preceding description come from more than just skepticism. For, like all good theories, our theory of the interior of the Sun should be able to make predictions that we can test. And current theory fails at least one test in a surprising manner.

The fusion reactions described above can actually be carried out in a laboratory on Earth, as well as be described theoretically. And these experiments confirm what theory predicts: that one of the minor byproducts of fusion is the production of a remarkable kind of subatomic particle called a **neutrino**.

Neutrinos are particles with no charge and no mass, but they can carry energy and momentum away from a nuclear reaction. They are virtually featureless, and they very rarely interact with matter. The probability that a neutrino reacting with the material it travels through is so small that more than 10 billion neutrinos could pass completely through Earth for every one that is absorbed.

From the energy that we see emitted from the Sun,

we can estimate the rate at which nuclear reactions are occurring inside the Sun and from that calculate how many neutrinos ought to be emitted. Unlike gamma rays, neutrinos are not scattered by material in the Sun's core but rather they move directly, at the speed of light, out into space. It is estimated that neutrinos should be reaching Earth at the rate of one trillion per square meter every second!

Neutrino interactions are exceedingly rare. But they do happen. A detector consisting of a huge vat of carbon tetrachloride (dry-cleaning fluid), buried deep in a South Dakota gold mine to protect it from cosmic and other radiation, was built in 1970 to try to capture solar neutrinos. The estimate was that roughly nine **solar neutrino units** (**SNU**; 1 SNU $= 10^{-36}$ events per target atom per second) a year should be detected from the Sun, if all our theories about solar fusion were correct. Even now this prediction has not been proved. The detectors are consistently seeing only two or three SNU per year.

Something is wrong. Either we don't understand fusion, or we don't understand neutrinos, or we don't understand what's really going on in the Sun. The neutrino emission from fusion has been measured since the 1950s, and that agrees with theory. The detectors have seen neutrinos from other events—most notably a supernova in a nearby galaxy, the Large Magellanic Cloud, in 1987—showing that our detectors are working as expected. It is possible that some arcane mechanism is transforming the neutrinos emitted by the Sun's core into some undetectable particles before they reach Earth. However, there is still the very real chance that some piece of our picture of the Sun may be wrong.

## SUMMARY

The Sun, the center of our solar system, is a star with a radius of 695,000 km and a mass of $1.99 \times 10^{30}$ kg. It contains 99.9% of the mass of the solar system. The Sun shines with a power output of $4 \times 10^{26}$ watts.

The energy source that powers the Sun is the fusion of hydrogen nuclei—protons—into helium nuclei. This fusion occurs in the core of the Sun, releasing energy in the form of gamma rays. These energetic photons are scattered by protons and electrons and may take a million years to travel from the core to the upper regions of the Sun. In the upper 20% of the Sun, the photons are absorbed by the solar gas, which convects their energy, as heat, to the Sun's surface.

The surface of the Sun has two notable features. The first are the patterns of granulation, which are the tops of the convecting cells. The second are the dramatic sunspots, regions that are over 1200 K cooler than the surrounding gases, but supported by intense magnetic fields. The number of sunspots is observed to vary in cycles, with a period of approximately 11 years.

Our understanding of the physics occurring in the interior of the Sun has been challenged by the results of experiments looking for neutrinos that should be produced by the Sun's fusion. Such experiments consistently show that fewer neutrinos are seen than are predicted. Thus we recognize that our understanding of the Sun's interior may be still incomplete.

## STUDY QUESTIONS

1. The nucleus of an atom is made of two kinds of particles. Name them.

2. Carbon-12 is an isotope of carbon. What does the "12" mean?

3. Which statement about neutrinos is false?
   (a) Neutrinos take a million years to get out of the Sun into space.
   (b) Billions of neutrinos pass through your body every second.
   (c) Neutrinos appear to have no mass and no charge.

4. How long does it take for a gamma ray to travel from the center of the Sun to the surface: a few seconds, a few hours, or millions of years?

5. What on the Sun varies with an 11-year cycle? A 22-year cycle?

## 2.2  SUNLIGHT AND SPECTRA

The study of the solar system can be divided into two questions: What is in the solar system? And, how did it get to be the way it is today? To understand the Sun and the planets we have to start with an understanding of what they are made of. That's the first step in finding out "what's in the solar system."

But how can we determine the compositions of places we've never been to, places we've only seen from a distance? The answer lies in the light we use to see the planet. The sunlight that bounces off those planets and goes into our telescopes carries information about the places where it's been. To see how, we must first understand the nature of light.

### Light Waves

Start with an electron. You can picture this electron as a tiny particle, with a charge attached to it, and this charge creates an electric field. A common way of picturing an electric field is simply to think of lines radiating away from

**FOR THE RECORD...**

# Star Types

The Sun is but one of an estimated hundred billion stars in the Milky Way galaxy. Among these stars there is a wide range of color. A view of a starry night shows stars like Arcturus in the summer and Betelgeuse in the winter that are distinctly reddish in color, especially compared to blue-white stars such as Vega or Rigel. Furthermore, there is a wide range of brightness among the stars. Clearly, some of this variation comes because some stars are closer to Earth than others, but for the nearest stars it is possible to actually measure their distance and so take this factor into account to compute their luminosity, or how bright each star would appear if they were all equally distant from us.

Through the late nineteenth and early twentieth centuries, the precise colors and luminosities of the nearby stars were recorded and tabulated, mostly by Harvard astronomers Antonia Maury and Annie Jump Cannon. Cannon alone was responsible for the classification of 225,300 stars. Gifted with enormous powers of concentration, it was said that she could classify stars at a rate of three per minute. Stars were classified into categories that were simply given the letters A, B, C, and so on. (Further work showed that some of these categories, such as C, D, and H, were redundant and could be eliminated.) The original sequence was based on individual absorption lines; if the groups are reordered in terms of the effective temperatures of the stars, equivalent to their predominant colors, then the sequence becomes "$O, B, A, F, G, K, M, R, N, S$" (forever immortalized in the mnemonic "Oh, Brutal And Fearsome Gorilla, Kill My Roommate Next Saturday").

When ordered this way, the luminosity can be plotted against the temperature (or color), and a striking pattern is seen (Figure 2.6). For most stars, the rule is the brighter, the hotter. The Sun (a G-type star) fits well in this **main sequence** of stars. More interesting are the stars that do not fit. Some red stars seem very luminous, even though they are relatively cool. This can be explained if they are very large; we call these stars **red giants**. At the other extreme, hot but dim stars must be physically small; we call them **white dwarfs**.

Why do these stars exist? Assuming that all stars shine from the power of fusion, like our Sun, we can predict from theory why stars should have different temperatures and colors and how they evolve with time. According to

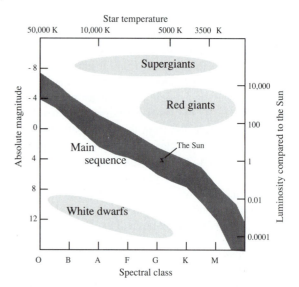

**FIGURE 2.6**  A plot of star temperature (or color, or star type) versus luminosity. Such a plot is called a Hertzsprung–Russell diagram.

theory, the important variable is the star's mass. Massive stars fuse hydrogen faster and thus burn hotter and brighter than less massive stars. They also burn more quickly and exhaust their fuel first. When the star's fuel is gone, it cools off and collapses. The sudden energy of collapse can shock the outer layers off into a huge, but cool, red cloud; hence red giants. Collapses can occur several times, becoming ever more violent until the outer cloud is completely blown off. What remains is, in most cases, a small but very dense hot star, a white dwarf.

Plots like Figure 2.6 are called **Hertzsprung–Russell** diagrams. Ejnar Hertzsprung, a Danish astronomer, used the Harvard catalogs of Maury and Cannon to construct the first diagram in 1906, from which he deduced the existence of giant and dwarf stars. In 1913 the American astronomer Henry Norris Russell, who was unaware of Hertzsprung's unpublished work, presented a similar diagram and showed its implications for stellar evolution theories. Russell went on to observe detailed spectra from the Sun and from them produce the first estimates of the cosmic abundances of the elements, a topic discussed later.

the electron. The lines represent the path that some other charged particle would follow as it was attracted, or repulsed, by our electron; and the closer together the lines are, the stronger the attraction or repulsion is. (Recall that electrons have a negative charge; they repel other negative charges and attract positive charges.)

Now move the electron back and forth (Figure 2.7). What happens to the field lines in such a case? They'll develop kinks. It takes a certain amount of time for the field far away from the electron to find out that the electron has moved. The kinks in the field lines travel down the field lines like waves travel across the surface of a pond. (The moving electron also creates a magnetic field that also "waves" as the electric field "waves.")

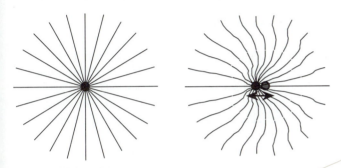

**FIGURE 2.7** The lines radiating from a point represent the electric field radiating from a point charge. If the charge is stationary, as on the left, then the lines diverge uniformly. But if the charge moves back and forth, as on the right, then the lines develop wiggles, or kinks.

Electrons usually don't exist all by themselves in nature, of course; usually they are part of atoms that also have protons, positively charged particles. And these protons also have an electric field. The field lines tend to bend from the proton around to the electron, producing a dipole field; the proton and the electron are the two poles of the dipole. If the proton and electron change places (for instance, by spinning the dipole around its center) or even if the electron just moves from one side of the proton to the other, then the direction of the field lines off to the side of the dipole will also switch directions. If the dipole keeps spinning or the electron constantly bounces back and forth, then a very special kind of wave is set up (Figure 2.8). Instead of having tiny kinks in the field, the field completely changes direction as a wave crest passes by.

(Why do we only talk about electrons moving, not protons? Because electrons do not weigh as much as protons and so are more likely to move around. Also, the electrons are found on the outside of the atoms, where they can get disturbed by outside forces; the protons are all clustered together in the center of the atom.)

The important thing about this wave is that it tends to reinforce itself. This kind of waving electric field sets

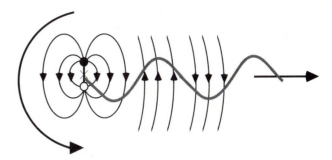

**FIGURE 2.8** A rotating dipole creates an electric field whose direction alternates up and down. That's an electric wave.

up the same sort of waving magnetic field that in turn produces a waving electric field, and so forth. This sort of wave is called an **electromagnetic wave**. What we have just described is, in fact, a light wave.

## Temperature and Wavelength

It takes energy to move the electron back and forth; the electrons don't move that way by themselves. To keep the electron moving back and forth we find that we have to apply energy constantly. If we have to keep adding energy, then we know that the energy must be going somewhere. It must be carried away by the wave.

The length of the wave depends on how fast we're moving the electron. If we wiggle it quickly, the waves are shorter and therefore the **frequency** of the waves (that is, how frequently we'd see a wave crest passing by us) is higher. To make a high-frequency, short-wavelength wave requires more energy than to move the electron sedately back and forth and make a lower-frequency, longer-wavelength wave. The energy of the wave is related to the wavelength of the wave.

Heat makes the electrons vibrate. Figure 2.9 shows how much energy, in theory, should be emitted at each wavelength of light for a simple **blackbody** (that is, one without an intrinsic color) at a given temperature. Notice that there's a peak in this curve. There is some wavelength that puts out the most energy. Another way of saying this is that if you raise a substance to a temperature $T$, it may glow in many colors, but there is one specific color in which it glows most brightly. (The color of light is the same thing as its wavelength.)

Can we find what that brightest color should be for any given temperature? In the early twentieth century Max Planck worked out the mathematical equation that produces the curve in Figure 2.9. We can plug in various values for the temperature and find out for each temperature the value of the wavelength $\lambda$ that gives the largest energy. (The more elegant way is to use calculus to find where the change in the brightness is zero, the point where the flux for a given

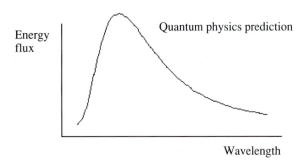

Energy flux

Quantum physics prediction

Wavelength

**FIGURE 2.9**    A graph of energy versus wavelength, as predicted by quantum physics and as actually observed.

wavelength stops increasing and starts to decrease again.) When we do this, we find that the temperature $T$ times the wavelength $\lambda$ obeys an equation $\lambda T = 2900$, for $T$ in degrees Kelvin and $\lambda$ in microns ($\mu$m, one millionth of a meter). This formula is called the **Wien displacement law**.

This is a tremendously powerful law. It explains why a campfire burns with a bright yellow light, while its cooler dying embers glow dull red. But more quantitatively than that, it says that if you split the light coming from any blackbody into its various colors and carefully measure which color is the brightest, you can tell from the wavelength of that color the precise temperature of the object that's producing the light.

## The Temperature of the Planets

Let's try applying this to the planets. Jupiter is very bright, and so its spectrum should be easy to measure. A typical spectrum is shown in Figure 2.10. The wavelength with the highest energy is easy to find; it's around $0.50 \times 10^{-6}$ m, or $0.50$ $\mu$m, giving a temperature of 5770 K. This is awfully hot; are we doing something wrong?

Obviously, we aren't looking at light produced by Jupiter, we're looking at sunlight reflected off of it. It's the Sun, not Jupiter, that has a temperature of 5770 K.

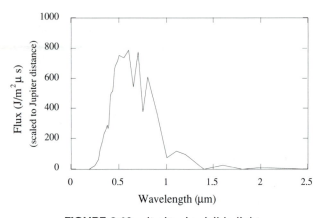

**FIGURE 2.10**    Jupiter in visible light.

If we were to keep observing into longer and longer wavelengths, in infrared light invisible to our eyes, we would find a second peak. This time it really is light radiated by the planet itself. For Jupiter, this peak is at about 23 $\mu$m. We can calculate from it a temperature of 125 K. (This is degrees Kelvin, remember; room temperature is about 300 K.) Jupiter has a thick atmosphere and clouds; this temperature is the temperature at the tops of those clouds, high in the atmosphere. This spectrum has two peaks because it is the spectra of two different bodies, the Sun and the planet itself, superimposed on one another. Jupiter is not a simple blackbody. And it differs from the pure theoretical blackbody in other ways, too.

## Absorption Lines

In addition to the two peaks, this spectrum looks very ragged. The curve in Figure 2.9, what we called a **blackbody** curve, was smooth. But the blackbody curve assumed a body that had no inherent colors of its own, hence its name. After all, an object that radiates primarily blue light may be 1000 degrees hotter than one that radiates mostly red, but you don't raise the temperature of a rock 1000 degrees just by painting it blue. But what makes a substance have a particular color?

Consider a very simplified view of an atom, with a positive charge in the nucleus and electrons orbiting around this center. Each electron has a certain orbiting frequency. When a light wave comes along, carrying the electric field past the electron in the atom, it's going to try to make that electron vibrate at the frequency of the wave. When the frequency of the wave and the frequency of the electron's orbit match, the electron in the atom absorbs the light wave. This sort of absorption tends to be very sharp; it is what makes the spectra in Figures 2.10 and 2.11 appear jagged instead of smooth. Light from the Sun's surface has passed through an atmosphere of colder hydrogen and helium gas just above the Sun, and this colder gas has absorbed certain colors out of the light. Likewise, as this light travels through Earth's atmosphere, still more colors can be absorbed. Sorting out the effect of Earth's atmosphere from that of the star or planet we wish to study is one of the challenges of spectroscopy.

Electrons that bind atoms together into molecules can also absorb light. Here, the possible electron vibrations are much more complicated, especially if the molecule is complicated. Light waves can cause the atoms within the molecule to vibrate, or they can make the whole molecule rotate. Because molecules are bigger than atoms, their resonant frequencies are much lower, and so they tend to absorb light of much longer wavelengths. Most of this absorption occurs in the infrared part of the spectrum (the part to the right as seen in Figure 2.10).

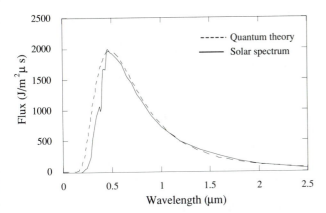

**FIGURE 2.11** The actual solar spectrum as seen from Earth is the solid line; the dotted line represents the blackbody curve that an object at the Sun's temperature should emit. The difference occurs because the light has travelled both through the Sun's atmosphere and Earth's atmosphere, where gases can scatter and absorb the light.

Gases give a series of very sharp absorption lines; from the precise measurement of the positions of the lines in such a spectrum one can deduce which gases are present and how hot they are. Sometimes you can even tell the pressure of the gas. Thus it is possible to see absorption lines in the infrared spectrum of Jupiter that we can identify as methane at 100 K, so it must be high up in the cold part of the atmosphere; and we can also see lines due to water vapor at 300 K, so it must be absorbing light much deeper in the atmosphere where the temperature is warmer. (Of course, some gases are easier to see than others, or more than one gas may absorb nearly the same wavelength of light, thus making the problem of interpreting spectra quite tricky.)

Rocks tend to give much broader absorption features because the rigid crystal structure of rocks changes the way the molecules can absorb light. By measuring the positions of the broad bands, however, it is still possible to deduce which minerals are present in a rock.

Thus one can see why astronomers spend so much time analyzing the light from planets. The general shape of the spectral curve tells us the temperature of the planet, and the specific lines in the spectrum tell us what chemicals are present and under what conditions.

In this way the abundances of the elements in the Sun are determined and the compositions of the atmospheres of Venus, Mars, Titan, and the gas giant planets are found. This technique told us that there was ice covering the moons of Jupiter and Saturn even before we sent a spacecraft to photograph those moons. The spectra of the Moon, Mercury, and the asteroids can tell us what minerals are present on these rocky bodies. For solid objects, like the rocks on the surfaces of these bodies, the reflectance spectrum is simply a very precise measure of the color of the rock. For instance, rocks that look reddish compared to sunlight (which our eyes interpret as colorless) mostly reflect longer-wavelength red light and have a wide absorption band in the blue part of their spectrum. The subtle shades of brown or olive green that rocks exhibit are mirrored in their reflectance spectrum.

## Cosmic Abundances of the Elements

The vast bulk of the mass of the solar system (99.9%) is in the Sun; of the remainder, more than 70% is in Jupiter; and what remains after Jupiter is found mostly in the other gas giant planets: Saturn, Uranus, and Neptune. The Sun, Jupiter, and the other gas giants are mostly hydrogen and helium. Looking at other stars, we can see that most of them have compositions similar to our own Sun. By far the most important features in the spectra of any star are the sharp visible and ultraviolet absorptions due to hydrogen and helium.

In fact, by carefully studying the spectrum of light that comes from the Sun it is possible to identify the relative abundances there of the elements hydrogen (H), helium (He), oxygen (O), carbon (C), neon (Ne), nitrogen (N), magnesium (Mg), silicon (Si), iron (Fe), sulfur (S), aluminum (Al), calcium (Ca), sodium (Na), nickel (Ni), and a few others. Similar observations can be made of Jupiter and Saturn and of other stars as well. Unfortunately, most of the other naturally occurring elements are so scarce that it is impossible to determine their relative abundances in the Sun with any accuracy.

A class of meteorites called **carbonaceous chondrites** have ages of 4.6 billion years (see Chapter 9), as old as any rocks seen anywhere; they are chemically and physically so fragile that it appears no chemical changes have ever occurred in them, going back to the time they were first formed, probably at the beginning of the solar system. Most intriguing, the relative abundances in carbonaceous chondrites of the eight "rocky" elements given above is very close to the solar relative abundances of these elements. (The elements Mg, Si, Fe, S, Al, Ca, Na, and Ni are the ones that tend to form rocks; H, He, O, C, Ne, and N occur in nature most often as gases.) For example, the proportion of silicon to aluminum is the same in these rocks as it is in the solar atmosphere.

Unlike the Sun, it is possible to take a sample of a carbonaceous chondrite into the laboratory and measure the abundance of every element that can be found in it. One can then combine this list of abundances with the known solar abundances. Usually the element Si, which is common to both lists, is chosen as the element against which all other abundances are normalized. This combined list is usually called a table of **cosmic abundances**, because it seems to be valid for the Sun, the gas giant planets, primitive meteorites,

**TABLE 2.1**

Cosmic Abundances of the Elements (in Number of Atoms Relative to Si)

| Element | Z | Abundance | Element | Z | Abundance | | Element | Z | Abundance | |
|---------|---|-----------|---------|---|-----------|---|---------|---|-----------|---|
| H | 1 | 26,600 | Li | 3 | 60 | ppm | Ag | 47 | 0.46 | ppm |
| He | 2 | 1,800 | Kr | 36 | 41 | ppm | Gd | 64 | 0.42 | ppm |
| O | 8 | 18.4 | Ga | 31 | 38 | ppm | Rh | 45 | 0.40 | ppm |
| C | 6 | 11.1 | Sc | 21 | 31 | ppm | Cs | 55 | 0.39 | ppm |
| Ne | 10 | 2.60 | Sr | 38 | 23 | ppm | La | 57 | 0.37 | ppm |
| N | 7 | 2.31 | Zr | 40 | 12 | ppm | Dy | 66 | 0.37 | ppm |
| Mg | 12 | 1.06 | Br | 35 | 9.2 | ppm | Sb | 51 | 0.31 | ppm |
| Si | 14 | $\equiv 1$ | B | 5 | 9.0 | ppm | W | 74 | 0.30 | ppm |
| Fe | 26 | 0.9000 | Te | 52 | 6.5 | ppm | Sm | 62 | 0.24 | ppm |
| S | 16 | 0.5000 | As | 33 | 6.2 | ppm | Er | 68 | 0.23 | ppm |
| Ar | 18 | 0.1060 | Rb | 37 | 6.1 | ppm | Au | 79 | 0.21 | ppm |
| Al | 13 | 0.0850 | Xe | 54 | 5.8 | ppm | Hg | 80 | 0.21 | ppm |
| Ca | 20 | 0.0625 | Ba | 56 | 4.8 | ppm | Yb | 70 | 0.20 | ppm |
| Na | 11 | 0.0600 | Y | 39 | 4.8 | ppm | In | 49 | 0.19 | ppm |
| Ni | 28 | 0.0478 | Mo | 42 | 4.0 | ppm | Tl | 81 | 0.19 | ppm |
| Cr | 24 | 0.0127 | Sn | 50 | 3.7 | ppm | Pr | 59 | 0.18 | ppm |
| Mn | 25 | 0.0093 | Pb | 82 | 2.6 | ppm | Hf | 72 | 0.17 | ppm |
| P | 15 | 0.0065 | Ru | 44 | 1.9 | ppm | Bi | 83 | 0.14 | ppm |
| Cl | 17 | 0.0047 | Cd | 48 | 1.6 | ppm | Eu | 63 | 0.094 | ppm |
| K | 19 | 0.0035 | Pt | 78 | 1.4 | ppm | Ho | 67 | 0.092 | ppm |
| Ti | 22 | 0.0024 | Pd | 46 | 1.3 | ppm | Tb | 65 | 0.076 | ppm |
| Co | 27 | 0.0022 | I | 53 | 1.3 | ppm | Re | 75 | 0.051 | ppm |
| Zn | 30 | 0.0013 | Be | 4 | 1.2 | ppm | Th | 90 | 0.036 | ppm |
| F | 9 | 0.0078 | Ce | 58 | 1.2 | ppm | Tm | 69 | 0.035 | ppm |
| Cu | 29 | 0.00054 | Nb | 41 | 0.90 | ppm | Lu | 71 | 0.035 | ppm |
| V | 23 | 0.00025 | Nd | 60 | 0.79 | ppm | Ta | 73 | 0.02 | ppm |
| Ge | 32 | 0.00012 | Ir | 77 | 0.72 | ppm | U | 92 | 0.010 | ppm |
| Se | 34 | 0.00006 | Os | 76 | 0.69 | ppm | | | | |

and other stars. A recent compilation of cosmic abundances is given in Table 2.1.

If any rock or any planet has a composition different from the cosmic abundances in this table, these differences can be attributed to some process of chemical fractionation. Much of the work of planetary geochemists is trying to figure out exactly what those chemical fractionation processes are, to answer the second question posed at the beginning of this chapter, "How did the planets get to be the way they are today?"

## SUMMARY

The light of the Sun is the basic tool we have to explore the surfaces of other worlds. Light is made up of electromagnetic waves, the transmitted energy of electrons in motion. Every substance emits some sort of light; the peak color of its emis-

sion is directly related to the temperature of the body, according to the Wien displacement law: $\lambda T = 2900$ where $\lambda$ is the wavelength of the light (in $\mu$m) and $T$ is the temperature in kelvins. The Sun is 5770 K hot and radiates in visible light with a blackbody peak at a wavelength of $0.50 \times 10^{-6}$ m; Earth and most things on its surface are at about 300 K and radiate in infrared light at roughly $10^{-5}$ m.

Along with radiating light, objects also reflect and absorb sunlight. Each atom, and each collection of atoms that make up a molecule, selectively absorbs certain precise wavelengths of light. By carefully observing the sunlight reflected from a distant planet and seeing which colors have been absorbed, we can tell what substances are on the surface of that planet. In this way, we can also determine the composition of the Sun and other stars. We find that most stars in the universe, including our Sun, have the same proportions of the elements, which we call the **cosmic abundances**. They are made mostly of hydrogen and helium, with the other elements together equalling less than 1% of the mass of the cosmos.

## STUDY QUESTIONS

1. Why do we talk about electrons moving, and not protons moving, when we discuss the origin of light?

2. What does the symbol "$\lambda$" stand for in the context of this chapter?

3. What is the most abundant element in the universe?

4. What quantity is measured in kelvins?

5. What is a typical wavelength for visible light:
   (a) 0.5 microns (half of one millionth of a meter)
   (b) 0.5 nanometers (half of one billionth of a meter)

## 2.3 NUCLEOSYNTHESIS

Ay, for 'twere absurd
To think that nature in the Earth bred gold,
Perfect i' the instant: something went before.
There must be remote matter.

*—Ben Jonson, 1610*

Where did the elements come from? What is it about the process of **nucleosynthesis** that makes an element like silicon more common than an element like gold and makes hydrogen and helium far and away the most abundant of all?

### Hydrogen and Helium

According to the widely accepted **big bang theory**, the universe started from a state of pure energy, compressed in a single point, that exploded. In this explosion, much of that energy was turned into matter, mostly protons, neutrons, and electrons. (Energy and matter are interchangeable, according to Einstein's famous equation, $E = mc^2$. In this equation, $E$ stands for a certain amount of energy, $m$ is the equivalent amount of mass that can be formed by that energy, and the factor $c^2$, the speed of light squared, is the constant that determines how much mass can be made from a given amount of energy.)

These subatomic particles soon formed themselves into the simplest types of atomic nuclei: hydrogen, H, consisting of one proton (in some hydrogen atoms, there is a neutron attached to the proton), and helium, He, with a pair of protons and (usually) a pair of neutrons.

Protons and neutrons are more stable when they come in pairs; notice several trends in Figure 2.14. First, the gray isotopes mark out a sort of "valley of stability" among all possible isotopes. Stable isotopes seem to have the same, or slightly greater, number of neutrons as protons. Atoms with an even number of protons tend to have a greater number of stable isotopes than atoms with an odd number of protons. And atoms with an odd number of protons tend to make stable isotopes only when they have an even number of neutrons. In all nature there are only four stable isotopes with both an odd number of neutrons and an odd number of protons.

Some time after the big bang, these gases of H and He cooled down enough to form star-sized masses. Eventually the pressure and temperature inside these **protostars** became high enough that individual H nuclei, the proton and neutron, got squeezed together and fused, forming He nuclei and releasing energy in the process. This **fusion** is the basic source of energy in stars, the source of their light and heat.

### Carbon to Iron

The big bang is the source of hydrogen and helium in the universe, and simple hydrogen fusion inside stars can make more helium. But, if you notice, even the CNO cycle described below does not turn hydrogen or helium into carbon or heavier elements; it just uses whatever elements may be around to make more helium. Where might these heavier elements come from? Is it possible that continued fusion can make them?

As it turns out, it's not easy. Say we try to fuse two $^4$He's together:

$$^4\text{He} + {}^4\text{He} \rightarrow {}^8\text{Be}$$

The trouble is, $^8$Be is an unstable isotope; it has a half-life of only $10^{-16}$ s. That means, before we can use it to make some more complicated isotope, it will have already decayed back into two helium nuclei. If we could fuse three $^4$He nuclei together to make $^{12}$C, that would be perfectly stable, but it means we would have to get the three nuclei together in the same place within $10^{-16}$ of a second. Only when the temperature reaches 100 million K, and the density is very high, is it likely that three $^4$He nuclei will encounter each other within such a short period of time. This is much hotter than most normal stars get in their interiors.

If that's the case, then the only fusion that can take place in most stars involves burning hydrogen into helium. What happens when all the hydrogen in the star's core is consumed and no more fusion can take place there? Once fusion stops, the core of the star—the hottest and densest place, where the fusion has been occurring—cools off, and as it cools it contracts. The energy of the outer material falling in towards this cooler center is enough to heat the outer parts of the star for a while. These outer layers start to fuse; the very outermost, being heated now, expand into giant size, with radii of several AU. Stars in this phase turn red in color—after their expansion, they are cooler at their

**FURTHER INFORMATION...**

# The Blackbody Curve and Wien Displacement Law

Heat makes the electron vibrate. In classical nineteenth-century physics, it was possible to show mathematically that the energy flux, $u_\lambda$, in a wave of a given wavelength $\lambda$ was equal to

$$u_\lambda = \frac{2\pi k T c}{\lambda^4}$$

where $k$ is **Boltzmann's constant**, which is just a number to convert units of temperature into units of energy; $T$ is the temperature of the system that is radiating the light; and $c$ is the speed of light. In other words, if you have a substance at a temperature $T$, it radiates light waves with a variety of wavelengths, and the energy carried by a given wave with wavelength $\lambda$ is the energy $u_\lambda$.

Figure 2.12 shows what happens when you use this equation to plot this energy against the wavelength. (This sort of graph is called a **spectrum**.) As the wavelength shrinks to zero, the energy becomes infinite! This result is absurd to begin with; a finite temperature shouldn't give rise to infinitely energetic light waves. Furthermore, it's not at all what is actually observed. The light radiated from a hot, glowing substance produces the spectrum seen in Figure 2.13.

Something was terribly wrong with classical physics. Its laws broke down when dealing with short-wavelength, ultraviolet light. The physicists of the nineteenth century called this the **ultraviolet catastrophe**. **Quantum mechanics** was invented in the early twentieth century to solve problems like this. Out of quantum mechanics a more precise equation for $u_\lambda$ was developed:

$$u_\lambda = \frac{2\pi h c^2}{\lambda^5 (e^{\frac{hc}{\lambda k T}} - 1)}$$

When the term $hc/\lambda$ is much less than $kT$, that is, when $\lambda$ is large and we have long wavelengths of light, then this equation reduces to the classical expression. (The term $h$ is called **Planck's constant**.) This equation produces the curve seen in Figure 2.13, in agreement with observations, called a **blackbody** curve.

The equation given above gives energy as a function of temperature and wavelength. One can then ask, What is the peak wavelength for a given temperature? To find that, one takes the derivative of that function with respect to wavelength—that gives the slope of the curve—and sets it to zero. When the appropriate values for the constants are inserted, one arrives at the **Wien displacement law**.

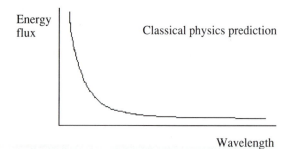

**FIGURE 2.12** A graph of energy flux versus wavelength, under the assumptions of classical physics.

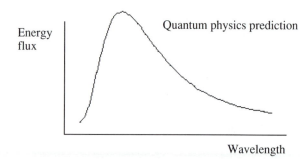

**FIGURE 2.13** A graph of energy flux versus wavelength, as predicted by quantum physics and as actually observed.

surfaces—and so these stars are called **red giants**. Our Sun will go through this phase about 5 billion years from now.

Meanwhile, the energy of this collapse is also heating up the centers of these stars. The more massive the star, the

hotter the center gets. For a star with the mass of our Sun, a temperature of $10^8$ K will be reached and at that point the formation of $^{12}$C by the fusion of three $^4$He can finally take place. More massive stars can reach higher tempera-

The following chart is organized by neutron number N (columns, N = 0 to 15) and proton number Z (element rows). Stable isotopes are shaded gray; long-lived isotopes (marked with slanting bars in the figure) are indicated here with their half-lives.

| Element (Z) | N=0 | N=1 | N=2 | N=3 | N=4 | N=5 | N=6 | N=7 | N=8 | N=9 | N=10 | N=11 | N=12 | N=13 | N=14 | N=15 |
|---|---|---|---|---|---|---|---|---|---|---|---|---|---|---|---|---|
| Si (14) | | | | | | | | | | | Si24 0.15 s | Si25 0.22 s | Si26 2.2 s | Si27 4.14 s | Si28 (stable) | Si29 (stable) |
| Al (13) | | | | | | | | | | Al22 0.07 s | Al23 0.47 s | Al24 2.07 s | Al25 7.17 s | Al26 7e5 yr | Al27 (stable) | Al28 2.25 m |
| Mg (12) | | | | | | | | | Mg20 0.1 s | Mg21 0.12 s | Mg22 3.86 s | Mg23 11.3 s | Mg24 (stable) | Mg25 (stable) | Mg26 (stable) | Mg27 9.45 m |
| Na (11) | | | | | | | | | Na19 0.03 s | Na20 0.45 s | Na21 22.5 s | Na22 2.6 yr | Na23 (stable) | Na24 15 h | Na25 60 s | Na26 1.07 s |
| Ne (10) | | | | | | | Ne16 4e-21 s | Ne17 109 ms | Ne18 1.67 s | Ne19 17.2 s | Ne20 (stable) | Ne21 (stable) | Ne22 (stable) | Ne23 37.2 s | Ne24 3.38 m | Ne25 0.61 s |
| F (9) | | | | | | | F15 5e-22 s | F16 e-20 s | F17 64.5 s | F18 1.83 h | F19 (stable) | F20 11 s | F21 4.16 s | F22 4.23 s | F23 2.2 s | F24 0.3 s |
| O (8) | | | | | O12 1e-21 s | O13 8.9 ms | O14 70.6 s | O15 122 ms | O16 (stable) | O17 (stable) | O18 (stable) | O19 26.9 s | O20 13.5 s | O21 3.4 s | O22 3 s | |
| N (7) | | | | | | N12 11 ms | N13 9.97 ms | N14 (stable) | N15 (stable) | N16 7.1 s | N17 4.2 s | N18 0.62 s | N19 0.3 s | N20 0.1 s | | |
| C (6) | | | C8 2e-21 s | C9 127 ms | C10 19.3 ms | C11 20.3 ms | C12 (stable) | C13 (stable) | C14 5730 yr | C15 2.45 s | C16 0.75 s | C17 20 ms | C18 0.07 s | | | |
| B (5) | | | B7 4e-22 s | B8 770 ms | B9 8e-19 s | B10 (stable) | B11 (stable) | B12 20.2 ms | B13 17.4 ms | B14 14 ms | B15 9 ms | | | | | |
| Be (4) | | | Be6 5e-21 s | Be7 53 d | Be8 7e-17 s | Be9 (stable) | Be10 1.6e6 yr | Be11 13.8 s | Be12 24 ms | | Be14 4 ms | | | | | |
| Li (3) | | | Li5 3e-22 s | Li6 (stable) | Li7 (stable) | Li8 0.84 s | Li9 177 ms | | Li11 8.7 ms | | | | | | | |
| He (2) | | He3 (stable) | He4 (stable) | He5 7e-22 s | He6 807 ms | He7 3e-21 s | He8 122 ms | | | | | | | | | |
| H (1) | H1 (stable) | H2 (stable) | H3 12 yr | | | | | | | | | | | | | |

**FIGURE 2.14** The beginning of the chart of the nuclides. Stable isotopes are shaded gray; isotopes with very long half-lives are marked with slanting bars. The half-lives of the unstable isotopes are given; s stands for seconds, ms for milliseconds (one thousandth of a second), m for minutes, h for hours, and yr for years. The terminology "$3e-21$" is a modified form of scientific notation, standing for "$3 \times 10^{-21}$."

tures, and heavier elements can be formed by fusing He to these C nuclei. Adding $^4$He to $^{12}$C forms oxygen, $^{16}$O; another $^4$He makes neon, $^{20}$Ne; another makes magnesium, $^{24}$Mg; another, silicon, $^{28}$Si; and so on to make even heavier elements.

Once we have $^{12}$C present, an interesting new kind of reaction can start to take place, called the **CNO bi-cycle**:

$$^1\text{H} + {}^{12}\text{C} \rightarrow {}^{13}\text{N} + \gamma$$

$$^{13}\text{N} \rightarrow {}^{13}\text{C} + \beta^+ + \nu$$

$$^{13}\text{C} + {}^1\text{H} \rightarrow {}^{14}\text{N} + \gamma$$

$$^{14}\text{N} + {}^1\text{H} \rightarrow {}^{15}\text{O} + \gamma$$

$$^{15}\text{O} \rightarrow {}^{15}\text{N} + \beta^+ + \nu$$

$$^{15}\text{N} + {}^1\text{H} \rightarrow {}^4\text{He} + {}^{12}\text{C}$$

or

$$^{15}\text{N} + {}^1\text{H} \rightarrow {}^{16}\text{O} + \gamma$$

$$^{16}\text{O} + {}^1\text{H} \rightarrow {}^{17}\text{F} + \gamma$$

$$^{17}\text{F} \rightarrow {}^{17}\text{O} + \beta^+ + \nu$$

$$^{17}\text{O} + {}^1\text{H} \rightarrow {}^4\text{He} + {}^{14}\text{N}$$

and the cycle goes on. This produces lots of energy (the star shines) and lots of $^4$He's, by a process that is easier to accomplish than simply fusing protons together. The $^{12}$C promotes the reaction without being depleted—or created—itself. It also promotes the production of lots of other isotopes, such as $^{13}N$, $^{13}C$, $^{14}N$, $^{15}O$, $^{15}N$, $^{16}O$, $^{17}F$, and $^{17}O$, that can take part in the other nuclear equilibrium processes.

Once you have a large number of these heavier seed nuclei inside a star, all sorts of complicated nuclear "burning" processes can occur. Nuclides are destroyed and new ones made as they combine with protons, He nuclei, or each other, decaying if they are unstable, or serving as targets for further fusion if they do not decay. Eventually a balance is achieved among the abundances of relatively rare isotopes such as $^{17}O$ or $^{21}Ne$ and the more abundant He-multiples such as $^{16}O$ and $^{20}Ne$.

## Elements Beyond Fe

Up to this point, all the reactions we've talked about have released energy. The formation of these elements contributes to maintaining the high temperature inside a star and continuing nucleosynthesis. But as bigger and bigger nuclei are built, the amount of energy released becomes less and less. Finally, once you get to iron, $^{56}Fe$, you've gotten all the energy out that you're going to get. It actually absorbs energy inside a star to make heavier elements.

The sort of equilibrium balance between common and rare isotopes means that every now and then one of these heavier isotopes will be formed anyway, even though energy is absorbed in the process; but beyond copper, Cu, and zinc, Zn, only a few elements past Fe, the odds of getting any heavier nuclei are very small. So how are the rest of the elements made? There seem to be three important ways to make such elements.

## S-Process

In an older star, which has had time to manufacture many elements heavier than H and He, the equilibrium "burning" we described above may make any number of isotopes around iron. And as we saw, the elements that are made by adding alpha particles (helium nuclei) together tend to be common. Thus inside such stars, there tends to be plenty of carbon-12, oxygen-16, neon-20, and so forth. But still, by far, the most abundant nuclei inside these stars are simple hydrogen-1 and helium-4.

An interesting reaction can take place between hydrogen, helium, and these heavier isotopes. Say a hydrogen nucleus combines with a neon-20 nucleus:

$$^{1}H + {}^{20}Ne \rightarrow {}^{21}Na$$

The sodium-21 is unstable and tends to decay:

$$^{21}Na \rightarrow {}^{21}Ne + \beta^+ + \nu$$

Now if the neon-21 encounters an alpha particle:

$$^{21}Ne + {}^{4}He \rightarrow {}^{24}Mg + n$$

This reaction tends to produce lots of neutrons, $n$, that are quite energetic.

These neutrons deposit their energy when they collide with some other particle. Nuclei such as $^{66}Zn$ are quite massive—they make a big target—and they'll absorb both the neutron and the energy. Considering that there are some 20,000 times more $^{20}Ne$ nuclei than $^{66}Zn$ nuclei, we can see that many neutrons can be available; after making $^{67}Zn$ by adding a neutron, another neutron can turn that into $^{68}Zn$, and so on. This process continues through $^{69}Zn$, which is unstable. But it has a reasonably long half-life of 14 h, and some of these nuclei might survive until they are hit by another neutron to make $^{70}Zn$, which is stable. $^{71}Zn$, the next isotope, is unstable again, and there are no further stable isotopes of Zn to be made.

What happens when these unstable isotopes decay? They decay by emitting $\beta^-$ particles (that is, electrons). This happens when a neutron turns itself into a proton. But, an extra proton means that we've turned the nucleus into a different element; by this decay process, new elements with higher atomic numbers are formed. In this case, an isotope of gadolinium, $^{71}Ga$, is formed and turns out to be stable. But neutrons hitting it produce $^{72}Ga$, which is unstable, and decays—by emitting a $\beta^-$ particle—into germanium, Ge.

Thus it is possible to form isotope after isotope in this fashion, slowly adding neutrons until you get to an unstable isotope that decays into the next element, whereupon the process continues. This process is called the **s-process**, for the slow addition of neutrons, and is illustrated in Figure 2.15.

## Supernovas

All these reactions can go on deep inside red giant and supergiant stars. But our Sun has all these elements, and it's not a giant. How did these heavier elements get into the Sun and its planets?

These elements have to get taken out of a star's interior and spread into space. This happens when a star dies. Stars like our Sun that are rich in these heavier elements are thus second-generation stars, made up of gases and dust from the debris of old, dead stars.

Observations of stars show that they seem to come in two distinct populations. Those stars whose spectral lines show that they are made of almost pure H and He are called **Population II** stars. In many other ways, they appear to be very old stars. **Population I** stars, such as our Sun, are younger and have the spectral lines of heavier elements like oxygen, carbon, and iron. They appear to be made out of recycled material.

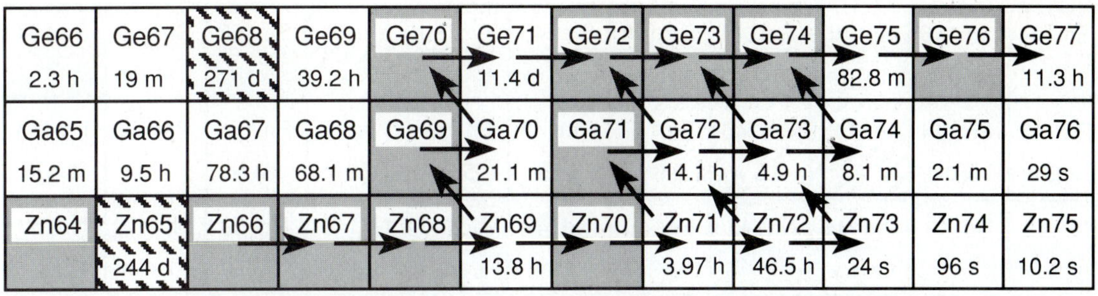

| Ge66 | Ge67 | Ge68 | Ge69 | Ge70 | Ge71 | Ge72 | Ge73 | Ge74 | Ge75 | Ge76 | Ge77 |
|------|------|------|------|------|------|------|------|------|------|------|------|
| 2.3 h | 19 m | 271 d | 39.2 h | | 11.4 d | | | | 82.8 m | | 11.3 h |
| Ga65 | Ga66 | Ga67 | Ga68 | Ga69 | Ga70 | Ga71 | Ga72 | Ga73 | Ga74 | Ga75 | Ga76 |
| 15.2 m | 9.5 h | 78.3 h | 68.1 m | | 21.1 m | | 14.1 h | 4.9 h | 8.1 m | 2.1 m | 29 s |
| Zn64 | Zn65 | Zn66 | Zn67 | Zn68 | Zn69 | Zn70 | Zn71 | Zn72 | Zn73 | Zn74 | Zn75 |
| | 244 d | | | | 13.8 h | | 3.97 h | 46.5 h | 24 s | 96 s | 10.2 s |

**FIGURE 2.15** A section of the chart of the nuclides including isotopes of zinc (Zn), gadolinium (Ga), and germanium (Ge). The slow addition of neutrons to the stable isotopes produces a new isotope, as indicated by the arrows pointing to the right. When an unstable isotope is formed, it will decay into a new isotope with one neutron being turned into a proton; this is indicated by the arrows pointing up and to the left.

How does a star die? How does its death spread materials into space? To understand this, we must briefly examine the physics inside a star. Once a star enters the red giant phase, it starts fusing the hydrogen in its outer layers. As this fusion gets closer and closer to its surface, more and more of the surface gases get hotter and boil off from the star, carrying with them the heavier elements that have been mixed into the outer parts of the star from the center when it first collapsed. As this gas is blown off, these heavier elements cool and condense into solid grains of dust a few tens of nanometers (1 nanometer = $10^{-9}$ m) in diameter, which eventually can find their way into newly formed star systems. This is one source of heavy elements, but there is another, more spectacular source.

When even the outer layers of a star have also run out of hydrogen to fuse, the star contracts again, forming a white dwarf. The temperature inside a white dwarf is so high that all its atoms are completely ionized. There are no electrons attached to any of these atoms. Instead, the electrons flow freely, and it is the mutual repulsion of these like-charged electrons that provides the pressure to "hold up" the inside against further collapse. In this state, with the electrons and protons completely independent and the electron pressure holding up the star, matter is said to be **degenerate**.

If the star's He-rich core has less than 1.4 times the mass of our Sun, then the pressure of the electrons stops this collapse and the star eventually just cools off, becoming a lump of cold helium. However, if the star's core has more than 1.4 solar masses, the density of matter in its center is great enough that electrons and protons can combine to form neutrons. This process absorbs energy and it absorbs electrons; but the outward pressure of the electrons prevents the star from collapsing under its own weight. So the pressure inside the star suddenly drops, and the star starts to collapse again. This collapse is very sudden and very violent. The energy of the collapse is sufficient to explode the star completely. This explosion is called a **supernova**.

This makes quite a spectacular sight. Astronomers have observed supernovas in distant galaxies that are as bright as the rest of the stars in their galaxies combined, some 10 billion times as bright as our Sun. A supernova can also leave behind strange things in its debris, such as pulsars, neutron stars, and possibly black holes. But the most important result, for us, is that this explosion takes the elements created deep inside a star and spreads them out into the galaxy, where they can get incorporated into new stars. And it provides the energy to make isotopes that could not be formed in any other way.

## Other Nuclides

The s-process accounts for 75% of the isotopes heavier than Fe. But, looking at a table of the nuclides, one can spot perfectly stable isotopes that do not lie on the zigzag path of the s-process. Look at the section of the "Chart of the Nuclides" reproduced in Figure 2.16. Some isotopes, like $^{144}$Sm, are to the left of the zigzag path; they are too light to be reached by the s-process and do not have any unstable isotope in the row below them that can decay into them. Others, like $^{154}$Sm, are to the right of the zigzag path and are separated from it by a highly unstable isotope (such as $^{153}$Sm, with a half-life of only 47 h) that is likely to decay before additional neutrons can be captured. (True, this 47-h half-life is longer than the 14-h half-life of $^{69}$Zn, which we used before; but there's 10,000 times more Zn available than Sm, so it is that much more likely that one of them will capture a neutron.)

It is not surprising that the abundance of such isotopes is low. But they do exist. We've never made a stable isotope of any element in the lab that hasn't occurred in nature already, someplace. How are they made?

| Sm144 | Sm145 | Sm146 | Sm147 | Sm148 | Sm149 | Sm150 | Sm151 | Sm152 | Sm153 | Sm154 |
|---|---|---|---|---|---|---|---|---|---|---|
|  | 340 d | 1e8 yr | 1e11 yr | 7e15 yr |  |  | 90 yr |  | 46.7 h |  |
| Pm143 | Pm144 | Pm145 | Pm146 | Pm147 | Pm148 | Pm149 | Pm150 | Pm151 | Pm152 | Pm153 |
| 265 d | 360 d | 17.7 yr | 5.5 yr | 2.6 yr | 5.4 d | 53 h | 2.7 h | 28 hr | 4 m | 5.4 m |
| Nd142 | Nd143 | Nd144 | Nd145 | Nd146 | Nd147 | Nd148 | Nd149 | Nd150 | Nd151 | Nd152 |
|  |  | 2e15 yr |  |  | 11 d |  | 1.72 h |  | 12.4 m | 11.5 m |

**FIGURE 2.16** A section of the chart of the nuclides. Notice the position of the stable elements $^{144}$Sm and $^{154}$Sm, which cannot be made by the s-process. The elements with very long half-lives, such as $^{144}$Nd, decay by emitting an alpha particle, in essence a $^{4}$He nucleus with two protons and two neutrons.

## R-process

Consider a supernova again. Detailed physical models have predicted that the violent explosion rips apart certain nuclei, freeing up a large number of very energetic neutrons. All the other stable nuclei, rather than being in the neutron "bath" of a normal star, are in a regular torrent of hot neutrons and may be hit with these neutrons repeatedly, without having time to $\beta$-decay into new elements.

After the torrent is over, the decay begins, and the nuclei keep decaying until they reach a stable isotope (see Figure 2.17). This process makes all the stable nuclei on the right-hand side of the s-process zigzag that the s-process couldn't get to, such as $^{150}$Nd and $^{154}$Sm. This way of making nuclides, by the rapid addition of neutrons, is called the **r-process**.

## P-process

The process that makes the isotopes to the left of the s-process zigzag on the chart of the nuclides is not well understood. At one time it was proposed that a torrent of protons, as well as neutrons, could be hitting the original s-process nuclei, thus forming these isotopes that are rich in protons. But detailed models of such a process have not been successful in matching calculated with observed abundances. Another theory suggests that it's not protons, but photons of high energy light, that are disrupting the original nuclei, changing neutrons into protons. The source of the photons would, again, be the supernova. Either theory could be called the **p-process**, which is convenient, if confusing. In any event, this process—if indeed it is one process—is still far from well understood.

## SUMMARY

The most abundant elements in the universe, by far, are hydrogen and helium. They were made in the big bang that began the universe. Inside giant stars, fusion reactions formed more helium and other atoms, primarily nuclides that were multiples of $^{4}$He nuclei. These include the next most abundant elements, such as O, C, Ne, Mg, and Si; they, and N, were also byproducts of simple fusion within these stars. This fusion can produce elements only a bit past $^{56}$Fe. Beyond this, the rest of the elements were built up either by the slow addition of neutrons inside stars (the s-process) or by the rapid addition of neutrons and other particles, followed by decay, in supernovae (r-process and p-process).

Because each element has to be made from a lighter element, it is not surprising that, as a general trend, the farther away from iron an element is, the less abundant it is. Also, because nuclides with even numbers of protons

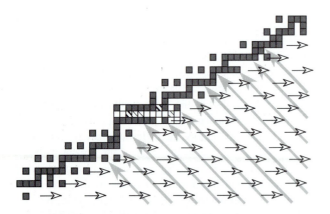

**FIGURE 2.17** A section of the chart of the nuclides, illustrating the r-process. The open arrows represent the direction in which new isotopes are formed by a flood of neutrons hitting stable isotopes. These unstable isotopes decay, following a path to the top and left of the chart, as indicated by the gray arrows, until a stable isotope is reached.

and neutrons are less likely to decay or capture a neutron, they should tend to be more abundant; checking back to Table 2.1 confirms that elements with even atomic numbers tend to be more abundant than those with odd atomic numbers.

The eventual blowoff of a supernova spews the products of these nuclear furnaces out into space, where eventually they can form second generation stars, rich in heavy elements, which are capable of having rocky and icy planets.

## STUDY QUESTIONS

1.  What elements were created in the big bang?
2.  What do we call a star that explodes?
3.  In the s-process, what does the letter "s" stand for?
4.  True or False: Our Sun will become a red giant some day.
5.  A popular science writer once said, "We are made of star-stuff." What was he talking about?

## 2.4  PROBLEMS

1.  In the text we discussed three forms of energy transport inside the Sun: conduction, convection, and radiation. Which of these three describes how the energy is transported from the Sun to Earth?
2.  Four hydrogens fuse to make one helium, but the mass of four hydrogen atoms is greater than the mass of the resulting helium atom. Where does the rest of the mass go? Is the Sun actually losing matter and becoming less massive, as it fuses hydrogen into helium?
3.  In the nineteenth century, astronomers believed that the Sun was slowly shrinking and that the compression of its internal gas was the source of its heat. What is wrong with this theory? Why did they believe it?
4.  The surface of Mercury is estimated to be at a temperature of 700 K in the daytime. Draw a rough sketch of what you expect a spectrum of Mercury to look like. Label the visible and infrared portions of this spectrum. At what wavelengths do you expect to find peaks in the spectra?
5.  A telescope with an infrared detector is pointed towards a spot in the sky where an asteroid should be visible. But when the astronomer looks through the eyepiece, she sees two points of light. One is the asteroid, and the other is a distant star. To her eye, they appear to be equally bright, but which is which? How can she tell the asteroid from the star?
6.  Restate the Wien displacement law for $\lambda$ in meters, $T$ in degrees Kelvin. What would the Wien displacement law be for $T$ in degrees Fahrenheit and $\lambda$ in feet?

7.  A gray asteroid orbits in space out beyond Pluto, where the temperature is only 40 K. Give a rough sketch of what its spectrum should look like (as in Problem 4). An interstellar imp splashes red paint all over this poor, cold asteroid. Now what does its spectrum look like?
8.  What are the chemical elements important to life? Where are they made?
9.  Take the numbers given in Table 2.1 and make a plot of abundance versus atomic number. You'll want to use semilog paper. What trends do you see? How could you explain these trends?
10. Name six ways elements are made, and for each process list an element with nuclides that could be made by that process.
11. Look at the chart of the nuclides section in Figure 2.16. What are the stable isotopes of Sm? What processes made each isotope?
12. A light element such as He gives off energy when it is fused into a heavier element such as C to make an even heavier element such as O; but a very heavy element such as U (uranium) gives off energy when it fissions into lighter elements (ultimately, He and Pb, lead). Why can't we get endless energy simply by fusing the He back into the Pb to make U, then letting the U decay back into Pb and He?
13. How far away from Earth would a typical supernova have to be to be as bright as our Sun? (The brightness dims as the square of the distance. If Earth were 2 AU from the Sun, the Sun would look one quarter as bright.) Chinese observers record seeing a supernova 1000 years ago that appeared $10^8$ times less bright than the Sun. (This is still bright enough to be visible in the daytime!) How far away was this supernova?

## 2.5  FOR FURTHER READING

A very readable and complete book on the Sun, written for a popular audience, is Robert W. Noyes, *The Sun, Our Star* (Cambridge: Harvard University Press, 1982). For a good introduction to solar physics, try the opening chapters of John C. Bradt, *An Introduction to the Solar Wind* (San Francisco: W. H. Freeman, 1970). A collection of technical review papers about the Sun has been assembled by the University of Arizona Press as part of their Space Sciences series: *The Sun in Time*, edited by C. P. Sonett, M. S. Giampapa, and M. S. Matthews (Tucson: University of Arizona Press, 1991).

The science of nucleosynthesis is described with wonderful detail in D. D. Clayton's classic text, *Principles of Stellar Evolution and Nucleosynthesis* (Chicago: University of Chicago Press, 1983). Perhaps the best general introduction to nuclear chemistry is Gerhart Friedlander and others, *Nuclear and Radiochemistry* (New York: Wiley, 1981).

**FIGURE 3.1**   The Moon.

# CHAPTER 3

# The Moon

## 3.1 ON THE MOON

What is it like to stand on another planet? The Moon is the one place, other than Earth, about which a human being can actually answer that question firsthand, because it is the one other planet where people actually have stood, picked up samples, and walked on the surface (Figure 3.2).

The pull of gravity one feels is one sixth that on Earth. Not only do you feel lighter, but things fall more slowly. Astronauts tended to travel across the lunar surface by hopping. Because the gravity is much less, it took less effort for them to get off the ground and they seemed to float for a longer time before they settled back down to the Moon.

The soil is white and powdery, a very fine sand, with the consistency of the scouring powder you might use to clean a kitchen sink. There are larger rocks on the surface as well, but many of them turn out to be a kind of rock called a **breccia**, a conglomeration of unrelated bits of stone and sand that have been packed together into boulder-sized chunks.

The terrain comes in two distinct types. Most of the Apollo landings took place on **maria** regions. These are remarkably flat plains that stretch for hundreds of kilometers, interrupted only by an occasional crater or by long cracks,

| The Moon's Vital Statistics | |
|---|---|
| Radius | 1738 km |
| Surface area | $3.79 \times 10^7$ km$^2$ |
| Mass | $7.35 \times 10^{22}$ kg |
| Density | 3.36 g/cm$^3$ |
| Local gravity | 1.62 m/s$^2$ |
| Escape velocity | 2.38 km/s |
| Albedo | 0.07–0.24 |
| Surface temperature | 380 K (day) |
| | 120 K (night) |
| Sidereal period | 27.32 days |
| Distance to Earth | 384,401 km |

called **rilles**, that can range from a few hundred meters to several kilometers wide and run for tens to hundreds of kilometers along the flat plains. The other kind of terrain is called the **highlands**, mountainous regions of bright white rock towering over the maria and covered with craters.

Because the Moon is smaller than Earth, the horizon appears closer. The eye, trained to Earth perspectives, has a hard time judging the size or distance of these soft, round-topped hills. Another factor confusing the perspective is the lack of atmosphere. There's no hazy air to obscure far-off mountains. The sky is black and stars are visible,

**FIGURE 3.2**   *Apollo 15 buggy on the Moon.*

even during the daytime. At night, when you don't have to worry about the glare of the Sun reflecting off the grayish white rocks, you can see the stars more brightly and more clearly than anywhere on Earth. The familiar constellations of bright stars can get lost in a background of fainter stars many times more numerous than you are used to seeing on Earth.

Because there's no air, you'll have to be inside a spacesuit, of course. The suit also has to have a heating and cooling system, because the Sun is relentless during the day but its heat is quickly lost at night. Temperatures range from 380 K during the day (water boils at 373 K on Earth) to 120 K at night. Your cooling system has a special difficulty on the Moon. Like all refrigerators, it has to dump the excess heat it has removed from your spacesuit to some place outside the suit. On Earth, air conditioners just let air outside the space being cooled carry the heat away. But on the airless Moon the heat has to be radiated away as infrared photons, which is a much less efficient process.

This brief picture of the Moon's surface raises many major questions we would want to answer concerning the Moon. Why is its gravity so low? What is the soil made of, and why is it powdery? What scattered the rocks, and what compacted them back together into breccias? How and when were the highlands made? How were the maria made? Why are they low and flat and darker than the highlands?

## Global Properties of the Moon

First, let's look at the general properties of the whole Moon: its size and shape, its gravity field, and its magnetic field.

To the naked eye, the Moon seems perfectly round; and indeed, it is rounder and smoother than any billiard ball. Its radius is 1738 km; in comparison, a mountain or

crater wall 2 km high is less noticeable than a scratch a thousandth of an inch deep on a billiard ball.

But by carefully measuring the position of the Moon in its orbit, we can determine the position of the center of mass of the Moon, and the center of the mass is not located at the center of the 1738-km ball. There appears to be more mass in the half of the Moon facing Earth than in the far hemisphere; enough that the center of mass is 3 km closer to Earth than is the **center of figure** (the geometrical center).

The Apollo spacecraft landed seismometers on the Moon's surface. Using data they collected on "moonquakes," we can determine the structure of the lunar interior (Figure 3.3). What we find is that the Moon has a thick crust of low-density rocks, presumably similar in composition to the samples returned from the highlands regions, made of **feldspars** and **pyroxenes**. This crust is up to 120 km thick on the far side, but only 60 km thick on the side near Earth. The far side is nearly all highlands (Figure 3.4), while virtually all the mare regions are on the near side, and the rocks from these maria are basalts rich in **olivine**, much denser than highland rocks.

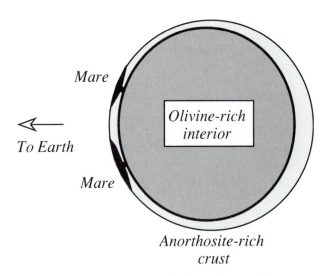

**FIGURE 3.3**   The distribution of material inside the Moon. The thickness of the crust is exaggerated.

One result is that over the mare basalts, the local force of gravity is found to be slightly stronger than the average gravity on the surface. This is caused by the extra concentration of mass that the thick, dense mare basalts have placed on the lunar surface. These **mass concentrations** are called **mascons**. By measuring the inferred extra mass of each mascon, one can determine that the mare basalts range from 0.5 km thick over the Oceanus Procellarum area to more than 5 km thick in the smaller round basins such as Imbrium.

**FIGURE 3.4** The far side of the Moon, photographed by the *Galileo* space probe.

The magnetic field is unusual in the vicinities of the mare basalts. The Moon as a whole does not have any magnetic field, but individual Moon rocks from the mare regions are slightly magnetic, typical of rocks that froze out of a lava in the presence of a magnetic field. The mare rocks are over 3 billion years old, and so it seems likely that 3 billion years ago the Moon did have a magnetic field, even though it does not have one now.

Planetary magnetic fields are thought to be generated in a region of molten metal, the **core**, that forms the central part of a planet. A planet without such a metal core would not have a magnetic field; or conversely, a planet with a magnetic field must have a metal core. The evidence for early lunar magnetism implies that the Moon must have a core, which was once molten but now is frozen.

If it does have such a core, it must be relatively small, less than a few hundred kilometers in radius, because the Moon's average density is low, lower than any other terrestrial planet. The core of Earth, by comparison, extends out to nearly half the distance from the center to the surface of Earth.

The Moon as a whole has the same density as an ordinary rock, 3.36 g/cm³. But the average density of the other planets is far greater than the density of the rocks we see on their surfaces. This means that the other planets must have some dense material in their cores, covered over by ordinary rock, to raise up their average densities. From this one can conclude that some dense core component is present in other planets, but not in the Moon. The only material that is both dense and plentiful deep inside planets

is metallic iron; thus it follows that the Moon does not have the metallic iron that other planets have and so cannot have a large metallic core.

The Moon is both smaller than the other terrestrial planets and made only of low-density material, and so it has less total mass than any other terrestrial planet. That is why the acceleration of gravity on the Moon's surface is so low: The less mass a planet has, the less its gravitational pull.

## The Moon's Surface: Physical Aspects

The most numerous surface features on the Moon are craters. These craters come in all sizes, ranging from 100-km-wide basins like Orientale to tiny microscopic pits one millionth of a meter in diameter. The lunar craters are formed by meteorites hitting the surface of the Moon; the large basins are formed by the impact of huge rocks, kilometers in size, while the microscopic pits are made by grains of dust hitting a Moon rock.

Besides peppering the surface with round, bowl-shaped features, meteorite impacts have another effect. They are the primary way that the surface features of the Moon become eroded. The little impacts by tiny meteorites, which occur continuously, unimpeded by any atmosphere, slowly grind down the rough edges of the larger craters and break up the Moon rocks into ever smaller pieces. The result is that the surface of the Moon today is covered with a sandy "soil" tens of meters deep, called the **regolith**.

The continued impacts tend to mix up both dark mare rocks and light-colored highlands material and keep mixing the soil already on the surface constantly. This process, called **gardening**, means that over a few million years material from the bottom of the regolith is eventually uncovered and brought up to the surface by the force of an impacting meteorite; then, as more impacts occur nearby, other debris eventually buries this material back deep into the regolith. In this process, fragments of rock from originally far-flung parts of the Moon may be compressed together into the conglomerates called **breccias** (Figure 3.5).

The rate at which material impacted the Moon's surface was much greater in the first billion years of the Moon's existence than it is now, as debris left over from the formation of the solar system was slowly swept up by the newly formed planets. During this period the highland regions became heavily cratered and the large circular basins were formed. The energy of the impacts that made these basins was on the order of $10^{24}$ J, comparable to billions of nuclear explosions set off simultaneously. Debris blasted out from such impacts cut scars across the highlands mountains. These basins are often surrounded by rings, marking the collapse of material in the region of the basin after the impact, as shown in Figure 3.6.

After such a basin was made, often long afterwards, molten rock from the interior of the Moon erupted to flood

**FIGURE 3.5** This conglomeration of rock was formed by bits of other rocks that were broken into shards when hit by meteorites and then glued together as further impacts piled up rocky debris on top of it. It is an example of a *breccia*.

the bottoms of these basins. This molten rock froze to make the dark mare areas of the Moon. We can tell that these areas are younger than the surrounding highlands because the lava lies on top of the heavily cratered material (which sometimes pokes up through the mare, as seen in the photographs) and because the mare regions are not nearly as heavily cratered as the highlands.

Even submicroscopic impacts can have an effect on the surface of the Moon. On airless bodies with no magnetic fields, high-energy particles from the solar wind or lo-

cal planetary magnetospheres can impact directly onto the surface materials, a process called **sputtering**. These particles are usually protons, electrons, or the nuclei of atoms such as sodium or sulfur that are travelling at extremely high velocities.

Upon impact a number of different reactions can occur. The particles, although tiny, can break up chemical bonds as they plow into a rock. Cosmic rays, the most energetic (but rarest) of these particles, leave trails in the upper few millimeters of each rock or meteorite they hit, providing a convenient way of dating just how long that particular rock has been exposed to space. (The more trails, the longer the rock has been sitting on the surface of its parent planet.)

The surface of the Moon tends to get darkened, in a way not completely understood, as it sits exposed to the solar wind. A combination of mechanical effects and actual chemical reactions between the rocks and the impacting hydrogen nuclei may be responsible.

## The Moon's Surface: Chemical Aspects

The energy of the impacts early in the formation of the Moon would have been sufficient to melt a considerable fraction of the Moon's surface. The chemistry of the Moon rocks reflects the outcome of such global melting.

Samples of highlands rocks returned by the Apollo astronauts showed several intriguing features. First, pristine highlands rocks are extremely rare; most are breccias. The ages of these rocks were difficult to measure because they had been so heavily battered by impacts, but all techniques indicated that they are at least 4 billion years old. But, finally, the very nature of the types of rocks they turned out to be was very surprising.

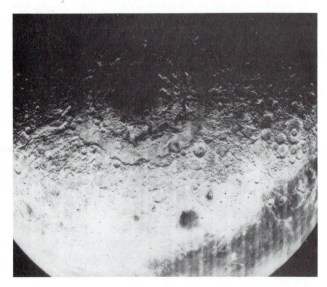

**FIGURE 3.6** Mare Orientale. This **multi-ring basin** on the eastern side of the Moon was formed by the impact of a large asteroid-sized meteoroid early in the Moon's history.

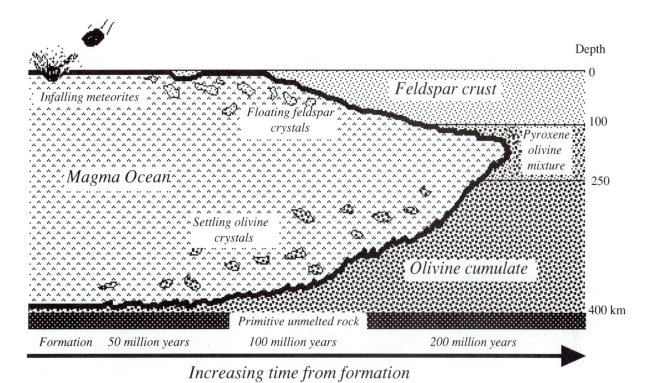

**FIGURE 3.7**  The Magma Ocean. Meteorites hit the early Moon with enough energy to melt the rocks, down to a depth of 400 km or so. In this ocean of magma, floating "rockbergs" rich in feldspar (lower density than the magma) gather on the surface to form an anorthosite crust, while higher-density crystals of olivine sink to the bottom. A mixture of different crystals, including all the incompatible elements, eventually freeze in a layer between the olivine and the anorthosite.

Igneous rocks made by lava flowing onto a planet's surface and then freezing have a specific chemical composition and mineralogic texture; basalts are rocks of this sort. Indeed, the rocks from the dark mare regions are classic examples of such basalts. But the highland rocks, for the most part, are not; rather, they belong to a class of rock called a **gabbro**. These are more typical of rocks crystallized slowly, deep inside a planet. So what are they doing on the surface of the Moon? Gabbros on Earth are brought to the surface with the mountain-building processes associated with plate tectonics, but there's no evidence for plate tectonics anywhere on the Moon.

One clue comes from the composition of these gabbros. They are rich in **plagioclase**, more specifically the mineral anorthite. This mineral is of a lower density than most other minerals, and indeed can be lower in density than the molten lava from which it was formed, if that melt also includes melted pyroxene or olivine. Thus, in a lava that was slowly cooling, the first mineral to crystallize would be plagioclase. Once formed, these crystals could tend to float upwards to the top of the lava, forming **anorthosites**, rocks made up almost completely of plagioclase.

Apparently the highlands rocks formed out of a large region of molten lunar material. How large might this region have been? The seismic detectors on the Moon indicated that the lunar crust, made up of gabbroic rocks, is 60 km to 120 km thick. Another sharp boundary in composition may exist about 400 km deep into the Moon. So one possibility is that the top 400 km of the Moon was completely molten at one time and the anorthosites crystallized out of this **magma ocean** to form the crust of the Moon (Figure 3.7).

This magma ocean was most likely melted by the energy of the large numbers of impacting rocks that hit the Moon as it was forming. It was probably completely frozen within 200 million years after the Moon was formed. As time went on and radioactive isotopes decaying in the lunar material caused the interior to warm up, temperatures eventually reached the the melting point of rock and basaltic lava could be formed. These basalts erupted to the surface and flooded the lowest areas of the Moon, creating the mare plains and their associated mascons. Judging from the ages of the mare rocks, this occurred between 3.9 and 3.0 billion years ago.

**TABLE 3.1**

Spacecraft That Have Orbited or Landed on the Moon

| Name | Launch Date | Type | Name | Launch Date | Type |
|------|-------------|------|------|-------------|------|
| Luna 1 | 01/02/59 | Unmanned flyby | Surveyor 5 | 09/08/67 | Unmanned lander |
| Luna 2 | 09/12/59 | Impact | Surveyor 6 | 11/07/67 | Unmanned lander |
| Luna 3 | 10/04/59 | Unmanned flyby | Surveyor 7 | 01/07/68 | Unmanned lander |
| Ranger 3 | 01/26/62 | Unmanned flyby | Luna 14 | 04/07/68 | Unmanned orbiter |
| Ranger 4 | 04/23/62 | Impact | Zond 5 | 09/14/68 | Unmanned flyby |
| Ranger 5 | 10/18/62 | Unmanned flyby | Zond 6 | 11/10/68 | Unmanned flyby |
| Luna 4 | 04/02/63 | Unmanned flyby | Apollo 8 | 12/21/68 | Manned orbiter |
| Ranger 6 | 01/30/64 | Impact | Apollo 10 | 05/18/69 | Manned orbiter |
| Ranger 7 | 07/28/64 | Impact | Luna 15 | 07/13/69 | Impact |
| Ranger 8 | 02/17/65 | Impact | Apollo 11 | 07/16/69 | Manned lander |
| Ranger 9 | 03/21/65 | Impact | Zond 7 | 08/08/69 | Unmanned flyby |
| Luna 5 | 05/09/65 | Impact | Apollo 12 | 11/14/69 | Manned lander |
| Luna 6 | 06/08/65 | Unmanned flyby | Apollo 13 | 04/11/70 | Manned flyby |
| Zond 3 | 07/18/65 | Unmanned flyby | Luna 16 | 09/12/70 | Unmanned lander |
| Luna 7 | 10/04/65 | Impact | Zond 8 | 10/20/70 | Unmanned flyby |
| Luna 8 | 12/03/65 | Impact | Luna 17 | 11/10/70 | Unmanned lander |
| Luna 9 | 01/31/66 | Unmanned lander | Apollo 14 | 01/31/71 | Manned lander |
| Luna 10 | 03/31/66 | Unmanned orbiter | Apollo 15 | 07/26/71 | Manned lander |
| Surveyor 1 | 05/30/66 | Unmanned lander | Luna 18 | 09/02/71 | Impact |
| Lunar Orbiter 1 | 08/10/66 | Unmanned orbiter | Luna 19 | 09/28/71 | Unmanned orbiter |
| Luna 11 | 08/24/66 | Unmanned orbiter | Luna 20 | 02/14/72 | Unmanned lander |
| Luna 12 | 10/22/66 | Unmanned orbiter | Apollo 16 | 04/16/72 | Manned lander |
| Lunar Orbiter 2 | 11/06/66 | Unmanned orbiter | Apollo 17 | 12/07/72 | Manned lander |
| Luna 13 | 12/21/66 | Unmanned lander | Luna 21 | 01/08/73 | Unmanned lander |
| Lunar Orbiter 3 | 02/05/67 | Unmanned orbiter | Luna 22 | 05/29/74 | Unmanned orbiter |
| Surveyor 3 | 04/17/67 | Unmanned lander | Luna 23 | 10/28/74 | Unmanned lander |
| Lunar Orbiter 4 | 05/04/67 | Unmanned orbiter | Luna 24 | 08/09/76 | Unmanned lander |
| Explorer 35 | 07/19/67 | Unmanned orbiter | Galileo | 10/18/89 | Unmanned flyby |
| Lunar Orbiter 5 | 08/01/67 | Unmanned orbiter | Hiten | 01/24/90 | Unmanned flyby |

## The Origin of the Moon

Humans always have wondered where the Moon came from. For a long time, we believed that the Moon was formed much as other planets were, out of a condensing nebula about the Sun. But there have always been some difficulties with this hypothesis. The Moon is *extremely* dry, much drier than Earth, which one might think would be most like it in composition, being so close. Even though it has been differentiated, if it has a metallic core it must be very small. The uncompressed density is wrong; it also should be much like Earth's, if they were both formed simply by condensation from the solar nebula.

A theory that is gaining in popularity is that a proto-Moon (perhaps originally more the size of the present Mars) and Earth underwent a giant collision with each other, after they were condensed but still very early in their history. This impact completely disrupted the proto-Moon, melting or even vaporizing much of it. Part of it was completely ejected from the system, but the rest formed a cloud of debris around Earth.

This collision had the effect of removing a lot of the volatile material originally present in the Moon, changing its density and chemical composition. The metallic material in the cores of both planets was combined into that of Earth. The material that had been in the mantle of the proto-Moon recondensed from the debris in orbit around Earth to form the present Moon.

## SUMMARY

The Moon, a small, rocky, airless body orbiting Earth, is the closest, most visited, and best explored extraterrestrial body (Table 3.1). Its surface is covered with impact craters, scars left over from the accretion of the Moon 4.5 billion years ago. Continued impacts to this day tend to wear down the walls of these craters and churn up the soil to a depth of several tens of meters. These impacts also tend to mix up, and compress together, fragments of different rock to form the conglomerate type rock called breccia that is the most typical type of boulder seen on the Moon today.

Early in its history these impacts melted the Moon to a depth of at least 400 km. Out of this magma ocean, a crust roughly 100 km thick of anorthositic gabbros was formed. Roughly half a billion years later, internal heating caused basaltic lavas to form; this molten rock erupted onto the surface, filling the low-lying areas in the bottoms of the largest impact basins, forming the dark mare regions. These eruptions went on for a billion years. For the past 3 billion years, however, little more has happened than the occasional new crater being formed to change the surface of the Moon.

The center of mass of the Moon and most of the mare regions are in the hemisphere facing Earth. Many mare basins have mascons associated with them, gravity anomalies due to the extra mass of the dense mare basalts that fill the basins.

## STUDY QUESTIONS

1. The Moon's surface is divided into two main types of terrain. What are the dark, flat areas called? What are the bright, rough areas called?

2. What is the most common feature seen on the Moon's surface?

3. For an astronaut standing on the Moon, it is hard to judge by sight just how far away features such as mountains are. There are several reasons for this difficulty. Name one.

4. One theory for the formation of the lunar crust proposes that it was frozen from a layer of molten rock 400 km thick. What is this molten layer commonly called?

5. The daytime temperature of the Moon is:
   (a) hot enough to boil water
   (b) hot enough to melt lead
   (c) not that different from Earth

## 3.2  MINERALS AND ROCKS

We have learned, by going there, that the surface of the Moon is covered with specific rock types. To understand better how the Moon was formed and how it produced the surface we see today, we have to know something about minerals and rocks.

There are three major issues we need to address. First, how are minerals and rocks made? Second, what can their composition tell us about the interior of the planet where they are found? And third, how can we discover how long ago the formation of these rocks took place?

## Nucleosynthesis and Rocks

Nuclear fusion in the centers of stars combines protons and neutrons to make the nuclei of the elements. When a supernova explodes, the elements are scattered to space. Out of these elements, the planets are formed.

But places like the Moon and the other rocky planets are not simply chaotic blobs of chemical elements. Rather, the elements are in chemical compounds, and these compounds form themselves into ordered crystal structures, which we call **minerals** (Figure 3.8). Finally, in the processes that form a planet's surface, several different kinds of minerals may be formed together and mixed into a conglomerate we call a rock.

**FIGURE 3.8** Minerals are chemical compounds with a regular crystal structure. The regular ordering of atoms in a crystal can create large, well-formed crystals, as seen in this sample of quartz ($SiO_2$).

Given the cosmic abundances of the elements, we should be able to predict which chemical compounds are the most plentiful. It turns out that virtually all the rocks in the solar system can be characterized as mixtures of just a few general types of minerals. Likewise, the most important types of rocks can be characterized by the way they are formed, which is limited to a few basic processes.

## The Primary Rock-Forming Minerals

Of the 83 naturally occurring elements, only a handful are abundant enough to be important in most minerals. In Earth and most of the other planets, these elements react chem-

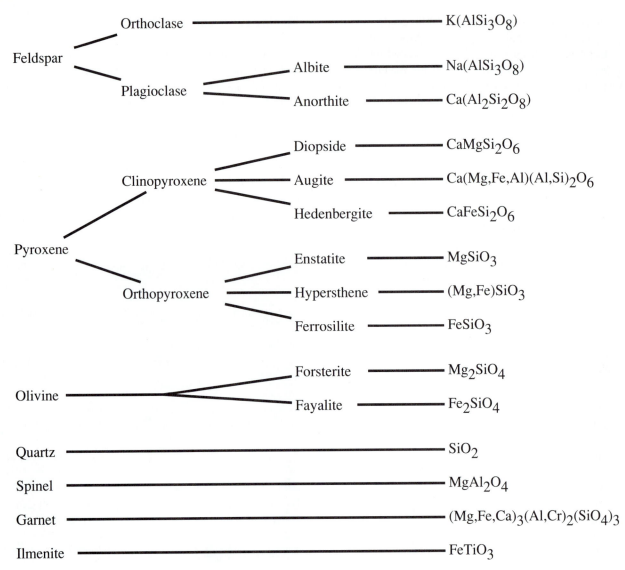

**FIGURE 3.9** The major rock-forming minerals.

ically to form oxides. These oxides, in turn, combine to form compounds that are the basis of the minerals.

*Mineral* refers not only to a chemical formula but also to a crystal structure. Two minerals can be very different, even if they have the same chemical composition, if those elements are arranged in different positions. On the other hand, two minerals might be very similar to each other, even if they're made of differing chemical elements, if they share the same crystal structure. The important classes of minerals are shown in Figure 3.9. The three most abundant types of minerals on the surfaces of planets are olivine, pyroxene, and feldspar. They are all examples of **silicate** minerals. This means that they are built from a crystal structure of silicon and oxygen atoms, with the other elements fitting in wherever their size and electric charge allows (Figure 3.10). The elements magnesium (Mg) and iron (Fe) both occur

with a +2 charge and have similar sizes, so they can substitute freely for each other in many crystals. That's why there's a range of olivines, from the iron-rich olivine (called **fayalite**) to the magnesium-rich olivine (**forsterite**). Likewise, calcium (Ca), sodium (Na), and potassium (K) all substitute for each other in pyroxenes and feldspars. But calcium has a +2 charge, while sodium and potassium are +1, so to balance out the charge, the abundance of aluminum (Al) and silicon (Si) has to be juggled. (Aluminum in minerals normally carries a charge of +3, while silicon carries +4.)

**Olivine** is a dense, usually olive-green mineral (hence its name). It is often found in the interiors of planets or in surface rocks made from a lava formed deep within the planet. It consists of silicon (Si), magnesium (Mg), and iron (Fe) with two atoms of Mg or Fe for each silicon

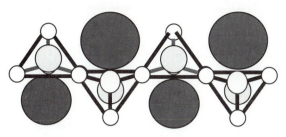

Pyroxene

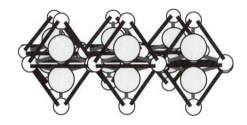

Olivine

**FIGURE 3.10**   The positions of the atoms in a mineral determine its crystal structure; examples of pyroxene and olivine are shown.

atom. These are the three most abundant rock-forming elements, and so we expect olivine to be the most common mineral in planets. Rocks rich in olivine are of necessity rich in **ma**gnesium and iron (**Fe**); these are often called **mafic** rocks.

**Pyroxene** is also generally greenish colored. Its chemical formula is quite similar to olivine, except that it has twice as much $SiO_2$, **silica**. Because the cosmic abundances of silicon and magnesium are roughly equal, a rock made with cosmic abundances of these elements but without iron oxide (FeO) would be mostly pyroxene instead of olivine. Pyroxene too, therefore, is a major constituent of planetary interiors, especially where iron is present only as metallic iron, and not iron oxide.

**Feldspar** is a general term given to the class of silicate minerals that are rich in aluminum and have an open, "framework" sort of crystal structure. Because of this open structure, there's room in feldspar crystals for larger-sized atoms, such as sodium (Na), calcium (Ca), and potassium (K). Many trace elements (any of 70 or so less-abundant elements that might be found in a rock) can also fit into this structure. Also, because the aluminum atoms are less massive than Fe or Mg, and because of the open structure of the crystal, feldspar minerals tend to be less dense than the other minerals and even less dense than most lava. As a result, when planets melt and refreeze, this lighter mineral tends to float to the crust of the planets. That is how we explained why the surface of the Moon is rich in feldspar.

The other types of minerals listed in Figure 3.9 can be important under certain circumstances. Deep inside planets

high pressures force the crystals into densely packed structures such as **garnet** and **spinel**. Compared to Earth, the mare rocks from the Moon show a high abundance of titanium (which does not fit easily into the crystal structure of the other minerals); on the Moon, the titanium-bearing **ilmenite** is an important mineral.

## The Formation of Rocks

Minerals are found combined together into rocks. Each rock type can have any number of minerals, usually two or three major minerals.

Traditionally, scientists have talked about three types of rock on Earth. These are **igneous** rocks, formed from molten lava that cools and crystallizes; **sedimentary** rocks, formed from the accumulation and conglomeration of weathered bits of other rocks; and **metamorphic** rocks, rocks that have been subjected to heat and pressure sufficient to alter their internal structure without melting them completely.

Not all these types are found on every planet. But it is clear that both sedimentary and metamorphic rocks are just altered forms of some other rock; and, because it appears that every terrestrial planet was melted at some time in its history, it is reasonable to assume that all rocky material on these planets started out as igneous rock. Thus it is the formation of igneous rocks that we will explore here.

The different types of igneous rock can be characterized by the process that made the rock. **Basalt**, an **extrusive** or **volcanic** rock, is made by lavas erupting onto the surface of a planet, then cooling; **granite**, **gabbro**, and **peridotite**, are formed from the slow freezing out of crystals from a melt below the surface of a planet (to form **intrusive**, or **plutonic**, rocks).

Basalts tend to have many tiny crystals, because the lava froze too quickly to organize itself (Figure 3.11, upper). One typically needs a microscope to easily see the individual crystals in such rocks. By contrast, the minerals in plutonic rocks had time to form large crystals, several millimeters across at least, as the lava slowly cooled (Figure 3.11, lower). Thus we call basalts "fine grained" while the plutonic rocks are "coarse grained."

Molten lava, from which igneous rocks are derived, is a part of a planet that has become hot enough to melt. A pure substance has a specific temperature at which it melts, called the melting point. But rocks are not made up of pure substances; they are mixtures of different minerals. A mixture of two (or more) substances often starts to melt at a temperature much lower than the melting point of either substance (Figure 3.12). This temperature is called the **eutectic temperature**.

For example, the mineral pyroxene melts at 1650 K and the mineral plagioclase melts at 1825 K, but if the two are mixed together, a molten lava starts to form when the

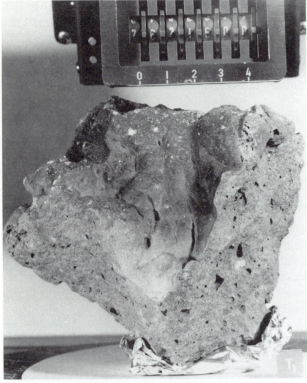

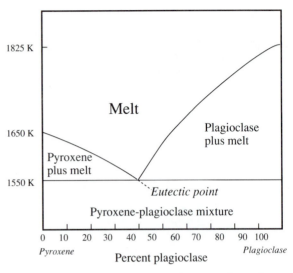

**FIGURE 3.12** A **phase diagram** for the pyroxene–plagioclase system.

**FIGURE 3.11** A lunar basalt (upper), and a terrestrial gabbro (lower).

temperature reaches only 1550 K. (The precise temperatures depend on the compositions of the minerals.)

The first lava to form when a rock starts to melt has a very specific composition, called the **eutectic composition**. Any mixture of these two minerals starts to melt with the same eutectic composition. You get the same composition whether you start out with almost all plagioclase and just a little pyroxene, or the other way around. In fact, substantial amounts of other minerals, such as olivine or metallic iron, can be present without changing the composition of the eutectic melt.

The composition of this eutectic melt is generally close to 60% pyroxene and 40% plagioclase. Although it does not depend on the proportions of plagioclase or pyroxene present, the precise composition does depend on the pressure. High pressures, such as might be seen deep inside a large planet, can cause a shift in the eutectic composition.

By precisely measuring the proportions of each mineral in a basalt, it is sometimes possible to determine how deep in the planet the lava was when it melted. This technique is called **geobarometry**. Naturally, nature is never quite this simple. Other complications, such as the iron content of the pyroxene, can also change the proportions of the minerals in a basalt.

The place where the lava is formed is called the **source region** of the lava. Using the techniques of geobarometry, we can infer that the basalts on the Moon came from several hundred kilometers below the surface of the Moon, while very similar-looking basalts seen in certain meteorites must have been formed inside much smaller planets where the pressure was very low.

## Trace Elements in Igneous Rocks

As we saw above, only a few elements are abundant enough to be important constituents of planets. Common minerals are formed from the arrangement of silicon, iron, magnesium, aluminum, calcium, and oxygen atoms into only a few configurations, a few crystal structures. Some other atoms, such as potassium and titanium, are abundant enough to make their own minor minerals. But where do rare elements such as hafnium or praseodymium fit in?

The answer is, quite literally, they fit in wherever they can. Most of these rare elements are more massive and physically larger than the major elements we listed above.

**FURTHER INFORMATION...**

# Ternary Diagrams

Understanding how mixtures of minerals behave is the key to understanding rocks. Minerals with similar structures, such as iron-rich olivine (fayalite) and magnesium-rich olivine (forsterite), can form a **solid solution** because the two minerals are intimately mixed and behave as if they were dissolved in each other. Other very different minerals, such as pyroxene and plagioclase, can form eutectic melts when heated. We saw (in Figure 3.12) how to represent a rock made of two different minerals. But how can we examine a system made of three minerals?

The answer is a three-sided figure called a **ternary phase diagram**. Assume that our rock is made of three minerals: quartz, olivine, and the calcium-rich plagioclase, anorthite (see Figure 3.13). The less anorthite there is in our rock, the farther away our composition is from the anorthite corner. If no anorthite is present at all, our composition lies on the line between olivine and quartz, and we are back to a two-mineral situation. Notice that the mineral pyroxene has a composition equivalent to one part olivine, one part quartz; thus it, too, can appear on our ternary diagram.

Look closely at the ternary phase diagram, Figure 3.13, for minerals under lunar conditions. The "v" markings along each axis mark off the composition scale, in intervals of 10%; each side of the "v" lines up with a "v" side on another axis. Thus, for example, the lunar basalt 60315 has a composition of about 45% olivine, 30% quartz, and 25% anorthite.

A melt of this composition, as it cooled, would first form olivine crystals (hence the point lies in a region marked "olivine"). The composition of the melt, once the olivine is removed, moves towards the plagioclase region. When the composition hits the boundary between the regions, both plagioclase and olivine crystallize together; the melt composition follows the arrow until it reaches the junction between the pyroxene, olivine, and plagioclase regions.

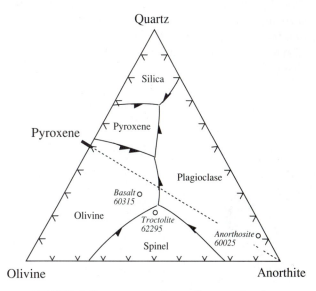

**FIGURE 3.13** **ternary phase diagram** for the olivine–anorthite–quartz system.

Melts whose initial composition lie below the dotted line, like our rock, will not evolve any further. However, more quartz-rich melts (like the anorthosite 60025) could continue to evolve up to the junction between the pyroxene, plagioclase, and silica regions. The boundary between olivine and pyroxene has two arrows, signifying that chemical reactions between the melt and the quartz take place as the melt evolves along this line.

Rocks that melt evolve in the opposite way; the first melt has a composition matching one of these two junctions; which junction depends on the overall composition of the rock. In effect, they mark the eutectic melting points for this system.

For them to fit into a crystal, the whole crystal has to be distorted in shape just a bit to make room for them.

Thus it is not surprising that once a crystal starts to melt, among the first bonds to be broken are those distorted bonds holding the large, incompatible elements into the crystal framework. The result is that molten lava tends to be relatively enriched in these trace elements. Thus basalts,

which are frozen lavas, are rich in these trace elements, while gabbros, which are formed of crystals that slowly settled out of molten rock, leave those elements behind in the melt and thus are poor in them.

That is the general pattern. Getting to specific cases, we find that each mineral has its own affinity, or disaffinity, for any given trace element. The **rare-earth** elements, the series from lanthanum (La) to lutetium (Lu), are a good case in point. (These are the elements that are usually found off by themselves in the periodic table.) These elements are almost chemically indistinguishable from one another. Experiments have shown, however, that plagioclase rejects lighter rare-earth elements, like lanthanum and cerium (Ce), more than the heavier rare earths, like ytterbium (Yb) and lutetium. But the opposite is true for pyroxene, which rejects the heavier ones more. Olivine and metal tend to reject all rare earths more uniformly. These trends are shown in Figure 3.14.

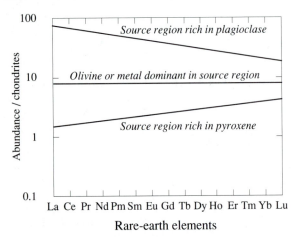

**FIGURE 3.14** Comparison between three simple types of rare-earth patterns. All values are normalized to chondrite composition.

Thus we can take a basalt and analyze the rare-earth content. The abundance pattern of the high-titanium (Hi-Ti) basalt in Figure 3.15 is mostly flat but slightly richer in lutetium than in lanthanum. The flatness suggests that olivine or metal was the dominant material in the source region, even though neither is present in the basalt itself, and the slight tilt towards lutetium suggests that the source region was richer in pyroxene than in plagioclase. The dip at europium is also significant, as we shall see.

The troctolite and anorthosite rocks shown in Figure 3.15 are not basalts, and so modelling their origin is not quite so straightforward. But it is still possible using their rare-earth element abundances and arguments similar to those described for basalts to learn much about the magma from which these rocks crystallized. For instance, the large superabundance of europium is characteristic of plagioclase (an important constituent of anorthosite in highlands rocks), whose crystal structure can readily accept this particular rare-earth element. The negative europium anomaly in the Hi-Ti basalt probably occurred because this rock was melted out of a region that crystallized after most of the plagioclase in the original magma ocean had formed the highlands anorthosite. On the other hand, apparently there was very little plagioclase crystallized when the troc-

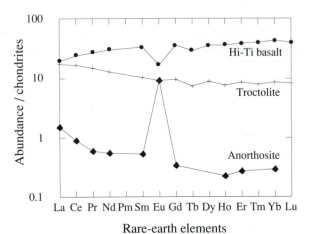

**FIGURE 3.15** Rare-earth element abundances for three lunar rocks. Trace element abundances give us information on the environment where a rock was formed.

tolite was formed, because it does not have a large europium anomaly.

## Dating Rocks

Is it possible to measure how old a rock is? First we must agree as to what we mean by a rock's age. Generally, the age of a rock is defined as the time that a rock has existed without being chemically disturbed. But there are different degrees of disturbance, so there may well be different ages for the same rock.

For instance, a lava on the Moon may have frozen into rock 4 billion years ago; but that rock may have been hit by an incoming meteorite some 500 million years later, which would significantly rearrange some of the atoms inside that rock. Such a rock could show two different ages, depending on how we measured the age.

The usual way to date a rock is to look at some radioactive element and measure how much of its **daughter** (decay product) is present in the rock. An example is an isotope of rubidium, $^{87}$Rb, which decays into strontium, $^{87}$Sr. The problem is how to sort out the daughter $^{87}$Sr from the $^{87}$Sr that was present in the rock originally ($^{87}$Sr$_{orig}$).

The amount of $^{87}$Sr present today ($^{87}$Sr$_{now}$) depends on three unknown quantities: the amount of $^{87}$Sr originally in the rock, the amount of original Rb ($^{87}$Rb$_{orig}$), and the time, $t$. (The age also depends on the rate at which Rb decays in Sr. However, this can be measured in the laboratory.) The amount of original Rb can be found by measuring the amount of Rb present today ($^{87}$Rb$_{now}$) and adding to it the excess Sr that's been made from the original time until now, the difference between $^{87}$Sr$_{now}$ and $^{87}$Sr$_{orig}$. We can also measure $^{87}$Sr$_{now}$ today. That still leaves us with

$^{87}Sr_{orig}$ and $t$ to find. In terms a mathematician would use, we have two unknowns; to solve for them, we must find two equations.

To do this, we look at two different minerals in the same rock that were presumably made at the same time, but which started off with different Sr and Rb contents. By comparing the two, we can solve for both the different initial Sr contents and for the time. (See Figure 3.16. The steeper the slope of the final abundance line, the older the rock is, while the point where this line intercepts the $^{87}Sr$ axis denotes the original abundance of $^{87}Sr$.)

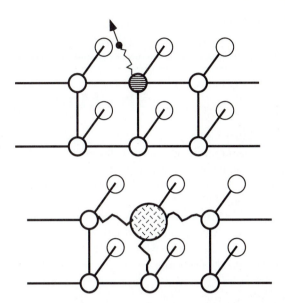

FIGURE 3.17 A radioactive element decaying while sitting in a crystal. By emitting a subatomic particle, the striped element transforms itself into a new element, with a different size and charge. The new element does not necessarily fit well into the old element's spot in the crystal structure; when $^{40}K$ decays into $^{40}Ar$, for instance, the $^{40}Ar$ is not chemically bound at all to its site in the crystal.

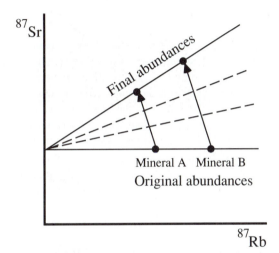

FIGURE 3.16 Dating a rock by looking at two minerals with different initial amounts of radioactive Rb. The more Rb the mineral started with, the more Sr it will have later on. The older the rock is, the steeper the slope of the line is.

Another atomic clock is provided by an isotope of potassium, $^{40}K$, which decays into argon, $^{40}Ar$. We can use a similar technique to Rb–Sr dating to find an age for the rock, but sometimes the two clocks arrive at different results. This happens because the daughter isotope of potassium decay, $^{40}Ar$, is a gas. If the rock is hit very hard—for example, if a Moon rock were hit by a meteorite forming a crater on the surface of the Moon—then some of the $^{40}Ar$ could get shaken loose and escape. In this way, the K–Ar clock would be set back to zero.

Argon is a noble gas and does not chemically combine with any atom in the mineral, but merely sits in the place where its parent $^{40}K$ used to sit (Figure 3.17). $^{87}Sr$, by contrast, does make a chemical bond with the same surrounding atoms that had been bound to $^{87}Rb$, and so it won't be dislodged by a shock. Thus the same rock can tell

us both the age when it was frozen from a lava and also the last time when it was struck by a meteorite.

## SUMMARY

Rocks are collections of minerals. Minerals are specific chemical compounds with specific crystal structures. On Earth and most other planets, the large abundance of oxygen ensures that most minerals are compounds of oxygen and other elements; each mineral can be considered as a collection of the oxides of various elements. Thus, by understanding the chemistry of the mineral classes olivine (two parts magnesium or iron oxide, one part silica), pyroxene (one part magnesium or iron oxide, with an occasional calcium oxide, and one part silica), and feldspar (one part calcium or sodium oxide, one part aluminum oxide, and one part silica), one has mastered the most overwhelmingly abundant of the minerals found in planetary rocks.

Likewise, the most common types of rock themselves are basalts (pyroxene and feldspar, made from lavas erupted on the surface of a planet), gabbro (pyroxene and feldspar, slowly cooled and crystallized inside a planet), peridotites (pyroxene and olivine, cooled inside a planet), and granite

**FURTHER INFORMATION...**

# The Math Behind Dating Rocks

In this chapter we have discussed the physical ideas behind how the measured abundances of radioactive trace elements in a rock can be used to determine its age. The mathematics behind this theory is a straightforward application of Newton's calculus.

Assume that we have an isotope with an abundance of $N$, which decays at a rate $\lambda$ decays per second. By the definition of a derivative, we can write the rate at which the abundance changes as

$$\frac{dN}{dt} = -N\lambda \tag{3.1}$$

(the minus sign indicating that the abundance is decreasing with time). This can be integrated to show that after a time $t$,

$$N = N_{orig}e^{-\lambda t} \tag{3.2}$$

where $N_{orig}$ is the original abundance of the isotope. The age of the rock, $t$, is ultimately what we're attempting to find, but to use this equation we also need to know $N_{orig}$. Furthermore, in a practical sense it is much easier to measure ratios of isotope abundances than to count the exact number of atoms of a given isotope in a particular rock.

In the Rb/Sr system, all abundances are measured relative to the stable isotope $^{86}$Sr. We measure the abundance of $^{87}$Rb/$^{86}$Sr; call that $N$. As the amount of $^{87}$Rb decreases, the amount of $^{87}$Sr increases, and we can also measure the amount of the daughter isotope, $^{87}$Sr/$^{86}$Sr. Call this ratio $D$.

The amount of $D$ present in a mineral depends on the initial abundance, $D_{orig}$, the amount of $N$ present originally, and the amount of time that the rubidium has had to decay into strontium. Initially, when $t = 0$, we know that $D =$ $D_{orig}$; after an infinite amount of time, all the initial $N_{orig}$ is converted into $D$ and so $D_{max} = D_{orig} + N_{orig}$. In between, the $N$ that is present at time $t$ represents material that is not yet converted into $D$, and so we can write

$$\begin{aligned} D &= D_{orig} + N_{orig}\left(1 - e^{-\lambda t}\right) \\ &= D_{orig} + N\left(e^{+\lambda t} - 1\right) \end{aligned} \tag{3.3}$$

Consider two minerals, $A$ and $B$. If they formed at the same time, when strontium isotopes were well mixed, then even though the amount of strontium in one mineral may be very different from that in the other, the ratio of strontium isotopes, $D$, starts out exactly the same in both minerals. Hence $D_{orig}$ of mineral $A$ equals $D_{orig}$ of mineral $B$. After formation, both minerals evolved for a time $t$ (again, $t$ is the same for both minerals). Thus we can write

$$D_A = D_{orig} + N_A\left(e^{+\lambda t} - 1\right) \tag{3.4}$$

$$D_B = D_{orig} + N_B\left(e^{+\lambda t} - 1\right) \tag{3.5}$$

Solve the second equation for $D_{orig}$ and substitute into the first equation:

$$\begin{aligned} D_A &= \left[D_B - (N_B)\left(e^{+\lambda t} - 1\right)\right] \\ &\quad + (N_A)\left(e^{+\lambda t} - 1\right) \end{aligned} \tag{3.6}$$

This can be solved for the age of the rock, $t$:

$$t = \frac{1}{\lambda}\ln\left(\frac{D_A - D_B}{N_A - N_B} + 1\right) \tag{3.7}$$

(mostly feldspar and quartz, pure silica, cooled beneath the surface of a planet). By studying the trace element abundances of a rock, one can deduce the conditions under which its parent lava was formed. Likewise, if the trace element is the daughter of a radioactive decay process, one can estimate the time at which the lava froze and thus find the age of the rock.

## STUDY QUESTIONS

1. For each term, state whether the material named is an element, a mineral, or a rock:

   | | | |
   |---|---|---|
   | titanium | plagioclase | granite |
   | gabbro | pyroxene | basalt |
   | silicon | olivine | peridotite |

**FURTHER INFORMATION...**

# Modelling the Origin of a Rock

One can mathematically model, in detail, the evolution of the trace elements in a geochemically evolving system as melting goes on inside the planet. From such a model, one can determine precisely how much plagioclase, pyroxene, or other minerals were present in the original source region (which may be hundreds of kilometers below the surface and thus otherwise impossible to sample directly). Furthermore, from these models one can determine how much the rock had melted when the basaltic lava was erupted, which tells us how much heat was present at the time and how thorough was the evolution of the interior.

We noted that every mineral has a certain affinity for each trace element. When a mineral begins to melt, the ratio of the concentration of a trace element in the molten, liquid phase to that in the solid phase is constant; every time we start to melt that mineral we get the same ratio of trace element concentrations. We call this constant the **partition coefficient**, and it is often given the symbol $D_{\ell/s}$.

But the ratio of the concentration of trace element in the liquid phase to the total initial concentration changes continually as the melting proceeds; so does the solid phase/initial concentration ratio. When melting just begins, most of the trace element is in the solid; when melting is almost complete, most of it is in the liquid. It is obvious that if the system is closed, that is, if the lava hasn't started to erupt yet but is still sitting with the unmelted portions of the rock, then the total amount of trace element is not changing. The total amount of trace element present is equal to however much is in the lava plus however much is in the rock. Mathematically, we can write

$$c_o = c_\ell F + c_s (1 - F) \qquad (3.8)$$

where $c_o$ is the original concentration of the trace element, $c_\ell$ is the concentration of that trace element in the lava, and $c_s$ is its concentration in the solid rock; $F$ is the fraction of the original rock that has melted into lava. But our definition of the partition coefficient, $D_{\ell/s}$, is the ratio of the liquid to solid concentrations, so $D_{\ell/s} = c_\ell/c_s$. So we can write the equation above as

$$c_o = c_\ell \left[ F + D_{\ell/s} (1 - F) \right] \qquad (3.9)$$

and solve for the ratio

$$\frac{c_\ell}{c_o} = \frac{1}{\left[ F + D_{\ell/s} (1 - F) \right]} \qquad (3.10)$$

In a typical lunar rock model, $c_\ell$ may be the observed concentration of the different rare-earth elements measured in a certain lunar sample, $c_o$ is the abundance of that trace element in the source region, and $D_{\ell/s}$ is the average value of the $D_{\ell/s}$ for the various minerals suspected of being in the source region. We measure $c_\ell$ in the lab; if we have a good guess for the composition of the source region, we can solve for amount of fractionation, $F$.

Alternatively, if we can estimate the original rare-earth element abundances, $c_o$ (for instance, by assuming that the different trace elements originally occurred in the same relative proportions as are seen in our table of cosmic abundances), and the amount of fractionation, then we can solve for $D_{\ell/s}$ and determine what combination of minerals would provide an average $D_{\ell/s}$ value consistent with the trace element abundances we observe in the rock.

---

2. We can measure precisely the amounts of rubidium and strontium isotopes in the minerals of a rock. What can this information tell us about the rock?

3. Argon is a gas. Do we ever find it in rocks?

4. True or False: If two rocks have identical amounts of the same elements present, then we can say that they are made of the same minerals.

5. The following is a list of qualities about an igneous rock. Name the one that cannot be determined from studying the geochemistry of that rock:

(a) age of the rock
(b) temperature at which it was formed
(c) pressure at which it was formed
(d) other minerals (not in the rock now) present when it was formed
(e) the depth inside the planet where it was formed
(f) none of the above

## 3.3 CRATERS AND VOLCANOES

Every surface in the solar system is peppered with round features. It was easy for early astronomers to see them on the Moon, which is thoroughly covered with these pockmarks (Figure 3.18). We now know that the moons of the other planets are covered with them; on Venus and Earth and Mars they are rarer, but still exist. Where do these round forms come from?

**FIGURE 3.18** Small craters on the Moon tend to be bowl-shaped.

Many bizarre theories were proposed in the nineteenth century. One scientist speculated that bubbles percolated up through a molten Moon, and the round features were what they left behind as they reached the surface and burst. But by the turn of the twentieth century only two theories were in serious contention.

One group of astronomers noted that the most common form of round **crater** feature (*krater* is a Greek word meaning "bowl") seen on Earth were volcanoes (see Figure 3.19). Because volcanoes leave craters on Earth, they should also leave craters on the Moon.

The argument against this theory was the size and number of the lunar craters. Some lunar craters are several hundred kilometers in radius. What sort of volcano would make so large a hole? And volcanism would have to have occurred on a scale unknown anywhere on Earth to leave so many craters.

In addition, the more detail that could be seen in lunar craters, the less they looked like terrestrial volcanoes. No lava flows could be seen coming from them, and the crater walls seemed rough and abrupt. Worst of all, by carefully

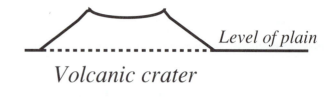

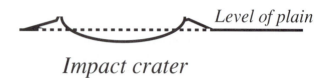

**FIGURE 3.19** A comparison of impact versus volcanic craters, in cross section.

measuring the lengths of shadows on the Moon's surface, astronomers could measure the relative heights of different areas on the surface, and it always turned out that the floors of the craters there were lower than the surrounding region. Volcanic craters on Earth are usually at the tops of mountains, and the bottoms of the craters are elevated above the surface of the surrounding plains (Figure 3.19).

To get around these problems, another theory was proposed suggesting that the craters were holes made by the impact of large meteorites striking the surface of the planet. Barringer (Meteor) Crater in Arizona, a round feature with no associated volcano and with a floor that was indeed lower than the surrounding countryside, was held up as an example of such a process occurring on Earth (another such crater is illustrated in Figure 3.20). To account for the large size and number of these craters on the Moon, it was pointed out that the Moon has no atmosphere and no running water; hence there should be no erosion, and so the surface could be extremely old, showing all the scars it

**FIGURE 3.20** Wolfe Creek Crater, an impact crater in Australia.

had accumulated over the past 4 billion years. By contrast, weather on Earth would soon wipe away any craters.

Still, this would demand an extremely high cratering rate for the Moon; and because there wasn't any evidence for such cratering in Earth rocks going back 3 billion years, even the youngest parts of the lunar surface would have to be older than virtually all the surface of Earth.

Another point that seemed intuitively wrong was that the craters on the Moon were all round. Our experience in digging holes by throwing things on Earth—snowballs thrown into a snowy field or rocks tossed onto a sandy beach—is that the objects tend to make oblong-shaped pits, unless the rocks fall straight down. It seemed unlikely that all the meteors on the Moon struck with such an absolutely precise straight-down direction.

This controversy has been solved. The craters on the Moon are caused by impacts (and we know from radionuclide age dating that the surface of the Moon is, indeed, older than any surviving surface on Earth). But it nonetheless is true that most of the crater-shaped features on Earth are caused by volcanoes. When we look to an unexplored planet and also see round features, how can we tell the difference? And why are the craters of the Moon so round? To understand we must examine the way both craters and volcanoes are formed.

## Impact Craters

The key to understanding impact craters is to realize that they are not the same as holes dug by a rock dropped onto sand. A rock falling onto sand is only travelling at a few meters per second; it pushes material out of its path as it hits the sand. Physically, what happens is that an impulse, travelling at the speed of sound, passes from sand grain to sand grain, moving each grain out of the path of the incoming rock and leaving a (noncircular) trail of moved sand grains behind the rock.

A large meteorite crashing into a planet, however, is travelling at perhaps 10,000 m/s. Its velocity is faster than the speed of sound in the rock. (The impact of such a projectile is called a **hypervelocity impact**.) The surface material does not have time to move out of the path of the meteorite. Because the surface does not yield at all, the meteorite must suddenly stop. All its kinetic energy must be instantly turned into heat. The meteorite explodes, with a force comparable to that of a nuclear bomb. The explosion is perfectly symmetrical, and so the crater is round.

Exactly what happens next? Geologists have made detailed studies of impact craters on Earth, such as Meteor Crater in Arizona, and a special gun has been built that shoots rock balls at hypervelocities into specially prepared targets, while the whole process is filmed with a high-speed camera. From this work, we get the following picture of how a crater forms (Figure 3.21).

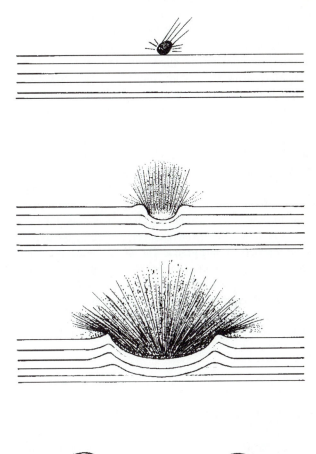

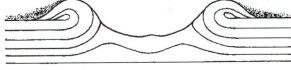

**FIGURE 3.21** The several states in the formation of an impact crater. From impact to final crater may occur in a few seconds; slumping and erosion continues for the life of the crater.

When the projectile explodes, it sends a shock wave into the surface, compressing the rock initially, and jetting, or "squirting out," a small amount of mass at the point of impact. The meteorite itself is totally destroyed, although small fragments may be found far from the crater. (Not realizing this, miners spent years looking for the large iron meteorite they thought must be buried underneath Meteor Crater.)

As soon as the explosion is over, the rock, which had been compressed by the shock wave, is suddenly released from this compression as the wave passes on. This release causes a general expansion of the rock, which "burps up" out of the surface and folds itself over onto the side of the crater. The resulting walls can be quite steep; as things settle down, they tend to slump back into the crater.

Meanwhile, small rocks thrown out by the initial jetting and the excavation of the crater can be thrown a considerable distance. These produce numerous smaller craters near the main crater and streaks of craters that often run in lines for thousands of kilometers away from the impact site. Material newly dug up from below the surface of an airless body like the Moon tends to be lighter in color than the surrounding rock. Radiation from the Sun and micrometeorite impacts tend to make the soil darker as time goes on. As a result, these streaks of freshly overturned soil are often very visible as the bright "ray" patterns seen around fresh craters on the Moon and Mercury.

Craters come in all sizes. The 500-km-wide basins on the Moon, filled with mare basalts today, were initially huge impact craters. Moon rocks have been found with tiny pits only a few microns in diameter, caused by the hypervelocity impacts of dust grains. The smaller impacts are more numerous and have the effect of slowly eroding the surface of any planet that is not protected by an atmosphere. It is these 4 billion years' worth of small impacts, in a process called **gardening**, that have rounded and softened the contours of the mountains on the Moon and turned the lunar soil into sand.

## Central Peaks and Central Pits

As we described above, the impact of a meteorite into a planet's surface imparts a sudden burst of energy into the surface. The energy travels through the crust of the planet in a sudden shock wave. The wave travelling up to the surface of the planet ejects material, forming the crater; however, part of the wave can also travel downward, into the lower regions of the planet (Figure 3.22).

Large impacts can be hundreds of kilometers in diameter. If we realize that the shock wave has roughly the same strength in all directions, this means that a large impact can send a very strong shock wave hundreds of kilometers into the planet. By travelling to such a depth, the shock wave will almost certainly pass completely through the planet's crust and into new material with a different composition deeper inside the planet. When a wave travels from one kind of material into a different type, some part of the wave is reflected (just as a pane of glass reflects light). The reflected wave then travels back up to the surface of the planet. On the Moon, such reflected waves tend to throw material up from the crater floor into mountains in the center of the crater, called **central peaks** (Figure 3.23). On other planets where fluids such as water may be present, the energy of this rebounding wave may cause the center of the crater to collapse into a **central pit**.

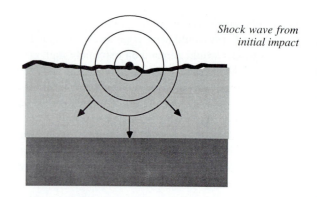

*Shock wave from initial impact*

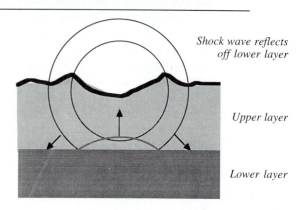

*Shock wave reflects off lower layer*

*Upper layer*

*Lower layer*

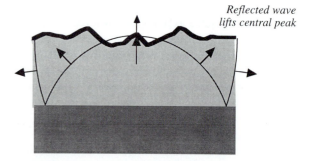

*Reflected wave lifts central peak*

**FIGURE 3.22** When a crater is formed, a shock wave is sent through the crust of the planet. If the wave is strong enough, it may be reflected off the bottom of the crust and may rebound to the surface, causing a mountain to form in the center of the crater.

## Volcanoes

When molten lava rises to the surface of any planet, it can form a variety of surface features. The final shape the lava takes depends on two factors: how viscous the lava is and whether it encounters any unusual conditions when it hits the surface.

**FIGURE 3.23** Tycho, one of the most prominent craters on the Moon, is 85 km in diameter. Notice the central peak and the slumped crater walls.

The **viscosity** of a lava, how sluggishly it flows, depends on its composition and temperature. In general, the more silicon there is in the lava, the more sluggish it behaves, that is, the higher its viscosity is. Simple lavas, like those that made the mare basalts on the Moon, are made of minerals such as olivine that are relatively poor in silicon, and so they are very low in viscosity. Such material can erupt through long **fissures** (cracks in the surface) and quickly flow onto the surface, filling in low-lying areas and travelling hundreds of kilometers before they freeze. These sheets of rock are called **flood basalts**. Besides the maria on the Moon, a good example is the flows in the Columbia River plateau in the northwestern United States. Once frozen, flood basalts often cover over their source fissures, and thus do not leave obvious volcanoes behind. However, hollowed-out tubes that the quickly flowing basalt left behind often collapse, leaving a chain of round holes on the surface that look not unlike impact craters.

**Shield volcanoes** occur when the eruption of this sort of material comes from one spot, rather than from many long fissures. Eventually a very large, shallow cone can be built up, with a slope less than 10° or so. Standing on such a mountain, you might barely be able to tell which direction is uphill. Yet such volcanoes can be very large; the islands of Hawaii were all built this way, including the main island where Mauna Loa (see Figure 7.8) is still erupting today. Mount Kenya and Mount Kilimanjaro, two shield volcanoes near the African Rift Valley, are high enough that their peaks are snow-capped year round, even though they lie on the equator. Olympus Mons on Mars is a shield volcano; at 26 km high, it may be the tallest mountain in the solar system. But its slopes are very gentle; the whole volcano

is over 1000 km across. Large crater-shaped **calderas** are often seen at the tops of these volcanoes.

**Composite volcanoes** are the classic cone-shaped mountains, like Mount Fuji in Japan (see Figure 6.11), or Mount Shasta in California, which one thinks of when one pictures volcanoes. The lava that makes such mountains is very rich in silica. Thus it is very viscous and tends not to travel too far from its source before it freezes. The slope of such a mountain may be 40°. The sources of this silica-rich lava are often sea sediments, pulled below Earth's crust by the motions of **plate tectonics** (see Chapter 6), and so these volcanoes tend to come in chains along the boundaries of plates. The material also tends to be rich in gases, such as steam, which drive vigorous eruptions.

All the eruptions described above assumed that the lava encountered only rock when it reached the surface of the planet. However, if the hot lava hits a surface material with a low boiling point, other odd configurations can result. Consider what happens when hot lava hits groundwater. The water, suddenly turned to steam, explodes, and this leaves a crater behind that can often be confused with an impact crater. Such volcanoes are called **maars** (named after the fine example near the town of Maar in Germany. Oddly enough, there also appear to be maars on Mars as well!). On Earth, the groundwater that led to the maar's formation may fill the crater after it is formed.

One theory for the volcanic plumes of Io (one of the moons of Jupiter) suggests that much the same process occurs there, only it is sulfur, not water, that is vaporized. The material in the plumes of Io travel in ballistic paths, as if they were shot out of a tube by an explosion. Lava flows of liquid sulfur, reminiscent of shield volcanoes, are also seen on Io. Sulfur is also seen in Earth volcanoes, although not in the same quantities as on Io.

## SUMMARY

Craters and volcanoes are two of the most common features on the surfaces of planets. Because both tend to make round bowl-shaped features, it is sometimes difficult to tell them apart. However, certain distinguishing features such as rough sides and crater floors deeper than the surrounding countryside are characteristic of impact craters. Volcanic craters tend to occur at the tops of mountains and usually have visible lava flows.

Meteorite craters are made by the impact of a meteorite into the planet at a speed greater than the speed of sound in rock. The meteorite explodes upon impact, compressing the ground where it hit. The release of this compression causes the ground below the impact to be thrown up and out of the surface, creating a pit with rough, overturned walls. Debris sent out of the pit by this event can make smaller, secondary craters around the main impact, including rays of craters that may extend for 1000 kilometers.

Volcanoes are made by lava erupting onto the surface of a planet. Shield volcanoes are caused by low-viscosity lava, rich in molten olivine and pyroxene. They tend to be very large but not very steep. Composite volcanoes are made of more viscous, silica-rich lavas. They tend to be steeper, cone-shaped mountains. Other features associated with volcanic eruptions include flood basalts and maars.

## STUDY QUESTIONS

1. What is the simplest way to tell, by appearance, an impact crater from a volcanic crater?

2. There are several reasons why impact craters are more abundant on the Moon than they are on Earth. Name one.

3. The mare regions of the Moon are examples of:
   (a) flood basalts
   (b) shield volcanoes
   (c) maars

4. Olympus Mons on Mars is an example of a:
   (a) flood basalt
   (b) shield volcano
   (c) maar

5. If you dug beneath Meteor Crater, would you find the meteor that dug out the crater?

## 3.4   PROBLEMS

1. A nuclear bomb is set off on the Nevada desert. What sort of feature do you expect this to make? Describe, step by step, the evolution of this feature.

2. The southern hemisphere of Mars is almost as heavily cratered as the lunar highlands, while the northern hemisphere of Mars has much fewer craters. Guess the relative ages of each hemisphere of Mars, compared with the lunar highlands and compared to the surface of Earth. What extra effects must you consider on a planet like Mars that you don't have to worry about for the Moon, which might complicate your estimates of relative ages?

3. Name two reasons why secondary craters are found farther from a crater of given size on the Moon than on Mars.

4. You are planning an expedition to explore the Moon. Which of the following supplies would be useful, and which would not? Give your reasoning.

   | compass | sextant | shovel |
   | flashlight | telescope | parachute |
   | flares | mirror | radio |

5. Gold is an example of an element called a "siderophile," which tends to be found with other metals. Do you think gold prospecting would be a profitable enterprise on the Moon?

6. How does the surface area of the Moon compare with the surface area of North America? Compare the difficulties of settling North America in the 1500s with the likely problems of colonizing the Moon in the 2000s.

7. How high an athlete can jump over a hurdle depends on how much energy she puts into the jump; the potential energy at the height of the jump must equal $E$, where $E = mgh$, $m$ is the mass of the jumper, and $h$ is the height of the jump. A good athlete can do a high jump of 6 ft on Earth. How high can she jump on the Moon? What if her spacesuit weighs as much as she does?

8. Given the cosmic abundance of the elements, which do you think is more abundant, plagioclase or pyroxene? Orthopyroxene or clinopyroxene? Plagioclase or ilmenite?

9. Two basalts with identical major element compositions are measured for their trace elements. Rock A has uniformly 10 times the cosmic abundances of the rare-earth elements. Rock B has rare-earth element abundances ranging from 15 times cosmic for La to 5 times cosmic for Lu, with the other rare-earth elements falling between these extremes. How can you explain the difference in rare-earth element patterns?

10. A single crystal of pure orthoclase is found. What age-dating technique might you try?

11. Radioactive age dating gives some rocks on Earth ages older than 3 billion years. However, creationists believe that Earth is only a few thousand years old. What assumptions did we make in the radioactive dating scheme that would be thrown into question if it turned out that the creationists were right?

12. No composite volcanoes have been seen on the Moon. Why do you suppose this is?

13. Which sort of volcanoes are likely to have explosions associated with their eruption? Mount St. Helens erupted with a large explosion in May 1980. What sort of volcano do you think this was? What sort of volcano is Mount Vesuvius? What composition of rocks would you expect to see around these volcanoes?

## 3.5   FOR FURTHER READING

Of the many excellent books written about the Moon since the Apollo era, perhaps the best place to begin exploring detailed lunar science is Stuart Ross Taylor, *Planetary Science: A Lunar Perspective* (Houston, Tex.: Lunar and Planetary Institute, 1982). Another interesting source is *Lunar*

*Sourcebook: A User's Guide to the Moon*, edited by Grant Heiken, David Vaniman, and Bevan M. French (New York: Cambridge University Press, 1991). The Arizona Press, whose Space Science series books make up the backbone of planetary science reference libraries, has not produced any book on the Moon, curiously enough. However, in many ways this gap has been filled by the compilation of review papers, *The Origin of the Moon*, edited by W. K. Hartmann, R. J. Phillips, and G. J. Taylor (Houston, Tex.: Lunar and Planetary Institute, 1986).

Many good introductory textbooks in geology discuss the basics of minerals and rocks. Among the classics are Cornelius S. Hurlbut, Jr., and Cornelis Klein, *Manual of Mineralogy*, 21st ed. (New York: Wiley, 1993), which is the rewriting of James D. Dana's classic 1848 mineralogy manual; and W. A. Deer, R. A. Howie, and J. Zussman, *An Introduction to the Rock Forming Minerals*, 2d ed. (New York: Wiley, 1992).

The authoritative book on craters is *Impact Cratering: A Geologic Process*, by H. Jay Melosh (New York: Oxford University Press, 1989). Many excellent books have been written about volcanoes; one favorite, written at an introductory level, is by Cliff Ollier, *Volcanoes* (Cambridge, Mass.: MIT Press, 1969).

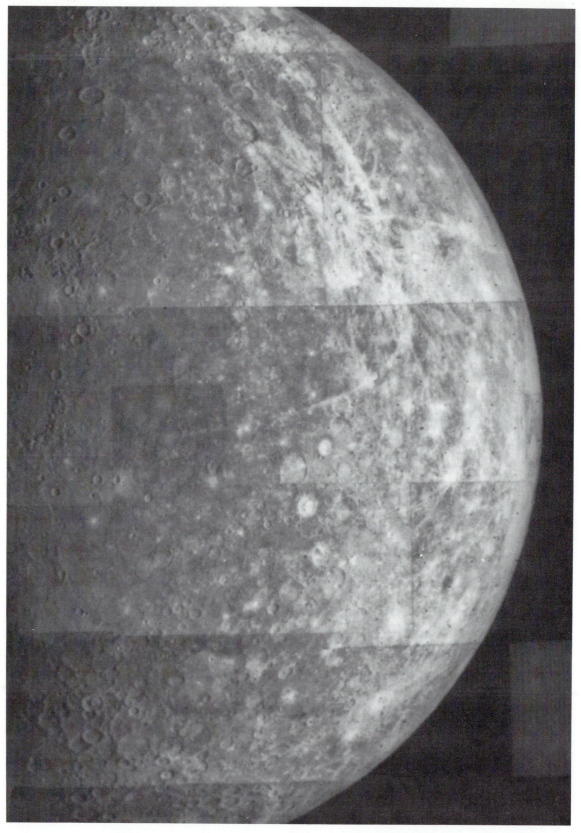

**FIGURE 4.1** Mercury, as seen by *Mariner 10*. The giant Caloris Basin is just visible on the left.

# CHAPTER 4

# Mercury

## 4.1 ON MERCURY

If you were to show the photograph on the opposite page (Figure 4.1) to someone and ask what planet this is, the answer you'd most likely get would be "the Moon." It's not; it is Mercury, as photographed by the *Mariner 10* spacecraft. But this initial reaction would not be an unreasonable one. Indeed, the first, most important thing you can say about Mercury is that it does look remarkably like the Moon. It is heavily cratered, just like most of the Moon's surface, and it appears that very little has happened on the surface of Mercury to erase the craters, except for the emplacement of yet more craters. A survey of the entire side of Mercury that *Mariner 10* saw continues the analogy. There are regions that look like large circular basins, just like those we see on the Moon, and there appear to be regions where the craters have been filled in or covered by flows of lava, similar to the maria regions of the Moon. Even the general color and brightness of Mercury are very similar to the Moon's.

But there are also differences between the surfaces. The shapes and spacing of the craters are slightly different on Mercury. The material filling the basins isn't as dark on Mercury as it is on the Moon. And if you look at the photo of Mercury, you may notice long jagged cracks in

| *Mercury's Vital Statistics* | |
| --- | --- |
| Radius | 2435 km |
| Surface area | $7.5 \times 10^7$ km$^2$ |
| Mass | $3.3 \times 10^{23}$ kg |
| Density | 5.43 g/cm$^3$ |
| Local gravity | 3.7 m/s$^2$ |
| Escape velocity | 4.25 km/s |
| Albedo | 0.12 |
| Surface temperature | 700 K (day) |
| | 100 K (night) |
| Length of year | 87.97 days |
| Length of day | 58.646 days |
| Distance from the Sun | 0.30–0.46 AU |

the surface, making steep cliffs, not like anything seen on the Moon.

From a practical standpoint there is one important difference in our views of Mercury versus the Moon. Mercury is a lot farther away. For the Moon we have a wealth of photographs, from Earth-based telescope views to orbiter pictures to pictures taken at the surface itself by the astronauts, not to mention all the other data and the rocks themselves acquired by the Apollo program. By contrast, practically all we have for Mercury are the fleeting glimpses of a spacecraft that passed by briefly and radioed back pictures of what it could see. As a result, we've really seen only about half the planet.

Furthermore, we often have only one view of a given region. For the Moon we are able to look at many pictures taken at different times of the lunar day to watch how shadows move across the surface, allowing us to estimate the heights and depths of mountains and valleys and compare colors with surface features. We can't do that on Mercury. This all adds to the challenge of studying Mercury.

## Comparison of Mercury and the Moon

Because we only have a limited amount of knowledge about Mercury, it's natural for us to start with the one similar place we do know a lot about, and then ask two complementary questions: How is Mercury like the Moon? How is it different? This approach to studying a planet is sometimes called **comparative planetology**.

We'll start by listing a few basic facts about Mercury, things we can learn by measuring its size and position with our earthbound telescopes. It's the closest planet to the Sun, on average a little more than a third of an **AU** (the distance from the Sun to Earth and the Moon) away. As a result, it receives nearly 10 times as much heat and light.

Mercury is 40% larger than the Moon in radius, giving it nearly twice the surface area. Thus it has 2.75 times the volume of the Moon. But it has 4.5 times the mass of the Moon. This means that on average there must be nearly twice as much mass packed into each cubic meter of Mercury than there is on the Moon. Therefore, it must be made up at least in part of some dense material that is not commonly found on the Moon. The one element that is both dense and abundant enough to account for this difference is metallic iron, so it is generally assumed that Mercury is made up of about 70% iron. Because there is no evidence for this iron on the surface—the surface looks just like the Moon, free of metal—we assume that the iron is buried deep in Mercury's interior, probably making up a large iron core (Figure 4.2).

One other difference, undetectable from Earth but seen by the instruments on *Mariner 10*, is the presence of a small magnetic field around Mercury, about 1/3000 the strength of Earth's magnetic field. The Moon has no detectable magnetic field. Mercury's field may be related to its iron core, but exactly how is something we still don't understand.

## Evolution of the Surface

### Similarities between Mercury and the Moon

Why should Mercury look like the Moon? The obvious answer is that Mercury, at least on its surface, must be made

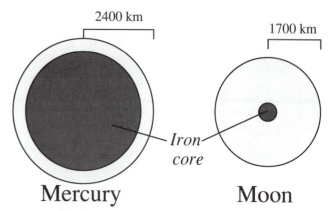

**FIGURE 4.2** Mercury is larger than the Moon, and its higher density implies that it has a substantial iron core.

of the same material as the Moon and must have undergone the same kind of geological history. So, in particular, we can guess that the rocks that make up the surface of Mercury are like highland Moon rocks, which were formed by crystallizing aluminum-rich feldspar minerals out of a molten lava that covered the initial surface like an ocean.

At the same time that these rocks were freezing on Mercury's surface, there also must have been a large population of planetesimals and meteoroids travelling in the vicinity of Mercury, which could crash onto its surface to make the craters. The proportion of the number of larger craters to smaller craters on Mercury is very similar to what has been seen on the Moon. (This is not true of the craters on other places, such as the moons of the outer solar system planets.) This implies that both the Moon and Mercury encountered a similar collection of planetesimals. Indeed, it seems reasonable to assume that they were peppered by exactly the same collection and, therefore, most of the cratering was occurring at the same time for both Mercury and the Moon. We can then guess that the age of the surface of Mercury ought to be close to the age of the Moon's surface, which has been dated at 4.0 billion years (from the age of the samples of lunar highland rocks returned by the Apollo program).

Thus we can work out a plausible scenario for the early history of Mercury. We assume that the surface was molten early in its history, while it was still accreting mass from the infall of thousands of planetesimals, and that the surface froze and continued to collect planetesimals and craters during the first half-billion years of its existence up to about 4.0 billion years ago. Large basins were formed during this time as well, and they were probably filled in with lava from the interior at about the same time as the lunar maria flooded their basins, 4 to 3 billion years ago.

After they froze, little further evolution occurred. Neither Mercury nor the Moon ever evolved water or a substan-

**FOR THE RECORD...**

# Solar and Sidereal Days

When we refer to the spin rate, we are referring to the rate the planet rotates once relative to the fixed stars. The length of time for one such rotation is called a **sidereal day**.

However, a solar day—sunrise to sunrise—as seen on Mercury is quite another matter. If it had been locked into an 88-Earth-day spin rate, as astronomers once thought, the solar day would be infinitely long, because the Sun would never appear to move in the Mercurian sky. However, the 59-day sidereal day, coupled with the uneven rate that Mercury moves through its orbit, results in a very peculiar sunrise–sunset phenomenon.

It takes two Mercury years (176 Earth days) from sunrise to sunrise. Right at perihelion, the closest approach of Mercury to the Sun, Mercury travels slightly faster than it spins. As a result, if you were watching sunrise on Mercury at this time you might actually see the Sun come above the horizon, then stop, reverse direction, and set again briefly, before finally rising again. Then the Sun would stay in the sky for a whole Mercury year. When it finally set it would rise again briefly, before setting for a second time. The night then would last for another Mercury year, until the phenomenon repeated itself.

---

tial atmosphere, so their surfaces would be eroded only by the occasional new crater formed over the past 3 billion years.

### Differences between Mercury and the Moon

We assume that the similarities between Mercury and the Moon mean that they had similar geological histories. What about the differences between the two surfaces?

For one thing, the craters seem slightly different in shape (Figure 4.3). If you look at any given crater on Mercury, you can see that the ejecta blanket—the dirt and rocks

**FIGURE 4.3** Craters on Mercury. Compare with the lunar surface, as seen in Chapter 3. Also notice the scarp in the center of the picture.

thrown out when the impacting meteorite made the crater—did not travel as far away from the impact as it would have for a similar-sized crater on the Moon. This we can explain as a consequence of Mercury having a stronger gravity field than the Moon. Ejected material with a given amount of energy won't travel as far on Mercury before it falls back to the surface as on the Moon.

One consequence of this greater gravity is that the areas between the craters appear to be smoother on Mercury than on the Moon. On the Moon, every time an impact makes a crater, the debris that flies out of the crater travels some distance before it hits the surface again, filling the regions between the craters with secondary impact craters. Thus every impact event on the Moon gives rise to many more small craters as well as the initial larger crater. On Mercury, the secondary debris does not travel as far, and so the regions between the craters are not so readily cratered themselves.

An interesting feature, so far seen only on Mercury, is called **hilly and lineated terrain** (Figure 4.4). This terrain is thought to have formed by the seismic waves from the impact that formed the huge Caloris Basin, converging on the opposite side of the planet.

Another difference between Mercury and the Moon is that the lava flows on Mercury do not seem to be significantly darker than the surrounding cratered terrain, while on the Moon the mare regions are much darker. We can measure the chemical content of the dark mare rocks from the Moon that the Apollo astronauts sampled, and in general they tend to be very rich in titanium and iron oxides, which is one reason that they are dark. This may suggest that Mercury is not as rich in these oxides as the Moon.

**FIGURE 4.4**   Hilly and lineated terrain.

Finally, one of the most important differences between Mercury and the Moon are the **scarps**, cliffs several kilometers high and thousands of kilometers long snaking across the surface of Mercury. Recall the example seen in Figure 4.3, stretching across the craters. There is nothing on the Moon even faintly resembling these cliffs. They represent a fundamental difference between Mercury and the Moon. To understand them, we have to look deeper into Mercury.

## Evolution of the Interior

Where did these scarps come from? Why should there be a chemical difference between lunar and Mercurian basin-filling lavas? And why should Mercury be so much denser than the Moon? All these differences may be tied together.

If you look at the cosmic abundances of the elements, it is not surprising that the majority of rocks everywhere in the solar system should be made of silicate (Si- and O-bearing) minerals also containing Mg, Ca, and Al, such as olivine, pyroxene, and feldspar. This is, in fact, what is seen in lunar rocks and most meteorites. However, iron is nearly as abundant as silicon and it should also be a major component of any rocky planet. Iron oxides can be found in many silicate minerals, as indeed they are in meteorites and in some lunar rocks. Iron can also be metallic, as in iron meteorites or (we believe) in Earth's core.

In either case, the presence of iron in a planet is obvious because it is a much heavier element than the other common rock-forming elements. Planets made with a cosmic abundance of iron along with the other rock-forming elements should have densities somewhere around 4 g/cm$^3$. The densities of Venus and Earth are higher than this, but these planets are large enough that the very high pressures deep in their interiors squeeze ordinary rock crystals into denser crystal phases. It is possible to calculate "uncom-

pressed" densities of these planets; for Venus and Earth, these densities are in fact near 4 g/cm$^3$.

However, the density of the Moon is 3.4 g/cm$^3$ and that of Mercury is 5.4 g/cm$^3$. We can only conclude that for some reason the Moon has less iron and Mercury more iron than cosmic abundances would predict.

Furthermore, the iron in Mercury must be metallic, not iron oxide. Iron oxide is less dense than iron metal, because not only does oxygen, a light element, dilute the density but also the metal has a more densely packed crystal structure. Therefore, even pure fayalite, the most iron-rich of the common silicates, has a density of only 4.2 g/cm$^3$, which is much less than the density of Mercury. And as we saw before, the color and darkness of Mercury's surface seem to indicate that little iron oxide is present there. Thus both the high density of Mercury and the absence of dark basalts there indicate that most of the iron in Mercury is metallic.

Why should Mercury be super-rich in metallic iron even above cosmic abundances, while the Moon is so poor? Planetary scientists have been debating this issue intensely since the early 1950s, but no single theory has yet solved this question to everyone's satisfaction. One possible explanation, that the Moon underwent a giant impact with Earth a very long time ago, was discussed in the last chapter. This scenario has also been proposed for Mercury, with the difference being that in Mercury's case, some of its

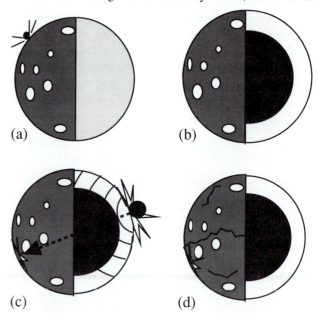

**FIGURE 4.5**   A possible series of steps in the evolution of Mercury: (a) a warm, still-accreting Mercury starts to melt in its interior; (b) the iron moving into the center of Mercury releases a pulse of heat as it falls, warming the whole planet and causing it to expand; (c) a large impact forms the Caloris Basin and fractures the opposite side of the planet; and (d) as the planet cools, it shrinks, forming the cracks on the surface we see as scarps.

**FOR THE RECORD...**

# Reflectance Spectroscopy

Reflectance spectra were mentioned briefly in Chapter 2, but not very much. As a reminder, a reflectance spectrum is a measure of how light is reflected back from some object. It is often shown as a plot of intensity versus wavelength.

The atoms in a crystal lattice vibrate all the time, and the frequency of this vibration is determined by the atomic weight of the elements making up the lattice, the structure of the lattice (the geometry of how the atoms are bonded together), and the temperature and pressure (which together affect the distance the atoms are apart and even what crystal structure the material is). When light hits the lattice, the wavelength characteristic of that particular lattice will be absorbed.

When the light is reflected back to your eye or to a spectrometer (the device that records the spectrum), the wavelengths that were absorbed will have a low reflectance while the wavelengths that were not absorbed will have a greater reflectance. By careful comparison between reflectance spectra and known materials, it is possible to identify the minerals on a distant planet by their spectra alone (Figure 4.6).

But, as an added complication, spectra are additive. By this we mean that if your field of view includes more than one material (whether because two materials are side by side or whether one is actually on top of the other but not obscuring it completely, perhaps because it is translucent), the spectrum you see is the sum of the spectra of the component parts, weighted according to the fraction of each component present. Because so much in geology is mixtures of many different materials, this effect can be very important.

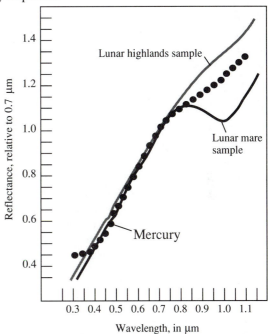

**FIGURE 4.6** These curves compare a spectrum of Mercury with samples of the lunar surface. The absorption at about 0.95 $\mu$m in the lunar mare sample is caused by iron-rich basalts, not present in either of the other samples.

---

mantle was stripped away by the impact. The question of elemental abundances in the planets will be explored more thoroughly in Chapter 14.

Regardless of why it is so, it remains that Mercury is much richer in metallic iron than the Moon. How would this change the way they evolved? Look at Figure 4.5.

When Mercury was partly molten, early in its history, it was inevitable that any metallic iron would settle downwards towards the center of the planet, because iron is denser than molten rock. The driving force moving all this iron to the center of Mercury is gravity, and in the process of settling down, large amounts of gravitational potential energy would be turned into heat. One can think of this,

in a crude sense, as a friction process: As the iron slides toward the core of the planet, it rubs against the rock, generating heat. Gravity keeps forcing the iron downward, in spite of this friction, as long as the surrounding rock is molten enough to let it pass. This process is known as **core formation**.

The result of this process is that the interior of Mercury would heat up to a degree far greater than the Moon ever did, because the Moon has hardly any metallic iron while Mercury must be nearly 70% metal to account for its density. As Mercury heated up, it must have expanded, because most rocks expand when heated. All this heating and expansion must have occurred very early in Mercury's

## FURTHER INFORMATION...

# Mercury's Orbit and Spin

Despite the advantages of comparative planetology, it can be terribly misleading at times to assume that Mercury is just like the Moon. The classic example of this mistake is the story of Mercury's spin.

Before spacecraft, the only way Mercury could be studied was by telescope, and such observations were very difficult for several reasons. Mercury is a very small planet, as planets go; of the nine only Pluto is smaller. This means that even in the most powerful telescope, Mercury appears only as a small, fuzzy disk.

An even greater problem is Mercury's closeness to the Sun. Mercury can't be observed during the daytime because sunlight scattered through Earth's atmosphere, making our sky blue, tends to overwhelm the light coming from Mercury. However, after the Sun sets, Mercury is also close to setting, and observing a planet or star when it is near the horizon means looking through much atmospheric dust and turbulence (see Figure 4.7).

But one factor that helps is that Mercury is in a very elliptical orbit. Its distance from the Sun ranges from 0.3 to 0.46 AU, which is quite a variation. Obviously, the best

time to observe Mercury is when it is at its farthest distance from the Sun, and when the Sun-Mercury-Earth angle is 90° (see Figure 4.8). At this time, it will be the highest above the horizon.

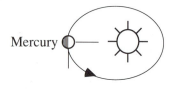

**FIGURE 4.8**   This is the best orientation of Mercury and Earth for astronomers to observe the surface of Mercury. The shape of Mercury's orbit has been exaggerated, as have the relative sizes of the Sun, Mercury, and Earth.

The nineteenth-century astronomer Giovanni Schiaparelli spent several years observing Mercury at such configurations. Despite all the difficulties of such observations, he was able to see distinct spots from observation to observation. This in itself was quite a feat; recall that Mercury doesn't have the same contrast between light mountains and dark plains that the Moon has. He noted that, during the few days he could observe each time the planets were correctly lined up, the spots hardly moved. And when he looked the next time the planetary alignment was repeated, he would see the same spot in the same position.

He also realized that Mercury was quite close to the Sun, and so it was subjected to solar tides just as Earth raises small but important tides on the Moon. The result of these tides on the Earth–Moon system has been to slow down the Moon's spin until it was locked, with the same side always facing Earth. By analogy, Schiaparelli reasoned, the Sun ought to keep the same side of Mercury facing

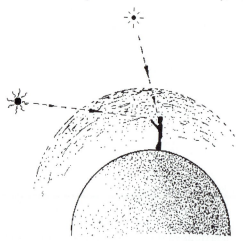

**FIGURE 4.7**   An astronomer looking at a star or planet low on the horizon has to peer through much more dust and turbulent air than when looking at a star straight overhead.

it at all times. This would be completely consistent with his observations. For the next 80 years, astronomers were convinced that Mercury kept the same face to the Sun, with a spin rate the same as its year, 88 Earth days.

In 1964, however, scientists were able to bounce radar waves off Mercury and measure the spin rate directly (by measuring how the spin of Mercury changed the waves that bounced back to Earth). The result was astonishing: The spin rate was 58.6 days, not 88 days!

What went wrong? Schiaparelli was misled by the lunar analogy and trapped by an odd coincidence of orbital periods.

Recall that the best time to observe Mercury is the time in its orbit when it appears from Earth to be farthest from the Sun. Consider a time when Mercury is at its **aphelion**, the point in its orbit farthest from the Sun, and when Earth is in a favorable viewing position (Figure 4.8). One Mercury year later, 88 Earth days, Mercury will be back at the same spot, but Earth will have moved. Because 88 days is nearly a quarter of an Earth year, we can see that at this time Mercury will be hidden from Earth by the Sun. Another 88 days later Mercury and Earth will be lined up favorably again (with Earth now seeing the other side of Mercury relative to the Sun), and 88 days past that Mercury will again be blotted out by the Sun, as seen from Earth. Finally, after another 88 days, we'll be almost exactly back to our original positions.

How has Mercury been spinning in the meantime? The spin rate of Mercury is just two thirds the rate it goes around the Sun. This means that one aphelion will see one side of Mercury facing the Sun, then 88 days later the opposite side will be sunward; and these two sides will alternate every other orbit. But from Earth, we can only observe Mercury with ease once every other orbit, so we would keep seeing Mercury in the same configuration (Figure 4.9). It's no wonder that Schiaparelli was fooled!

But what about the theory that explained why Mercury should be locked by tides so that one side always faced the Sun? Ironically enough, that theory was correct, but with an important variation.

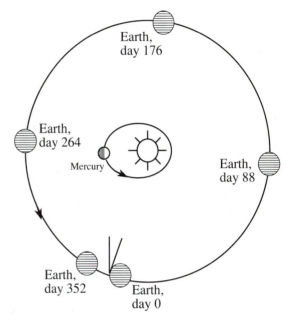

**FIGURE 4.9** During successive Mercury years, Earth orbits at such a rate that the ideal observing orientation illustrated here will occur only every other Mercury year. But after 2 Mercury years, the planet will have rotated exactly three times. Thus an observer on Earth will see Mercury always turned with the same face to the Sun.

Recall that Mercury has an eccentric orbit. It moves closer and then farther from the Sun as it goes through 1 year. The tidal forces that affect the spin are strongest when the Sun and Mercury are closest together. In addition, planets in eccentric orbits do not move at a constant speed, but move faster when they're closer to the Sun and slower when they're farther away. It turns out that when Mercury is closest to the Sun, it is moving in its orbit at such a rate that if it were to continue at that rate for its whole orbit, its year would be just under 59 days. This is precisely when the tidal forces are strongest. Hence, this is why the tidal forces lock Mercury into a 59-day spin rate, rather than an 88-day rate.

history, before the surface had completely frozen, because the surface today has no sign of any iron metal.

After core formation was over and the surface was frozen, the planet would slowly cool off over the next 4 billion years. As it cooled, it would contract. As the total volume of the planet shrank—current estimates are that the radius must have changed by about 20 km—the surface area would likewise get smaller, and so parts of the surface would have to be thrust over one another. The result would be the long scarps that are seen all over Mercury. Because the Moon did not have the surplus iron, it never formed a core, never heated up to such an extent, and never cooled off and contracted to the same degree. Thus, scarps would never be formed on the Moon.

## SUMMARY

From the *Mariner 10* photographs of Mercury, we know that it has a surface similar in many ways to the heavily cratered highland areas of the Moon. Thus we can infer that the surface of Mercury is made of very old rocks, rich in aluminum feldspars, formed from the freezing of an ocean of lava that originally covered the planet. Intense cratering of the surface has ground these rocks into a powder probably similar to the sandlike surface of the Moon. However, Mercury lacks the iron oxide and titanium oxide minerals seen in the Moon's mare basalts.

Mercury is a larger and denser planet than the Moon, with a consequently higher gravity. This has changed the shape and distribution of craters on its surface compared with the Moon. Mercury also apparently underwent a much more extensive period of heating and cooling, leading to a period when the planet shrunk by several kilometers and resulting in great cracks and cliffs called scarps.

Mercury, like the Moon, has probably been geologically quiet for the past 3 billion years. Without water or an atmosphere, the only changes to the surface come from the occasional impact of a meteoroid to form a new crater and rocks cracking from the relentless heating and cooling of the surface, which is more than three times closer to the Sun than the Moon is and so absorbs roughly 10 times more heat.

## STUDY QUESTIONS

1. Name one kind of feature seen on Mercury's surface that is also seen on the Moon's surface.

2. Name one kind of feature seen on Mercury's surface that is not seen on the Moon's surface.

3. What is present deep inside Mercury that is *not* present on the Moon?

4. How long is it from sunrise to sunrise on Mercury?

5. If you were on Mercury watching the sunrise at the time when Mercury was closest to the Sun, what peculiar phenomenon would you observe?

## 4.2   INSIDE PLANETS: CHEMISTRY

Mercury and the Moon have surfaces that look remarkably similar, and yet we have argued that their interiors are very different. But how can we determine the composition of an entire planet? We have literally only scratched at the surface of our home planet, Earth; how do we know what is deep in its interior, and how can we possibly guess the compositions of planets we've never been to?

It is a difficult problem, and until the day when we can actually take apart a whole planet and analyze it piece by piece we'll probably never know for certain if our solutions to this problem are correct. But there are several specific ways we can use a variety of tricks to guess at a planet's bulk composition. In this section we concentrate on how we can use a planet's density and the composition of its surface materials to infer what materials it is made from.

### Mean Density

The first, most important criterion for the bulk composition of a planet is its mean density. To measure the density, first we measure how large the planet is. Then, by observing the effect of this planet on the motions of other planets, moons, or spacecraft over a long period (which is difficult to do, but possible) or by timing the orbital periods of its moons (much easier, if it has moons) we can find out how massive the planet is. Then we can divide the mass by the volume and determine the mean density. This density can be compared with the densities of common substances, and thus we can start to guess what material, or mixture of materials, could go into making up the planet.

For example, we can see from Table 4.1 that water has a density of 1.0 grams per cubic centimeter. An ordinary rock has a density of about 3 $g/cm^3$, and metallic iron is close to 8 $g/cm^3$. So a planet like Earth, with a mean density of 5.5 $g/cm^3$, probably has both iron and rock in its composition, certainly more iron than water. Mercury's density, at 5.43 $g/cm^3$, is clearly different from that of the Moon, 3.36 $g/cm^3$. So we must conclude that Mercury's interior must be largely made up of some dense material lacking in the Moon, probably iron.

(For consistency with the SI units, we really ought to list densities in terms of $kg/m^3$, so that water has a density

**TABLE 4.1**

Densities of Some Common Planet-Forming Materials (in g/cm³)

| | |
|---|---|
| Water ice | 0.94 |
| Water liquid | 1.0 |
| Carbonaceous chondrite | 2.5 |
| Plagioclase | 2.7 |
| Pyroxene | 3.3 |
| Ordinary chondrite | 3.5 |
| Olivine | 3.3 |
| Iron sulfide | 4.8 |
| Iron | 7.9 |

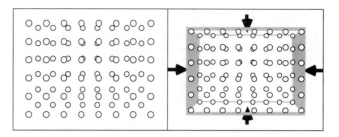

**FIGURE 4.10** The pressure deep inside a planet can compress the materials there; this can significantly alter the density of minerals deep within the planet and the density of the planet as a whole.

of $1 \times 10^3$ kg/m³; however, this is one case where the cgs units are clearly more convenient to use, so we'll continue to use them. For SI units, just multiply all these densities by $10^3$.)

In the same way, the very low densities of Jupiter, Saturn, Uranus, and Neptune tell us that these planets are made mostly of gas, with at most only a small fraction of rocky material in them. We call them the **gas giant planets**. The moons of these planets are obviously solid but also of low density; we infer that they are made of a mixture of ices and rock. They are often referred to, as a class, as the **icy satellites**.

The four inner planets and the Moon all have densities roughly the same as Earth. From their densities we can infer that they too are mixtures of rock and iron. We call them **terrestrial** (from the Latin for "Earthlike") or just the **rocky** planets.

There are two major difficulties, however, that must be overcome before we can use the mean density to determine the planet's composition in any quantitative manner. First, the density of materials deep inside a planet may be different from the density of the same materials measured in a laboratory. Second, we have to have a reasonable idea of what components are likely to be present.

A cubic centimeter of rock 2000 km deep inside a planet is squeezed by the pressure of 2000 km of rock sitting on top of it. Its density, and hence the mean density of the whole planet, is greater because of this pressure (Figure 4.10). On the other hand, the temperature deep inside a planet may reach 1500 K or more. Rocks, like most other substances, tend to expand as they get warmer. This could have the opposite effect, tending to make the rock, and the planet, seem less dense.

Which effect wins? To find out, experiments are made on many substances to measure precisely how much they change with temperature and pressure, and calculations can be made to estimate what the temperature and pressure might be inside each planet. From these measurements we've been able to estimate uncompressed densities for the terrestrial planets. As Table 4.2 shows, the uncompressed density of

the Moon is virtually the same as its observed density. Here, the temperature and pressure effects appear to balance out. All the other planets are larger than the Moon, however, and so inside them the pressure is more important.

**TABLE 4.2**

Compressed and Uncompressed Densities of the Terrestrial Planets (in g/cm³)

| *Planet* | $\rho_{comp}$ | $\rho_{uncomp}$ |
|---|---|---|
| Mercury | 5.43 | 5.3 |
| Venus | 5.24 | 4.0 |
| Earth | 5.515 | 4.1 |
| Moon | 3.36 | 3.3 |
| Mars | 3.94 | 3.7 |

The other problem comes in actually choosing which components are likely to exist inside a planet in the first place. A planet of density 2.5 g/cm³ could be made of pure carbonaceous chondrite material; or it could be a mixture of water and iron; or it could be made of any number of other materials that, in the proper combination, would give that density. Clearly, the density itself is not sufficient to determine the composition of the planet completely.

Our first guess might be to assume that planets have the same composition as the Sun. Cosmic abundances would make a mostly gaseous planet that, if it were big enough, would indeed look much like Jupiter or Saturn. However, the terrestrial planets are obviously not made up of mostly hydrogen and helium.

Our next guess would be to assume that these rocky planets are made of cosmic abundances of the rock-forming elements. This is probably not a bad guess; but we know that, in detail, this can't be precisely true of all the terrestrial planets, simply because (as seen in Table 4.2) the terrestrial planets don't all have the same densities. They are different from one another. Finally, even if a rock of cosmic abundances did match the density of the terrestrial planets, we'd like to find some way of confirming the composition of the planet through actual observations and experiments.

## FURTHER INFORMATION...

# Modelling the Densities of Planets

If a planet is made of one component, the simplest way to model its density is to find some substance with the same density as the planet. But what if the body in question is partly made of rock and partly of metal? This is thought to be the case for Mercury (and the other terrestrial planets). Its uncompressed density is 5.3 g/cm$^3$, between that of iron and rock. Clearly, the simplest way to make a moon with this density is to mix rock with iron.

How much of each do we need? It is simple enough to reason that the total density, $\rho_{total}$, is just the weighted average of $\rho_{iron}$ and $\rho_{rock}$, the densities of our two components. The more iron we have, the closer $\rho_{total}$ will be to $\rho_{iron}$. So we might write

$$\rho_{total} = \rho_{iron} X_{iron} + \rho_{rock} X_{rock} \qquad (4.1)$$

where $X_{iron}$ is the volume fraction of the planet that is iron and $X_{rock}$ is the volume fraction that is rock. Because the two fractions must add up to one, we can say that $X_{rock} = 1 - X_{iron}$ and so

$$\rho_{total} = \rho_{iron} X_{iron} + \rho_{rock}(1 - X_{iron}) \qquad (4.2)$$

and then use algebra to solve for $X_{iron}$.

Using the numbers in Table 4.1 for $\rho_{iron}$ (= 7.9) and $\rho_{rock}$ (= 3.5, for ordinary chondrites), and the uncompressed density of Mercury, $\rho_{total}$ = 5.3, we then do the algebra. We find that $X_{iron}$ in this case is 0.41; that is, Mercury is 41% iron by volume (and, therefore, 59% chondrite).

What if we want the mass fraction ($Y_{iron}$), which may be quite different? With a little bit of thought, you should be able to convince yourself, using the same reasoning as above, that the proper equation is

$$\frac{1}{\rho_{total}} = \frac{Y_{iron}}{\rho_{iron}} + \frac{1 - Y_{iron}}{\rho_{rock}} \qquad (4.3)$$

We solve this equation for $Y_{iron}$, the mass fraction of iron in Mercury, and find that it is 61% iron by mass.

## Surface Compositions

Knowing what the planet's surface is made of can help in both a general way and a specific way. If a planet has a predominantly icy surface, like the moons of Jupiter and Saturn, then one is justified in including ice into the density calculation.

The most detailed observation, of course, comes from an actual sample of rock returned from the planet in question and analyzed in a laboratory. It's easy to find the composition of any rock we can get our hands on; however, it's not as easy to deduce the composition of an entire surface from just one rock.

Certain general characteristics, such as the presence of water or methane ice, the relative proportions of major mineral types such as olivine, pyroxene, and plagioclase, and the presence or absence of iron and carbon can all be determined from Earth-based telescopic observations. Each of these substances reflects some colors of sunlight while absorbing other specific colors. So, by splitting the reflecting light into its component colors (for example, with

a prism) and comparing this light with sunlight, an astronomer can deduce which of these materials might be present on a planet's surface. (This is called **reflectance spectroscopy**.)

Much more detailed information can be determined by a satellite orbiting close to the planet's surface. In this case, reflectance spectroscopy can give us information, not just about what minerals may be present, but even how they are distributed over the surface, the *geology* of the surface.

Another tool we can use is **gamma-ray spectroscopy**. Gamma rays are very short wavelength photons of light, more energetic than ultraviolet light or x-rays. If the planet has no atmosphere, then the surface is exposed directly to gamma rays from the Sun, as well as to cosmic rays. Individual elements such as iron and titanium are excited by these rays and respond by emitting gamma rays at a characteristic energy. Thus a gamma-ray detector in a satellite orbiting near the planet can map out the abundances of these elements over an entire surface. In addition, some radioactive elements such as potassium and uranium emit gamma rays as they decay, so such a satellite can look for these

**FOR THE RECORD...**

# Angular Momentum and the Moment of Inertia

Every moving object has associated with it an **angular momentum**, **L**. The angular momentum of a particle is equal to the momentum of the particle (the mass times the velocity) multiplied by the position vector of the particle (with respect to some coordinate system, usually the center of mass of the system).

All these terms are vectors, that is, they have not just a magnitude, but a direction also. The momentum and position vectors are multiplied together using the vector **cross product**:

$$\mathbf{L} = \mathbf{r} \times \mathbf{p} \qquad (4.4)$$

This means that the directions of the three vectors follow the "right-hand rule": If you hold your right hand with the thumb up, first finger straight ahead, and middle finger pointing left, and say that **r** is in the direction of your first finger and **p** in the direction of your middle finger, then **L** will be in the direction of your thumb.

Now imagine that we have a solid body such as a planet. It has associated with it an angular momentum. Because angular momentum is a vector, it has components in the $x$-, $y$-, and $z$-directions. If the coordinate system is chosen with the origin at the planet's center of mass and the $z$-direction along the axis of rotation, then the $x$- and $y$-components of the angular momentum will be zero. So let's only look at the component in the $z$-direction, the axis of rotation. The $z$-component of the total angular momentum, $L_z$, is the sum of all the individual $z$-components for all the particles in the body:

$$
\begin{aligned}
L_z &= \sum_i L_{zi} \\
&= \sum_i r_i m_i v_i \\
&= \sum_i r_i^2 m_i \omega \qquad (4.5)
\end{aligned}
$$

where $r_i$ are the radial distances to the particles, $m_i$ are the masses of the particles, and $v_i$ are the velocities of the particles (which equal $r_i \omega$, the distance times the angular velocity, $\omega$). The angular velocity is a constant throughout the body (otherwise the body would tear itself apart!).

We can now introduce the **moment of inertia**, **I**, which is defined

$$I \equiv \sum_i m_i r_i^2 \qquad (4.6)$$

so that

$$L_z = I\omega \qquad (4.7)$$

The moment of inertia is a way of describing the distribution of mass in a body. For a uniform sphere of mass $M$ and radius $a$,

$$I = \tfrac{2}{5} M a^2 \qquad (4.8)$$

---

gamma rays and map out the regions rich in these radioactive elements. (Gamma rays are absorbed by atmospheres, and so these observations can only be made for places like the Moon or Mars that have little or no air.)

In any case, we must keep in mind the difference between a surface composition and the planet's bulk composition. Recall that three quarters of Earth's surface is covered by oceans, but clearly, Earth is not made of three quarters water and one quarter rock.

## SUMMARY

Without literally taking a planet apart, there is no direct way to measure what elements are present. But there are several techniques that can be used to deduce the abundances of the elements within a planet.

The first, fundamental clue to a planet's composition is its density. A very large planet with a low density, such as Jupiter or Saturn, must be made primarily of low-density gases such as hydrogen and helium. A small, dark asteroid with the same density as a carbonaceous meteorite probably is made out of carbonaceous meteorite material.

However, the density does not give a unique answer to the composition of a planet that might be made of a mixture of different types of materials. We can use our knowledge of cosmic abundances to make educated guesses of the compositions of such bodies. For instance, many moons in the outer solar system have densities between 1.5 and 2 $g/cm^3$; from cosmic abundances, we know both water ice (density 1.0 $g/cm^3$) and rocky meteorite material (density 3.0 $g/cm^3$) should be available, and so we can conclude that these moons are made of a mixture of ice and rock. Given the density, we can even calculate the specific proportions of rock and ice present.

Knowing the composition of the surface also helps us to tell which materials are likely to be present inside the planet. Any bulk composition that we might guess for a planet should be able to produce the sort of surface we actually see on that planet. If we have actual rock samples from the surface, we can be more specific; for instance, knowing the trace elements in a basalt from a planet's surface lets us determine what minerals must have been present deep inside the planet where the lava originally came from.

However, no technique gives a complete, unique determination of precisely what material exists deep inside a planet. Instead, what we find is a composition that is most consistent with all the different techniques we can use. The more different ways we have of determining what's inside the planet, the more confident we can be that we know its bulk composition.

## STUDY QUESTIONS

1.  A typical density of an ordinary rock, compared with water at 1 $g/cm^3$, is:
    (a) 0.3 $g/cm^3$
    (b) 3 $g/cm^3$
    (c) 30 $g/cm^3$

2.  Jupiter is made from the same material as Saturn, but it is denser. Why?

3.  Why is a gamma-ray detector orbiting Mercury able to obtain surface composition data, while one orbiting Earth is not?

4.  Which planet's composition is better determined by the composition of its surface rocks, an asteroid or Earth?

5.  Why do we say that it is iron, instead of some other dense material, that must be present in Mercury's core to explain Mercury's high density?

## 4.3   INSIDE PLANETS: PHYSICS

Our understanding of planets has begun with knowing the ingredients of the planets. However, it's one thing to know what pieces are in the puzzle; it's another to know how the puzzle is put together. What can tell us how the chemical constituents of the planets are organized?

### Moment of Inertia

One of the most basic things we can determine about a planet is if it has ever differentiated. Did it ever get warm enough to melt, so that the heaviest materials, the iron, sank to the center to form a core? This is an important question; answering it doesn't just tell us about the history of the planet, it also tells us about what it is like today.

An **undifferentiated** body is the same on the surface and inside. If we determine what the surface of the body is made of, we know all about the interior, too. Such bodies tend to be very small, because only small bodies were able to radiate heat away fast enough while they were forming so as not to melt.

A **differentiated** body, on the other hand, has been through a period of melting and may still contain melted regions. When a core is formed, iron isn't the only material that is separated out. Lighter compounds rise to the top to form the crust. This means that when we look at the surface of a differentiated body, we may see quite a different composition than that of the body as a whole.

We can tell just from the density of a planet that there must be heavier material on the inside than is visible on the surface. But how do we know that it is concentrated into a blob at the center? For this we go back to some basic physics (see the box on angular momentum and moment of inertia).

A spherical, undifferentiated planet would have a **moment of inertia** equal to 0.4 $Ma^2$ (mass times radius squared). The moment of inertia of the Moon is 0.39 $Ma^2$, indicating that it has only a small core. The moment of inertia of a planet with a core, for example, Earth, is considerably less, only 0.33 $Ma^2$.

So, the moment of inertia can tell us something about the mass distribution inside a planet. But how can we measure the moment of inertia? We measure it by determining the gravitational field of the planet. The harmonics of the gravity field (discussed below) may be used to determine the moment of inertia. This is not a simple calculation, however. We will explore this subject further in Chapter 11.

### Isostasy

In the 1800s, British surveyors mapping the locations of towns near the Himalayas were faced with an odd discrepancy. The distance from the town of Kalianpur, some 700 km south of the mountains, to Kaliana, a mere 100 km from them, was measured in two different ways. One way simply used chains along the ground; the other compared the altitude of various stars and calculated how much Earth curved between the two towns, much like Eratosthenes measured the size of the whole Earth. The two methods disagreed by 170 m, a small but troubling difference.

After a little thought, it was realized that the mass of the Himalayas could account for the difference. The altitudes of the stars were being measured against a plumb bob, which was presumed to hang in a straight vertical, pointing towards the center of Earth; but the extra mass of the mountains would be enough to pull slightly at the plumb bob in Kaliana. The surveyors knew how big the mountains were, and so they could guess at how much extra mass there was.

But that calculation gave an even bigger surprise. Rather than predicting that the astronomical method should be off by 170 m, the calculation said the error should have been closer to 500 m! The mountains were only a third as massive as they thought. Were the mountains hollow? Where did the mass of the mountains go? In 1865 the English astronomer George B. Airy suggested that the mass was missing from underneath the mountains.

Mountains on Earth are made of rocks like granite, rich in plagioclase. They are lower in density than rocks from deep inside Earth. In general, the crust of Earth has an average density of about 2.9 g/cm$^3$. Rocks below this crust in the mantle (such as **xenocrysts**, rocks that have been brought up from the mantle by volcanoes) are mostly olivine, with a density of 3.3 g/cm$^3$. Piling a mountain of rock several miles thick onto the surface of Earth creates a tremendous load on Earth's crust.

As a result, the low-density crust sags down into the mantle, displacing the denser rocks of the mantle. Every cubic centimeter that used to be mantle rock and contained 3.3 g of material is now crustal rock and contains only 2.9 g. So for every cubic centimeter of mantle that is displaced, the total mass of the region is 0.4 g lighter than it used to be. In fact, this "sagging" continues until the total change of mass in the mantle region is the same as the mass of the mountains that caused the sag in the first place. In effect, the mountain "floats" in the mantle, just like an iceberg floats in water (Figure 4.11).

Say, as in this case, that the difference in density between mantle and crust (called the **density contrast**) is 0.4 g/cm$^3$. It can be shown that, for the mass loss in the mantle below the crust to compensate for extra mass piled on top of the crust, an amount equal to 2.9/0.4, or about seven times as much crustal material, must extend into the mantle below the mountains as extends into the air above the surface of the surrounding crust.

This principle is called **isostasy**. In the absence of other forces, surface features on a planet would try to flow so as to maintain a constant gravitational potential every-

**FIGURE 4.11** The crust under these mountains extends down into the mantle, displacing denser mantle rocks with lighter crustal rocks. In effect, the mountains float in the mantle. This is an example of **isostasy**.

where across the surface of a planet. There are always other forces, however. The crust resists deformation, and the mantle resists the downward pressure of the "roots" of mountains.

The surface of a planet appears to be **isostatically compensated** when the "sagging" of the crust into the mantle balances the mass of the mountain above the surface. This does not mean that a constant gravitational potential is achieved, however. Because the mass of the mountain above the surface of the planet is always closer to an observer than the mass of the "root" below the surface, its mass will always dominate the total gravitational force from the mountain.

Rocks do not flow nearly as fast as fluids, and so young mountains may not be isostatically compensated while the older ones have had time to "settle in." Some features may not ever become isostatically compensated. If the crust is strong enough (by being very thick, or very cold, or by the feature being very small in area), the mass of the feature may be supported by the strength of the lithosphere.

For planetary exploration, the importance of isostasy is that it provides another tool to examine the structure and evolutionary age of a planet. First, we can assume (from cosmic abundance arguments) that most planets have this sort of density contrast between crustal and mantle rocks. (All rocky planets seem to have plagioclase-rich crusts, but cosmic abundances tell us that olivine should be far more abundant than plagioclase). That means that simply by looking at the topography on a planet's surface, we have a rough idea of how thick its crust must be. Whatever "relief" we see on a planet—the difference in elevation between the highest and lowest points on the planet—the crust under the highest point must be at least seven times thicker than under the lowest point, if the mountain is isostatically compensated. Of course, it may not be completely isostatically compensated.

By measuring the details of the gravity field of a planet, described below, and comparing gravity anomalies with surface topography, we can often tell if a planet is isostatically compensated or not. There is a certain uncertainty here. For a given gravity field and topography, a feature could be fully compensated at some depth, or it may be partially compensated at some shallower depth. By comparing the results from features of different sizes (areas) or adding in data about crustal thickness from other sources (such as seismology; see Chapter 6) we can often resolve this uncertainty.

From our experience on Earth of how long it takes for rocky material to flow, we can then estimate the conditions inside that planet. How easily do the rocks in that planet's mantle flow? Warm rock flows more easily than cold, so the amount of isostatic compensation tells us something about how warm the interior of the planet is. Depending on how warm the mantle is, compensation can take millions or even billions of years to achieve, so from this information we can also learn something about how old the mountains are or how quickly the planet has cooled off.

## The Gravity Field

Planets are made of matter, and all matter is subject to the laws of gravity. We usually think of gravity as the force that pulls us towards Earth, but in fact, every bit of matter has its own gravity field and exerts a force on every other bit of matter in the universe. While Earth is pulling on us, we are also pulling on Earth and on every other bit of stuff around us. The Earth is $10^{23}$ times more massive than we are, however, which is why its force seems so much stronger than any of the other pulls.

Recall that the strength of the force of gravity depends on two things: the masses of the two objects attracting each other and the distance between those objects. The direction of the force is important, too; the force on each object is directed towards the other object. The two objects want to pull themselves towards each other.

Consider now a person standing on a planet. Not only is the planet much more massive than the person, it's also much larger. Its mass is spread out over a wide area. All the bits of the planet are attracting the person. Some are pulling her one way, some another. Some bits are just under her feet, some are clear on the other side of the planet. And what if some bits are denser, and hence more massive, than other bits? If we want to determine the total force acting on our subject, which masses and which distances do we use?

The total force is equal to the sum of all the little forces between our subject and all the bits of the planet. The resulting force field is often so complicated that it can be difficult to see general trends or specific details.

To avoid these complications, one maps out the gravitational potential energy, rather than the gravitational force. To separate special anomalies from general trends, we divide the potential field into harmonics.

## Harmonics

If the surface of a planet were completely fluid, as it is for a planet like Jupiter that is made of gas and clouds, then its surface would flow until it felt the same gravitational potential everywhere. Even for planets made of rock, which do not have fluid surfaces, one can measure the gravitational potential and think of an imaginary surface of constant potential. The level of the seas on Earth (ignoring waves) is this kind of surface. Such a surface is called the **geoid**.

If a planet were smooth and uniform, if there were no mountains and the density of the material in the planet stayed constant at any latitude or longitude, if it were perfectly still and unaffected by any other body in the universe its shape would be that of a perfect sphere. Real planets, of course, are somewhat different from this case. We sort out these differences in terms of how large an area of the planet they cover, decreasing in size in regular intervals. Each interval is a **harmonic** of the gravity field. (See the box at the end of the chapter for more on how we define harmonics.)

For instance, if our perfectly spherical planet were somehow squeezed, like a balloon, so that the fluid surface bulged out of the northern hemisphere and was pushed in at the southern hemisphere, this would represent the **first harmonic** of the gravity potential field. Such a situation doesn't usually happen to planets, however. But if the planet spins, then its surface would be a smooth **oblate spheroid**, different from a perfect sphere by bulging out at the equator. This bulge represents the **second harmonic** of the planet's gravity field. All spinning planets have a significant second harmonic. The second harmonic is often referred to as the $J_2$ term of the gravity field.

The **higher harmonics** consist of all further differences between the real geoid and such a spinning, oblate spheroid. They are called **gravity anomalies**. They are places where the local gravity is slightly larger or smaller than you would expect for a smooth, spinning planet.

**Gravity anomalies** are usually caused by local concentrations of extra-dense matter below the surface of the planet. For example, such gravity anomalies are seen on Earth in areas where there is a concentration of metal ores, because the ores are much denser than typical surrounding rock and so have a stronger gravitational field. By mapping the gravity field one can thus get more detailed clues as to how mass is distributed inside a planet.

The motion of a moon or satellite orbiting a planet can be subtly affected by the precise distribution of mass inside the planet. In this way we were able to estimate how thick the dense mare basalt flows were on the Moon's surface. Although the mare basalts cover a substantial part of the surface, it turns out that most of them are only a few kilometers thick; the bulk crust of the Moon has a composition much closer to that of the highland rocks.

**FOR THE RECORD...**

# Potential Energy and Kinetic Energy

A 1-kg body sitting at some point feels a certain gravitational force. The amount of force it feels doesn't depend on how it got to that point or whether it's moving or standing still. We can assign a number to that point, which represents the force felt by any kilogram of stuff that happens to find itself there. And that number never changes. (Notice that we said "by any kilogram"; if you put 2 kg of stuff there, you get twice the force.) In fact, every point in space can be represented by such a number. This collection of numbers assigned to points is called a **gravitational field**.

Gravity is a well-behaved force, but not every force in nature is like that. A force like friction, for instance, depends on how the object that feels the force is moving. If a box slides along a floor to the right, the force of friction opposes its motion and pushes to the left. But if the box moves to the left, the friction pushes to the right. If the box isn't moving, then it feels no friction at all. So the force of friction, unlike gravity, depends on something more than just the position of the box.

There's another number we can assign to each point in a gravity field, the **potential energy**. The potential energy represents how much motion—how much **kinetic energy**—it is possible to get from the force of gravity.

For example, a rock at the top of a hill has a certain amount of potential energy. If it starts to fall down the hill, it will pick up speed. The faster it moves as it falls downhill, the more kinetic energy it has, until it reaches its fastest speed at the bottom of the hill. The kinetic energy the rock acquires by falling down the hill is the same amount as the difference between the potential energy it had at the top of the hill and the potential energy it has at the bottom of the hill. (It still has potential energy at the bottom of the hill because it is not at the bottom of the gravitational field of Earth. If you dug a hole and pushed the rock into it, it would gain some more kinetic energy.)

The **total energy**—that is, the kinetic energy at any point plus the potential energy at that point—is **conserved**: The sum of kinetic energy plus potential energy is a constant. That is why we call gravity a **conservative force**. A force like friction, on the other hand, wastes energy by turning it into heat. If the rock hits a wall at the bottom of the hill, it's lost both its potential and kinetic energy, but it will be a little bit hotter.

---

A spinning planet that has much of its mass concentrated at its center has a slightly different gravity field from one where the mass is evenly distributed. The size of the equatorial bulge is different if a core is present (Figure 4.12). The more material concentrated in the core of a planet, the smaller the bulge. Thus the second gravity harmonic is closely related to the moment of inertia of the planet.

## Putting It All Together

As we have seen, each method of finding clues to the interior of a planet has its advantages, but it also has its drawbacks. Knowing the bulk density of a planet can let you make some guess as to its bulk composition, but there are many different ways to combine different types of material and still arrive at the same density. The surface composition can be determined with some accuracy, but relating the surface to the interior can be tricky. Where the data are

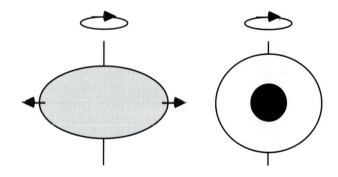

**FIGURE 4.12** As a planet spins, centrifugal force causes the equator to bulge out. The more mass there is in the outer parts of the planet, the bigger the bulge will be. Thus the size of the bulge gives a clue as to whether a planet is uniform in density or if it is differentiated into a dense core and a less dense mantle.

**FURTHER INFORMATION...**

# Spherical Harmonics

Using the potential solves the problem of having to deal with vector addition. But the mathematical equation that describes the gravitational potential of a planet can still be quite complicated. To break down these complicated functions into understandable bits, we use a very useful mathematical trick.

In the nineteenth century, the French mathematician Jean Baptiste Joseph Fourier discovered that any function, no matter how complicated, can be written as an infinite sum of powers of sines and cosines. An example is shown in Figure 4.13. If the function is nearly the same as a sine or cosine in the first place, then often just a few terms of that infinite series will come very close to describing the whole function. To get greater accuracy, more terms can be included. Each term represents a sine wave with a progressively shorter wavelength.

A note from a musical instrument is a good example of such a function. Every note has a fundamental pitch and a variety of other pitches (overtones and so forth) that give that instrument its particular sound. Each one of these sine waves is called a **harmonic**.

Planets are nearly spherical. If a planet were perfectly spherical and completely uniform in its composition, then perhaps you can intuitively see (as, in fact, one can prove, using calculus) that the force of gravity someone would feel, standing anywhere on the surface, is exactly the same

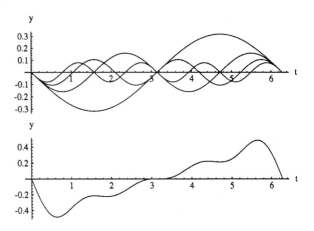

**FIGURE 4.13** An example of how a Fourier series can approximate an arbitrary curve. The bottom graph shows a Fourier approximation of a sawtooth wave. Above this curve is another graph, showing the first four terms individually. When they are added together, they produce the wavy solid line of the lower graph.

as if all the mass of the planet were concentrated at the center of the planet. Thus the force would be $F = GMm/r^2$,

available, geophysical information, such as the gravity harmonics, can describe the structure and distribution of matter inside a planet; but again it does not tell us exactly which elements or minerals are present. Only through putting all these bits of the puzzle together can we begin to find a consistent picture of the composition of a planet.

Consider Earth. We know that the uncompressed density is a bit denser than rock, but less than iron; probably Earth is a mixture of both. The geophysical data tell us that Earth has mass concentrated in a core; presumably that's where the iron is.

But earthquake data (described in Chapter 6) tell us how big the core is, and a solid iron core that big would be too dense to match the moment-of-inertia data. Furthermore, the speed at which certain earthquake tremors pass through the core tells us that the core is partially molten.

Most geochemists think that the solution to these problems is to include sulfur in the core. Sulfur is lighter than iron, so it solves the density problem, and experiments have shown that sulfur and iron make a eutectic melt at relatively low temperatures, resulting in a core that would be partially molten. And the amount of sulfur we'd need to do the job is consistent with the cosmic abundance of sulfur.

The seismic data tell us that Earth is divided into a roughly 30-km-thick crust and a 2900-km-thick mantle. The crust, which is what we see at the surface, seems rich in plagioclase and other low-density rocks. But lava from volcanoes suggests that the rocks below the crust are rich in olivine. Indeed, occasionally these lavas bring up with them **xenocrysts**, odd rocks that clearly aren't made of the lava material but were broken off and carried up to the surface by the flow of the lava, and these xenocrysts are where $m$ is the subject's mass, $M$ is the total mass of the planet, and $r$ is the distance one would be from the center of the planet, the planet's radius. $G$ is known as the **universal gravitational constant**. The potential, $V$, for such a planet also depends only on the distance from the center of the planet; it is given by $V = -GM/r$.

Real planets are not perfectly uniform or spherical, however, which is why their gravity fields cannot be described by such a simple equation. But they are close enough to spherical that we can write out the gravity field in terms of a series of sine-and-cosine functions of the latitude and longitude of the planet, a Fourier series. These terms are the spherical harmonics of the gravity field.

Figure 4.14 shows how these harmonics are related to the shape of the potential field. The **zonal harmonics** represent bulges in latitude. For example, when a planet spins, centrifugal force tends to make the equator bulge out; this is the simplest zonal harmonic. Bulges along lines of longitude are called **sectoral harmonics**; they remind you of the sections of an orange. A combination of both kinds of bulge leads to the complex **tesseral harmonics**. An example of the latter is the bulge in a planet from tides raised by another passing moon or planet.

The equatorial zonal harmonic in a rapidly spinning planet is the largest and most important of these bulges. How large this bulge is depends on how the mass is dis-

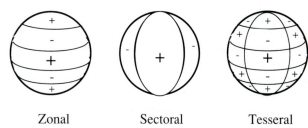

| Zonal | Sectoral | Tesseral |

**FIGURE 4.14** Zonal, sectoral, and tesseral harmonics. The sectoral harmonic is third order; that is to say, in going around the planet once, one sees three complete sets of bulges and dips, or plus and minus signs.

tributed inside the planet. The bulge is caused by centrifugal force; the more mass there is in the outer parts of the planet, where the spinning material moves the fastest, the stronger this force and so the larger the bulge will be. Thus by measuring the zonal harmonic of the gravity field we can estimate whether dense material is distributed throughout a planet or packed into a dense core. This ultimately gives us a value for the moment of inertia of the planet, the number that describes how much of the mass of a planet is concentrated at its center.

indeed almost pure olivine. Thus we infer that the mantle is made of a rock type, **peridotite**, that is mostly olivine.

From these arguments we can eventually put together an estimate of the composition of each part of Earth. In fact, it is now known that the mantle and core themselves have very complicated structures.

By knowing how big each part of Earth is, we can add up the contribution of each part to the whole and come up with a bulk composition of the whole planet. The elements that are in Earth now are, we assume, the same elements that it started with. When a planet evolves, those elements get sorted into different layers, such as a core, a mantle, and a crust.

For Mercury, our situation is much different. We do not have samples of its crust, seismometers on its surface, or a detailed model of its gravity. Because Mercury spins

slowly it does not have much of a bulge at its equator; furthermore, we've never put any satellites into orbit around Mercury. Thus we have very little information about its gravity field. All these difficulties mean that our conclusions about its interior structure are very tentative.

We do know its density, which implies a large iron content, and we do see a surface with a reflectance spectrum not too different from the lunar highlands. This implies that Mercury is probably differentiated into a large iron core and a plagioclase-rich crust. Is there an olivine mantle between core and crust? Our best estimate of topographic relief, from the limited set of photographs available, is about 1 km; thus if Mercury is isostatically compensated and has a denser mantle (neither of which is known for sure), the plagioclase crust must be at least 7 km thick. Judging by analogy with the Moon, it's probably much thicker than that.

# SUMMARY

The internal structures of the planets can be explored in several ways. The size of mountains on the surface of a planet gives one clue as to its internal structure. If these mountains are **isostatically compensated**, then they must have "roots" that extend into the mantle of the planet. Thus we can put limits on the relative densities of the crust and mantle and on the thickness of the crust under the mountains. If the mountains are not isostatically compensated, then we will see a gravity anomaly in the potential field, which may tell us something about the temperature and composition of the mantle below the mountains.

By studying the gravity field of a planet and expanding the potential of this field into its harmonics, one can determine the moment of inertia of the planet (a measure of how much mass is concentrated at the center of the planet) and whether odd lumps of unusual density lie buried under its surface.

# STUDY QUESTIONS

1. We say that mountains are isostatically compensated when they are:
   (a) floating in the mantle
   (b) surrounded by valleys
   (c) young and newly formed

2. A spinning planet bulges at the equator. If it has a massive core, does this make the bulge bigger or smaller?

3. True or False: "The force of gravity for a planet is always directed exactly towards the geometrical center of the planet."

4. Is Mercury a differentiated or undifferentiated body?

5. The moment of inertia of a planet tells us:
   (a) the mass of the planet
   (b) how mass is distributed inside the planet
   (c) the speed at which the planet spins

## 4.4   PROBLEMS

1. Mercury has the most extreme range of temperatures of any planet. Why? There are two spots on Mercury that get hotter than any other part of the planet. Describe where these spots are. Why are there two spots? (Hint: Read the "Further Information" box on Mercury's orbit and spin, and consider Mercury's eccentric orbit and unusual spin).

2. We know that the Moon has never had an atmosphere, going back to 4.6 billion years ago. What about Mercury? Do you think it ever had an atmosphere? State your reasons.

3. Assume that metallic iron has a density of 8 $g/cm^3$ and that Mercury rocks are similar to Moon rocks, with a density of 3 $g/cm^3$. If Mercury has a core of pure iron, what fraction of Mercury's mass would it represent? How big would the radius of the core be?

4. Callisto, a moon of Jupiter, has a density of 1.85 $g/cm^3$. The surface appears to be made of a mixture of ice and carbonaceous chondrite material. Given the densities of the materials as listed in Table 4.1, what fraction of Callisto's mass is icy?

5. A thin atmosphere of $^4He$ has been observed around Mercury. Where might this helium come from?

6. A spaceship is put into orbit about a small, rocky, airless planet. What information about the structure and composition of this planet can we get by observing the orbit of this spacecraft?

7. "Studying the internal structure of a planet is like trying to guess what's in a present before it's unwrapped." What "techniques" might a small child use on Christmas morning that have analogies to the way we try to study planets?

8. The density of ice is 0.94, while the density of water is 1. How much of an iceberg is below the surface of the water when it is floating?

9. The gravity field of a planet can be expanded into harmonics fairly easily because it tends to have regular, periodic variations from a smooth surface. What other periodic phenomena occur in nature that might be analyzed in terms of harmonics? (Hint: We already mentioned one in this chapter.)

10. Do you have any potential energy when you are asleep? What about kinetic energy?

11. The density of Mercury indicates that it is roughly two thirds iron metal, one third rock. Describe in a general way, perhaps with a drawing, the difference in the strength and direction of gravity that a person would feel if she were standing at the north pole of Mercury and:
    (a) the iron is distributed evenly throughout the planet;
    (b) the iron is concentrated at the south pole of the planet;
    (c) the iron is concentrated on the side of the planet nearest the Sun;
    (d) the iron is concentrated in the core.

## 4.5   FOR FURTHER READING

The primary sourcebook for Mercury studies is *Mercury*, edited by Faith Vilas, Clark Chapman, and Mildred Shapley Matthews (Tucson: University of Arizona Press, 1988). Robert Strom has produced a popular, readable introduction to Mercury, *Mercury: The Elusive Planet* (Washington, D.C.: Smithsonian Institution Press, 1987).

Good introductory books on the interiors of planets include William Hubbard's *Planetary Interiors* (New York:

Van Nostrand Reinhold, 1984) and Alan H. Cook's *Interiors of the Planets* (Cambridge, England: Cambridge University Press, 1980); you may also wish to consult advanced geophysics texts, such as Frank D. Stacy, *Physics of the Earth* (New York: Wiley, 1977), and Donald Turcotte and Gerald Schubert, *Geodynamics* (New York: Wiley, 1982).

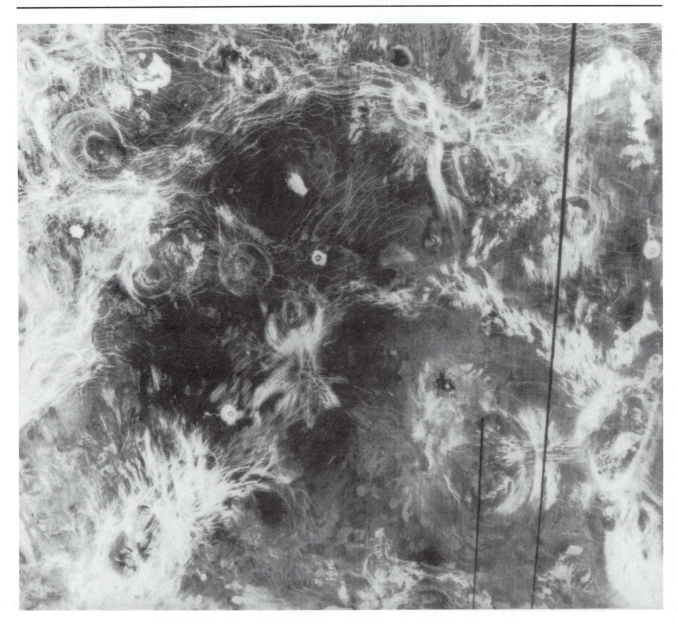

**FIGURE 5.1** Bereghinya Planitia. A wide variety of features are visible, including arachnoids, a corona, small shield volcanoes, steep-sided domes, impact craters, and lava flows. The area shown is 1843 km by 1613 km.

# CHAPTER 5

# Venus

## 5.1 ON VENUS

On October 18, 1967, a Soviet spacecraft called *Venera 4* arrived at Venus, the first spacecraft to pass through its clouds. (That it was number 4 is testimony to the difficulty of just reaching another planet. In fact, the Soviets had tried 10 times to send a spacecraft to Venus before finally succeeding.) It carried a simple detector to report the chemical composition of the atmosphere and a radar detector to locate the surface of the planet. However, neither device survived the conditions in Venus's atmosphere long enough to send back useful results.

Three years later *Venera 7* endured those conditions to reach the surface. In 1972 *Venera 8* followed. The surface pressure of the Venus atmosphere was found to be $9.2 \times 10^6$ Pa, or 92 Earth atmospheres, and the temperature a hot 740 K. The *Venera 7* and *Venera 8* landers survived for a matter of minutes before their radios failed under the intense heat.

Later Soviet landers succeeded in sending back color photographs from the surface and through heroic engineering efforts were able to survive on the surface for a little over an hour. The only American mission to penetrate the atmosphere of Venus, on December 9, 1978, included a probe with a complicated atmospheric chemistry sampler. While going through the clouds of Venus, a droplet of sulfu-

ric acid apparently clogged the inlet to the detector, leading to bizarre chemical readings.

For as long as people have tried, Venus has strongly resisted all attempts to penetrate its secrets. The relentless cloud cover never parts to show a glimpse of the surface, nor do the clouds themselves have any markings observable in visible light. Venus has no moon, no natural satellite for us to use to probe its gravity field. The enormous pressure

| Venus's Vital Statistics | |
|---|---|
| Radius | 6052 km |
| Surface area | $4.6 \times 10^8$ km$^2$ |
| Mass | $4.87 \times 10^{24}$ kg |
| Density | 5.24 g/cm$^3$ |
| Local gravity | 8.87 m/s$^2$ |
| Escape velocity | 10.3 km/s |
| Albedo | 0.75 |
| Surface temperature | 740 K |
| Surface pressure | $9.2 \times 10^6$ Pa |
| Length of day | 243.02 days (retrograde) |
| Length of year | 224.7 days |
| Distance from the Sun | 0.723 AU |

and temperature have completely disabled every probe that has reached its surface.

Our most complete knowledge of the surface of Venus comes not from sending landers to its surface, but from sending radar waves there. Earth-based radar can easily see roughly half the surface of Venus, and it gave us our first clues to the geography of Venus. The American *Pioneer Venus* probe made the first nearly complete map of the topography of its surface. The later Soviet *Venera* probes in the 1980s mapped its northern hemisphere in much greater detail. And most recently, in the early 1990s, *Magellan* arrived at Venus and made the most detailed mapping yet of its surface, in fact, more detailed than has yet been made of Earth, including all ocean basins (Figure 5.2).

**FIGURE 5.2** A global mosaic of Venus, centered at 180° longitude.

## Before the Space Age

Some clues to the nature of Venus were available before the space age. Carbon dioxide was first detected in the Venus atmosphere by spectroscopy in 1932. Little or no water was seen, implying that the clouds were not water clouds like Earth's. Faint markings in the clouds could be seen, but only when photographed with ultraviolet-sensitive film.

In 1940 Rupert Wildt, the German–American astronomer, theorized that the carbon dioxide could create a **greenhouse effect** on Venus to make the surface very hot. His reasoning went this way: Every planet stays in thermal equilibrium with the Sun by absorbing sunlight (which is strongest in visible wavelengths) and reradiating the energy

of this sunlight in infrared wavelengths. But in the case of Venus, although some visible light can eventually pass through the clouds and reach the surface, the infrared light from the surface cannot escape back out into space. Instead, it is absorbed by the carbon dioxide. The surface keeps receiving light and energy from the Sun, getting hotter and hotter because this energy cannot get out. Eventually the temperature becomes high enough that the atmosphere itself glows in an infrared wavelength close enough to visible light that sufficient energy can finally get past the carbon dioxide absorption to escape to space.

Wildt turned out to be right. Everywhere on the surface of Venus, day side or night side, equator or pole, the temperature is almost exactly the same, near 740 K.

The composition of the clouds themselves was a puzzle for many years. Finally, in the early 1970s teams of Earth-based observers and theorists determined that the cloud drops bent sunlight with a refractive index like that of concentrated sulfuric acid.

To peer through these clouds to the surface was a hopeless task until the advent of radar. For example, telescopic observers had noted that the ultraviolet markings on the clouds seemed to take about 4 Earth days to circle the planet. But Earth-based radar discovered that the surface moved at a much slower rate, and retrograde: backwards from the direction that the clouds moved and from the direction that most other planets spin. Each Venus sidereal day takes 243 Earth days, longer than a year on Venus. From sunrise to sunrise on Venus is 117 Earth days; but the Sun rises in the west on Venus!

To further confound the radar astronomers, Venus always shows the same hemisphere towards Earth whenever Venus and Earth are closest together. This is the only good time for radar studies, because a radar astronomer is in effect shining her own light on the planet and needs to be as near as possible for the "radar light" to be as bright as possible. Thus it is impossible to get a good radar map from Earth of more than part of one side of Venus.

## A Map of Venus

The *Pioneer Venus* probe of 1979 finally got around this problem. This satellite carried a radar device to measure the height of the terrain below its path. From this, the first radar maps of the surface could be made. These maps were superseded first by the Soviet *Venera 15* and *Venera 16* probes, which mapped the surface in 1983–1984, and most recently by the Magellan radar orbiter in 1990–1993 (Figure 5.3).

The surface of Venus can be divided into two general units. The most common type of terrain, covering about 85% of the surface, is the **lowland plains**, relatively smooth plains at close to the planetary average radius, dotted with thousands of individual volcanic features. These regions appear to be covered with basaltic lavas.

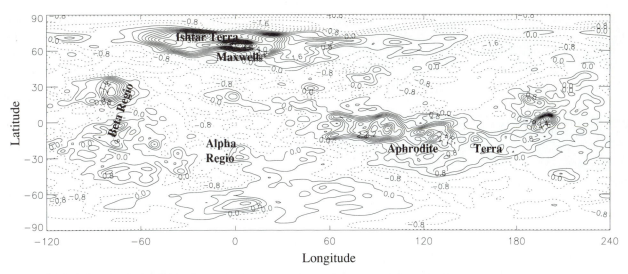

**FIGURE 5.3** A topographic map of Venus, based on radar data from *Magellan* and *Pioneer Venus*. The elevated region in the upper left is Maxwell; the one just to the right of center is Aphrodite Terra.

Most of the Soviet Venera spacecraft landed on this type of terrain and sent back data on the radioactivity (which gave information on the uranium, thorium, and potassium abundances) and density of the surface. Of the five Soviet landers that attempted to measure chemical abundances, four of them found that the rocks around them were basaltic. The results from *Venera 8* implied that the rocks in its region were more like terrestrial granites. In every case, the rocks were much more evolved than the basalts from a planet like the Moon; complex rock formation is probably occurring on Venus, just as it is on Earth.

The second type of terrain is found in the **highlands**, tectonically deformed plateaus several kilometers higher than the lowlands. This terrain, which appears to be formed by compression of the crust, contains many fascinating geological features. The two largest examples have been named **Ishtar Terra** (or Ishtar Land, in everyday English) and **Aphrodite Terra**. (Ishtar and Aphrodite are the Babylonian and Greek equivalents of Venus, the Roman goddess of beauty). Ishtar is in the northern hemisphere, about 20° from the north pole. It's an area as large as Australia, with a large, high plateau (called **Lakshmi Planum**) on the west, surrounded by mountains on all sides. The highest feature on Venus, named **Maxwell Mons** (mountain), stands to the east, 12 km higher than the lowland plains. At first, it was thought that Maxwell might be a very old volcano because there was a circular feature near its summit. Later, more detailed radar mapping of the area has shown that the circular feature, called in error **Cleopatra Patera** (a **patera**, or "shallow saucer," is a type of low-relief volcano), is actually an impact feature (although it may have undergone later volcanic modification).

Aphrodite is near the equator of Venus and a third of the way around the planet from Ishtar. It's larger than Ishtar, too, about the size of Africa. It lies only a few kilometers above the surface of the lowland plains and contains a variety of terrains. To the west are the plateaus **Ovda** and **Thetis**; the east, called the **chasmata** region, contains a lot of ridges and troughs, including the large **Diana** and **Dali** troughs (or **chasmata**). Farthest to the east is **Atla Regio** (region), containing the volcanic peaks **Ozza** and **Maat**.

Along with these two "continents," there are several other highlands regions named **Beta Regio** (there is also an Alpha Regio, but it is not a true highland feature), **Eistla Regio**, and **Bell Regio**. Of interest is that all features on Venus, except for Alpha, Beta, and Maxwell, are named after women (both fictional and real), as is the planet itself. These three exceptions were discovered before this naming convention was begun, and their original names were never changed.

Beta Regio is a large (2000 km by 3000 km) area of rifted and uplifted terrain containing two mountains, **Theia Mons**, a volcano, and **Rhea Mons**, whose origin is uncertain. Beta Regio may be the site of an upwelling in the mantle of Venus. Eistla and Bell are also regions of faulting and uplift, with many volcanic features.

The *Venera 13* and *Venera 14* landers were set down around Beta Regio. *Venera 13* landed on the surface of Beta, while *Venera 14* was placed in a valley just to the south of the first lander. These probes carried devices to measure the composition of the rocks in each place. They also took photographs of the surface (Figure 5.4).

*Venera 13*, in the highlands, landed on soft ground on which were scattered heavily cracked boulders. The rocks

**FIGURE 5.4** The surface of Venus, as photographed by the Soviet lander *Venera 14*.

had the chemical content of a weathered basalt, such as those found on Earth in the west rift zone of Africa. The *Venera 14* lander, in the valley, saw no dust and a very flat surface to the horizon. Most of the rocks in this area appeared orange colored, probably because only orange light could penetrate through the thick atmosphere (much like sunlight on Earth at sunrise or sunset).

Another type of terrain on Venus needs to be mentioned, the highly deformed **tesserae**, or **complex ridged terrain**, such as **Fortuna Tessera**, to the east of Maxwell. (The word *tessera* means "a small tile used in mosaic work" in Latin.) What kind of rock this area is made up of is still uncertain. The whole region appears very rough, covered with boulders on the order of a tenth of a meter or larger in radius. Tessarae are characterized by ridges and troughs, with roughly equal spacing (about 10–20 km), often with other **graben** superimposed. Most, but not all, are found in the highlands.

## Active Geology on Venus?

In size and density, Venus is quite similar to Earth; thus it seems reasonable to expect that it might have a composition similar to Earth's. On Earth we know that radioactive nuclides such as $^{40}K$ and $^{238}U$ create a considerable amount of heat inside the planet; this heat provides the energy to break the surface of Earth into **plates**, which can spread apart, push together, or slide past each other. Plate tectonics produces mountain belts and composite volcanoes on Earth, and at the plate boundaries most of this **radiogenic** heat (produced by radioactive decay) escapes from inside Earth. This will be discussed more thoroughly in Chapter 6.

Does Venus also have plate tectonics? This became a controversial question following the *Pioneer Venus* mapping. If it does not have plate tectonics, so the argument went, then there would have to be some other way of getting the heat out of Venus; for instance, it might be necessary for Venus to have intensive volcanism occurring all over

its surface. Because large areas of flood basalts or large shield volcanoes were not obviously visible (except possibly near Maxwell) in the *Pioneer* data, there would have to be very many small volcanoes present to account for the heat loss. This seemed quite unlikely; instead, one group of researchers suggested that Aphrodite Terra might actually be a ridge similar to the mid-ocean ridges on Earth where Venus plate tectonics was occurring.

Scientists realized then that only detailed radar maps could tell the difference. They would show if Aphrodite really did resemble a spreading center; they would show if Venus really was covered with numerous small shield volcanoes. Thus stood the debate after *Pioneer Venus*.

### *Magellan*'s Maps

In August 1990, the *Magellan* orbiting radar mapper arrived at Venus and was able to give the highest resolution images of the surface of Venus yet seen. It has begun to answer some of the debates about the nature of Venus geology, but has started new debates.

Although *Magellan* saw lots of tectonic features, showing that the crust of Venus has been under tremendous stress, it did not find confirmation of terrestrial-style plate tectonics. Aphrodite Terra does not actually seem to be a mid-ocean ridge. And Venus *is* covered with flood basalts and numerous small shield volcanoes.

In addition to the small shield volcanoes (< 20 km diameter), *Magellan* saw a host of other volcanic features, of which many were types not seen anywhere before (although coronae and arachnoids were seen first by *Venera 15* and *Venera 16*). Some examples are seen in Figures 5.5

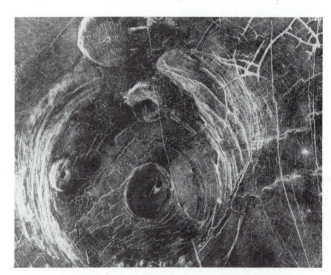

**FIGURE 5.5** Aine corona, about 200 km in diameter. Also visible in the upper part of the photo is a steep-sided dome, about 35 km in diameter. Complex fracture patterns (upper right) are often associated with coronae.

and Figure 5.6. These features, so new that not just the genesis but the names of many types are still under discussion, include such things as **coronae** (crowns), **arachnoids** (spiders), **novae** (stars), **anemones** (flowers), **ticks**, and **steep-sided domes** (at first called "pancake" domes). Of these, anemones, steep-sided domes, and ticks are considered to be intermediate-sized volcanoes (20–100 km diameter). In addition, there are large (> 100 km diameter) volcanoes, calderalike structures 60–80 km in diameter, flood basalts, and lava channels.

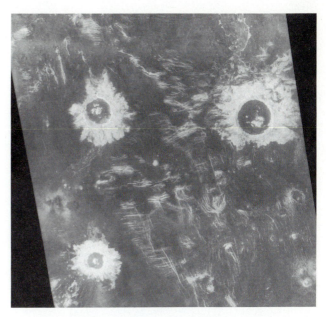

**FIGURE 5.7** This region of Venus, seen by the *Magellan* radar orbiter, has been called the "Crater Farm" for the large number of impact craters visible here. The area shown is 550 km by 500 km.

**FIGURE 5.6** This volcano, on the plains between Artemis Chasma and Imdr Regio, consists of a dome, 100 km across, with many lava flows.

Less than 1000 impact craters have been seen on the surface of Venus (Figure 5.7), evidence that some process is occurring to erase older craters, and that process is probably at least locally catastrophic. There are hardly any partially flooded or destroyed craters; either they are in good condition or they have been erased entirely. From the distribution of craters, one can infer an average surface age of half a billion years.

Also of interest about the craters on Venus is that there appear to be no craters of a diameter smaller than about 10 km. This is thought to be because the atmosphere of Venus is so thick that meteoroids that would have formed craters of this size are destroyed in the atmosphere, before they reach the surface. However, there are odd blotches that appear to be scars from atmospheric detonations, similar to the Tunguska event on Earth.

Curiously, smaller craters on Venus have not been preferentially destroyed, as they have on Earth by weathering. And the craters are distributed over the surface of Venus randomly, indicating that the entire surface of the planet is the same age. Two theories have been proposed to explain this: that the entire planet was completely resurfaced (by some catastrophic event) about half a billion years ago, or that active, random resurfacing is presently occurring, such that the average age of the surface is half a billion years. We don't know yet which of these two models is the more likely.

## Clouds, Winds, and Weather

Just like the Venus rocks, the clouds of Venus are complex in their composition and evolution. The air on Venus is 96% carbon dioxide and 3.5% nitrogen, with the rest consisting of traces of neon, argon, and sulfur compounds. There is essentially no water. From the ground to about 30 km above the surface, the air is clear; but above 30 km, a thin haze of sulfuric acid droplets appears (Figure 5.8). This cloud layer thickens at 48 km above the surface and then gradually thins out again over the next 20 km. The top of the cloud layer is 90 km above the surface of the planet. The clouds, even at their thickest, are not made of big drops like the water clouds of Earth. Instead, they are very fine droplets (called **aerosols**), 1 $\mu$m to 10 $\mu$m in diameter (about as fine as flour particles), and the effect is much more like a

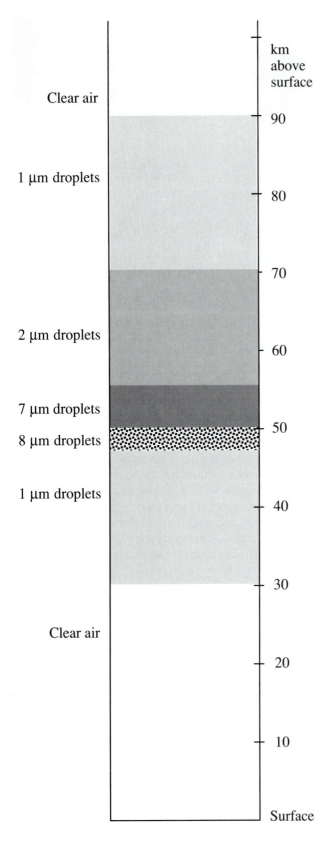

Clear air

1 µm droplets

2 µm droplets

7 µm droplets

8 µm droplets

1 µm droplets

Clear air

km
above
surface

90

80

70

60

50

40

30

20

10

Surface

**FIGURE 5.8** Layers in the atmosphere of Venus.

haze of pollution over a big city than like a solid cloud layer with a distinct bottom and top. Sitting in the cloud, you might not even recognize the cloud was there, except that your vision of distant objects generally would be obscured.

The weather on Earth, especially to those of us living in the middle-latitudes, is confusing and ever-changing. By contrast, the weather on Venus is downright boring. Every spacecraft landing on Venus has recorded a temperature of 740 K, within 5 K (corrected for altitude variations). This includes both daytime and nighttime landers at a variety of latitudes, spread over a period of nearly 15 years.

Likewise, the winds at the surface of Venus are quite low, ranging from 0.3 m/s to 1 m/s (less than 2 mph). In the upper atmosphere, however, where the clouds lie, the winds get much stronger, ranging up to 100 m/s at an altitude of 50 km above the surface. But even there, the pattern of the winds is quite regular, always blowing westward and rarely changing.

Indeed, the remarkable thing about the weather on Venus is that it so closely approximates the very simple theory put forth by George Hadley in 1735. **Hadley cells** are a convection pattern where air heated at the equator rises and flows to the poles, where it cools and descends. Thus there should be gentle winds at the surface blowing from pole to equator.

The reason this simple mechanism fails to explain Earth's weather but works so well for Venus is that many of the forces acting on the atmosphere of Earth do not occur on Venus. Venus is not tilted on its axis. This means that unlike Earth or Mars, Venus does not have any seasons. The planet's northern and southern hemispheres always see the same amount of sunlight. And again, unlike Mars, Venus has very little eccentricity to its orbit; its orbit is even closer to a perfect circle than Earth's. So there's very little change in the amount of sunlight Venus receives during the year.

Also, Venus spins very slowly. This means that the effect of the **Coriolis force** (which will be discussed further in Chapter 10) is negligible. This force tends to break up the flow of air towards the equator on Earth, turning this flow into the familiar cyclone (counterclockwise in the northern hemisphere) and anticyclone (clockwise in the northern hemisphere) whirling masses that we associate with low and high pressure centers. None of this happens on Venus.

Finally, the temperature of Venus is determined by the greenhouse effect of the carbon dioxide in the atmosphere described above. This means that the temperature at any given spot does not depend on how much sunlight that spot receives at any time. Instead, it depends solely on what temperature can radiate sufficient heat to maintain a thermal balance over long periods. Furthermore, with such a dense atmosphere, there's simply much more gas to cool off. Thus even though the night side can be in shadow for nearly 59 Earth days, the temperature will not drop significantly during that time. (Recall the difference between solar and sidereal days described in Chapter 4. The sidereal period of Venus is 243 days, but the solar period is only 117 days.)

Despite the monotony of the atmosphere, there is one exciting aspect to the weather. The *Venera* spacecraft recorded several bursts of radio noise that could be explained as powerful strokes of lightning in the atmosphere. Thus, if acid and heat were not enough, electrocution might also strike a visitor to Venus!

## Evolution of Venus

What is the interior of Venus like today? How has it evolved with time? These are questions to which we cannot give definitive answers yet; we just don't have enough detailed information about the surface of Venus to make educated guesses about what geologic processes are going on. We still don't have enough detailed information on the chemistry of the surface rocks. We have not landed seismometers on the surface of Venus (nor indeed can we make seismometers that could operate for very long at the high temperature of the Venus surface).

The *Pioneer Venus* satellite was carefully tracked to determine the gravity field of the planet. It found gravity anomalies on Venus similar to those on Earth, much less than had been seen on Mars or the Moon. Its few large "continents" are probably supported by convection of the Venus mantle. But because Venus spins so slowly, it has no bulge at its equator; thus one cannot directly estimate the moment of inertia or the size of the core. We do know that it has no magnetic field.

In other words, it is difficult to put strict limits on our ideas of the possible ways that Venus could have evolved or even what the interior of Venus looks like today.

The possible models show the degree of our uncertainty. On the one hand, Venus could have started much drier and much less oxidized than Earth. In this case, Venus would have a metallic iron core, but no liquid core of iron sulfide, and the mantle rocks would be rich in magnesium and low in iron oxide. The temperature of the mantle rocks in Venus by this model would be higher than on Earth, both because the surface started out hotter and also because the lack of iron oxide raised the viscosity of the mantle rocks, making solid-state convection less likely. The water and sulfur seen in the atmosphere are only a tiny fraction of the mass of the planet, and they might have been added by an occasional infalling comet.

On the other hand, Venus might have started with as much water as we see on Earth today. Over time, solar radiation broke this water up into hydrogen, which escaped to space, and oxygen, which reacted with the rocks. Thus the mantle of Venus would be rich in iron oxide and convect easily, cooling off the interior. If there was water like Earth's, there might also be sulfur in the metallic core, like Earth probably has. In this case, the reason Venus

doesn't have a magnetic field may be that it doesn't spin fast enough; both a liquid core and a spinning planet seem to be necessary to make magnetic fields. Alternatively, there may be so much sulfur in the core that there's no metallic iron left, or at least enough that the melting point is lowered below the current core temperature. Some theorists believe that magnetic fields need condensing metallic iron to provide their energy. In other words, the lack of a magnetic field might be telling us that there's no sulfur in the core, or it might be telling us that there's too much sulfur there!

## SUMMARY

Venus, although the planet closest to Earth, is covered with such a thick atmosphere that studying its surface and interior is very difficult. The atmosphere, mostly carbon dioxide, has a surface pressure 92 times that of Earth. The surface temperature is an unpleasant 740 K. The thick atmosphere contains several haze layers, in total 60 km thick, made of droplets of sulfuric acid.

This haze completely obscures the surface. Only radar can give us a picture of the topography of Venus. Radar maps have shown us that most of Venus is covered with plains. Some continentlike highlands are also seen on the planet. The crust of the planet is highly cracked, indicating that it is under a lot of stress. Recent landers have found the composition of Venus rocks to be similar to basalts on Earth. The vast numbers of volcanoes seen by *Magellan* fits in with the volcanic rocks seen by *Venera*; both imply a surface that may still be constantly evolving, and volcanoes, rather than plate motions, may be the dominant way of getting radiogenic heat out of Venus.

Because Venus has no seasons and spins slowly, the winds and weather do not change much. The upper clouds tend to travel in a westerly direction, moving about the planet once every 4 Earth days. At the surface, wind speeds are very low.

## STUDY QUESTIONS

1. Are basaltic rocks on Venus closer to Earth rocks or to Moon rocks in composition?

2. Standing on the surface of Venus, what color does the sunlight make everything look? Why?

3. Venus is covered with clouds. What are the clouds made of?
   (a) water
   (b) sulfuric acid
   (c) carbon dioxide

4. How long is a solar day on Venus?

5. Does Venus have an iron core?

## 5.2  ATMOSPHERIC CHEMISTRY

As you sit right now reading this chapter, you are (we can presume) on the surface of the planet Earth, surrounded by a gas at a pressure of $10^5$ Pa (Pa stands for **pascal**, the SI unit of pressure) and a temperature of 290 K, with a composition roughly 80% nitrogen and 20% oxygen, with other components such as argon, $CO_2$, and water in trace amounts. If you are fortunate, you may be sitting outside right now while this gas is gently moving past you. Maybe you're getting a sunburn; maybe you're getting wet because it's raining. In either case, every breath you take brings oxygen into your lungs where it undergoes a chemical reaction with your blood.

Your suntan, or soggy clothes, and your warm $CO_2$-rich breath are all examples of the way the chemistry of our atmosphere affects our lives. Chemical reactions with ultraviolet light at the top of the atmosphere (and the little bit that reaches the surface to cause suntans and sunburns), condensation of solid material within the atmosphere forming clouds and rain, and chemical reactions between the air and the surface of the planet are all covered in this chapter.

As important as these topics are on Earth, they play an even more dramatic role for Venus. The high temperature and very high pressure of its atmosphere means that it is much more chemically reactive than the atmosphere of Earth. Its location, closer to the Sun than Earth is, means that solar ultraviolet (UV) radiation can play an even more important role in its chemistry. And the eternal shroud of sulfuric-acid clouds has been the side of Venus which most terrestrial astronomers have known (Figure 5.9).

What controls the chemical composition of its atmosphere? Why are its clouds made of sulfuric acid? And what role has its atmosphere played in the evolution of the surface of Venus?

### Composition of Atmospheres

Before we attempt to understand how the chemicals in any planetary atmosphere behave, we must first ascertain just what those chemicals are. Consider our best estimates for the composition of the atmospheres of Venus, Earth, Mars, and Jupiter, as given in Table 5.1.

There is an obvious difference between Jupiter and the terrestrial planets: Jupiter's atmosphere is dominated by hydrogen and helium. The cosmic abundance of H and He is 99 times that of the other elements, so a planet that keeps its cosmic share of these gases, as Jupiter did, will have a very large atmosphere.

The abundant hydrogen has an effect on the chemical nature of its atmosphere. Chemists tend to think of a chemical system in terms of the amount of free oxygen available to react with the other species. This might seem

**FIGURE 5.9**  The clouds of Venus are seen in this ultraviolet photograph, taken by *Mariner 10*.

like an Earth bias at first, because our atmosphere alone has much free oxygen. But in fact the difference between oxidizing (oxygen-rich) and reducing (oxygen-poor) systems has a more general significance. When hydrogen is available in cosmic abundances, it forms water with the oxygen, removing it from other reactions. This tends to make an extremely reducing chemical climate. When hydrogen is absent, however, the next most abundant element to react with other species is in fact oxygen. Thus a hydrogen-free system tends to be oxidizing.

Jupiter's atmosphere is very reducing. Its carbon and sulfur compounds tend to be methane ($CH_4$) and hydrogen sulfide ($H_2S$). On the terrestrial planets, however, carbon is oxidized as $CO_2$, and sulfur can exist in its elemental form, S, as $SO_2$, or as $SO_3$.

A more subtle difference is also seen in not only the contrast between the terrestrial planets and Jupiter, but also in "gas-rich" versus "gas-poor" meteorites. When you look at the abundances of various isotopes in these meteorites, especially the inert gases like neon and krypton, you find that there are two distinct types of gas present.

Most meteorites have small amounts of **adsorbed** gas, gas that has been physically trapped onto its surface. The isotopes in this gas are quite similar to the isotopes seen in Earth's atmosphere. This component of gas is thus called the planetary component, because it seems to be similar to that seen in terrestrial planets. However, gas-rich meteorites have another component that is quite distinctive and that seems to have similar isotope abundances to those seen in

**TABLE 5.1**

Atmospheres of Four Planets

| Planet | g | P | T | $H_2$ | He | $H_2O$ | $N_2$ | $O_2$ | Ar | $CO_2$ |
|--------|------|-------|-----|-----|-----|------|-----|------|-------|--------|
| Venus | 8.87 | 92 | 740 | — | — | 0.01 | 3.5 | — | 0.007 | 96 |
| Earth | 9.87 | 1 | 290 | — | — | 1 | 77 | 21 | 0.93 | 0.0033 |
| Mars | 3.72 | 0.006 | 220 | — | — | 0.03 | 2.7 | 0.13 | 1.6 | 95 |
| Jupiter | 26.6 | 0.5 | 125 | 75 | 25 | 0.02 | — | — | — | — |

Notes: Gravitational acceleration at the surface, g, is in units of m/s$^2$; pressure, P, is relative to Earth's atmosphere, which is $10^5$ Pa in SI units; T is in kelvins. Figures for Jupiter, which has no surface, are for the cloudtop level. Abundances are in weight percent.

the **solar wind**, the ionized gas that is blown off the surface of the Sun and fills interplanetary space.

It seems reasonable to conclude, therefore, that these meteorites have been exposed to the solar wind for a long time and have collected much gas from it and that this solar component in fact reflects the abundances of gases in the Sun. Because Jupiter is also of solar chemical composition, we might assume that it too has solar isotopic composition (although we won't know for certain until we go there and sample it). The terrestrial planets, on the other hand, seem to have only those isotopes that could most easily diffuse into the matrix of a rock.

From this, we come to the following conclusion about the atmospheres of the terrestrial planets: Unlike Jupiter, which apparently captured part of the same gas that formed the Sun, *the terrestrial planets were formed without any atmospheres at all.* It is only after they were formed that the heating of the interior of these planets drove gases out of the rocks. These gases gathered on the surfaces of the larger planets, forming atmospheres on Venus, Earth, and Mars, and escaped the gravitational pull of the smaller ones like Mercury and the Moon, which remain airless to this day.

This is an important first step in understanding the atmosphere of Venus.

## Chemical Cycle on Venus

### Reactions with the Surface

If the atmospheres of the terrestrial planets are made of gases that were driven out of rocks and up through these rocks to the surface of the planet, then this provides a ready situation for chemical reactions between the gas and the surface rocks. One effect is to destroy surface rocks; this process is called **chemical weathering**. As time goes on, any other changes to the atmosphere will result in chemical reactions to try to keep the air and the surface in chemical equilibrium.

Let's look at the chemical system on Venus. The clouds tend to be rich in sulfuric acid, and the radar response of certain parts of the surface indicate that the mineral pyrite, $FeS_2$, may be present on part of the surface

of Venus. How does sulfur interact between surface and atmosphere?

The reactions that occur to pyrite include

$$FeS_2 + 2CO_2 \rightarrow FeO + 2COS + \tfrac{1}{2}O_2 \quad (5.1)$$

$$FeS_2 + 2H_2O \rightarrow FeO + 2H_2S + \tfrac{1}{2}O_2 \quad (5.2)$$

Thus the presence of carbon dioxide and water in the atmosphere tends to destroy pyrite on the surface. Rocks that originally had iron sulfide will have iron oxide instead, and this iron oxide may eventually become a part of any of a number of iron-bearing minerals. The other chemicals on the right-hand side of these equations are gases; these gases have indeed been observed in the atmosphere of Venus.

### Reactions with Solar Ultraviolet Light

What then happens to the chemicals COS and $H_2S$ in the atmosphere of Venus? Convection eventually carries these chemicals into the upper atmosphere; once they diffuse above the cloud layer they can be broken apart by high-energy bits of ultraviolet light. Recall how a light wave is an oscillating electric and magnetic field and that this oscillating field can make other electrons in its path vibrate. If such a wave encounters a chemical compound and its energy is high enough, it will shake the compound apart. Ultraviolet wavelengths of sunlight are high energy, and high-energy light tends to break up compounds.

In particular, reactions that occur in the atmosphere of Venus include

$$3CO_2 + h\nu \rightarrow 3CO + \tfrac{3}{2}O_2 \quad (5.3)$$

$$\tfrac{3}{2}O_2 + H_2S \rightleftharpoons SO_3 + H_2 \quad (5.4)$$

$$\tfrac{3}{2}O_2 + COS \rightleftharpoons SO_3 + CO \quad (5.5)$$

Note what happens. Carbon dioxide makes up 96% of the atmosphere of Venus, and so it is the most likely target of a UV photon. It is broken up by a high-energy, high-frequency photon of light traditionally called $h\nu$; $\nu$ stands for the frequency of the light wave and $h$ is Planck's constant, which converts units of frequency into units of energy. The first reaction produces free oxygen in large amounts, which then combines with $H_2S$ and COS to remove them from the atmosphere. Because there is no iron present in the upper atmosphere to take up the sulfur, the sulfur forms a sulfate, $SO_3$.

Other reactions that occur include the direct breakup of $H_2S$ and COS by UV radiation:

$$H_2S + h\nu \rightarrow H_2 + \tfrac{1}{n}S_n \qquad (5.6)$$

$$COS + h\nu \rightarrow CO + \tfrac{1}{n}S_n \qquad (5.7)$$

In other words, the sulfur produced this way gets formed into chains of sulfur atoms. Here, $n$ stands for the number of sulfur atoms in a chain. There might be many short chains or a few long ones, hence the $1/n$ term in the equations above. Typical values of $n$ range from 3 to 8. These chains tend to be solid specks of dust, not gases.

What happens to the $SO_3$ made in the reactions described above? There's also much CO made in these reactions, and so in the lower atmosphere this can react:

$$SO_3 + CO \rightleftharpoons SO_2 + CO_2 \qquad (5.8)$$

This replenishes the amount of carbon dioxide in the atmosphere, destroyed by the UV reactions in the upper atmosphere. Thus, by a complicated chain of reactions, we eventually convert sulfur in the form of iron sulfide, $FeS_2$, from the surface of the planet into sulfur dioxide, $SO_2$, in the lower atmosphere.

### Condensation Reactions and Cloud Formation

That's not the only reaction that can take place with $SO_3$. In the upper atmosphere, the temperature is low enough to favor the reaction

$$H_2O + SO_3 \rightleftharpoons H_2SO_4 \qquad (5.9)$$

This tends to build up the amount of sulfuric acid, $H_2SO_4$, in the upper part of the atmosphere, and liquid droplets of $H_2SO_4$ start to form. These aerosol droplets are so small that random motions of the molecules in the air tend to keep them aloft and prevent them from falling through the atmosphere to the surface for a long time. Thus clouds of droplets are formed, like the clouds made of water that form in Earth's atmosphere.

Recall the sulfur particles formed by direct UV destruction of $H_2S$ and COS. These sulfur particles can get incorporated, like little bits of dust, into the droplets of sulfuric acid. Because the particles absorb UV light they make the clouds appear dark when viewed in ultraviolet light, as in the "barber-pole" stripes in UV photographs of Venus.

Despite the small size of the droplets, they eventually collect together into a size large enough to rain down onto the surface. But as they reach the lower parts of the atmosphere, below the clouds, where temperatures and pressures are much higher, the sulfuric acid tends to break down into its components

$$H_2SO_4 \rightleftharpoons H_2O + SO_3 \qquad (5.10)$$

followed by

$$SO_3 + 4CO \rightleftharpoons COS + 3CO_2 \qquad (5.11)$$

$$SO_3 + H_2 + 3CO \rightleftharpoons H_2S + 3CO_2 \qquad (5.12)$$

Finally, the product gases react to form $SO_2$:

$$H_2S + \tfrac{3}{2}O_2 \rightleftharpoons H_2O + SO_2 \qquad (5.13)$$

$$COS + \tfrac{3}{2}O_2 \rightleftharpoons CO_2 + SO_2 \qquad (5.14)$$

### Finishing the Cycle

What happens to the $SO_2$ formed by all the above reactions? It can react with surface rocks again:

$$2SO_2 + O_2 + 2CaCO_3 \rightleftharpoons 2CaSO_4 + 2CO_2 \qquad (5.15)$$

Calcium carbonate, $CaCO_3$, is one of the minerals involved in the reactions that regulates carbon dioxide pressure on Venus. Furthermore, the mineral $CaSO_4$, called **apatite**, can react with other surface rocks and atmospheric carbon dioxide:

$$2CaSO_4 + FeO + 2CO_2 \rightleftharpoons$$
$$FeS_2 + 2CaCO_3 + \tfrac{7}{2}O_2 \qquad (5.16)$$

We've made pyrite again, and completed the cycle.

For every chemical species in every atmosphere, some sort of cycle like this must go on. The rate at which each step can take place determines how much material is tied up in any one form. For example, if droplets grew much faster on Venus, then the clouds would rain out sooner. Thus less sulfur would be tied up in the clouds, and more would exist in other forms elsewhere in the atmosphere.

## SUMMARY

Most theorists believe that the planets were formed from a cloud of gas and dust with the same chemical composition as the Sun. Large planets, like the gas giants Jupiter and Saturn, were able to hold on to some of this gas while they

**FURTHER INFORMATION...**

# Chemistry of Earth's Atmosphere

We've used one specific chemical cycle on Venus to illustrate the various ways that chemical reactions in the atmosphere interact with sunlight, surface rocks, and clouds. How do these processes work on Earth?

Clouds here are formed mostly by water vapor. Water is not constantly reacting with other chemicals in the air to the extent that sulfuric acid reacts on Venus. Reactions involving water do take place, most importantly forming carbonic acid, $H_2CO_3$, by reaction with the small amounts of $CO_2$ that exist in Earth's atmosphere. This results in chemical weathering reactions on the surface of Earth. This weathering helps break rocks up into soils, where plants can grow. Animals like human beings (or other plant-eating animals) eat the plants and breathe, producing more $CO_2$ to keep the cycle going.

In the clouds themselves, the principle reaction that occurs is a change in the vapor pressure of water (a measure of the total amount of water in the air) that varies both as the temperature changes and as rivers or winds bring water from one part of the planet to another. For instance, winds and currents from a moist and hot region, the Gulf of Mexico, hit cold, drier northern Europe, producing the famous London fogs and other types of weather.

The free oxygen (made by the plants, growing from weathered soil and fed by rainfall) that eventually gets into the upper atmosphere of Earth is broken up by solar UV radiation into energetic single atoms of oxygen. These single atoms react with $O_2$ molecules to form $O_3$, ozone. The ozone is then broken up by solar UV radiation back into oxygen atoms and molecules, and the cycle continues.

The importance of the ozone is that every time it is broken up, it prevents the light that broke it from reaching the surface of the planet. The photons of light stopped by the ozone are those UV photons that are not energetic enough to be absorbed by other gases, but that still pack enough energy to do serious damage to living species like plants and people. As long as there is ozone in the upper atmosphere, it absorbs this UV light. The little that gets through is generally just enough to give us suntans instead of skin cancer.

---

were forming, and so they have atmospheres of solar composition. The terrestrial planets, however, formed without atmospheres initially. Later, as they heated up, the small amounts of gas trapped in the rocks were driven out and migrated to the surface of these planets.

Chemical reactions between this gas and the rock determine the composition of both surface rocks and the lower atmosphere. In the upper atmosphere, solar UV radiation can break up chemical compounds, which can react to form new compounds. Along with forming new compounds, these reactions also absorb the energy of the UV light, preventing it from reaching deeper into the atmosphere, down to the surface of the planet.

If the abundance of a component of the atmosphere increases or if the temperature drops to the point where individual molecules are likely to stick together and form droplets, then clouds will condense in the atmosphere. Small droplets can be supported by winds. As the drops grow, they tend to fall back towards the surface of the planet as rain. When the droplets hit the ground, they come in chemical contact with the surface rocks and the chemical cycle starts over again.

## STUDY QUESTIONS

1.  What do we call the energetic photons of sunlight that cause chemical reactions at the top of a planet's atmosphere?

2.  At the surface of Venus, $CO_2$ reacts with iron sulfide to make COS. Why does this reaction occur only at the surface of Venus, not in the upper atmosphere?

3.  The clouds of Venus consist of droplets of acid. What is it about the droplets that prevents the clouds from falling (as rain) down to the planet's surface?
    **(a)** small size
    **(b)** chemical composition
    **(c)** peculiar color

4.  Where did the atmosphere of Venus come from: gases from the Sun trapped by Venus's gravity, or gases originally adsorbed in the rocks that made up Venus?

5.  What does ozone in Earth's stratosphere do to protect life on Earth?

## 5.3   TECTONICS: GLOBAL STRESSES

Beneath the clouds of Venus we now know there exists a rocky surface covered with volcanoes, a surface with endless miles of rolling hills, continents of rugged mountains, plains crisscrossed with cracks and scars. What are the forces that have shaped this surface?

Planetoids impacting a planet, massive volumes of lava piled high on its surface, convection of rock stirring the interior of the planet all lead to tremendous stresses on a planet's surface. And if there's enough stress, all solid materials will eventually move, crack, buckle, and break.

We are familiar with stresses on a small scale, on Earth. Stress in a plate of hot glass causes it to shatter when touched by a drop of cold water. It makes road pavements crack and heave in the winter, creating bumps and potholes. On a planet-sized scale, the cracks and stresses of a planetary crust create mountain chains and rift valleys. The study of how a planet's surface responds to stress is called **tectonics**.

### Stress on Rocks

To understand how surfaces are shaped by tectonics, let's start by looking at how ordinary solid materials act under stress. Imagine a block of wax being squeezed in a vise. Because the wax is soft, it might slowly bulge in the middle as the two ends are pressed together. This is called **plastic**, or **viscous**, behavior. Most materials, even stone, will flow like this if they are warmed up to close to their melting points, or if the stress is applied very slowly, over a very long period of time. The **asthenosphere** of a planet such as Venus is a region, beneath the surface, in which the response to stress is plastic.

However, most rocks at temperatures typical of planetary surfaces, well under 1000 K even on Venus, will not relieve the stress of the squeezing by flowing. Instead, the stress inside the rock will continue to build up, until the rock suddenly cracks and breaks. This is called **brittle** behavior. The parts of the crust and uppermost mantle of a planet that behave mostly in a brittle manner, including the rocks at the surface, are called the **lithosphere**.

In what direction does the rock crack? Say we had our rock on a table, with a weight pressing down on it. A horizontal crack, parallel to the surface of the table, does not release any stress; nor does a vertical crack. Neither type of

crack lets the rock move sideways (the way the bulging wax moves). However, a crack at an angle between these two extremes will allow the top part of the rock to slide both down and sideways, thus releasing the stress. (See Figure 5.10.)

All things being equal, one might expect the crack to run at 45° from the direction of the vertical. However, this assumes that once the rock has cracked, the pieces can slide easily past each other. If there is friction between the pieces, however, some of the energy that might move the rock pieces sideways is needed instead to overcome the friction between the two pieces. The crack tends to occur at a steeper angle. Thus a more typical angle of fracture, for rocks with internal friction, is about 30° from the direction of the stress. From this we can conclude that, in planetary features, the direction that a stress is applied is usually at roughly 25° to 40° from the direction of the cracks it produces.

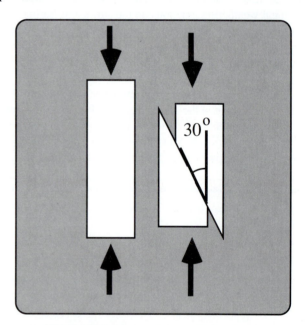

**FIGURE 5.10**   Stresses make rock crack; the direction of the crack is at an angle to the direction of the stress.

Of course, in nature all things are usually not equal. Mineral crystals tend to have certain directions that break more easily than others (hence the beautiful smooth facets of a diamond). A rock, or a segment of a planet, that already has a crack tends to release its stress along that crack. In general, stress causes the surface to break where it is thinnest or weakest. A planetary surface, like the proverbial chain, is only as strong as its weakest link.

By talking of rocks under pressure, we have been discussing **compressional** stresses. But an opposite kind of stress can also occur, called **extensional** stress, when pressures are arrayed to try and pull the rock apart. A pleasant example can be seen in the surface of a cake taken from a hot oven. As the cake cools, the top starts to contract faster than the center. Because the whole cake can't shrink at the same rate, this leads to extensional cracks in the top of the cake.

On planets, we find both compressional and extensional stresses, some occurring just in small parts of the surface, or **local stresses** (more about these in Chapter 7), and others happening everywhere on the planet, or **global stresses**.

## Global Stresses

Virtually all materials change density as their temperature changes. Most substances expand as they warm up and melt and contract as they cool. This expansion and contraction means that the whole volume of the planet will be growing and shrinking and the surface will have to be stretched or compressed accordingly.

Global expansion upon heating leads to extensional stresses. Cracks on planets that have experienced such stresses tend to look like grooves where the main pieces of the old surface have been pulled apart, exposing fresh (and warmer, more ductile) material from below, which slowly oozes up to partly fill the cracks. Such features are often seen in the icy moons of Jupiter, Saturn, and Uranus. Valles Marineris on Mars may be a case where it has occurred on a terrestrial planet.

Once a planet starts to cool off, however, it will contract and the pieces of the old surface will be compressed onto each other. Long chains of mountains, or high cliffs called scarps, can result. The best example of these can be seen on Mercury. Planetary cooling tends to be more important on smaller planets, like Mercury or the icy satellites. Convection and topographic stresses are more important on larger planets, like Earth and Venus.

## The Energy Source

Where does a planet get the energy to fold or crack its surface?

We saw in Chapter 4 how the formation of a large core in Mercury might have added so much heat to its interior that the whole planet expanded by several tens of kilometers. After core formation, the planet would have cooled off rather rapidly, leading to Mercury-wide contraction. The result, it is thought, are the long scarps seen nearly everywhere on the surface of Mercury. These scarps are excellent examples of compressive global tectonics.

No other planet is thought to have such a proportionally large core, however. But the cracking of the surface, and the formation of volcanoes, is strong evidence that considerable amounts of heat are generated inside these planets. Where does this heat come from?

### Accretional Heating

Assume that a planet was formed by bits of matter that fell from infinitely far away onto its surface. As these bits were falling, they converted gravitational potential energy into kinetic energy; they moved faster and faster as they got closer to the center of the gravitational attraction, the center of the growing planet. When they hit the planet, their energy had to go someplace. That someplace is heat.

But how much heat? Obviously, the bigger the planet, the more bits of matter we have falling into the planet, so the more energy we'd expect to have available for heating the planet. But the source of this energy is the gravitational potential energy of the planet itself, which also gets bigger as the planet gets bigger. The total amount of energy ought to look something like the potential energy of two spheres, each of mass $\frac{1}{2}M$, separated by a distance of $\frac{1}{2}R$, or $\frac{1}{2}GM^2/R$. In fact, a more careful derivation yields

$$E_{max} = \frac{3}{5}\frac{GM^2}{R} \qquad (5.17)$$

Let's say that all this energy was available to heat the body, that none was lost to space. Let $C$ be the heat capacity of the material that makes up the body, the amount of heat it takes to raise the temperature of a unit of mass by 1 K, and let $\Delta T$ be the number of degrees that the mass $M$ actually heats up. A typical value of $C$ for a rock is about 100 J/kg·K. (For large planets, $C$ may change considerably as the pressure and temperature of the material deep in the planet increases; on the other hand, for small moons and asteroids we can neglect this effect.) We set

$$\frac{3}{5}\frac{GM^2}{R} = CM\,\Delta T \qquad (5.18)$$

and solve for

$$\Delta T = \frac{3}{5}\frac{GM}{RC} \qquad (5.19)$$

If we assume that our planet is a sphere with a constant density $\rho$, then we can rewrite this equation by substituting $\frac{4}{3}\pi\rho R^3$ for the mass $M$ and get

$$\Delta T = \frac{4}{5}\frac{G\pi\rho R^2}{C} \qquad (5.20)$$

For the various planets, we arrive at the temperatures listed in Table 5.2. These are the maximum temperatures, as if all the matter arrived instantaneously and didn't have a chance to radiate away any of this heat before more rock, and heat, was deposited. In fact, planets may take up to

# Stress and Strain

A **viscoelastic** material is a material, such as a rock, a golf ball, or Earth's mantle, that behaves elastically on short time scales and viscously on long time scales. But what do we mean by "elastic" and "viscous," and how do we determine what time scales are important?

An **elastic** material deforms when a force is applied to it and returns to its original shape when the force is removed. If the force is too strong, however, the material will break. In the case of planetary surfaces, this breaking can be seen as faults. The force pushing on the material, divided by the area over which it pushes, we call the **stress**. The force can push directly against the surface like pressure, or it can drag along the surface like friction; both are examples of stresses. How the material responds to stress—how much it stretches—we call the **strain**.

An elastic material deforms according to **Hooke's law**:

$$\sigma = \epsilon_e E \tag{5.21}$$

where $\sigma$ is the **stress**, or force per unit area on the material; $\epsilon_e$ is the **elastic strain**, or deformation of the material; and $E$ is a constant, called **Young's modulus**. (Hooke's law is strictly true only for uniaxial stresses, stresses applied in one direction only. Stress and strain are actually **tensors**, arrays of numbers relating the response of the material to both compression and sliding forces in every direction.)

A **viscous** material also deforms when a force is applied to it, but it stays deformed after the stress is removed. In a simple, **Newtonian** fluid, the rate of strain is directly proportional to the applied stress, such that

$$\dot{\epsilon}_f = \frac{d\epsilon_f}{dt} = \sigma/2\mu \tag{5.22}$$

where $\epsilon_f$ is the fluid strain and $\mu$ is the **viscosity**.

In a viscoelastic material, the strain, $\epsilon$, is the sum of the viscous and elastic terms:

$$\epsilon = \epsilon_e + \epsilon_f \tag{5.23}$$

$$\dot{\epsilon} = \frac{1}{E}\frac{d\sigma}{dt} + \frac{1}{2\mu}\sigma \tag{5.24}$$

Suppose a strain, $\epsilon_0$, is applied to a viscoelastic material for some period of time, then released. The behavior of the material may be derived by integrating Equation 5.24 to get

$$\sigma = \sigma_0 e^{-Et/2\mu} \tag{5.25}$$

where $\sigma_0$ is the initial (elastic) strain, and $t$ is time. The stress relaxes to a value of $1/e$ times its original in a characteristic time scale, the **viscoelastic relaxation time**,

$$\tau_{ve} = \frac{2\mu}{E} \tag{5.26}$$

A viscoelastic material responds elastically to a stress applied on a time scale short relative to the viscoelastic relaxation time, and viscously to a stress applied on a time scale long compared to this time scale. The temperature of a material is important in determining how it will react to stress. The same material at low temperature may be brittle while at high temperature will flow readily.

The **lithosphere**, consisting of the crust and the topmost part of the mantle, is the part of a planet that behaves elastically in response to typical stresses. The **asthenosphere**, farther down in the mantle, behaves viscously.

---

100 million years to form. Furthermore, note that the heat from an impact is preferentially deposited at the surface of the growing body (although in fact seismic waves from the impact can carry some of the heat into the body's interior) where the heat can radiate out to space and never affect the interior. Bodies heated by accretion may have relatively cool interiors and be heated only in their outer regions. In any event, the temperatures in Table 5.2 are maximum values, much higher than we would realistically expect. Clearly the accretion temperatures predicted by this simple model for the larger planets are absurdly large. On the other hand, even these maximum accretional energies for the small moons and asteroids are negligible.

The heat pulse associated with the formation of a core, such as for Mercury, is just a variation on accretional heating. Instead of material falling from infinity onto the

**TABLE 5.2**

Maximum Temperatures Attainable by Accretion from Infinity

| Object | Radius (in km) | Density (in g/m³) | ΔT (in K) |
|--------|----------------|-------------------|-----------|
| Earth | 6,378 | 5.5 | 375,000 |
| Venus | 6,052 | 5.24 | 325,000 |
| Mars | 3,397 | 3.9 | 75,000 |
| Ganymede | 2,631 | 1.94 | 23,000 |
| The Moon | 1,738 | 3.4 | 7,000 |
| Ceres | 512 | 3 | 1,300 |
| Vesta | 277 | 3 | 360 |
| Enceladus | 250 | 1.13 | 120 |
| An asteroid | 50 | 3 | 13 |
| Thebe | 50 | 1.6 | 4 |

surface of a planet, denser constituents (such as iron) fall through a planet's molten interior onto a growing core. The bigger the planet, or the bigger the core, the more heat will result. This calculation suggests that Mercury's internal temperature could have been raised 700 K through core formation.

Besides Mercury, core formation could also contribute to the heating of other rocky planets with iron cores, such as Venus or Earth. Estimates suggest that the interior of Earth could have heated up by 2300 K, and although we know less about the size of Venus's core, the total heat should not be too different. Mars is a smaller planet with a smaller core; heating by formation of a Martian core probably produced a temperature rise of less than 350 K.

Finally, it is important to remember that accretional heating and core formation warm a planet only at the time when accretion or formation occurs. We know that most of the craters on the Moon, the scars from the tail end of its formation, are at least 4 billion years old. Thus accretional heating for a Moon-sized object might provide a pulse of heat early in its history (possibly producing the **magma ocean** discussed in Chapter 3), but it is not responsible for remelting, resurfacing, or tectonics after the first half billion years of a moon's existence. Likewise, core formation could have resulted in the heating and expansion that created the scarps on Mercury, but it could not keep the core molten for the next 4 billion years. Only the gas giant planets are large enough to still retain some of their accretional heating.

### Radioactive Heating

A number of radioactive isotopes occur naturally in rock. As they decay, they give off heat. How much heat? It depends on the isotope.

To be an important heat source, the isotope must be reasonably abundant; it must give off a significant amount of energy while it decays; and it must have a half-life short enough so that it decays fairly frequently, but long enough so that the material continues to give off heat long after the planets have been formed. The most important isotopes, and the amount of heat they produce, are given in Table 5.3.

The first four isotopes are known to be present in rocks today. The next three are short-lived isotopes that would be expected to have almost completely decayed away within the first 10 million years after they were formed. However, they might have played an important role in heating rocky bodies when the solar system was first forming. The presence of $^{26}$Al in certain meteorites has been inferred from the presence of its decay product, $^{26}$Mg; however, no good experimental evidence has yet been found that $^{36}$Cl or $^{60}$Fe existed in the early solar system.

The power per kilogram for even the most active isotopes in this list looks very small, a few trillionths of a watt. But there are many kilograms of matter inside a large planet and billions of years for that heat to build up. As a result, it is clear that bodies the size of the Moon will melt at some point in their history from the heat of radioactive materials alone. Where accretional heating probably produced the lunar **magma ocean** at the start of its history, this heating is probably responsible for the formation of mare basalts a billion years after the magma ocean froze.

A planet's interior heats up as radioactive materials release energy while they decay. Yet, by the very process of decay, the amount of radioactive isotopes inside a planet is constantly decreasing. As a result, a newly formed planet will stay warm or even get warmer from this decay process, but gradually cool off as time goes on and the abundance of heat sources declines. The rate at which the planet cools off depends on its size. A small body like the Moon is expected to be completely frozen today.

However, in Venus and Earth the heat from long-lived radioactivity is enough to keep their interiors partially molten even to this day. It is this heat that produces volcanoes, complicated tectonics, and internal convection on Earth. Thus we should expect to see some expression of this energy on the surface of Venus, too.

## Tectonics of Venus

Almost the entire surface of Venus is covered with tectonic features. In fact, most of the surface has been faulted or folded during the last 500 million to 1 billion years. We see slopes of over 20° to 30° over scales of tens of kilometers; such high slopes are not likely to be stable over very long periods. Thus Venus is almost certainly tectonically active today.

Many tectonic features that appear to be related to local stresses indicate common strain patterns over distances of hundreds of kilometers, reflecting global-style tectonics, influenced by mantle convection. The crust of Venus is being pulled, pushed, and compressed by convection inside

**TABLE 5.3**

Radioactive Nuclide Heat Production

| Nuclide | Abundance (ppm) | Isotope fraction | Decay Rate $\times 10^{-10}/yr$ | Energy/Decay $\times 10^{-13}J$ | Heat Out $\times 10^{-12}W/kg$ Today | 4.6 $\times 10^9$ yr ago |
|---------|-----------|----------|----------|-----------|-------|------------|
| $^{40}$K | 815. | 0.0001167 | 5.543 | 1.14 | 2.9 | 36.7 |
| $^{235}$U | 0.012 | 0.0072 | 9.8485 | 72.4 | 0.05 | 4.6 |
| $^{238}$U | 0.012 | 0.9928 | 1.5513 | 75.9 | 1.12 | 2.3 |
| $^{232}$Th | 0.04 | 1. | 0.4948 | 63.8 | 1.04 | 1.3 |
| $^{26}$Al | 10,000. | 0.00001 | 9,500. | 3.46 | 0. | 25,000. |
| $^{60}$Fe | 200,000. | 0.014 | 4,600. | 0.2 | 0. | 8,000,000. |
| $^{36}$Cl | 1,000. | 0.015 | 23,000. | 0.34 | 0. | 600,000. |

Notes: Abundance columns are elemental and isotope fraction abundances for chondrites; final columns are heat production per kilogram of chondrite. Initial $^{26}$Al abundance assumed to be $10^{-5}$ as inferred from inclusions in the Allende meteorite. Initial abundance of $^{60}$Fe assumed to equal present day $^{60}$Ni abundance; the $^{36}$Cl abundance is 50 ppm, relative to Si.

the planet, driven from the heat of radioactive isotopes. This strain on Venus results in broad regions of deformed, cracked terrain, tens to a few hundreds of kilometers wide (Figure 5.11), separating somewhat larger relatively undeformed regions.

Although the surface of Venus has clearly experienced a lot of stress and although there is good evidence that the mantle is convecting, it shows no conclusive signs of terrestrial-style plate tectonics. How does the heat escape? Probably through the multitudes of small volcanic features that *Magellan* has seen.

Why would two planets so similar in size, history, and composition display different styles of tectonics? The answer to this question is not well understood and is an active topic of discussion among scientists who study Venus. It may have something to do with the lack of water in the rocks of Venus—water changes the **viscosity** of rocks—or it may be because the surface of Venus is so much hotter than the surface of Earth.

## SUMMARY

The stresses created as a planet evolves in time will tend to crack the material that makes up the surface of the planet, a process called **tectonism**. Prime causes of such cracks include the impact of meteorites to form craters and the growth or shrinking of a planet as it warms up or cools down, both especially important on smaller planets. On larger planets, the emplacement of large loads such as the lava from a volcano may crack the surface of the planet, or convection inside the planet may break its surface into individual plates, whose motions past one another can lead to mountains and trenches being formed.

Extensional stresses result in surfaces being pulled apart. Large rift valleys and grooves may result from this activity, and the cracks that open up will provide natural pathways for lavas to erupt onto the surface. Compressional stresses, pushing rocks together, can produce scarps and wrinkle ridges where material gets piled up.

Venus has active tectonics, just like Earth does, but it does not take the form of plate tectonics. Most of the interior heat of Venus escapes through scattered volcanoes. The highlands appear to be actively supported by mantle convection.

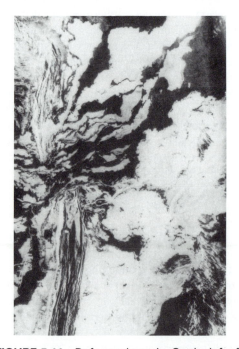

**FIGURE 5.11** Deformed terrain. On the left of the image, which is 550 km by 650 km in area, can be seen a ridge belt. Covering much of the image are lava flows, which have their source in Ammavaru caldera, 300 km to the west.

**FURTHER INFORMATION...**

# Energy of Accretion

A planet grows every time a little mass $m$ is added to its mass. If we assume these little masses fall from infinity, we can calculate the amount of gravitational energy contained in each mass, add up each bit of energy from each bit of mass that goes into our planet, and from this total up the maximum possible accretional energy available to heat up a planet.

The total energy that a little mass $m$ has as it falls must be constant as it falls, because energy is conserved. Assume that this total energy is zero (that is, the little mass $m$ began at infinity, where $U = 0$, and with zero speed, so that its initial kinetic energy $K$ was also 0).

At any position $r$ between the planet and the small mass, $m$ has a potential energy $U = -GMm/r$ (Equation 9.9), where $G$ is the universal gravitational constant and $M$ is the total mass of the body that is attracting the extra mass $m$. As the mass falls to the body, $r$ gets smaller and the potential energy becomes an increasingly large negative number, so the kinetic energy must be increasingly large and positive. (In other words, the object speeds up as it falls.) Thus the kinetic energy is always $K = -U = +GMm/r$, keeping the total energy constant and equal to zero. Thus, at the surface of the body, the amount of kinetic energy the small mass $m$ has when it hits is $+GMm/R$, where $R$ is the radius of the body. This kinetic energy is added to the heat budget of the body.

As the body grows, $M$ and $R$ increase; one must sum up over a large number of little $m$'s until the whole mass of the body $M_{\text{final}}$ is achieved. The amount of energy $dK$ which is gained when a mass $dm$ is added to some intermediate stage of the growing body, a stage where the body has a radius $r$ and a mass $M$ already, is given by

$$dK = \frac{GM}{r} dm \qquad (5.27)$$

But if we assume that the body is spherical and has a constant density $\rho$, then we can write the mass $M$ as

$$M = \frac{4}{3}\pi\rho r^3 \qquad (5.28)$$

and if the infinitesimal addition of mass $dm$ is done in a spherically symmetric shell, then

$$dm = 4\pi\rho r^2 dr \qquad (5.29)$$

Thus we can substitute Equations 5.28 and 5.29 into Equation 5.27 and integrate from a radius of 0 to a total radius $R$:

$$\Delta K = \int_0^R 3G\left(\frac{4}{3}\pi\rho\right)^2 r^4 dr \qquad (5.30)$$

$$= \frac{3}{5}G\left(\frac{4}{3}\pi\rho\right)^2 R^5 = \frac{3}{5}G\frac{M^2}{R} \qquad (5.31)$$

That's the total energy going into the body.

The amount of heat generated during core formation can be estimated using this same logic. Here it is important to realize that material is not falling from infinity and that as core material falls into the planet's center, releasing energy, some of that energy is needed to displace the less-dense rock and move up and out of the core. One estimate for the average temperature change when a core is formed is given by the equation

$$\Delta T \approx (1 - f)f\Delta\rho\frac{GM}{3R\rho C} \qquad (5.32)$$

where $f$ is the fraction of mass in the core, $\Delta\rho$ is the difference in density between the core and the mantle material, $R$ is the total radius of the planet, $\rho$ is its average density, and $C$ is its average heat capacity. $G$ is the universal gravitational constant, and $M$ is the total mass of the planet. This formula is only an approximation, but it gives a good order-of-magnitude estimate of just how important core formation is likely to be for any given solar system body.

## STUDY QUESTIONS

1. What happens to plastic material when stress is applied?

2. Give an example of something that could happen to a planet that would result in global tectonics.

3. Stresses make a rock crack. All things being equal, and assuming the rock is easily cracked, at what angle will the crack occur?

4. Name a place in the solar system (besides Earth) where you see tectonic features due to compression; name a place where you see tectonic features due to extension.

5. Name one element whose long-lived radioactive isotopes are important in heating up the interior of rocky planets.

## 5.4   PROBLEMS

1. What is the mass of the atmosphere of Venus? (Hint: What are the units of pressure, in terms of kilograms, meters, and seconds? Combine the surface pressure with the acceleration of gravity and any other terms you may need to end up with an answer that has units of kilograms.) What is the ratio of the total number of kilograms of atmosphere to the mass of the planet itself?

2. Do the same exercise for the atmosphere of Earth as you did for Venus in Problem 1. The atmosphere of Venus is 4% by mass nitrogen; that of Earth is 78%. Which planet's atmosphere has more kilograms of nitrogen?

3. An earnest young woman claims to have been taken to Venus by a flying saucer. To prove her story, she has photographs of the planet, taken with her pocket camera. The markings in the clouds look suspiciously like those seen in the *Mariner 10* photographs. How can you prove that her story is false?

4. Imagine an entire planet made of soft wax. Which of the following features might you expect to see?
   (a) craters
   (b) scarps
   (c) plate tectonics
   Explain your reasoning.

5. A small rocky planet is observed to have both extensional and compressional tectonic features on its surface. It is also fairly heavily cratered. The average number of craters on the surface of the extensional features is larger than the number of craters on the compressional features. From this information, give a brief description of the likely evolution of this planet:
   (a) Did it start out hot or cold in its interior?

(b) What changes in its internal temperature occurred?
(c) Is it hot or cold today?

6. The planet in Problem 5 has a twin that underwent exactly the same sort of evolution. However, this twin planet has a bizarre composition: It is made up mostly of rubber, not rock! Rubber has the unusual characteristic that it contracts when it gets warmer and expands when it cools. What sort of tectonic features do you expect to see on such a planet?

7. The daytime and nighttime temperatures on Venus are virtually the same. On the other hand, the most extreme difference in day and night temperatures on Earth occurs in desert areas. What principle do these two examples illustrate?

8. Moist winds that hit a mountain will be pushed higher into the atmosphere to get over the mountain. Sometimes clouds form over mountains as a result. Why?

9. It has been proposed that the sulfur on Venus originally came from comets, dumping water and sulfur-rich ices into Venus's hot atmosphere. A rival theory says that the pyrite on the surface is the original source of the sulfur, that it came from deep inside the planet and was erupted onto the surface by volcanoes. For the sulfur cycle described in this chapter, does it matter where the sulfur is introduced into the system?

10. Most of the carbon dioxide on Earth is tied up in limestones, formed from the shells of small ocean creatures, and is not in the atmosphere. If all the carbon dioxide on Venus were removed in a similar fashion, how would the sulfur cycle described in the chapter be changed?

11. If Venus had no surface, like Jupiter, how would this change the chemistry of its atmosphere?

12. It has been suggested that Venus could be **terraformed** (turned into an Earth-like planet) by such steps as directing water-rich comets onto the planet and introducing genetically engineered microbes to alter its atmosphere. What properties of Venus need to be altered before it can become habitable? How might these algae and comets help matters? What are the ethical questions involved in irreversibly transforming a planet to suit our needs?

13. On Earth, very cold weather can freeze out almost all the water from the air. The coldest winter days are usually dry and cloudless. But water is only a fraction of Earth's atmosphere, so the total pressure of the atmosphere hardly changes. On Mars, the winters get cold enough to freeze carbon dioxide. But carbon dioxide makes up most of the Martian atmosphere. What should happen to the pressure of the atmosphere on Mars? What stops the whole atmosphere from freezing?

## 5.5  FOR FURTHER READING

The University of Arizona Press book, *Venus*, edited by Don Hunten and others (1983), is a collection of research papers following the Pioneer mission, and it centers mostly on the atmosphere of Venus. For a post-*Magellan* view of Venus's surface features, consult two special issues of the *Journal of Geophysical Research (Planets)*, volume 97, issues E8 and E10 (August 25 and October 25, 1992).

A good introduction to the chemistry of atmospheres for upper-level undergraduates is Murray McEwan and Leon Phillips's *Chemistry of the Atmosphere* (London: Edward Arnold, 1975). It concentrates on the theory as applied to Earth's atmosphere, but it also devotes a chapter to the atmospheres of other planets. An advanced look at atmospheric chemistry can be found in John Lewis and Ronald Prinn's *Planets and Their Atmospheres* (New York: Academic Press, 1984).

For more background on geologic landforms, an excellent introduction is Michael A. Summerfield, *Global Geomorphology* (New York: Longman, 1991). It is well-written and beautifully illustrated. More detailed references for tectonics and structural geology are the textbooks by Donald Turcotte and Gerald Schubert, *Geodynamics* (New York: Wiley, 1982), and John Suppe, *Principles of Structural Geology* (Englewood Cliffs, N.J.: Prentice Hall, 1985).

GOES-7 21 Sep 1989 19:01 UTC 3800 x 3000 pixels
9 Hours before Hurricane Hugo landfall

**FIGURE 6.1**    The *GOES* satellite took this (visible and infrared) photo of Earth on September 21, 1989. Hurricane Hugo can be seen approaching the coast of South Carolina in the upper right.

# CHAPTER 6

# Earth

## 6.1 ON EARTH

For many thousands of years, the only planet that human beings had really studied was Earth. This planet, alone (as far as we know) in the solar system, is the home of living creatures, which take on an astonishing variety of forms. Among these myriad creatures, from viruses and one-celled organisms to giant redwoods and great blue whales, Earth is the home of human beings and as such will always be special to us, no matter where our future travels take us.

But in our realization of how special Earth is because it is the home of living things, let us not forget how much it has in common with the other planets, nor how it is unique in ways that have little to do with the presence of life.

| Earth's Vital Statistics | |
|---|---:|
| Radius | 6378 km |
| Surface area | $5.1 \times 10^8$ km$^2$ |
| Mass | $5.974 \times 10^{24}$ kg |
| Density | 5.515 g/cm$^3$ |
| Local gravity | 9.87 m/s$^2$ |
| Escape velocity | 11 km/s |
| Albedo | 0.36 |
| Surface temperature | 290 K |
| Length of day | 23 h 56 min |
| Length of year | 365.25 days |
| Distance from the Sun | $1.496 \times 10^8$ km |

## The Structure of Earth

The planet Earth is thought to have formed about 4.6 billion years ago, at the same time that the rest of the solar system was taking shape. As the planet accreted, its internal temperature rose enough for it to melt, so that it could differentiate according to the mass of the constituents. The heaviest parts, the iron and nickel metal, fell to the center of the planet, where they formed a **core**. The lighter material left over formed a **mantle** first, then as it cooled, a **crust** on the surface of the planet. Geologists today divide Earth into even more parts (Figure 6.2): The core is divided into two sections, an inner, solid core, and an outer, liquid one; the mantle is also divided into two sections, this time called upper and lower; and the crust is divided, not radially, but laterally, into regions of **oceanic** (thin, dense, and mostly basaltic) and **continental** (thick, light, and very diverse, but generally granite-rich) **crust**.

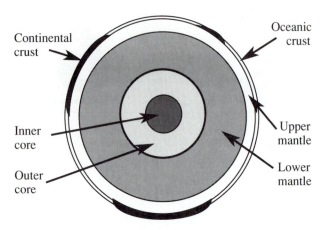

**FIGURE 6.2** Cross-section of Earth, showing the major parts of its interior.

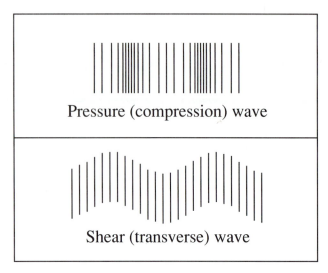

**FIGURE 6.3** P and S seismic waves.

## Seismology

Most of the information we have on the interior of Earth comes from the science of **seismology**, which began as the study of natural **earthquakes**, or **seisms**, but has since branched out into the study of both natural and manmade vibrations in Earth. When an earthquake happens somewhere on Earth, say in Japan (which gets a lot of earthquakes, for reasons we will discuss later), it starts waves, like sound waves, moving though the entire planet.

Sensors, called **seismometers**, can be placed on different parts of the planet to time the arrival of these waves from a distant earthquake and to record how strong the quake felt at that spot. These records can be used to determine what the material is like between the sensor and the source of the earthquake, just as one might hear a noise in the next room and be able to tell if the door is open or closed. With enough sensors, a general picture of the structure of the whole planet can be made.

There are two basic kinds of seismic waves. The difference between these two types of waves is illustrated in Figure 6.3.

**Pressure waves**, or **primary waves**, called **P-waves**, are exactly like sound waves. If you take one end of a spring and push or pull it suddenly, then let go, a wave will travel down to the other end of the spring. If the other end of the spring is fastened down, the wave will reverse direction and travel back towards your end. This is what goes on when a P-wave happens. The earthquake causes the surrounding rock to be momentarily compressed; the material relieves this compression by compressing the neighboring material, and so the wave of compression is passed along.

This compression will pass through any sort of material, solid or liquid; how fast the wave travels depends on the composition of the material. An abrupt change in material can cause these waves to be reflected, resulting in a seismic echo that can be used to determine the location of such changes deep within Earth.

**Shear waves**, or **secondary waves**, called **S-waves**, are caused by the back-and-forth motion of an earthquake. Imagine the spring again, but this time grab the end and move it from side to side. This will also cause a wave to travel down to the other end of the spring and perhaps to reflect back. This jiggling gets transmitted along the bonds of atoms in crystals; the bonds act something like a string that has been plucked. Because liquids are not strongly bonded like solid materials, molten rock cannot transmit shear waves.

In Earth, these two types of waves travel at different speeds, so that a P-wave arrives first (explaining how it got the name primary wave) and an S-wave second. But the physical properties of Earth greatly affect the way these waves travel through it, so much so that by comparing the recordings of these vibrations made by a seismometer (Figure 6.4), we can reconstruct what happened to the wave on its way through Earth and what Earth must be like inside to affect the wave that way. For example, we can see reflections of these waves caused when they encounter a boundary of some sort, say between the upper and lower mantle, or we can see the S-waves disappear (but the P-waves continue weakly) as they pass through a liquid region, such as the outer core (which is the only large region of Earth that is liquid).

A third type of wave can be seen in seismograms, however. In the presence of a surface (that is, the surface of the ground), the combination of P- and S-waves can lead to other types of waves: Rayleigh and Love waves. These waves are **surface waves**, which travel only along

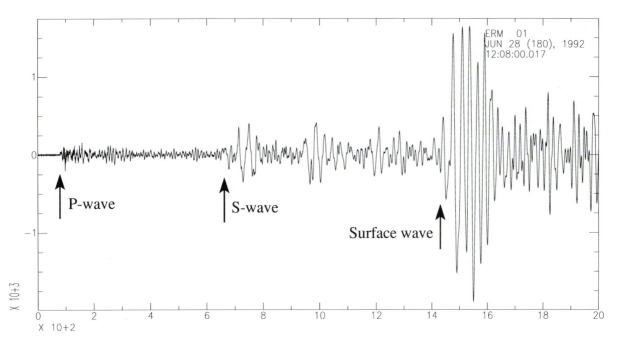

**FIGURE 6.4**   A seismogram.

the surface of the ground. Surface waves travel much slower than P- and S-waves and have a much larger amplitude. They give us no information about the interior of the planet.

If we know exactly when and where the earthquake took place, we can figure out how fast each type of wave was travelling through the rock between the quake and the seismometer. These velocities, in turn, depend on the density and compressibility of the rock. By measuring these velocities, we can learn something about the properties of the rock deep below the planet's surface.

Detailed study of seismic data from Earth and the Moon (the only two planets where it is available so far) has led to a reasonably good understanding of the internal structure of both planets. By seeing how tremors from earthquakes bounce off different layers of material inside the planet, we know that both Earth and the Moon are made up of layers of different kinds of rock; we can determine where inside Earth (or the Moon) one layer of rock ends and another begins; and we can even make estimates of the density of the rock in each layer and also the speed at which the earthquake tremors travel through the rock.

Of course, there might be several minerals, or mixtures of minerals, that would give the same densities and speeds. But by combining the geochemical modelling described earlier with this geophysical information, it is possible to make a good educated guess as to what minerals are likely to be found in each layer of the planet. The more we study earthquakes, the more rocks we analyze, and the more tests we make in the laboratory of different types of rocks, the closer we come to being able to define what each layer in Earth is made of. This is by far the best way to de-

termine the composition of a planet, but it requires years of detailed measurements from the surface of the planet itself.

## Plate Tectonics

Now let's go back and try to answer that question about why Japan gets a lot of earthquakes. We will take what might seem like a roundabout path. For many years, geologists believed that the surface of Earth was, in general, fixed and unmoving and had been since it first formed. Mountains might rise up and be worn down, volcanoes might spew forth lava, faults might even crack the surface and large blocks move relative to one another, but overall, things stayed basically the same. Starting in the twentieth century, however, geologists started to seriously consider other ideas.

Look at a globe of Earth. Notice how the coast of South America looks as though it could fit into the coast of Africa, assuming that it could travel several thousand miles to do so. And you could tuck North America up against the coast of Europe, with only a little twisting, and connect Arabia to the coasts of Africa and the Middle East, closing the Red Sea and the Persian Gulf. If you worked at it long enough, you could make most all the continents fit together on one side of Earth, leaving the Pacific Ocean to take up the whole other side (and more).

This idea of **continental drift** was first strongly advocated by Alfred Wegener in 1912. His arguments included not just that the continents *look* like they should fit together. He also found that fossil land animals common on one continent in the distant past were common on others too, as if these animals could move freely over the entire land mass.

Continental drift is a very interesting idea, if you think of continents as floating around in the ocean, but they don't. "How can you move continents through all the solid rock of the sea floor in between those continents?" traditional geologists argued back at the early proponents of continental drift. "What possible mechanism could move so much land around over the face of Earth?"

This question was not answered until the late 1950s, when the first magnetometer data were taken on the mid-Atlantic ridge, a chain of underwater mountains running down the middle of the Atlantic Ocean. (A **magnetometer** is a device that measures the magnetic field coming from Earth and from nearby rocks very accurately.) As the data from the ridge were examined, the magnetic fields coming from the rocks along the ridge were found to form a definite pattern. All the magnetic minerals (like magnetite) in the rocks had magnetic fields that were aligned in the same direction, indicating that the rock was exposed to a magnetic field when it cooled (all these rocks are volcanic). But that is not really so odd—lots of rocks on Earth have that property—because Earth itself produces a magnetic field strong enough to cause this effect. The odd part comes when you look at how the direction of this field varies in rocks at different distances from the center of the ridge (Figure 6.5). The field that the rocks were in when they cooled did not always point in the same direction; sometimes it reversed. Progressing away from the center of the ridge, the field, as recorded by the rocks, flips back and forth many times. If you think of rocks formed under a normal field as black and those formed under a reversed field as white, they will be seen to form a pattern of stripes of varying widths. If you look at the rocks on the other side of the ridge in the same manner, you can see the same pattern of stripes, but in mirror image to the first.

As you get farther and farther from the center of the ridge, the rocks get older and older, all the way to the edge of the ocean basin, where they are about 220 million years old.

These surprising data were used to develop the new theory of **plate tectonics**, because now we know what to do

with the rocks that form the sea floor. Plate tectonics gives us the following scenario for the history of the Atlantic Ocean: Imagine South America and Africa almost touching, about 220 million years ago, with just a narrow gulf separating them. In the center of this gulf, below the water, is a chain of mountains, the infant mid-Atlantic ridge. Lava seeps up from the interior of Earth along the ridge, as the two plates move apart, and solidifies. As time passes, the plates get farther and farther apart, and more and more oceanic crust is formed at the mid-Atlantic ridge. Depending on the direction of Earth's magnetic field at the time, either a positive or a negative "stripe" will be recorded as the rocks cool.

This explains nicely what happens as plates separate. But what happens on the other side of the plate? There must be some way of getting rid of plates as well, or Earth would have to grow larger with time, and there is no evidence of its having done so. As plates separate along one plate margin, they can collide or rub against each other along other margins. If two continental plates collide directly, mountains are formed. This is what is currently lifting the Himalayas higher and higher. If a continental and an oceanic plate collide directly, the denser oceanic plate is drawn under the continental plate (**subduction**), forming a deep trench, such as those seen along some of the boundaries of the Pacific Ocean. The oceanic plate is eventually drawn down, or subducted, deep enough into Earth for it to melt and become part of the mantle again. If the plates don't collide directly against one another, but merely slide along one another, you get a huge **transform fault** (see the box later in this chapter), like the infamous San Andreas fault in California.

Look at Figure 6.6. As a piece of the surface, such as the square labelled "A," moves relative to its neighbors, there are three kinds of interactions it will have with neighboring surface plates.

Plate A will be slipping past plates B and C on either side, resulting in **transform faults**. Where A and D are separating, lava is being brought up from the mantle and new crust is forming. Here, in places like the mid-Atlantic ridge, we find a long chain of very young mountains formed out of a basalt (called **MORB**, standing for **m**idocean **r**idge **b**asalt) that is derived from silicon-poor sources in the mantle.

What happens when plate A runs into plate E? If A is the surface underneath an ocean, the leading edge of plate A may get pulled underneath plate E, going down with a cooler part of the mantle returning back towards the center of Earth, or **subducting**. Sediments (such as sand and shells) pile up at the edges where the two plates meet, or they may just get carried down into Earth. Deep in Earth these wet, silicon-rich sediments will melt and erupt back up to the surface, forming a chain of **composite** volcanoes. Thus, in an area like Japan, we see a deep trench in the ocean bottom, followed by a string of volcanic islands.

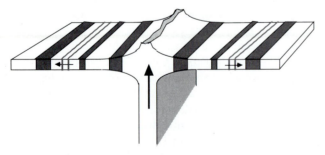

**FIGURE 6.5** Magnetic stripes on the sea floor, represented here by dark lines, spread out away from mid-ocean ridges symmetrically in both directions.

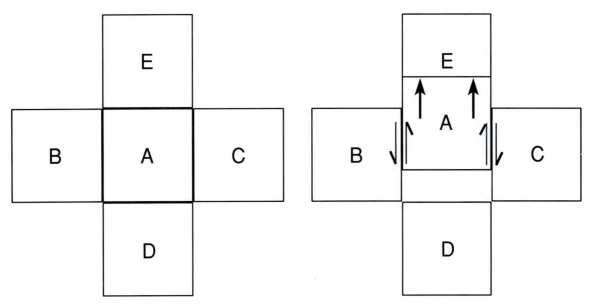

**FIGURE 6.6**   The various types of boundaries that a plate on Earth can have with its surrounding plates. Left: Consider plate A surrounded by plates B, C, D, and E. Right: As it moves towards plate E, it will slide past plates B and C, and separate from plate D.

If plate A is part of a continent, it will not be so easily carried below plate E. The jumble of two continental plates piling on top of each other results in the high plains and many high mountains of Tibet and the Himalayas.

As the plates collide, or rub against each other, they don't move smoothly. Instead, they stick and then move rapidly; this causes **earthquakes**. Most of the earthquakes occur along plate boundaries, shown in Figure 6.7.

Why should the various parts of Earth's crust be in constant motion over the surface of Earth? What could be driving them? The solid mantle of Earth, which is very hot, is too hot for its heat to escape merely by **conduction** (the type of heat transport that works in a frying pan), and so it **convects** (more like a pot of boiling water). The warm, ductile mantle brings heat up towards the surface by convection (convection will be discussed more in Chapter 10), carrying warm, partially molten rocks up in some parts of the planet and carrying down cooler material elsewhere (Figure 6.8). As cooler material on the surface descends into the planet (to replace the warmer material bubbling up) it tends to drag along behind it large sections of the rigid rock on the surface.

There are three ways that the **lithosphere** of Earth might be moved around by these motions inside Earth. One possibility is that the plates might be riding on the convection cells, moving as they move. However, there may not be enough friction between the convecting **asthenosphere** and the plates to couple their motions together. A more likely possibility is that hot lava erupting from the mid-ocean ridges is pushing the plates apart.

But the most widely held theory is that the cold, dense surface material subducting down into Earth where the convection cells are also moving material downward pulls the rest of the plates along with them.

## SUMMARY

Seismology, the study of vibrations within Earth, is one of the most valuable tools in our efforts to understand the structure of Earth.

Earth may be unique in the solar system in its possession of plate tectonics. Plate tectonics describes the movement of pieces of lithosphere—plates—over the surface of Earth. As the plates move with respect to each other, they form mid-ocean ridges where they separate, transform faults where they slide along each other, and deep trenches or huge mountain ranges where they collide. All these movements are in some way driven by convection in Earth's mantle.

Convincing evidence for plate tectonics has not been found on any other planet as yet, either presently occurring or in the past. Earth also may be the only planet that presently has oceans. The existence of plate tectonics, oceans, and life is what makes the surface of Earth very different from that of any other planet. It has very few craters (even though there is no reason to think that fewer meteorites have fallen on Earth than on other planets) because

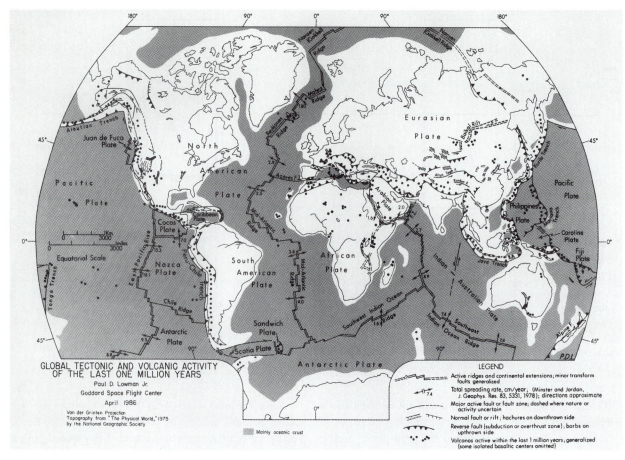

**FIGURE 6.7**   Plate boundaries on Earth.

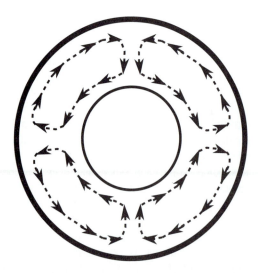

**FIGURE 6.8**   A schematic drawing of mantle convection. We actually don't know how many convection cells there are, or if they extend down to the core, as is implied here.

of the high degree of resurfacing that occurs, due largely to weathering but also to subduction. The oceans simply hide about 70% of the surface of the planet from view. And the presence of green plants on Earth gives us the oxygen in our atmosphere, which makes this planet such a pleasant place for creatures like ourselves.

## STUDY QUESTIONS

1.  Earthquakes produce P- and S-waves. What does the "P" stand for? What does the "S" stand for?

2.  We have detailed seismic data from only one other place besides Earth. Where?

3.  Give one piece of evidence that plate tectonics exists on Earth.

4.  What moves the plates around on Earth's surface?

5.  Give an example of a plate boundary on Earth that is a subduction zone.

**FURTHER INFORMATION**...

# Volcanoes on Earth

Volcanoes are often found along the boundaries of plates (called **plate margins**—see Figure 6.9). At mid-ocean ridges, lines of volcanoes thousands of miles long exude lava under the ocean, forming new oceanic crust as the plates move apart. As the lava is extruded underwater, it oozes out in the form of lumps and blobs, called **pillow lava**. Hot water (hotter than the boiling point would be on the surface, but it can't boil because of the great pressure at the bottom of the ocean) circulates through the new oceanic crust, taking elements, such as manganese and iron, out of the rocks. As the water cools, farther away from the new lava, the heavy elements are precipitated, often in the form of chimneys, which might have plumes of what looks like black smoke rising from the center (Figure 6.10). This "smoke" is made up of the heavy elements that are precipitating out of the water, and the ocean around these **black smokers** is very rich in nutrients that support a wide variety of strange life forms.

**FIGURE 6.10** Black smoker at the East Pacific Rise. The chimney is about 45 cm high.

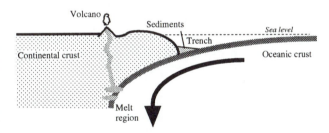

**FIGURE 6.9** A schematic drawing of a plate margin, with a volcano.

At **convergent margins**, where two plates are colliding, volcanoes can form too. When an oceanic plate is subducted beneath a continental plate, it carries with it a large load of sediments that have accumulated on top of it in the millions of years it was part of the sea floor. As the plate and its sediments are carried deeper and deeper down into Earth, they are heated hotter and hotter, until eventually they begin to melt. The melted material can then rise through the overlying crust (which will be full of cracks because it is under a lot of stress there at the plate margin) to erupt out onto the surface as volcanoes. The type of volcano formed by this sort of material (melted sediments and oceanic crust) tends to form exceptionally beau-

tiful, cone-shaped **composite** volcanoes, like Mount Fuji in Japan (Figure 6.11).

Sometimes, however, volcanoes form at other places, not plate margins at all. Some of the most famous of these **hot spot** volcanoes are the islands of Hawaii. These islands, which form a chain thousands of kilometers long in the Pacific Ocean, were formed by a hot spot in Earth's mantle. The movement of the Pacific plate over this hot spot caused the volcanoes to form in a chain, not all in one big volcano, as the volcanoes on Mars may have formed.

**FIGURE 6.11** Mount Fuji in Japan.

## 6.2 DINOSAURS, ICE AGES, AND CELESTIAL MECHANICS

In January 1991 a rock some 30 m across, travelling at roughly 30 km/s (more than 60,000 mph), was seen by the Spacewatch telescope in Arizona to pass within 100,000 km of Earth, less than a third of the distance from Earth to the Moon. It was the first such close-passing asteroid to be seen by Spacewatch, a newly constructed observatory designed specifically to look for asteroids passing near Earth. But almost certainly it was not the first such asteroid to pass by this planet; craters like Meteor Crater in Arizona are evidence that such objects have come much closer to Earth in its recent history.

For most of its history, humankind has thought of itself as inhabiting a universe bound by ground and sky. Even after Earth was understood to be a planet of the solar system, this knowledge was considered to be no more than an interesting fact of science, one with no obvious practical implications. Stars and comets might be considered omens by the superstitious, but most of us agreed with Shakespeare that our destinies were not controlled by the stars, but by ourselves.

That Earth is a planet in space, not a universe unto itself, really only entered the human psyche with the photographs of Earth from the Apollo missions (Figure 8.1). And only since the 1970s have scientists begun to realize just how much our life can depend, and has depended, on the interactions of this planet with its solar system neighbors.

### Impacts on Earth

No human being, as far as we know, has ever been killed by a falling meteorite (although a dog in Egypt was killed by the fall of the Nahkla meteorite in the nineteenth century). But as humanity continues to spread its cities across the surface of Earth, we are presenting more and more targets for the random falling of a rock from space. On November 30, 1954, a 9-pound meteorite crashed through a house in Sylacauga, Alabama, and bruised Mrs. Hewlett Hodges. It is estimated that a human-built structure somewhere on Earth is hit once every 3 years; defying all odds, two houses in the town of Wethersfield, Connecticut, separated by less than 1 mile, were struck in the span of 12 years.

What would happen if an asteroid like the close-approacher of January 1991 actually hit Earth?

To answer this question, we will first want to calculate how much energy a body has when it orbits the Sun near Earth. As we have mentioned previously, the kinetic energy of a body in motion is given by the formula $E = \frac{1}{2}mv^2$. Here, $m$ stands for the mass of the orbiting meteoroid. If we know its radius, we can guess at the mass by assuming it is

a sphere—so its volume is $\frac{4}{3}\pi r^3$—and then multiply this volume by a density typical for rock or iron. Let's assume that our asteroid was made of rock, with a density of 3000 kg/m$^3$; thus we find that if the asteroid is $r$ meters in radius, it has a mass of about $10^4 \times r^3$ kg.

Now, what about a value for $v$, its velocity? How fast do things orbit when they are close to Earth? In detail, it depends. Most comets are in elliptical orbits, literally falling in to the Sun from a great distance beyond Pluto, and so they will be travelling much faster than an Earth-crossing asteroid in a more circular orbit. But for both comets and asteroids, it's safe to say that most meteoroids that hit Earth will be travelling at least at a speed similar to the rate Earth travels around the Sun. That's $2\pi$ times 1 AU, $2\pi R$, in a year, or about 30 km/s.

When you put these numbers together into the energy equation, you find that the energy of a body of radius $r$ hitting Earth will be roughly $\frac{1}{2}10^{13} \times r^3$ J. Our 30-m-wide asteroid must have had an energy of something like $1.5 \times 10^{17}$ J of energy. Compare this with an atomic bomb. A 1-kiloton bomb produces $4.187 \times 10^{12}$ J of energy, so our asteroid's impact is equal to the power of a 37-megaton nuclear device! And even more important is to note that the energy increases as the cube of the asteroid radius. A meteoroid 10 times bigger carries 1000 times more energy.

How big a hole in the ground will an asteroid make when it hits? That depends on many factors. Obviously the crater will be bigger than the impactor itself; how much bigger depends on the size of the asteroid and the type of material it hits. It also varies from planet to planet, depending on how strong the local gravity is. With less gravity, you get a bigger hole.

People studying craters have performed different experiments, firing glass balls into water or sand at a variety of speeds and even looking at the craters left behind following the test explosions of nuclear bombs. From these data, a rough idea of the crater size expected for an impacting asteroid can be estimated; the results are shown in Figure 6.12. For a 30-m-diameter asteroid, we might expect to see a crater about half a kilometer (about a third of a mile) in diameter. That's just a bit smaller than the size of Meteor Crater in Arizona.

An even larger impact will leave an even bigger hole; a 10-km-radius comet (such as Halley) will make a crater 100 km wide. The crater would be as big as Rhode Island.

The impact shock blast will leave traces beyond the crater itself. The shock wave in the atmosphere could knock down trees (or buildings) for an area 10 to 100 times greater than the area of the crater itself; and it is calculated that 10 to 100 times the mass of the impacting meteoroid will be flung out of the ground and into the atmosphere by the impact.

The material blown into the atmosphere would in-

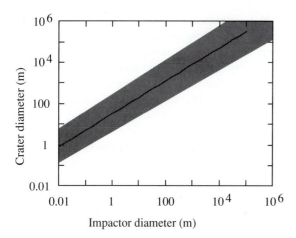

**FIGURE 6.12** The shaded region in this graph gives the range of crater sizes expected for given diameters of meteoroids hitting Earth.

clude boulders, shocked quartz, drops of melted glass (see the section on **tektites** below), and dust. **Shocked quartz** is quartz that has been in an explosion, so that it was under a very high pressure for a very short time. If you look at shocked quartz grains under a microscope, they show a pattern of intersecting lines, unlike quartz that has not been in an explosion. This pattern was first discovered in quartz grains from nuclear test explosions and has been seen in many terrestrial impact craters.

If fine dust reached the stratosphere, above the regions of the atmosphere that are well-mixed by convection currents, it might stay there for years before settling out. The shadow of this dust could shade Earth, causing temperatures to drop precipitously. We know that following the explosion of the Tambora volcano, Earth experienced a series of very cold winters and summers; following this volcano, in 1816 it snowed in New England on July 4! (A similar effect might occur after a nuclear war. The devastating cold that could result from dust lifted into the stratosphere by nuclear bombs is sometimes called **nuclear winter**.)

The dust from a large impact would raise much more dust into the atmosphere than Tambora did, causing larger and longer drops in temperature. We were able to measure the energy of the Mount St. Helens eruption in Oregon in 1980; it had roughly 10 megatons ($4.2 \times 10^{16}$ J) of energy. The Krakatoa eruption in 1883 was probably 10 times larger, and the Tambora eruption was probably not much bigger. But, as we calculated above, even a 30-m impacting asteroid would have nearly this much energy. Halley's comet, which is 30 times bigger, would have 27,000 times more energy!

Even before hitting Earth, some fraction of the meteoroid's energy would be dissipated as it passed through the atmosphere of Earth; probably not enough to slow it down very much, but almost certainly enough to heat up the atmosphere. The shock wave—in effect, the "sonic boom"—of a meteoroid passing through the atmosphere could induce chemical reactions between nitrogen and oxygen, making products that could eventually react with water to form nitric acid. This acid could spread across the continents, as acid rain, and also turn lakes and oceans acidic.

Finally, the foreign material of the impactor itself would become incorporated into the surface of Earth. Both meteorites and comets contain trace elements that normally are not found in crustal Earth rocks. Elements such as osmium and iridium were sorted out of crustal Earth rocks and into its core early in its history, as Earth melted and differentiated into a core and mantle. But these elements will occur in their full cosmic abundances in any rock or comet from space.

Two scientific problems have motivated the detailed study of such impacts. One is the question of the origin of **tektites**, strange blobs of glass that are found in distinct **strewn fields**. The other is the question of the **K/T event**, the apparently catastrophic event marking the change from the Cretaceous to Tertiary geologic eras, coinciding with the mass extinctions of huge numbers of marine animals and, most famously, the extinction of the dinosaurs.

## Tektites

There are four generally recognized tektite-strewn fields on Earth, places where tektites have been found. The tektites found in the North American field (which actually stretches across the Pacific Ocean to include parts of southeast Asia) are 34 million years old. Those found in Czechoslovakia are 14.7 million years old. Tektites from the Ivory Coast are about 1 million years old, while those in the Australasian field, covering Australia, southeast Asia, and most of the Indian Ocean, are about 700,000 years old.

Besides showing different ages, each type of tektite has a distinctive chemical composition. All the compositions, however, are similar to those of Earth rocks near the strewn field. This fact, plus the distinctive shape of the tektites, supports the theory that these rocks represent droplets of ordinary Earth rock that were melted and scattered by the impact of some large asteroid or comet.

Three of the four impact events seem to have occurred at the same time as Earth's magnetic field reversed direction. This may just be coincidence. Likewise, the chemistry of rock samples seems to indicate that the temperature of Earth as a whole dropped significantly following these impact events. But there are dozens of other occasions when the temperature of Earth appears to have dropped, so again it may just be coincidence that some of these drops occurred after tektite-forming impacts.

## The Case of the Dinosaurs: The K/T Transition

Geologists traditionally have divided the history of Earth into a number of ages, based (among other factors) on the kinds of fossils they find in rocks. Rocks that have been formed from the silt at the bottom of an ocean can be rich in marine fossils; furthermore, a relative time scale can be worked out for these rocks, because younger fossils will lie on top of older ones.

The **Cretaceous period** (abbreviated often by the letter **K** from the German spelling of the term; it also alleviates confusion with the **Carboniferous period**) ended 65 million years ago, and was followed by the **Tertiary period** (**T**). At this time, a significant number of marine animals became extinct; the temperature of the oceans dropped by several degrees; and—as every school child learns—the dinosaurs became extinct.

This was by no means the only period of mass extinction in Earth's history, nor even the most significant. It is, however, the most recent of the major mass-extinction events, and consequently the best documented in the fossil record. Harold Urey, in 1972, first proposed that an extraterrestrial impact may have been the cause of the mass extinctions. Urey, a Nobel chemist and pioneer planetary scientist, was by then elderly, and he was merely speculating about such events in a popular article in the *Saturday Review*. However, 8 years later, Luis Alvarez and his co-workers found evidence that such an impact may have occurred.

Looking at layers of rock in Italy whose fossils recorded the transition from the Cretaceous to Tertiary periods, they found a thin layer of clay at the boundary between the periods. Analyzing the clay from the K/T boundary, they discovered that it contained unusually high amounts of trace elements such as iridium. Similar iridium anomalies were found at the K/T boundary in three dozen other sites around the world. From this, and noting the arguments listed above concerning the effects of a large meteoroid impacting Earth, they felt that they had evidence that indeed an impact triggered the Cretaceous mass extinction.

This theory immediately caused a great controversy among both biologists and planetary scientists.

Most scientists would agree that such an impact could have occurred. Current estimates are that a 10-km-radius object ought to hit Earth about once every 40 million years, which is about the right time scale for the K/T event 65 million years ago. And most scientists agree that such an impact **did** occur, at a time close to the K/T boundary. The evidence of iridium anomalies in the K/T boundary clay has been supplemented by the discovery of shocked quartz and in some areas microtektites (very small tektites). There is even a very promising candidate for the crater that must have been formed.

In the Yucatan Peninsula of Mexico a 180-km-diameter crater, buried beneath sediments, has been discovered that may be the K/T crater. This feature, named Chicxulub after a nearby town, has been dated to be 65 million years old, the same time as the K/T extinction. In nearby areas of Mexico and Haiti, thick layers of tektites of the same age exist, forming the K/T boundary instead of a thin clay layer.

The effect of such a large impact is difficult to quantify, but there is no shortage of theories as to ways in which it might have led to wide-scale mass extinctions. Whether in fact any of those mechanisms actually caused the extinctions, however, or indeed if the extinctions were sudden and catastrophic or more gradual and produced by natural causes, remains controversial.

## Ice Ages

Not only impacts can change the climate of Earth. Ultimately, our climate is controlled by the amount of sunlight that reaches Earth, and the amount of sunlight reaching any part of Earth depends on the orientation of Earth and its distance from the Sun.

While studying how ice in the Alps cut grooves into the rocks beneath it and often left behind piles of rubble when the ice melted and retreated in summer, in 1842 Louis Agassiz noted that other parts of the Alps, too low to have ice at any time of the year nowadays, nonetheless showed similar scarring and piles of rubble. From this, he deduced that Earth was once considerably colder. Detailed study over the past 150 years has outlined both the magnitude and the dates of the **ice ages**. We now know that, at their peak, glaciers covered most of northern Europe; in North America, they extended as far south as Ohio.

The ice ages occurred relatively recently in Earth's history, within the most recent million years; indeed, the last significant ice age ended only 12,000 years ago. Thus it is possible to date with some precision both when the ice ages occurred, and how much ice there actually was during each ice age.

To do this, one looks at sediments on the ocean floor. Small marine organisms, called **foraminifera**, have shells made of calcium carbonate. As they die, their shells sink to the bottom of the ocean. By digging deeper into ocean sediments, one can look further back in time, in a way that is well-calibrated by other fossils and radiocarbon dating.

Carbonate rocks contain carbon, useful for $^{14}C$ dating, and oxygen. As we noted before, oxygen has several isotopes; the most common are $^{16}O$ and $^{18}O$. The sea animals making their shells get the oxygen for their shells from the water itself. But the ocean water is constantly evaporating into the atmosphere and then returning to the ocean as rain.

The exception to this process is during ice ages. When Earth is cold enough and a large ice cap is formed, this ice ties up much of the water that normally would be flowing back into the oceans. One obvious result is that the sea level drops; another, more subtle change is that the isotopic composition of the oxygen in the sea water also changes. Water containing the lighter isotope, $^{16}O$, is evaporated more easily; being lighter in weight, it takes less energy to move it away from the water and into the air. Thus the water that remains behind becomes enriched in the heavier isotope, $^{18}O$, and so do the carbonate shells made in this water.

Thus it is a relatively straightforward process to take a core of sediments from the ocean floor, measure the fluctuations in $^{18}O$ content, and from that deduce how much ice there was on Earth as a function of time. Then we can deduce the surface temperature of Earth at that time. The curve seems very irregular, at first glance. But it is almost periodic; that is, the temperature trend seems to repeat itself. Is there any way of analyzing this irregular curve to see if we can uncover any regular, periodic behavior?

That kind of analysis is called **Fourier analysis**. As we learned in Chapter 4, Fourier proved that any irregular curve could be broken down into an infinite series of sine waves, each sine wave with a different frequency and a different amplitude. What we can do with our irregular curve is apply the mathematical tools of Fourier analysis to find which sine waves, sorted by frequency, have the biggest amplitudes. They will be the frequencies that are the most important components to this curve.

That analysis is shown in Figure 6.13. We see that there are three periods that are very important in this curve. In other words, we could reproduce most of the wiggles in this curve just by adding together three sine waves, one with a period of 105,000 years, one with a period of 41,000 years, and one with a period of 23,000 years.

These periods all have an important astronomical significance. Earth orbits the Sun in an ellipse; but, from calculating the effect of the gravity of other planets on Earth's orbit, we know that the shape of this ellipse changes regularly. The time it takes for Earth's orbit to change from a nearly circular orbit to one of maximum eccentricity, and back again, is 105,000 years.

We also know that Earth is tilted on its axis, at about 23° from the plane of its orbit. But this tilt tends to rock back and forth, varying from 21° to 24°. This rocking motion, called **nutation**, can also be calculated, taking into account the effect of other planets' gravity on the harmonics of Earth's gravity field. It takes 41,000 years to complete one cycle.

Finally, the axis of Earth is oriented in a certain direction in space; at the moment, its orientation is such that the north pole happens to be pointing to a star we call Polaris, the North Star. But it wasn't always that way. In fact, we can tell from the records of the ancient Egyptians that, only a few thousand years ago, the north pole pointed to a completely different star, one we call Thuban. The reason

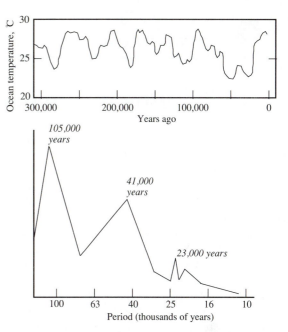

**FIGURE 6.13** Tropical ocean temperature records of Earth (upper). A Fourier analysis of these data show that three periods predominate (lower).

is that Earth tends to wobble as it spins, much like a spinning top that is losing its energy tends to wobble around and around before coming to rest. This wobbling motion is called **precession**. And the period of precession of Earth's axis is 23,000 years. (This period is measured relative to the location of the perihelion of Earth's orbit, which itself is precessing. The precession period relative to the fixed stars is 26,000 years.)

The shape of Earth's orbit and its orientation to the Sun clearly play an important role in how warm Earth gets. That's not surprising. Obviously, if Earth's orbit is eccentric, then Earth should be warmer during the times when it is closest to the Sun. Currently, this happens during the month of January. Because of our seasons, we may not notice it much in the northern hemisphere; but on the equator, the months of December through February are noticeably warmer than the months of June, July, or August. And the seasons in the southern hemisphere are much more intense than those in the northern hemisphere. The eccentricity of Earth's orbit makes winters milder in the north and summer hotter in the south during the month of January. When the eccentricity of Earth's orbit increases, this effect becomes even more pronounced.

Likewise, the intensity of the seasons depends on the tilt of the axis. The more the axis is tilted, the stronger the contrast in weather will be between winter and summer.

Finally, the orientation of the tilt determines just when the seasons occur. Our calendars are designed so that the months that occur during winter in the northern hemisphere

**FOR THE RECORD...**

# Why Is Summer Hotter than Winter?

Let's start with the obvious. Days are warmer than nights. That's because the Sun shines on the daytime side of Earth, and our atmosphere is thin enough to allow that heat to escape at night. (Humid or cloudy nights are warmer than nights when the sky is crystal clear; clouds and water vapor trap infrared light and prevent it from escaping to space, in an example of the **greenhouse effect**.)

Mornings and evenings are cooler than midday. You recall we learned that the energy output of the Sun is $1.4 \times 10^3$ J per second per square meter. In other words, in every second $1.4 \times 10^3$ J of solar energy will fall on a 1-m$^2$ area turned to face the sunlight head-on. But if the area where the sunlight falls is slanted away from the direction of the sunlight, that energy can be spread out over a much larger area, as shown in Figure 6.14. If, instead of being directly overhead, the Sun is only 30° above the horizon, the sunlight will be spread over 2 m$^2$, twice as much area. In that case, the amount of sunlight per square meter is only half as much as if the Sun were directly overhead.

Of course, only in the tropics does the Sun ever pass exactly overhead. Even ignoring the tilt of Earth (an assumption that is valid only at the spring and fall equinoxes, around March 21 and September 22), on average a place on Earth with a latitude of 60° will at best receive only half the solar heating that a place on the equator would get, because the Sun only gets to 30° above the horizon. Thus, no surprise, the North and South poles of Earth are much colder than the equator.

But Earth's rotation axis is tilted by 23°. At the height of summer—June 21 in the northern hemisphere—this tilt adds 23° to the altitude of the Sun, so at 60° N the Sun will reach a full 30°+ 23°, or 53°, above the horizon. At winter, however, we have to subtract 23°; the Sun only gets 7° above the horizon at latitude 60° N on December 21. North of latitude 67°, which is the Arctic Circle, the Sun never

rises on that day, and it never sets on June 21. This is what is meant by the "Land of the Midnight Sun."

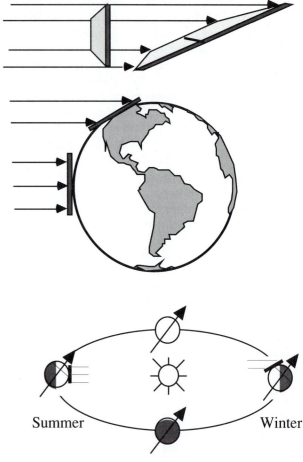

**FIGURE 6.14** An area of 1 m$^2$ captures less sunlight when it is tilted than when it is not tilted. This principle ultimately explains why winters are colder than summers in the northern hemisphere.

will always be called December, January, and so forth. But the position of Earth in its orbit during these months will change, as the axis precesses. Thus half a period from now (11,500 years) Earth will be aligned so that we will be farthest from the Sun during January; then it will be the northern hemisphere that has the more intense summers

and winters while the southern hemisphere has less intense seasons.

A connection between ice ages and changes in Earth's orbit and spin was first proposed in the 1920s by the Yugoslavian astronomer Milutin Milankovitch. But the oxygen isotope data confirming the fluctuations in Earth temperature were not obtained until the 1950s, by Cesare Emiliani and Harold Urey (the same Urey who proposed the impact hypothesis for the death of the dinosaurs).

How can eccentricity changes, nutation, and precession combine to make ice ages?

The basic idea is simple enough. Snow storms occur mostly when the temperature is just below freezing; if it's too cold, snow won't form. During mild winters in the northern hemisphere, therefore, we're more likely to have snow; and mild summers are less likely to cause the polar caps to melt. Because snow can only accumulate on land, we also need the majority of the land mass to be located in one hemisphere. Currently, the continents have drifted together in such a way that most of them are located in the northern hemisphere. This coincidence may explain why it is only recently (in geological terms) that we've had ice ages. There are several difficulties with this simple picture, however. The biggest problem is that, of all the periods, the 105,000-year eccentricity fluctuation is the one most strongly correlated with Earth's temperature history; but in terms of the relative heating of Earth's surface, it should have the smallest effect.

Thus several theories attempt to show why this period should be most dominant. One theory calculates that it takes about 100,000 years for the ice glacier to grow so big that its weight warps the crust down; at this point, the ice may find itself at a lower, warmer elevation and start to melt. Another theory postulates that the system of ocean currents can be significantly altered by minor changes in the ice sheet; thus the small changes resulting from Earth's eccentricity variations would trigger major changes in Earth weather.

## SUMMARY

Earth is not an isolated universe unto itself; it is a planet in the solar system and is subject to the events and forces that act on all planets. The impact of even small asteroids onto Earth has certainly happened in the past and will occur again in the future. The effect of such impacts could be catastrophic. In the past, such impacts have scattered molten rock (called tektites) for thousands of miles and left behind sizable craters. It has been proposed that especially large impacts may have led to mass extinctions.

The gravity of other planets causes the shape of Earth's orbit and the orientation of its spin to fluctuate. These fluctuations appear to have been the cause of the ice ages, but the exact mechanism is still not completely understood.

## STUDY QUESTIONS

1. When did the dinosaurs become extinct?
   (a) thousands of years ago
   (b) millions of years ago
   (c) billions of years ago
2. When did the ice ages occur?
   (a) thousands of years ago
   (b) millions of years ago
   (c) billions of years ago
3. If a 10-km-diameter asteroid hit Earth, how big a crater it would make?
4. What evidence suggests that a meteoroid hit Earth coincident with the death of the dinosaurs?
5. What data tell us the temperature of the oceans in past epochs?

## 6.3   LIFE

More than anything else, what sets Earth apart from the rest of the planets is the presence of life. Earth is, so far as we know, the only place in the universe on which life has developed. This doesn't mean that it *is* the only place, it's just that we have no evidence of its existence elsewhere.

But what is "life," anyway? That's a question for philosophers, but in the most simplistic sense, we might define it as collections of chemicals that are able to grow and reproduce. The difficulty in defining life is one thing that makes it so difficult to look for on other planets.

### *Viking* Looks for Life

On July 20, 1976, 7 years to the day from the first landing of men on the Moon, a robot spaceship called *Viking* landed on the plains of Mars. It carried, among numerous other experiments, three devices to test for the presence of life on Mars.

These were designed to look for "life as we know it," life based on complex carbon-based chemistry that we call **organic**. This sort of life would breathe gas or consume liquid nutrients and leave behind organic wastes. For the more unusual or unexpected types of life forms, the researchers could only depend on the cameras *Viking* carried; if any silicon giraffes walked past, they'd see them move.

The results of the experiments were mixed at best. The conclusion that most scientists have drawn is that there is no carbon-based life on Mars today. (No silicon giraffes were seen, either.) Every experimental result was either negative or else could be explained just as well by inorganic chemical reactions.

## Geologic Consequences of Life

To many scientists, the lifeless results of *Viking* were not too surprising. Each of the *Viking* experiments created a small environment of gas and food in a chamber and then searched to see how life would change that environment. But without sending *Viking*, we have still been able to see the general environment of Mars and contrast it with that of Earth. Life has completely altered the surface and atmosphere of Earth in ways unlike anything seen on Mars or anywhere else.

Life on Earth has changed the atmosphere completely. By looking at 3-billion-year-old rocks, we can see that the oxidation state of elements such as iron (the ratio of $Fe^{2+}$/$Fe^{3+}$) changed markedly as more and more free oxygen entered the atmosphere when life was beginning on Earth.

Free oxygen is an extremely reactive chemical. That's why fires are such a hazard on Earth. Anything that will burn, including metals that rust or oxidize, is out of chemical equilibrium with the atmosphere. It is life that keeps producing this free oxygen and the substances that would normally burn in its presence. Plants use sunlight to separate oxygen from carbon dioxide, making flammable substances such as trees and leaves in the process. Animals consume the plants, in effect slowly burning them inside their bodies. And animals (humans) create foundries and blast furnaces to complete the process of covering Earth with nonequilibrium chemicals like plastics or steel.

Venus and Mars both have $CO_2$ atmospheres. In the case of Venus, it's the $CO_2$ that traps infrared light and makes the surface temperature so hot. What happened to the $CO_2$ in Earth's atmosphere? Again, life is responsible. Small sea creatures use up the carbon dioxide to make hard shells of calcium carbonate. As they die, the shells fall to the bottom of the ocean where eventually they become limestone. Today there is enough $CO_2$ locked up in limestone on Earth to come close to matching the $CO_2$ levels seen on Venus. On Venus, the gas stayed in the air to smother the planet. On Earth, life turned it into rocks.

In more subtle ways, life continues to alter the chemistry of Earth. For example, Earth has liquid water oceans under an atmosphere that is nearly 80% nitrogen. Nitrogen and water react to make nitric acid. Why aren't the oceans dilute solutions of nitric acid? Because microbes in the ocean consume the nitric acid as rapidly as it is formed, bringing this nitrogen into the food cycle.

There is no evidence that anything like these processes has ever occurred on any other planet in the solar system.

## Life Elsewhere

Why do we concentrate so much on Mars in our search for life? It's mostly a matter of process of elimination. Life, as we know it, needs warm temperatures and an atmosphere to survive. Mercury and the Moon are airless; the satellites of the outer planets that do have atmospheres (Titan and possibly Triton) are too cold, and Venus is much too hot for any life forms we know of to survive. Earth and Mars are the only terrestrial planets left, and Mars is apparently lifeless.

It has been suggested that the colors of the clouds of Jupiter are evidence for the presence of complex organic chemicals, the kind of complicated chemicals, based on carbon, that make up living things on Earth. For example, the orange-red color of Jupiter's Great Red Spot might, it has been suggested, result from the presence of carotene. Carotene, the chemical that gives carrots their color, is the simplest colored organic chemical. Laboratory experiments, where a mixture of gases found in Jupiter's atmosphere was zapped with bolts of electricity to simulate lightning on Jupiter, have indeed produced colored substances with organic chemicals present.

However, theoretical calculations of how the observed gases in Jupiter (methane, with one carbon atom, $CH_4$; and ethane, with two carbon atoms, $C_2H_6$) might combine to make carotene (which has a chain of 40 carbon atoms, $C_{40}H_{56}$) indicate that all the intermediate substances (for instance, chains of 10, 20, or 30 carbon atoms) would be convected into the hot, deeper regions of Jupiter's atmosphere and destroyed long before they could be made in any quantity, much less survive to combine into any life-producing substances. And the very experiments that first produced these substances in a simulated Jovian atmosphere have shown that the most common coloring agent is, in fact, sulfur and not organic compounds. Thus the idea that lightning bolts could have produced chemicals that eventually might form living things on Jupiter has serious problems.

## Organic Materials in Meteorites

Other candidates have been suggested as possible homes for extraterrestrial life. Carbonaceous meteorites have been studied extensively for evidence of complex organic matter whose origin could be traced to some living system or that might be the precursors of life. Measurements so delicate that they can detect the residue of a human scientist's fingerprint on a rock have been made; in most cases, the complex organic materials found in these meteorites have been shown to be terrestrial contamination and not from outer space.

However, organic materials in the Murchison carbonaceous chondrite do appear to have originated in space. One strong bit of evidence is found in the structure of the organic molecules. Many complex organic molecules can exist in either of two possible structures, one being the mirror image of the other (see Figure 6.15). The two types of molecules are either "left-handed" or "right-handed."

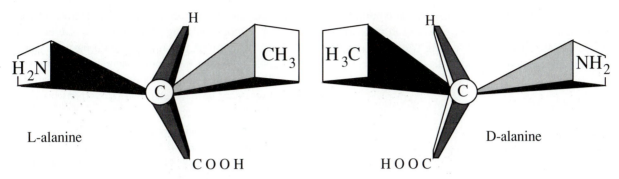

**FIGURE 6.15**   The left-handed and right-handed versions of alanine.

Terrestrial organisms produce only one version of such structures, the left-handed variety. The reason for this is that, if an organism can utilize left-handed sugars, for instance, then it will not be able to use right-handed ones; they will not be metabolized. So an ecosystem develops that depends on one handedness of molecule alone. However, in Murchison, both left-handed and right-handed amino acids were found in nearly equal abundances. This strongly indicates that these organic molecules did not come from terrestrial contamination.

In addition, in the meteorite are found a wide variety of other organic chemicals, including many amino acids not commonly found in ordinary life. Life on Earth commonly produces some two dozen different kinds of amino acids, but over 70 different kinds of amino acids have been found in the Murchison meteorite.

The processes that made these chemicals almost certainly had nothing to do with terrestrial life, nor does anyone think that life anywhere else was responsible for producing these acids. Any living organism on Earth would tend to make only the more common acids, and presumably life anywhere else would make either left-handed-only or right-handed-only versions of those molecules.

But the presence of these acids in a meteorite shows that such molecules can be made in space. Indeed, radio telescopes observing **cold molecular clouds** deep in space, the kinds of places at which it is believed stars (and their planets) are formed, have found radio spectra characteristic of a wide variety of relatively complex hydrocarbon atoms. Visible light spectra of organic molecules have also been observed in comet tails. The basic building blocks of organic life can be made, in space, without life as we know it present.

Is it possible that life on Earth got its start from these organic materials? One hint in favor of this idea is tied into the theory that dinosaur extinctions are caused by comet or asteroid impacts.

Recall that the dinosaurs lived during a geologic period called the Cretaceous; their bones are found only in rocks of a certain limited period of age. Younger rocks, lying on top of the Cretaceous rocks, have fossils of ani-

mals and plants that are more like modern species, and such rocks are said to represent the Tertiary geologic period.

Recently it has been discovered that there are amino acids, along with the iridium, in the rocks of the K/T boundary layer. As in the case of the meteorites, the types of acid present (a-amino-isobutyric acid and isovaline) are not typical of acids made by life on Earth, and both left- and right-handed versions of the isovaline are found. Thus there seems to be some evidence that organic material from space can somehow survive the trip through Earth's atmosphere and reach the ground, although it is hard to see how such molecules could survive such an impact.

Even if these organic acids could reach the surface of Earth in this way, it is unknown how they could chemically combine themselves into even the most primitive life forms. No one knows for sure if these extraterrestrial acids actually played any role in the formation of life on Earth. It is just possible. But if so, one would expect that the other planets should have their share of these building blocks, too. The absence of life on Mars or the other planets becomes all the more significant in light of this discovery. It takes more than just the right ingredients to allow life to flourish.

Does all this mean that Earth is unique and that there is no life anywhere else in the universe? Not necessarily. It just means that, if life does exist elsewhere, we haven't found it yet. It seems unlikely that, with all the myriad stars in the heavens, there is no other one that does not have planets harboring life. But the cold fact is that, so far, we have seen absolutely no evidence for that life.

## SUMMARY

Earth is the only planet known to harbor life. The existence of life on Earth has changed the climate, removing carbon dioxide and introducing free oxygen. We have looked for evidence of carbon-based life on Mars, the next-most hospitable place in the solar system, with the *Viking* lander experiments, but did not find it.

Organic materials have been suggested as coloring the clouds of Jupiter. They have been observed in meteorites and in interstellar space. But it's a long way from simple organic molecules to living organisms.

## STUDY QUESTIONS

1. Did *Viking* find life on Mars?

2. Water reacts with nitrogen in our atmosphere to make nitric acid. Why aren't the oceans on Earth full of nitric acid?

3. Why isn't Earth's atmosphere rich in carbon dioxide?

4. What is the evidence for complex organic materials in the universe outside Earth? Where are these organic chemicals seen?

5. Is there life anywhere else in the universe?

## 6.4  PROBLEMS

1. Earth is 6371 km in radius and its core is 3486 km in radius. If an earthquake occurs in Los Angeles, what region of Earth's surface will receive no S-waves from this quake? Assume for this problem that seismic waves run through a planet in straight lines.

2. The Richter scale is often used to measure the strength of an earthquake. It is a logarithmic scale, such that an earthquake of strength 4 on the Richter scale is 10 times as powerful as one of strength 3. How much stronger is an earthquake of strength 6.5 than one of strength 3?

3. Scientists have known for many years that the plates on Earth move relative to one another. Recently they have learned how to measure this directly. How do you think they might do this? Before we knew how to measure plate motion directly, scientists developed other ways in which to tell in what direction, and how fast, the plates might be travelling. What might these have been?

4. Imagine that you were the first person to get a good look at the surface of Saturn's moon, Titan. What would you look for, to tell you whether or not it had plate tectonics?

5. The impact trigger for the K/T event and the Milankovitch cycle theory for the origin of the ice ages are both still somewhat controversial. Why? For each theory, describe what you think needs to be done to make the theory more widely accepted. Try to imagine, and describe, a future discovery that would conclusively disprove either theory.

6. Imagine that a 30-m-radius asteroid struck the physics or geology building at your school. Describe in detail what effect this impact would have on your campus and on the surrounding neighborhood. Be specific. Where would the crater extend to? What nearby buildings might be knocked down? How many people might die?

7. Comet Swift-Tuttle has a period of 134 years; it most recently approached Earth in 1992 and it is due back in 2126. Because of its unusual orbit, inclined 117° from the plane of the other planets, it is unlikely to be perturbed out of its orbit. If its return date is 15 days later than calculated (and be aware that its return in 1992 was 17 days late), then there is a significant chance that it will strike Earth on August 14, 2126.

    (a) In general, the probability of hitting Earth on any one orbit is estimated to be 1 in 10,000. Show that the probability is 1 in 2 that the comet will strike Earth sometime in the next 1 million years. (Hint: What are the odds of it missing Earth? How many times does it encounter Earth over the next 1 million years?)

    (b) The speed of the comet relative to Earth is calculated to be 60 km/s. Assume that the radius of the comet is 5 km and its density is similar to that of ice, 1 g/cm$^3$. Calculate the energy of impact.

    (c) If all this energy went into heating the atmosphere, how much hotter would the air get? Assume that the atmosphere has a heat capacity of 400 J/K·kg and the mass of Earth's atmosphere is $5.2 \times 10^{18}$ kg.

    (d) Describe the consequences of this impact on Earth. Would Earth's spin or orbit change significantly? What effects would someone living far from the impact site feel?

8. What is life? Give two defensible, but clearly different, definitions. Only one of them should define life in a way that a spacecraft such as *Viking* could detect.

9. In the year 2525 an interstellar space probe discovers a solar system orbiting a G-type star much like our Sun. There are three terrestrial planets in this system. The first is rocky and airless. The second is about the size of Earth and has an atmosphere of 99% carbon dioxide, 1% water. The third is Mars-sized, with an atmosphere 40% argon and 40% neon, both inert gases, and 20% fluorine (the most chemically reactive element in the periodic table). Which planet do you think is most likely to have life? State your reasons.

10. The way life started is a question that is still hotly debated. One theory has it that life was carried to Earth in a form like "spores" from somewhere outside the solar system. Like all the other theories for the origin of life, this one has its problems. What problems do you see with it?

## 6.5 FOR FURTHER READING

Frank Press and Raymond Seiver's textbook, *Earth* (New York: W. H. Freeman, 1986), is one of the best of many good introductory geology texts available. A less formal book, written for the educated layperson, is Preston Cloud's *Oasis in Space* (New York: W. W. Norton, 1988), which treats Earth as a planet in space and discusses its early history in warm detail.

Sources on geology and geophysics have been mentioned in the Venus and Mars chapters; once again, we would especially call your attention to A. Summerfield's *Global Geomorphology* (New York: Longman, 1991). One delightful and personal introduction to structural geology as it is done by real people can be found in *Basin and Range* and its sequels, *In Suspect Terrain*, *Rising From the Plains*, and *Assembling California* by John McPhee (New York: Farrar, Straus and Giroux, 1981, 1983, 1986, 1992).

Earth science and its connection with life is a rapidly changing field; books on the topic are often out of date before they reach the libraries. If you're looking for more information, an intermediate step between introductory texts and advanced research articles are review articles in the journals *Science* and *Nature*. For instance, articles by C. F. Chyba, "Meteoritics: extraterrestrial amino acids and terrestrial life" (November 8, 1990, *Nature*, volume 348, p. 113), and Moustafa T. Chahine, "The hydrological cycle and its influence on climate" (October 1, 1992, *Nature*, volume 359, p. 373), include both useful overviews of current research, written at the level of the nonspecialist, and references to current work in the field. Try paging through this year's volume of *Nature* or *Science* for more recent review articles on the subject.

For a readable discussion about the question of impacts and their effects on terrestrial life, as well as some other exciting topics, try *Cosmic Catastrophes* (New York: Plenum, 1989), by Clark R. Chapman and David Morrison. More technical discussions can be found in articles by Digby J. McLaren, "Geological and Biological Consequences of Giant Impacts" (in *Annual Reviews of Earth and Planetary Science*, volume 18, 1990), and by Eugene M. Shoemaker and Ruth F. Wolfe, "Mass Extinctions, Crater Ages, and Comet Showers" (in another University of Arizona Press book, *The Galaxy and the Solar System*, 1986).

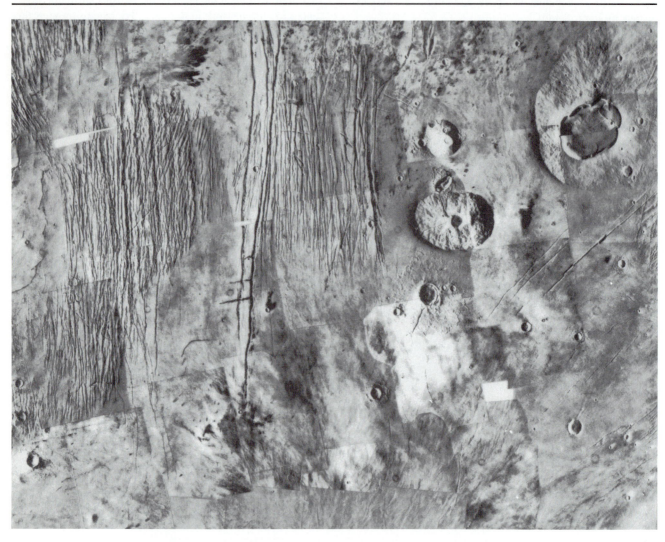

**FIGURE 7.1**  The Tharsis northeast quadrangle of Mars, showing Ceraunius Fossae and three volcanoes: Ceraunius Tholus, Uranius Tholus, and Uranius Patera.

# CHAPTER 7

# Mars

## 7.1  ON MARS

It is daybreak of an early summer day in the northern hemisphere of Mars. The temperature is a chilly 210 K, the wind is a light 2 m/s, and the atmospheric pressure is a relatively low 600 Pa (Earth's atmosphere has a pressure of roughly $10^5$ Pa, and 1 m/s is roughly 2 mph). The Sun just rising in the east is met by a tiny disk of light, the moon Phobos, which is racing from west to east. Three hours later it will be rising in the west again to repeat its path. The Sun seems only about two thirds as big as from Earth and feels as though it is giving off only half as much heat. The moon Phobos is only a quarter the size of the Sun's disk, much smaller in the sky than Earth's moon.

As the Sun climbs in the sky, the air takes on a reddish hue and everything seems bathed in a pink-colored light. Although the air is very thin, it supports a fine red dust, tiny as grains of flour, that also coats all the rocks. The force of the wind seems negligible. Not only is its speed gentle, but there is only 1% as much mass per cubic meter in the air as there is on Earth.

The field where we are sitting is surrounded by boulders, grayish pitted rocks capped by the fine red dust (Figure 7.2). As the day progresses, the rocks begin to warm up; eventually they will reach a temperature of 240 K. But

| Mars's Vital Statistics | |
| --- | --- |
| Radius | 3397 km |
| Surface area | $1.4 \times 10^8$ km$^2$ |
| Mass | $6.4 \times 10^{23}$ kg |
| Density | 3.94 g/cm$^3$ |
| Local gravity | 3.72 m/s$^2$ |
| Escape velocity | 5.1 km/s |
| Albedo | 0.25 |
| Surface temperature | 240 K (day) |
| | 210 K (night) |
| Surface pressure | 600 Pa |
| Length of year | 686.98 days |
| Length of day | 24 h 37 min 25 s |
| Distance from the Sun | 1.524 AU |

the air is so thin that it can hardly hold any heat. In fact, the air never gets much above 220 K. The winds pick up a bit, to about 5 m/s, but this still is not enough to stir the red dust on the ground. By evening, the rocks have cooled back down to around 210 K, and the winds die down as well.

By night, all is still. Sharp eyes might make out a point of light in the sky, slowly changing position from hour to hour. This is the other moon of Mars, Deimos.

**FIGURE 7.2** The Martian surface, as photographed by the *Viking 2* lander. The large rock in the center of the picture is about 60 cm long and 30 cm high.

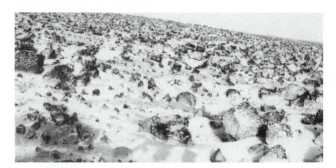

**FIGURE 7.3** White patches under the rocks indicate the presence of frost. This picture was taken by *Viking 2* during the Martian winter. Compare this with Figure 7.2.

It takes Deimos 5 days to move completely across the sky and back to its original position. However, the closer moon, Phobos, completes an orbit every $7\frac{1}{2}$ hours. It will rise and set and rise again before the night is out. Finally, 24 hours and 39 minutes from our last sunrise, it is sunrise again and a new day has begun on Mars.

As the seasons pass—each season takes 170 days, or about twice as long as an Earth season—some changes are seen. In the northern hemisphere fall, great dust storms sometimes arise, starting in the low-lying regions and growing until most of the planet is covered with dust. The winds rise to 20 m/s, with gusts up to 25 m/s. The temperatures stay much cooler during the day, hovering around 200 K. At the height of the storm it is as dark as night, even at noon, as the dust blankets the sky. The storms will take months to clear.

During the winter, a thin layer of water frost may appear on the ground at night (Figure 7.3), to disappear quickly at dawn. The pressure of the atmosphere is much higher now, nearly 1000 Pa (or 1% of Earth's atmospheric pressure), but the temperature stays cooler. Finally, after nearly 700 days it is early summer again. The atmospheric pressure drops down to almost half its wintertime value, and the days return to their quiet pattern.

This description is a summary of the weather as seen by *Viking Lander 1*, one of the two spacecraft that landed on Mars in 1976 and sent back information about the planet for the next 6 years. *Viking Lander 1* landed in the northern hemisphere, about 20° north of the equator, in a fairly low-lying region called Chryse. Its sister ship, *Viking Lander 2*, landed farther to the north, on a plain called Acidalia. (Actually, *Viking Lander 1* did not see frost in the winter; *Viking Lander 2* did.)

The two Viking Lander spacecraft were accompanied by two *Viking Orbiter* spacecraft, which took many pictures of the surface of Mars. From these pictures we have been able to map most of the surface of the planet to see craters, volcanoes, lava flows, huge valleys, dune fields, polar ice caps, and other features not so easily classified.

## The Geography of Mars

The surface of Mars can be divided into two parts (Figure 7.4). One part, covering most of the southern hemisphere and some of the northern, is heavily cratered, much as our Moon is. If the density of craters is any indication of age, then it appears that this area is nearly as old as the lunar surface, too. In general, the ground here is higher than in the other part of the planet, which covers most of the northern hemisphere. This area is often referred to as the **highlands**, while the northern part is referred to as the **lowlands**. But nestled within the highlands is the deepest impact basin on Mars, Hellas, where the atmosphere is almost thick enough to allow liquid water to exist and the temperature can sometimes get as warm as the arctic regions of Earth.

The lowlands are quite different. They are smoother and seem younger, with relatively few craters. Many parts of the lowlands are nearly featureless except in the highest-resolution photographs. It may be that these plains are covered with lava or a thick layer of dust, or that they used to be covered with a layer of water or ice; nobody really knows.

At the poles, there are caps made of water and carbon dioxide ice, with beautiful swirling stripes, perhaps caused by alternating layers of dust and ice (Figure 7.5). Near the ice cap, giant fields of sand dunes can be seen (Figure 7.6).

## The Volcanoes of Mars

Besides the major divisions of highlands and lowlands, there is one other great physiographic feature on Mars. This is known as the **Tharsis bulge**. It is a region where there are several gigantic volcanoes sitting on top of a gently sloping plateau (the "bulge"). The biggest known mountain in the solar system, **Olympus Mons** (Mount Olympus), is one of these volcanoes (Figure 7.7). It stands 26 km above the mean surface of the planet (dwarfing Mount Everest on

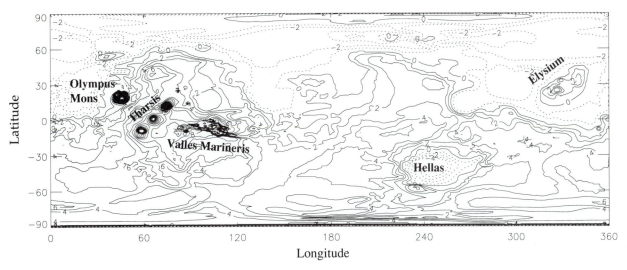

**FIGURE 7.4**  A topographic map of Mars. Notice that the northern hemisphere is generally lower than the southern. The Tharsis volcanoes are on the left.

Earth, which is less than 9 km above sea level) and covers an area comparable to the state of Arizona. The peak of this mountain is higher than 96% of the atmosphere. It is a type of volcano known as a shield volcano (its gently curving slopes make it look like a shield lying on the ground), the same type of volcano as Mauna Loa on the island of Hawaii (see Figure 7.8), the largest mountain on Earth (rising over 10 km from the surrounding sea floor, and therefore larger than Everest).

Some of the most outstanding features on Mars are the great shield volcanoes, but there are many other volcanic features as well. To start with, the plains of the southern highlands all appear to be made up of lava flows, much as the surfaces of the Moon and Mercury are. This area is very old, perhaps the oldest surface on Mars, and may have formed shortly after Mars accreted, over 4 billion years ago. There are also some very interesting volcanoes found in

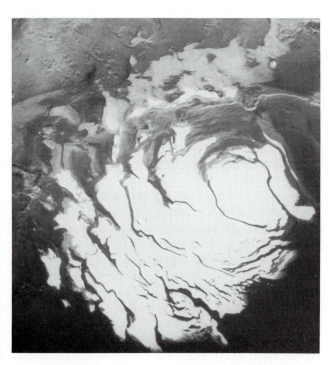

**FIGURE 7.5**  The southern polar cap of Mars.

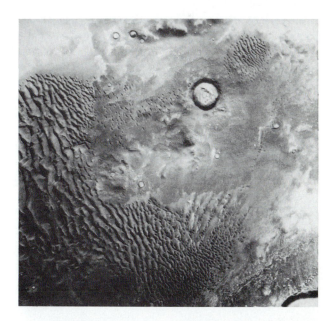

**FIGURE 7.6**  A sand dune field in the southern hemisphere of Mars. The dunes in the upper left are transverse dunes; those in the lower right are barchan dunes. The area shown is about 60 km across.

**FIGURE 7.7** Olympus Mons, with a diameter of 600 km, is the largest volcano in the solar system.

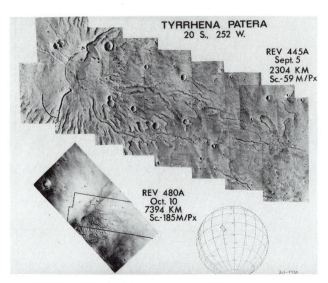

**FIGURE 7.9** Tyrrhena Patera, thought to be a pyroclastic volcano. The mosaic is roughly 500 km long.

the southern highlands. These volcanoes are thought to be **pyroclastic volcanoes** (Figure 7.9). Pyroclastic volcanoes don't ooze lava like the volcanoes in Hawaii; they explode! This behavior can be caused either by the presence of a lot of volatiles (such as water or carbon dioxide) in the lava itself or by the lava encountering a layer of water or ice (which is quickly melted) in the ground as it rises to the surface. These volcanoes—named Uranius Tholus, Ceraunius Tholus, Hecates Tholus, and Tyrrhena Patera— all appear to be surrounded by deposits of ash, that are highly eroded, possibly by running water. (*Tholus* is the Greek word for a circular building of beehive shape.)

Another astonishing volcano is known as **Alba Patera** (*patera* is the Greek word for "shallow dish"). This

feature, found near the Tharsis volcanoes, is only about 6 km high, but it has a diameter of 1600 km—1000 miles— giving it an area seven times that of Olympus Mons and 180 times that of Mauna Loa.

On the flanks of this volcano are large, beautifully preserved lava flows, some over 400 km long. This incredible length for a single flow indicates either that the lava was extremely fluid compared to typical Earth lavas, that it was flowing very fast, or both.

This sort of fluid lava seems similar to that which covers the southern plains of Mars. By contrast, the lavas produced in other volcanic regions of Mars, the Tharsis and Elysium volcanoes, do not appear to be nearly this fluid.

## Canyons and Channels

Next to the Tharsis region is a rift valley as long as the whole United States. One of the smaller side canyons is as large as the Grand Canyon on Earth. This rift valley, called **Valles Marineris** (Mariner Valley, after the spacecraft that discovered it), may have been formed when the Tharsis bulge first began to rise and stretch the crust around it.

Remember to keep the perspective of these giant surface features in mind. Mars is a much smaller planet than Earth. It has only a quarter the surface area, and its gravity is only about a third that of Earth. These volcanoes and the rift valley, big as they are, seem even bigger when you consider how much of the total surface area of the planet they represent.

But some of the most exciting features on Mars are the channels. Running down the sides of volcanoes, meandering through the plains, forming part of the Valles

**FIGURE 7.8** Mauna Loa, the largest shield volcano on Earth.

Marineris system, or debouching from the highlands onto the northern lowlands, channels are found on many parts of Mars. These features, which look like dried-up riverbeds (Figure 7.10), may actually have been formed by running water in Mars's distant past. There is no liquid water on Mars today; there is very little ice visible on the surface, and the atmosphere has only the thinnest of clouds. Even if there were water, it would be frozen, not liquid, perhaps existing only beneath the surface of the ground where evaporation would be extremely slow. But the presence of these channels is evidence that there may have been liquid water flowing on the surface of Mars in the past.

**Dendritic channels** are found all over the highlands, but are almost never seen on the lowland plains. These channels look very much like terrestrial river valleys, with a branching, treelike structure, except that they tend to just stop abruptly in low regions (Figure 7.10). Because these channels are found almost exclusively in very old terrain, they may indicate that there was running water on the Mars's surface in the distant past, but not very recently.

**Outflow channels** tend to start suddenly, full size, in the highlands (sometimes in regions of **chaotic terrain**) and travel for long distances before petering out, also rather abruptly, often just after reaching the lowlands (Figure 7.11). These channels are thought to be formed by catastrophic flooding. (A similar region on Earth, the Channeled Scablands in Washington state, was formed by this process.) One interesting feature of some of these channels are the teardrop-shaped islands found in them, which clearly show

**FIGURE 7.11**  Tiu Vallis, an outflow channel. There is an oval area of chaotic terrain, roughly 100 km across, in the center right.

the path of the water as it flowed through the channels (Figure 7.12).

This catastrophic flooding may have been caused by volcanism. Some scientists have suggested that subsurface ice was suddenly melted, perhaps by rising lava from a volcano, and that this flood of water roared downhill, digging channels as it went, until it evaporated into the thin air. The water possibly was lost from Mars forever as solar UV radiation broke the $H_2O$ molecules into oxygen and hydrogen, the latter escaping via Jeans escape (see the box). Meanwhile, the area where the ice used to be would have

**FIGURE 7.10**  A sample of the heavily cratered southern region of Mars. Notice the dendritic channel. The image is 500 km across.

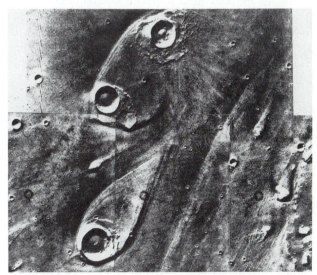

**FIGURE 7.12**  Teardrop-shaped rises in the floors of the channels may have been islands shaped by the flow of water. This area is at the mouth of Ares Vallis. Each island is about 40 km long.

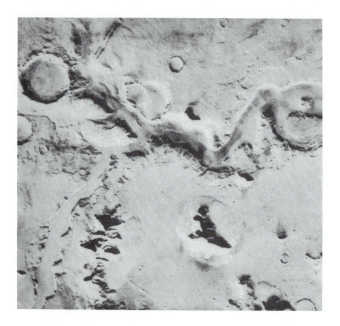

**FIGURE 7.13** Fretted channels are wide, steep-walled channels with flat floors. This image is 280 km wide.

collapsed as the water flowed away, leaving behind the jumbled, chaotic landforms seen today.

**Fretted channels** are wide, steep-walled channels with flat floors (Figure 7.13). They may have been formed by the enlargement of runoff channels by a process called **mass wasting**. Mass wasting is an erosional process in which the walls of a canyon slump down into the canyon, causing the walls to gradually recede. This can happen slowly; on the other hand, a quicker version of mass wasting results in a landslide. Valles Marineris, although thought to have originally formed as a large crack in the crust of Mars, has likely been enlarged greatly by mass wasting over time. Landslides are visible in some Viking photos of Valles Marineris (see Figure 7.22).

## Canals on Mars?

One of the reasons Mars has so fascinated humans is that it is the planet in the solar system whose conditions seem closest to those on Earth (although not nearly as close as people at the turn of the twentieth century believed or hoped). Another is that it is easily observable from Earth, its colorful and changing surface accessible to anyone with a telescope.

As the seasons on Mars progress—and not only is the Martian day nearly the same as Earth's, its seasons are similar as well, because its axis of rotation is tilted nearly the same as Earth's—an observer can see the Martian polar caps grow and shrink. The growth is most dramatic for the southern polar cap. Winter in the southern hemisphere is colder than in the north, because it occurs during the part of Mars's orbit that is slightly farther from the Sun. In addition, the elevations in the south are higher, and so the temperatures are colder than around the north pole.

One result is that the southern hemisphere winters are cold enough to freeze carbon dioxide, the main constituent of the Martian atmosphere. As a substantial fraction of the whole atmosphere gets frozen out at the south pole, the atmospheric pressure everywhere drops by about one third. That is why Viking saw such lower atmospheric pressures during the northern hemisphere's summer (the southern hemisphere's winter). And the enlargement of the south polar cap is visible even in small Earth-based telescopes.

That is not the only phenomenon seen in Earth-based telescopes, however. The surface of Mars has regions of strongly contrasting colors: a bright orange due to the dust (colored by various forms of iron oxide, similar to flakes of rust) and a dark gray from the basaltic rocks that lie beneath the dust. To the eye of an observer on Earth, this contrast in color can play peculiar tricks. The eye imagines that the dark spots are arranged in lines; hence the channels (or *canali*, in Italian, causing the misunderstanding that these features were canals as we would know on Earth) reported in the nineteenth century by the Italian astronomer Schiaparelli (the same astronomer who misinterpreted the rotation of Mercury; see Chapter 4). And the dark, grayish areas themselves can appear almost greenish when seen next to the bright orange regions.

Recall the dust storms that Viking saw in the (northern) fall. From Earth, an observer would see the south polar cap (more visible when Mars is closest to Earth than the north polar cap) retreating, the surface becoming obscured by clouds, and then, once the dust had settled, the dark areas (which the observer would think were greenish) would seem to have grown. It was natural for the observers of the nineteenth and early twentieth centuries to assume that the darkness represents vegetation, growing as the caps melt and water flows through the "canals" onto the surface of the planet. In fact, the red dust is being blown off the gray rocks, while evaporation of the polar cap (which is mostly carbon dioxide, not water) pumps up the atmosphere.

## SUMMARY

Mars is in many ways the planet with a climate most like that of Earth. The atmospheric pressure is 0.6% of Earth's, and the temperatures tend to be nearly 100 K colder. Surface rocks, for the most part, appear to be volcanic. A fine red dust, colored by iron oxides, covers most of the planet and can get stirred into the atmosphere during giant dust storms in the fall. Enough dust stays in the air year round for the sky to appear pink.

Roughly half the surface of Mars is very old cratered plains, while the other half is younger, with many volcanic features. Mars features large shield volcanoes, the highest mountains known on any planet in the solar system. Mars at one time may have had running water on its surface. The apparent lack of water on Mars today means either that it is trapped or hidden somehow beneath the surface or that it was lost to space.

There are many different types of volcanic features on Mars, including great lava plains, pyroclastic (exploding) volcanoes, shield volcanoes, and the unique structure of Alba Patera that has no analogy on any other planet.

Channels also come in many different forms on Mars. Some major types are runoff channels, outflow channels, and fretted channels. All these channels seem to indicate the presence of water on Mars. There is also a huge rift valley, Valles Marineris.

## STUDY QUESTIONS

1.  The terrain on Mars is divided into lowlands and highlands. Which type of terrain dominates the southern hemisphere of Mars?

2.  Which hemisphere is heavily cratered?

3.  Give one piece of evidence that Mars once had more water than we see there today.

4.  What may have triggered the catastrophic flooding that formed the outflow channels on Mars?

5.  At the turn of the twentieth century, some astronomers saw regions of Mars turn dark during the Martian spring season, which they interpreted as evidence of growing vegetation there. What turns out to be the real reason for this darkening?

## 7.2 TECTONICS: LOCAL STRESSES

Global stresses were mentioned in Chapter 5. Local stresses, which affect only some regions of a planet, are also important to the evolution of a planet's surface.

### Volcanic Loading

A planet is coldest on its surface, where heat can radiate to space, and warmest in its interior, where radioactive species such as uranium or potassium generate heat as they decay. As a result, the surface of a planet will tend to be more brittle, while the interior region may be warm enough to behave as a ductile material, or even be partly or completely molten. This can be pictured as a rigid layer called the **lithosphere** sitting on top of a flowing **asthenosphere**.

As rock in the interior melts, the melt is less dense and so more buoyant than the solid rock. As a result, it flows

up through the crust to the surface of the planet, where it forms a volcano. As the volcano grows, more and more mass gets piled onto this surface. If the surface is thick and rigid, it can support this mass for a while; but the weight of this material will eventually try to achieve **isostasy** by sinking downwards into the mantle. Another way of seeing this is to picture the rigid lithosphere sagging under the load of the mountain (Figure 7.14). This extensional stress will cause the crust to crack, as can be seen very clearly around the large shield volcanoes of Mars.

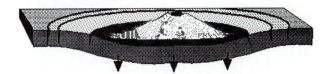

**FIGURE 7.14**  As a volcano grows, it places an ever-growing load on the crust of the planet. If the load is great enough, the surrounding crust may crack.

Other loads besides a volcano can cause a crust to crack. The ice caps on Greenland and Antarctica and the glaciers of the ice age create such cracks on Earth. The lava flows of the mare regions create similar local loading on the Moon.

### Basin Formation

The Hellas Basin, in the southern hemisphere of Mars, was probably formed by a giant impact early in the planet's history. If so, it could be similar to the basins on the Moon and Mercury. The making of these involved both local and global stresses.

These basins started forming early in the planet's history, when considerable accretion of planetesimals onto the body peppered its surface with craters large and small. Particularly large impacts left large circular depressions in the planet.

Because the planet was still young, internal heat sources (and the energy of the accretion itself) were still raising the temperatures in its interior. This means that the global stress felt on the surface at that time would be extensional. And the surface tended to pull apart and crack where it was thinnest and weakest, namely in the bottom of these basins.

# Faults and Folds

When you drive in the mountains, the road will pass through a number of road cuts. Sometimes you can see quite deep below the surface into the mountainside. You are likely to see parallel layers, or **beds**, of rock bent into wavy lines, or

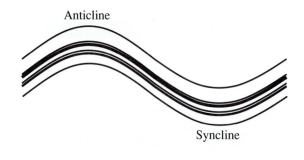

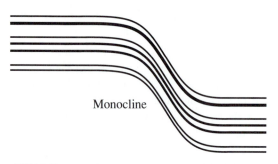

**FIGURE 7.15** Folds are caused by compression of the crust.

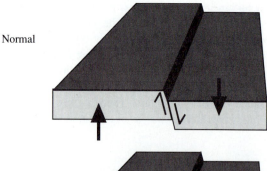

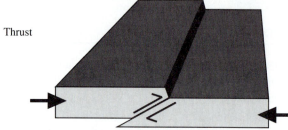

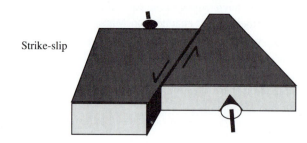

**FIGURE 7.16** The three basic types of faults.

Heating in the interior eventually progressed to the point where a molten lava might form. Following the path of least resistance, the lava flowed upwards through the cracks and out into the bottom of the basin, eventually filling a considerable fraction of the volume of the original crater.

As the planet started to cool, melting ceased. Contraction of the whole planet included compression of the surface, closing up the cracks and stopping the flow of molten lava onto the surface. The lava, now frozen, contracted as it cooled and cracked.

The mass of this lava can be considerable. It may have formed a disk hundreds of kilometers in radius and tens of kilometers thick. As the crust of the planet tried to adjust isostatically to this mass, the whole disk started to subside,

to settle in towards the center. The rigid lithosphere beneath this mass sags, and extension cracks can be seen ringing the outer edges of the mare. In the center of the disk of frozen lava, however, the stresses are compressional. The settling tends to pull material away from the edges and down into the center. Here, compressional ridges can be formed. All these processes thus can result in the multiple cracks seen on the surface of the basin.

## Tectonics of Mars

The surface of Mars shows an interesting intermediate case between the heavily cratered plains of Mercury and the Moon and the completely resurfaced topography of Earth, even broken and offset from one another. Geologists have names for all these features.

In places where the crust of the planet has been compressed, **folds** may form (Figure 7.15). Two terms are used to describe these folds: the crests of the folds, the "hills" (although they may not correspond to an actual hill on the surface), are termed **anticlines**, and the "valleys" are called **synclines**. If the fold is not complete, more like a slope between two flatter areas, it is called a **monocline**.

Often, instead of deforming like this, the crust will simply break. The broken pieces are then fairly free to move relative to one another. This is called **faulting**. There are three basic types of faults (Figure 7.16). **Normal faults** occur when the crust is pulled apart, or under **extension**. Two blocks move apart, and one travels down relative to the other. **Thrust faults** (also called **reverse** faults) occur when the crust is under compression. Two blocks move toward each other, and the vertical motion is opposite (reverse) to that of the normal fault. **Strike-slip faults**, such as the famous San Andreas fault in California, occur when two blocks move horizontally, relative to one another.

If a piece of crust is in between two normal faults, so that it drops down relative to the crust outside the faults, it is called a **graben** (Figure 7.17). An example of a terrestrial graben is the Red Sea. Graben are important features, because their appearance shows that that part of the crust must have been pulled apart.

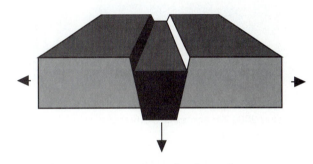

**FIGURE 7.17** A schematic of a graben.

which is nearly free of craters. On Mars, in general, the southern hemisphere is heavily cratered and looks quite Moon-like, while the northern hemisphere is completely resurfaced and reminiscent of Earth.

We have some information about the distribution of material below the Martian surface, by observing the orbits of Phobos, Deimos, and Viking spacecraft about Mars, to measure the gravity field. The moment of inertia factor is about $0.365MR^2$, which tells us that Mars has a core of higher density than its crust.

The southern hemisphere is isostatically compensated and seems to have a thicker crust than the northern hemisphere. The northern hemisphere is generally at a lower elevation than the southern hemisphere; there is a several-kilometer difference in elevation between the northern plains and the southern highlands, the **crustal dichotomy**.

Found primarily in the northern hemisphere, but extending across the equator, are shield volcanoes of immense size. These volcanoes are associated with the Tharsis bulge, as is the large rift valley of Valles Marineris.

All these facts can be combined to give us a picture of the tectonic history of Mars. The planet is differentiated into a core, mantle, and crust. And this differentiation must have

occurred early in its history to allow for a thick, heavily cratered but isostatically compensated crust to exist.

Mars probably started out hot, with an actively convecting mantle. Somehow, possibly through a fortunate superposition of large impacts or perhaps by some action of convection in the mantle, the northern plains were lowered, relative to the southern highlands. The formation of the crustal dichotomy is not yet well understood.

A **mantle plume**, or **hot spot**, may have caused the formation of the Tharsis bulge, pushing the still-thin crust upwards as the hot material rose through the mantle. Many of the tectonic features seen on Mars are associated with the Tharsis bulge and its associated volcanoes. Radiating away from the center of the Tharsis bulge are hundreds of simple graben, most of them a few kilometers wide and about a kilometer deep, but hundreds of kilometers long. The walls of these graben typically dip into the surface at an angle of 60 degrees; from this and the width of the graben, one can infer that the faults making up the two walls should intersect at a depth of less than 10 km below the surface, indicating that these cracks do not extend through the entire lithosphere but are indeed local features. By contrast, Claritas Fossae, a region south of the main Tharsis volcanoes, is a more complex set of graben 100 km wide, 1000 km long, and reminiscent of a continental rift on Earth.

The most prominent example of rifting on Mars is the massive set of troughs 4000 km long, at places 600 km wide, and up to 8 km deep, known as Valles Marineris (Figure 7.18). The morphology of this region is quite complex.

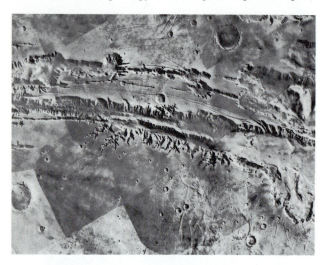

**FIGURE 7.18**   Valles Marineris. The image is over 2000 km across and covers less than half of the canyon.

At the western end are some relatively shallow graben, 3–20 km wide and 1–2 km deep. They form two sets that run from north to south and from east to west, crossing at roughly 90 degrees. This area is called Noctis Labyrinthus, the Labyrinth of Night. These channels open to the east onto two parallel valleys, roughly 700 km long. The northern one, Tithonium Chasma, starts with a width of nearly 100 km but narrows to less than 20 km wide at its eastern end. Opening out of Tithonium Chasma is Candor Chasma and north of it, Ophir Chasma. The southern channel, Ius Chasma, widens to the east into Melas Chasma, a roughly oval depression 400 km long and 250 km wide. At this point Melas, Candor, and Ophir together are more than 600 km wide. East out of Melas Chasma runs the longest and deepest of the troughs, Coprates Chasma. Near Melas, the walls of Coprates Chasma drop 8 km to the floor of this trough. As the surrounding plains gradually lose altitude, the height of the chasma walls falls to 4 km at the far end, some 1500 kilometers to the east. Opening out of Coprates is wide, shallow Capri Chasma and Ganges Chasma north of Capri.

The origin of this complex set of troughs is still not well understood. (The term *trough* instead of *canyon* is used, because canyons on Earth are formed by running water.) The direction that the valley runs seems to be controlled by the Tharsis bulge, and the most likely explanation for its formation is rifting of the lithosphere in response to the uplift of Tharsis. Other origins have been proposed as well, including that this might be a place where plate tec-

tonics on Mars tried to get started. The only comparable feature on Earth is the East African Rift, which runs from southern Africa up through the Red Sea. The African rift valley, however, is noted for the large numbers of volcanoes along the rift valley floor, whereas there is no unambiguous evidence for volcanic features on the floor of Valles Marineris. Regardless of its formation, it is clear that landslides and possibly water or ice erosion have substantially altered and widened this valley since its formation.

The heights of the Martian volcanoes seem to be related to the ages of the volcanoes. The oldest volcanoes are only a few kilometers high, while the youngest are about 20 km higher than the surrounding terrain. This may be because the source region of the magma has deepened with time as the planet cooled.

Judging from the cracks around the volcanoes it is possible to estimate the thickness of the rigid **lithosphere** at the time of formation of the cracks. This thickness varies greatly from area to area; it is as much as 120–150 km thick on some parts of Mars, but only 20–50 km thick near the Tharsis and Elysium volcanoes.

As Mars cooled, the crust thickened. The hot spot under the Tharsis bulge cooled and the bulge itself contracted, producing compressive ridges. The most common compressional features associated with Tharsis are wrinkle ridges (Figure 7.19). These look almost like graben in reverse: strips a few kilometers wide, raised a few hundred meters above the surrounding plains, and running for hundreds of

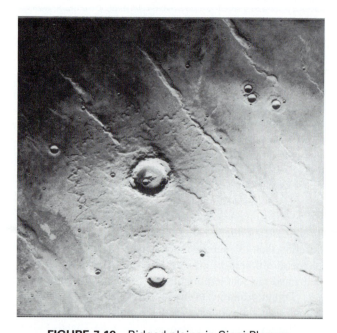

**FIGURE 7.19**   Ridged plains in Sinai Planum. These ridges are concentric to the Tharsis bulge. The craters in the center show the lobate ejecta structure common amongst Martian craters. The image is 250 km across.

kilometers along the surface. Wrinkle ridges are seen on the mare regions of the Moon, concentric to the centers of the basins—the compressional ridges described in the Basin Formation section above. On Mars, wrinkle ridges concentric with the Tharsis region extend over half the planet, some lying more than 2000 km away from Tharsis itself.

Olympus Mons, located on the edge of the Tharsis bulge, is not surrounded by concentric wrinkle ridges. This implies that the lithosphere under this volcano was already quite thick before the eruption occurred. However, within the caldera of Olympus Mons one does find a complex set of concentric fractures, probably related to the collapse of a magma chamber within the body of the volcano.

Though thick and relatively immobile at present, the Martian lithosphere may have been more active in the past. For instance, the Gordii Dorsum escarpment southwest of Tharsis contains possible anticlinal folding on the surface, with offsets in the folds suggestive of strike-slip faulting.

## SUMMARY

A load on a planet's surface will force the rigid lithosphere to sag into the more fluid mantle of a planet, producing extensional stresses and cracks around this feature.

Mars is intermediate in size between the Moon/Mercury airless planets and the Venus/Earth heavy atmosphere ones, and its internal and surface evolution also falls somewhere between them. Roughly half the surface is very old cratered plains, similar to the Moon, while the other half is younger. The division between the two is called the **crustal dichotomy**.

Mars probably started out hot, with an actively convecting mantle that produced major tectonic features on the surface. These features include the Tharsis bulge and its associated volcanoes, the great rift valley of Valles Marineris, radial graben, and concentric wrinkle ridges. Mars cooled and its crust thickened, however, and it does not seem to have active tectonics today.

## STUDY QUESTIONS

1. Name some geological feature, other than volcanoes, that could produce stress cracks.
2. What feature on Mars appears to be similar to the giant impact basins on Mercury and the Moon?
3. Give one piece of evidence that Mars is differentiated.
4. In which hemisphere are most of the large shield volcanoes on Mars located?

## 7.3 SURFACE PROCESSES ON MARS

Mars is an active planet. Although there does not seem to be life there, there are weather and seasons, dust storms and changing polar caps. Compared with a place like the Moon, Mars is warm and alive, and compared to a place like Venus, Mars is downright hospitable. The surface of Mars has undergone many changes since the planet first formed. The tectonic changes were discussed in previous sections. In this section we discuss the more subtle, but still quite important, modifications of the Martian surface.

### Mars: The Red Planet

The Martian surface is covered with red dust. In many places this dust is so thick that the rocks beneath are not visible at all. In other places there is only a thin coating, through which we can see the rocks.

How do we tell the difference between the dust and the rocks? This sounds like a simple question when you first hear it, but it isn't quite that simple. From a great distance, for instance viewing Mars through an earthbound telescope, a flat lava plain or a dust-filled valley might look very similar.

However, the dust and the rocks are different colors, and this means that they have different reflectance spectra. (For more about reflectance spectra, see the box in Chapter 4.) It is possible to tell something about the composition of a material by its spectrum.

When we look at the spectra of the red dust on Mars, we can tell that the red color comes from the presence of iron oxides. In a sense, Mars is rusty. We are not entirely sure exactly what this dust is made of, but theories include weathered basaltic glass or ash (called **palagonite**) and iron-rich clays. It helps greatly in this determination that we have some chemical analyses taken by the Viking landers. It is interesting that the chemistry of the two Viking lander sites seems to be almost identical, even though the sites were separated by over 1500 km. This could be because the global dust storms transport and mix the dust all over the planet. The Viking landers were not able to make any chemical analyses of the rocks beneath the dust, because every time they picked up what looked like a rock, it crumbled in their grasp.

### Erosion of Planetary Surfaces

After a surface has been created, after the mountains have been formed, the lava cooled, and the craters gouged out, then the slow but steady process of erosion sets in. Erosion is the process that levels the surfaces of planets, grinds down the mountains, and fills in the craters. It alters, softens, and reshapes the land. The way erosional processes act on a planet determines just how the surface of that planet looks today.

## FURTHER INFORMATION...

# Jeans Escape of the Atmosphere

One of the major changes that can happen to the chemical makeup of an atmosphere is for some component of the atmosphere to escape the planet's gravity and leave the atmosphere for the vacuum of space. How can we determine whether or not a gas will escape? Sir James Jeans presented an answer to this problem in his book *The Dynamical Theory of Gases* in 1904.

Any object can escape the gravitational pull of a planet if it is going fast enough. It doesn't matter if the object is a rocket or a molecule of gas; if it is travelling faster than the **escape velocity** it will overcome the attraction of gravity and escape to space. So, to find out if a molecule will escape, we must first estimate how fast it is travelling.

The temperature of a gas is a direct measure of how much energy it has. The energy measured by temperature is the kinetic energy of the molecules in the gas; if the gas is hotter, the molecules have more kinetic energy. Recall that the amount of energy in 1 mole of gas is $E = CT$, where $C$ is the heat capacity of the gas.

The equation for kinetic energy of any object is $E = \frac{1}{2}mv^2$, where $v$ is the speed of the object and $m$ is its mass. If a gas is made up of different sorts of molecules, all the molecules have the same temperature, so they all have the same kinetic energy. But because they have different masses, they must travel at different speeds.

Consider two gases, hydrogen and oxygen. A mole of either gas has a heat capacity of roughly 20 J/K, but a mole of $H_2$ has a mass of 2 g while a mole of $O_2$ has a mass

of 32 g. At room temperature, 300 K, a mole of either gas would have 6000 J of energy. Relate that to our equation for kinetic energy and we see that because hydrogen is $\frac{1}{16}$ the mass of oxygen, its value of $v^2$ must be 16 times greater, or $v$ itself four times greater than that of oxygen. Equating the two formulas above for $E$, we arrive at

$$v = \sqrt{\frac{2CT}{m}} \qquad (7.1)$$

From this we calculate that the speed of an $H_2$ molecule should be 2.5 km/s; the speed of an $O_2$ molecule should be 0.6 km/s.

In each planet's chapter, we give tables for the escape velocities of the various planets. We see that the Moon's escape velocity is about 2.4 km/s while that of Earth is closer to 11 km/s. It is easy to see why the Moon has no hydrogen atmosphere, but why can't it have an oxygen atmosphere? The trouble is that Equation 7.1 gives us an average velocity. There are some $10^{20}$ molecules in a liter of gas, and there's no reason why they should all be travelling at the same velocity. In fact, it turns out that their velocities are spread out in what's called a **Maxwell–Boltzmann distribution**. There's a whole range of velocities that the molecules might have, with most molecules travelling close to the average velocity but a few of them going much faster, or much slower. (The graph showing the percent of molecules travelling at any given speed is much like the bell-shaped curve that teachers use when

There are two distinct processes of erosion that we must keep in mind. One is the **chemical weathering** of rocks. This is the chemical reaction of surface rocks with the atmosphere of the planet (including water, if it is present) that weakens or dissolves rocks, making them more susceptible to the processes of **physical weathering**. Physical weathering includes the mechanical destruction of rocks by such things as meteorite impacts or thermal stresses, which act to break up the rocks into smaller pebbles, sand, or eventually dust; it also includes the transport of this material from higher regions of the planet to lower regions, which leads to the wearing down of mountains and the filling in of valleys.

## Chemical Weathering

Nature always attempts to maintain a chemical equilibrium between different chemical species. This is especially true of an atmosphere that contains mobile and potentially reactive chemicals such as carbon dioxide or oxygen acting on rocks that were formed deep inside the planet, under very different chemical circumstances than what is on the surface.

Consider a piece of granite on the surface of Earth. This rock is a collection of minerals, separate chemical species all intergrown and locked together. Two important constituents of granite are quartz, which has the chemical they "curve" a test around the average score of the class.)

This means that, at any temperature, some molecules will indeed be travelling faster than the escape velocity. Thus, at any moment, some molecules should be leaving the top of the atmosphere and escaping into space. When they leave this means that the hottest molecules in the gas are being lost, so you would expect that the average energy in the gas, which is to say the temperature, must drop.

However, the temperature of the atmosphere is fixed by how much sunlight it absorbs. That means that the molecules left over just keep getting heated up again until, once more, some of them are travelling fast enough to escape. The net result is that there will be a continual slow dribbling off of molecules from the top of a planet's atmosphere into space.

How long does it take to lose an entire atmosphere? After all, if the planet has been around for 4.5 billion years, even a very slow escape should have had an observable effect by now. It all depends on exactly how close the average speed is to the escape speed of the planet. For the escape of an atmosphere over 4.5 billion years, the general rule of thumb is that the average speed of the molecules must be greater than 20% of the escape speed.

Thus, in the case of the Moon, not only hydrogen but also oxygen should have completely escaped; oxygen molecules on average travel at 25% of the escape speed of the Moon. On the other hand, the escape of oxygen from Earth is not likely, but the escape of hydrogen is. So it's not surprising that Earth's atmosphere is lacking light gases such as hydrogen and helium while the Moon is completely airless.

You don't have to raise the temperature of an entire atmosphere to get molecules to escape, however. In the UV photochemical reactions described in Chapter 5, an energetic photon of light from the Sun hits a molecule and rips it apart. Each of the pieces has a lot of kinetic energy after this reaction.

These reactions tend to take place at the top of the atmosphere (because the air itself blocks the photons from getting very deep into the planet), and so it's quite possible for some piece of the original molecule to reach space before it has hit enough other air molecules to slow it down below escape speed. Such reactions may have been important in driving much of the nitrogen from Mars's atmosphere.

In addition, consider what happens if a water molecule gets ripped apart in this way. The hydrogen atom, even if it gets slowed down to its average speed in the atmosphere, will still be going fast enough on its own to escape relatively quickly from the atmosphere. That means that it's not likely to stay around long enough to get convected back into the deeper parts of the atmosphere, where it could recombine with the free oxygen to make water again. The result may be that much hydrogen is lost, and so water becomes very scarce in the planet's atmosphere. A process like this could be the reason why Venus no longer has much water in its atmosphere and why Mars seems to have lost much of its water.

---

formula $SiO_2$, and a feldspar such as orthoclase, with the formula $KAlSi_3O_8$. On the surface of Earth this rock can be swept by wind that contains oxygen and trace amounts of water and carbon dioxide, and it can be rained on by liquid water. The water on Earth has dissolved in it a small amount of carbon dioxide. It reacts with the water,

$$CO_2 + H_2O \rightarrow H_2CO_3 \qquad (7.2)$$

to form carbonic acid. In water, this acid is partially ionized into the species

$$H_2CO_3 \rightarrow H^+ + HCO_3^- \rightarrow 2H^+ + CO_3^{2-} \qquad (7.3)$$

What happens when this dilute acid hits the granite? The quartz is virtually impervious to the acid, but the feldspar will start to react:

$$2KAlSi_3O_8 + 2H^+ + H_2O \rightarrow$$
$$Al_2Si_2O_5(OH)_4 + 4SiO_2 + 2K^+ \qquad (7.4)$$

The feldspar gets turned into a clay mineral called kaolinite, plus quartz. The potassium is ionized and carried away by the rain water. This clay mineral is quite soft compared with feldspar, and the slightest mechanical weathering will turn what used to be solid granite into a pile of quartz crystals—sand!—and dust.

Carbonic acid is not the only acid that can drive this reaction. On Earth, acids produced by living organisms in the soil provide very strong drivers for weathering. It is also possible to see buildings with granite facings (or even worse, marble statues) that have begun to corrode away due to the acid-rich pollution that seems inevitably to accompany modern urban areas.

This sort of reaction, turning hard minerals into clays, is the most common sort of chemical weathering that occurs on Earth. By this process the basic fabric of rocks can be corrupted, allowing more mechanical processes to break them apart into boulders, pebbles, or sand. It also provides the source of those elements, especially sodium and chloride, which make the oceans salty.

This reaction depends on the presence of liquid water, and Earth is the only planet known to have liquid water present on its surface today. This sort of process could have played some role in the weathering of Mars in its past, however, and it might have some important role to play in the chemical composition of the rocky materials in icy satellites, especially if the ice was melted at any time in their histories.

## Water on Mars

There are many indications on the surface of Mars that perhaps much more water was once there than is visible today. This water may have escaped entirely from the planet's surface or it may merely be hidden beneath the surface.

If it is hidden, there are several possible places it might be. It may have reacted with the surface rocks to form clays and other hydrous minerals, and therefore be bound chemically to the surface. It may have just been absorbed onto the finest particles in the regolith, much the same way as particles of flour can clump together on Earth when they are exposed to damp conditions. Or, perhaps the most intriguing possibility, it may be present beneath the surface in the form of a layer of ice, or permafrost.

Some evidence seems to support this last possibility. On the northern plains of Mars there are many features that have puzzled scientists ever since they were first seen in the Viking photographs (Figure 7.20): polygonal networks of ridges and troughs, plains covered with small hills and ridges (called **patterned ground**), and chains of ridges that look, from above, like a human thumbprint. Some of these features bear strong resemblances to features seen in arctic regions on Earth, where they are formed by the action of ice in the soil. However, like the Martian volcanoes, on Mars these features are much larger than they are on Earth, and nobody knows quite why.

## Mechanical Weathering

Once chemical weathering has softened a rock, any of a number of mechanical processes can act upon it to complete its breakup. Even in the absence of chemical weathering, these mechanisms will have an effect on the sizes and shapes of rocks.

The most widespread mechanical weathering effect in the solar system is cratering. Not only do large impact events shatter the rocks they hit, but micrometeorites con-

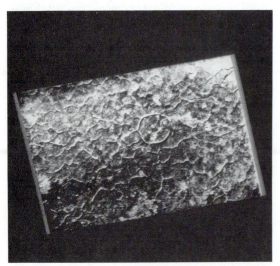
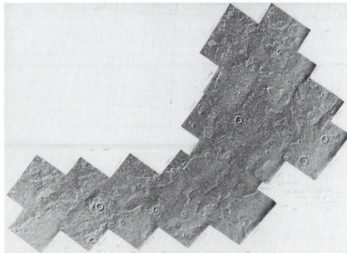

**FIGURE 7.20**   Possible permafrost features on Mars. On the left is an example of polygonally fractured ground; the image is 53 km across. On the right is a mosaic of a region of patterned ground; the image is nearly 600 km wide.

stantly bombard the Moon and Mercury and all the other airless moons and asteroids. The effect of these impacts, large and small, is to rub down any sharp surfaces on these planets into the soft, rounded mountains typical of the Moon and to **garden** the surface of the planet by grinding and overturning the rock into a compacted soil called a **regolith**. The regolith of the Moon is estimated to be tens of meters deep. Smaller bodies, such as the asteroids, should have shallower regoliths.

Another mechanical process breaking up rocks is thermal stress. As materials heat up and cool off, they tend to expand and contract. But just as planets crack when they expand, so do individual rocks crack. This effect is especially important on Earth where water can seep into cracks in rocks. Water has the unusual property that it expands when it freezes. During summer, liquid water can work its way into the cracks in rocks; then in the wintertime when the water freezes it will force the cracks apart, breaking up the rocks.

On rocky planets with atmospheres, wind can also act as a weathering agent. Small particles of rock, transported by the wind, strike larger rocks, breaking off more small particles. If there is running water available, it can also carry small rocks that impact other rocks and fracture them.

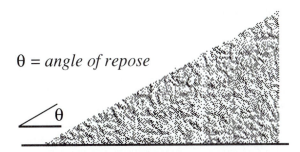

$\theta = angle\ of\ repose$

**FIGURE 7.21**  The angle of repose. Material piled at any steeper angle will slide down until this angle is reached. The precise angle of repose depends on the material involved and the local acceleration of gravity.

## Transport of Weathered Rocks

Once formed, the debris of weathering will tend to move about on a planet's surface, altering the shape of the features where it is found.

**Landslides** are the simplest form of motion. Loose solid material tends to lie piled up with sloping sides up to a certain angle, called the **angle of repose** (Figure 7.21). If the slope of such a pile is greater than this angle, the material will slide downhill until this angle is reestablished. The precise angle of repose will be that angle where the force of gravity pulling material downwards is matched by the friction of the individual particles sliding against each other.

A landslide can occur either due to a sudden change in slope (such as a crater suddenly being formed) or by a change in the frictional properties of the rocks (for example, if chemical weathering makes the particle size smaller or if liquid water lubricates the pile). Landslides can also be triggered by seismic activity, such as earth- (or Mars-) quakes. Some Martian landslides may be seen in Figure 7.22.

**Rivers** are probably the most common means of moving mass about on Earth. Rivers also seem to have played a role in shaping Mars at some time in its past.

A gently flowing stream will result in little erosion. The energy contained in the the motion of the fluid isn't enough to carry away boulders. Such flow is called **laminar**. However, the **turbulent** flow of a fast-flowing river is just the opposite. The energy in the motion of the water is sufficient to pick up and carry silt and to move boulders.

The amount of material that can be carried by a river will depend on both the size of the material and how fast the river is moving. The rate at which water moves down a river is called its **discharge rate**. A small stream may move less than 1 m$^3$ of water in a second, while the Mississippi

**FIGURE 7.22**  Several landslides can be seen in this *Viking* photo of Valles Marineris. The area shown is 70 km by 150 km.

has a discharge rate of 20,000 m³/s, and three times more than this during floods.

Even after a river has stopped flowing, the evidence of its stream bed will mark the surface of the planet. Such stream beds are common on Mars. There are several ways to tell a dried-up riverbed from a tectonic crack or a collapsed lava channel. In river networks, many small streams in the higher elevations merge into a few larger branches that eventually form a large river. This is called a **dendritic** pattern, after the Greek word for tree, because these channels spread out like the branches of a tree. And, at the downstream end of these channels, they tend to spread out into a characteristic delta-shaped alluvial fan.

**Wind** can also carry sediments, in many ways similar to the transport of silt by rivers. The processes of **suspension**, **saltation**, and **creep** that are active in running water also apply (see the box). However, because air is less dense and less viscous than water, the particles that can be carried by wind are much smaller.

On the other hand, winds are not confined to narrow riverbeds. The effect of wind, blowing dust over a large surface, will be to wear down mountains everywhere in its path and, when the conditions are right, it will build up sand dunes.

There are three major types of sand dunes (Figure 7.23). A **barchan** is a crescent-shaped dune of sand, standing alone. It is formed when the wind blows a limited amount of sand over a hard surface. The tails of the crescent are pointed away from the prevailing wind direction. With a sand-covered surface, long rows of barchan-shaped dunes will form together to make **transverse dunes**. These often run at an angle to the prevailing wind. If there is a prevailing wind from one general direction and only a moderate amount of sand, the dunes may line up parallel to the direction of the wind. Such dunes are called **longitudinal dunes**.

**Ice** can also transport debris and soil. Glaciers dig steep valleys in Earth mountains, flowing downhill and dragging loose boulders with them. The viscosity of ice is markedly larger than that of water, so that much larger stones can be carried. If a glacier melts or retreats, it will leave behind a large pile of glacial rubble called a **moraine**. Such features have been found so far only on Earth, although some scientists believe that they have seen evidence for them on Mars as well.

## A Warm, Wet Mars?

As was mentioned in Section 7.1, the presence of many types of channels on Mars is evidence that once water may have flowed upon its surface. Today, Mars is far too cold for liquid water to exist. Even water ice quickly sublimes away into the atmosphere. Could Mars once have had a climate more like Earth?

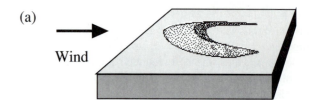
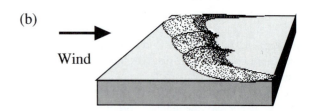
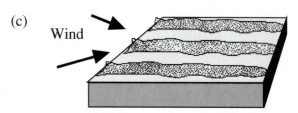

**FIGURE 7.23** Three different types of sand dunes: (a) a barchan dune, (b) transverse dunes, and (c) longitudinal dunes.

According to theories of stellar evolution, the early Sun should have put out 25–30% less energy than it does today, so one might expect that Mars (and the other planets) were significantly cooler several billion years ago. There is no evidence that the early Earth was 25% cooler then, but it is possible that the carbon dioxide content of the early Earth atmosphere was great enough to produce a greenhouse effect that made up for the weaker Sun. Having a weaker Sun in the past increases the difficulty in understanding how early Mars could have been warm.

On the other hand, one would expect that the interior of Mars was warmer in the past, before the abundance of radioactive nuclides had decayed to their present level (see the discussion in Chapter 5), allowing a greater amount of heat to flow out and melt subsurface water and carbon dioxide ice.

The most likely cause for an early warm Mars, however, is the greenhouse effect. Solar insolation can only warm Mars if there is a way to keep the heat in the atmosphere. If an atmosphere is rich in a greenhouse gas then visible light from the Sun warms the planet's surface and the infrared light emitted by the warm surface is absorbed by the gas before it can radiate into space. For a warm, wet early Mars, we need at least a carbon dioxide pressure of several times $10^5$ Pa, about 100 times the current pressure.

But even this might not have been enough. If you tried to put enough carbon dioxide to cause a significant greenhouse effect into Mars's atmosphere it would immediately condense out. The conclusion is that another greenhouse gas in addition to carbon dioxide must have been present in the early Mars atmosphere. Methane and ammonia have been suggested, but these are easily broken up by solar UV radiation and would not last long.

If Mars did have a dense carbon dioxide atmosphere at one time, what happened to it?

Perhaps, as happened on Earth, most of the carbon dioxide went into forming carbonate rocks. Today on Earth carbonates are formed primarily as the shells of sea creatures; however, before life developed on Earth carbonates were formed in the ocean by inorganic reactions. There is some spectroscopic evidence consistent with the presence of carbonates on Mars and certain landforms on the surface of Mars are similar to carbonate landforms on Earth, but the evidence is by no means conclusive. Even if there are carbonate deposits, there is no way to tell without going there just how widespread or how thick they are.

Carbon dioxide might also be physically bound to the surface. The polar caps could be holding an amount equal to about one quarter of the carbon dioxide seen in the atmosphere today. Water and carbon dioxide make a **clathrate** structure, where $CO_2$ molecules fit themselves into spaces between the $H_2O$ molecules in the loose framework structure of water ice crystals. If there is a large amount of water in a permafrost layer, a significant amount of carbon dioxide may be residing there, too.

The third possibility is that the carbon dioxide escaped to space. Jeans escape will not suffice; Mars is too cold and $CO_2$ too heavy. An alternate method is **hydrodynamic escape**: If the atmosphere is warm enough hydrogen can flow off a planet directly rather than slowly dribble off by Jeans escape, and such a flow can carry heavier gases with it.

Another suggestion is that large impacts could have blasted off much of the atmosphere. Certainly the crater record (and large features such as the Hellas basin) indicate that large impacts did occur early in Mars's history, but in detail this process is difficult to model mathematically to see if the numbers really work or not.

The final picture is still very unclear. Our best guess suggests that Mars started with an atmosphere rich in carbon dioxide and other greenhouse gases, nitrogen, and water. As the other greenhouse gases were destroyed by solar UV energy, or the carbon dioxide formed carbonates, or carbon dioxide was lost in the outflow of hydrogen, or an impact removed a significant part of the atmosphere, the greenhouse became less effective and the temperature dropped. Water ice forming a thick permafrost layer could have hidden even more carbon dioxide. Volcanic activity later in Mars's history may have led to additional episodes of a warmer, wetter Mars, but the continual escape of gas from the top of the atmosphere and the ever-cooling interior of Mars have combined to produce the thin, cold, dry atmosphere seen today.

## SUMMARY

In views from Earth, from orbiting spacecraft, and from the Viking landers on the ground, Mars looks red. It is covered with red dust, red from the iron oxides present in it. The dust is likely to be made of palagonite (a sort of weathered basalt), iron-rich clays, or both.

Erosion is the final force that shapes the surfaces of planets. Chemical reactions with liquid and atmosphere can break apart rocks. Impacts and thermal stresses are two mechanical processes that will also turn solid rocks into small pieces. Once rocks are broken up, the force of gravity will act in a number of ways to remove material from higher elevations and fill in lower areas. Landslides cause slumping of material, once the local angle of repose is exceeded. Rivers, winds, and glaciers can carry small debris over great distances, forming alluvial fans, sand dunes, or moraines when the material is deposited.

The channels on Mars suggest that its early atmosphere was once warmer and wetter than today. The original atmosphere may have had roughly 100 times the present abundances of carbon dioxide, nitrogen, and water. Although much water and nitrogen may have been destroyed by energetic solar ultraviolet light and lost from the planet, more may still be present in the soil and deep in permafrost layers beneath the surface. Most of the carbon dioxide might be tied up in carbonate rocks or frozen with the water permafrost as a clathrate ice, or it may have been lost by hydrodynamic escape or as the result of a large impact early in Mars's history.

## STUDY QUESTIONS

1. What is the red dust of Mars made of?
2. How do we know what the dust of Mars is made of?
3. What erosional process leaves behind a dendritic pattern?
4. What is a barchan?
5. What happens to granite when it is weathered by acid rain?

**FURTHER INFORMATION...**

# Suspension, Saltation, and Creep

There are three different ways that a river carries material, depending on the size of the grains of silt (Figure 7.24). If the silt is small, then the rate it settles out of the water will be slower than the rate at which turbulent eddies pick it up off the bottom of the stream bed, and it will be in **suspension** in the water. As a result, this material will be carried along with the water as the water flows downstream. The **settling velocity** depends on the size and the density of the particles; light, small fragments settle more slowly and so are less likely to settle out of the water. (This is the principle behind panning for gold. The lighter particles of dirt are more easily washed away by running water than the denser flakes of gold.) The settling velocity also depends on the force of gravity; thus, on a planet like Mars, a river could carry proportionally larger grains of silt than the same river could on Earth.

Larger grains of sand will be alternately picked up by turbulent eddies and then settled back down onto the stream bed. This process is called **saltation**. The average speed at which the material is dragged down the river in this manner depends on the particle size, the density, and the force of gravity. Again, the effect will be more important on Mars than on Earth, because Mars's lower gravity allows bigger particles to be carried this way.

Finally, this saltating sand will continually hit other, larger chunks of rock sitting on the river bottom. Each time

they're hit, they'll tend to move a little bit farther downstream. This process is called **sediment creep**.

It's not only rivers that can transport material in these three ways. Wind action also moves dust via suspension, saltation, and creep. The settling velocity is larger than in an atmosphere—material falls through air faster than it falls through water—and so only very small grains can actually be suspended in an atmosphere. However, as mentioned in the text, during seasonal dust storms a very fine dust is suspended in the thin Martian atmosphere and gets carried for thousands of kilometers about the planet.

How fast does a stream of air or other fluid have to be moving before material can be moved downstream? This "critical velocity," $v_{crit}$, can be estimated by the formula

$$v_{crit} \propto \left( \frac{\rho_f}{gd(\rho_p - \rho_f)} \right)^2 \qquad (7.5)$$

where $\rho_f$ is the density of the fluid, $g$ is the local acceleration of gravity, $d$ is the diameter of the particle and $\rho_p$ its density. From the formula, one can see that particles with smaller size or density can be carried by a fluid moving at a much lower speed than larger or more dense grains.

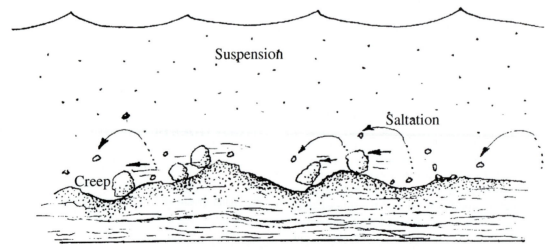

**FIGURE 7.24** Transport processes in rivers.

**FURTHER INFORMATION...**

# Isotopes in the Martian Atmosphere

*Viking* lander and groundbased spectroscopic observations have given us detailed elemental and isotopic abundances for the Martian atmosphere. From these we can compare the relative abundances of isotopes there with terrestrial values. Three elements in particular tell a complex story: nitrogen, hydrogen, and oxygen.

The ratio of $^{15}N/\ ^{14}N$ observed by *Viking* is 1.6 times the terrestrial value, implying that Mars is either enriched in $^{15}N$ or depleted in $^{14}N$. How is this possible? If energetic ultraviolet (EUV) radiation from the Sun breaks up $N_2$ molecules at top of the Martian atmosphere, the lighter $^{14}N$ is *slightly* more likely to be able to escape than the heavier $^{15}N$. It's a tiny effect, so to achieve the observed $^{15}N/\ ^{14}N$ ratio given present-day conditions it is estimated that Mars would have needed an original $N_2$ pressure of anywhere from 130 to 3000 Pa (0.0013 to 0.03 bars). The present pressure of $N_2$ is only 13 Pa (0.00013 bars), so Mars would have to have lost at least 90% of the nitrogen originally in the atmosphere to account for the shift in isotope abundance.

This has implications for the rest of the Martian atmosphere. The ratio of outgassed carbon to nitrogen on Earth is 24. If the ratio for Mars is similar, then the $N_2$ abundances derived above imply an original pressure between $7.8 \times 10^3$ Pa and $1.8 \times 10^5$ Pa (nearly 2 Earth atmospheres) of carbon dioxide. There are all sorts of uncertainties in this calculation, however. Nitrogen might be fixed in the soil rather than residing in the atmosphere; four billion years ago the Sun may have put out much more EUV radiation, so that Mars's nitrogen might have escaped faster; alternatively, a thicker carbon dioxide atmosphere would have tended to reduce the efficiency of EUV breakup, reducing the escape rate.

The difference between terrestrial and Martian hydrogen isotopic abundances is even more dramatic. The deuterium (D, or $^2H$) to hydrogen ratio on Mars is five times the value seen on Earth. Because meteorites and comets have the same D/H ratio as Earth we can assume that Mars started out with the terrestrial value too, and so it must have lost a lot of hydrogen. Indeed, the Martian D/H ratio implies that 99% of the hydrogen originally on Mars has escaped. Hydrogen on Mars occurs mostly in water, $H_2O$; thus we conclude that 99% of the water originally in the Martian atmosphere has escaped. Based on the present water abundance observed on Mars, this suggests that originally Mars had enough water to cover it with an ocean 3 m deep.

However, assuming Martian volcanoes emit water at a rate similar to Earth volcanoes, the volcanoes seen on Mars should have erupted much more water than this, enough to make an ocean 50 m deep. And some calculations of the erosion of the dendritic and outflow channels due to running water and flowing ice imply that there should have been even more water on Mars; the equivalent depth over the planet based on geological estimates ranges from 0.5 to 1 km.

Furthermore, losing so much hydrogen means that lots of oxygen, $O_2$, is released from water as well. Where did the oxygen go? Judging from oxygen isotopes, it did not escape. Viking found the ratio of $^{18}O/^{16}O$ at the surface of Mars to be the same as the terrestrial value, and Earth-based observations of Martian clouds imply that, if anything, the lighter isotope is enriched, not depleted. How could this be? Shouldn't the same processes that enrich hydrogen and nitrogen operate on oxygen? The answer apparently is that most of the oxygen in the atmosphere is not in the form of water, but in the form of carbon dioxide, $CO_2$. This implies that the original $CO_2$ abundance in the atmosphere was very large, at least $6 \times 10^4$ Pa (0.6 bars).

Thus the isotopic evidence alone gives a wide range of possible initial atmospheric pressures for Mars. There could have been enough carbon dioxide to have a significant greenhouse effect; depending on the calculation, it's estimated that as little as 1 to 5 Earth atmospheres of carbon dioxide may have been enough to warm the surface of Mars above the melting point of water.

## 7.4  PROBLEMS

1. What location on Mars has a climate closest to comfort for humans? What advantages would a colony of people on Mars have that a colony on the Moon would lack?

2. Olympus Mons, a shield volcano on Mars, is the highest mountain in the solar system. It rises 26 km above the surface and extends roughly 600 km in diameter. Assuming that this mountain rises steadily over its entire diameter, how steep are the sides of this mountain? What might this tell you about the viscosity of the lava from which it was made? Would this be a difficult mountain to climb?

3. Give a rough estimate of the volume and mass of Olympus Mons, stating whatever simplifying assumptions you've made.

4. Compare the acceleration of gravity on the surface of Mars to that on Mercury. They are almost the same, even though Mars is much bigger. How can this be? What does this tell you about the bulk composition of Mars, compared with Mercury?

5. Some scientists have proposed that the lowlands of Mars were formed when a giant meteor impacted the surface of the planet. How would you go about proving or disproving such an idea? Hints: What shape of a hole would such an impact be likely to leave? What other effects might this impact have? How long ago might this impact have happened?

6. Other scientists have suggested that the lowlands of Mars were at one time covered, or partially covered, with water. What sort of evidence would such a large body of water leave? Would you want to go sunbathing on a beach on Mars (assuming that it was warm enough)?

7. Judging from the age of the terrain where the dendritic channels are found, was there running water in Mars's recent past or distant past? Explain your logic.

8. In terms of the three types of Martian channels described in the text, which description best describes the Grand Canyon in Arizona? Why?

9. Martian astronomers (little green women?) observing Earth through a telescope have discovered that most of its surface is smooth and dark blue. A great argument has resulted: are these features vast plains of a strange blue lava or basins filled with an odd blue dust? Some radical astronomers have even proposed that they may be oceans of liquid $H_2O$! Short of sending a spaceprobe to Earth, how can the Martians resolve their dilemma?

10. Three types of surface alteration processes are chemical reactions, impacts, and thermal stresses. How important will each of these processes be on Mercury? Venus? Mars?

11. Three types of transport include landslides, winds, and glaciers. How important will each of these processes be on Mercury? Venus? Mars?

12. What process erodes a mountain the fastest on Mercury? Venus? Mars?

13. Which weathering processes discussed in this book are *not* important on Earth?

14. Which transport processes may have occurred on Mars? Which of them are still active today?

15. Assume that Mars is made up of rocks, density 3.5 $g/cm^3$, and a mixture of iron sulfide and nickel sulfide whose compressed density is 5.5 $g/cm^3$. What proportions of each component must be present to match the observed density of Mars? Given these proportions, assume that all the iron sulfide is in Mars's core. How big a core would this make? For extra credit, calculate the moment of inertia of a body with such a core and mantle, and compare it against the observed moment of inertia of Mars.

## 7.5  FOR FURTHER READING

*Mars*, edited by H. H. Kieffer, B. M. Jakosky, S. W. Snyder, and M. S. Matthews (Tucson: University of Arizona Press, 1992) is a massive tome (over 1500 pages) of research and review papers that thoroughly covers all aspects of Martian science. Two less-intimidating books that can serve as good starting points for studies of the Martian surface are M. H. Carr's *The Surface of Mars* (New Haven: Yale University Press, 1981) and Victor R. Baker's *The Channels of Mars* (Austin: University of Texas Press, 1982).

There also have been four International Conferences on Mars, since the first data were returned from the *Viking* spacecraft in 1976. The proceedings from these conferences were collected in special issues of the *Journal of Geophysical Research* (volume 82, pp. 3959–4681, 1977; volume 84, pp. 7909–8519, 1979; volume 87, pp. 9715–10305, 1982; volume 95, pp. 14089–852, 1990).

For further references on geophysics and geomorphology, see the reading list for Venus in Chapter 5. Along with those books, Ronald Greeley's *Planetary Landscapes* (Boston: Allen and Unwin, 1987) is beautifully written and illustrated, and the NASA special publication SP-469, *The Geology of the Terrestrial Planets* by M. H. Carr, R. S. Saunders, R. G. Strom, and D. E. Wilhelms, is a good introduction to geology from a planetary point of view.

The classic advanced text on the transport of dust and sand by wind action is R. A. Bagnold's *The Physics of Blown Sand and Desert Dunes* (London: Methuen, 1941). Ralph Bagnold was a British Army officer stationed in

Egypt in the 1930s who supplemented his careful observations of sand dunes with wind tunnel experiments. The modern extension of this work to situations on other planets is R. Greeley and J. D. Iverson, *Wind as a Geological Process on Earth, Mars, Venus, and Titan* (Cambridge: Cambridge University Press, 1985).

**FIGURE 8.1** This view of Earth from *Apollo 8* was important in creating humanity's present view of itself in relation to the universe.

# CHAPTER 8

# Moving Our Place

## 8.1 FINDING THE CENTER

"The paralysis of mind brought about by the rigid authoritarian tradition of the Dark Ages," Albert Einstein once wrote, produced "a naïve picture of the earth as a flat disk, combined with obscure ideas about star-filled space and the motions of the celestial bodies, prevalent in the early Middle Ages." Such a picture, he asserted, was the product of the "barren, primitive mentality... a petrified and barren system of ideas." ubiquitous in Europe at the time.

For all his genius, Einstein was no historian of science. Of course, medieval science was neither rigid, barren, nor obscure. Copernicus, Galileo, and Newton fit into a long-developing trend in scientific thought; to view them as lonely revolutionaries of science, as if they were twentieth-century men trapped in some alien world, does violence to the facts. It insults the work of the medieval and Renaissance scientific community.

It also misses a point of their true significance. The methods and world views of men such as Nicholas Copernicus or Johannes Kepler or Tycho Brahe do not contrast with the earlier scientists' nearly as dramatically as they do with the nonscientists' cosmology of the age, a world view of pervasive importance to the whole culture. Galileo made major contributions to our understanding of how na-

ture works; but perhaps even more, his achievement lay in bringing the fruits of a long-developing tradition of science to the attention of the nonscientific world.

### Before 1500

The idea of "two cultures," one of science and one of humanities, separate and rarely communicating, was articulated by C. P. Snow in the mid twentieth century. But it describes a state of affairs that has existed to some degree throughout human history. The late Middle Ages were no exception.

Scientists in the medieval centuries began with the ancient Greek and Roman teachings, but this was hardly a rigid authoritarian tradition. Even among the ancients, there were several different cosmologies. As we saw in Chapter 1, several Greek astronomers had developed the idea of a heliocentric universe. Aristarchus based his theory on observation and geometry. Other Greeks, including the mystical followers of Pythagoras, put the Sun at the center of the universe for purely philosophical reasons, and their estimates of the distances to the planets—wildly inaccurate—were based on "magic ratios" of numbers. They did not describe the universe as it was, but as they felt it ought to be.

Another scientific picture, one more widely held in the popular culture, was that of the geocentric universe. For instance, the Roman statesman and author Cicero, in his *Somnium Scipionis (The Dream of Scipio)*, described a dream where he imagined the great Roman general Scipio passing from Earth, through a series of concentric celestial spheres, to heaven. This literary device was loosely based on the cosmology of the fourth century B.C. Greek philosopher Eudoxus, who pictured each planet as a point of light encased in a transparent sphere that spun about Earth. Eudoxus postulated several "spheres within spheres" for each planet to account for that planet's observed complex motion, but this detail was ignored by the poets. The idea of the planets held in place by concentric crystalline spheres was too romantic an image to miss; it was developed and used by a long line of classical and medieval authors, from Lucian and Ovid to Chaucer and Dante.

The structure of the universe in these literary works served a metaphorical purpose, often to contrast the tiny and corruptible Earth with the immortal and unchanging heavens. However, there were also several good scientific reasons to believe that Earth, not the Sun, was the fixed center of the stars and planets.

*If Earth moves, why don't we feel its motion?* An arrow shot eastward travels just as far as an arrow shot westward; if Earth were spinning, west to east, you'd expect the surface of Earth to move underneath the eastward arrow and move away from the westward one, making the travel distances of the arrows very different.

*If Earth travelled around the Sun, while the stars stayed fixed, wouldn't we expect to see* **parallax***?* A star that is over the north pole in summertime should appear to be moved lower in the sky as Earth travels to its winter position (Figure 8.2). No such motion could be observed.

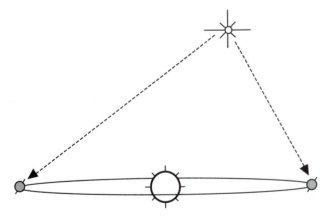

**FIGURE 8.2** If Earth moves around the Sun, shouldn't stars that are near the north pole at one point in Earth's orbit appear somewhat away from the polar direction at the opposite point in the orbit?

*If the planets moved around the Sun, not Earth, shouldn't they look different to us as they move closer to us, then farther away?* We would expect Mars to look much bigger when it is on our side of the Sun and much smaller when it is on the far side of the Sun. However, although Mars changes brightness, the naked eye can see no such change in its size. In addition, one would expect Venus and Mercury to show phases as they orbit the Sun, like the Moon does (Figure 8.3). Again, the naked eye can see no such phases.

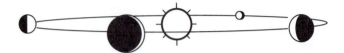

**FIGURE 8.3** If Venus orbits the sun, we would expect it to show phases, like the Moon does. Galileo finally saw this with his telescope in 1610.

Besides these observational problems, there were also theoretical difficulties. What force could possibly move something as big as Earth? And, assuming that space was filled with an **ether** (the supposed medium carrying light and heat from the Sun to us), wouldn't the drag of this æther tend to slow Earth down? Indeed, "nature abhors a vacuum," but if Earth ever tried to move from its spot, argued the philosophers, it would leave behind a vacuum in the æther that would suck Earth back to its original position.

A primary proponent for the geocentric universe was Aristotle, who outlined his version of eternal circular motion in his *Physics* and *De Caelo (On the Heavens)*. In the second century A.D., Ptolemy combined Aristotle's ideas with a wealth of Greek observational data in the *Almagest*.

But other cosmologies abounded, as well. For instance, the ninth century A.D. Irish philosopher John Eriugena believed that all the planets except Saturn orbited the Sun, which in turn circled Earth. (In his scheme, Saturn and the stars orbited Earth directly.) Eriugena was fluent in Greek, and from his writings it is clear that he was familiar with the work of Eratosthenes and other Greek scholars. However, he asserted (much like the Pythagoreans) that the ratio of distances from Earth to Moon and the other planets were simple small ratios of Earth's diameter; instead of dividing the circumference by $\pi$ to find the diameter, a value known by the Greeks, he simply divided the circumference by two! Clearly, the general structure of the universe was of more interest to him than any mathematical precision.

The debate concerned not only the heliocentric versus geocentric world views, but also the question of whether the daily motions of the stars and the Sun could be better explained by a starry sphere that spun around a stationary Earth once every 24 hours or a stationary sphere enclosing a spinning Earth. This argument was not confined to the West; for instance, a spirited debate occurred in Hindu astronomy in the sixth century A.D. between the schools of Brahmagupta, who had an essentially Ptolemaic view of the universe, and Aryabhata, who argued that Earth spun on its axis.

But perhaps the most curious argument concerning the spin of Earth was found in Bishop Nicholas Oresme's *Livre du Ceil et du Monde (Book of Heaven and Earth)*, written in 1377. In it he showed that no argument from reason alone could rule out the possibility of a spinning Earth. He pointed out (anticipating Einstein) that all motions are relative and that, if the eastward and westward arrows mentioned above also had a "natural circular motion" along with the spinning Earth, the arguments against spin would be answered. (This came close to anticipating Newton's concept of inertia.)

However, after all these powerful arguments, Oresme concluded his chapter by stating, "Nevertheless, everyone holds that the heavens move, not Earth; and so do I." The point of his exercise was to show the limits of reason: Rational thought, for all its power, could not even prove something as intuitively obvious as that Earth stood still!

## The Geocentric Theory

The great goal for ancient and medieval astronomy was to find a mathematically simple but reliable way of calculating phases of the moon, eclipses, and the positions of the planets. Various methods were accepted or rejected based on their ability to "preserve the appearances." To the scientist, this had a straightforward meaning: Did the calculation work? The philosophers, however, read into this phrase an implication that the astronomers did not see: Just because a system "preserved appearances," they realized, did not mean that it necessarily described, uniquely, the true situation of the orientation of the planets.

The scientific problem was this: If one plots the positions of a planet against the background stars, one finds that their general motion is west to east. If Mars is found in Gemini one month, night by night it can be found closer and closer to the constellation Cancer, to the east. But the rate at which it moves is by no means regular. Instead, it speeds up and slows down in a way very difficult to predict. Even worse, at times it will even appear to move "backwards" through the constellation (Figure 8.4).

The problem of calculating this motion was solved to some degree by Ptolemy and other geocentrists. Although Ptolemy wrote in the second century A.D., his work was first translated into Latin (from Arabic sources) in the late twelfth century. He started with two assumptions. The first one, that the observer was situated on a fixed Earth at the center of the system, is a natural starting position if all you are interested in is predicting the observations that the observer will see. It is still the starting assumption today for **celestial navigation**, finding one's position at sea or in the air based on the apparent position of the stars.

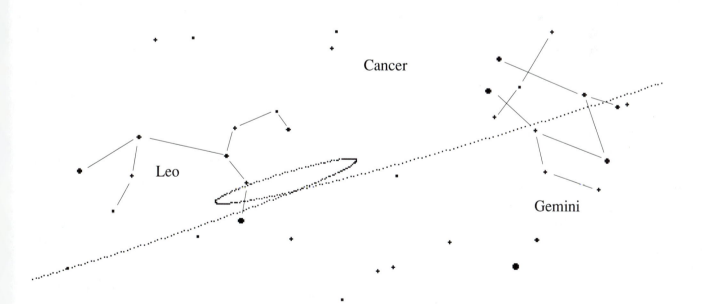

**FIGURE 8.4** The motion of planets against the background stars provided the most severe test for any theory of planetary motion. The position of Mars is plotted here over a period of nine months, October 1994 to August 1995, moving right to left against the background constellations Gemini, Cancer, and Leo.

The second assumption was based on a philosophical point. The ancients reasoned that all motion could be thought of as a mixture of straight-line and circular movements. Aristotle and other Greek philosophers differentiated between finite and eternal motion: Just as the gods lived forever while humans were born and died, they noticed that circles had no beginning or end while straight-line motion on Earth has a natural starting point and ending point. Thus it seemed perfectly reasonable for them to argue that the heavens, which seemed perfect and unchanging, must move in circular motion. (Copernicus and Galileo followed this argument as well.)

The first approximation for Ptolemy, therefore, was that the planets moved around Earth in perfect circles. This, however, failed to reproduce all the observed motions of the planets, mentioned above. He therefore postulated that this circular orbit actually marked the path of the center of a smaller circle, called an epicycle (Figure 8.5). By adding epicycles within epicycles, eventually a very close approximation to the observed paths of the planets could be obtained. (Ptolemy used up to 16 epicycles, circles within circles. In fact, we now understand that it takes an infinite number of circles to represent an arbitrary curve; this is the principle of Fourier analysis discussed in Chapter 4.)

Even 16 epicycles failed to completely reproduce the motions of the planets, however. Thus Ptolemy hypothesized that some of the planets' paths were "eccentric circles," that is, circles that were not centered on Earth but on some other point nearby.

This model worked reasonably well against observation, or, at least, Ptolemy claimed it did. (A modern analysis of his putative observations has led some astronomers to conclude that he lifted many observations from earlier works and perhaps juggled some numbers to make them fit his theory.) The success of his formulation led many educated people to conclude that there was scientific support for the world view described by poets such as Ovid and Dante. Needless to say, few poets had actually read the astronomy texts (just as today most people who talk about "quantum leaps" or "the uncertainty principle" have never really studied quantum mechanics).

By contrast, the astronomers of the time considered Ptolemy's model to be little more than a calculating device. Among those scientists arguing this point of view were the neoplatonist Proclus in the fifth century (A.D.), the twelfth-century Islamic physician Ibn Rushd, and the fourteenth-century scientists Jean Buriden and Henry of Hesse. Even those thinkers less engaged in natural science, such as the twelfth-century Jewish philosopher Maimonides and Thomas Aquinas in the thirteenth century (see his *Commentary on Aristotle's Heavens*), also understood this point. Furthermore, the combination of off-center circles and numerous epicycles described a universe that was in fact quite different from the simple series of concentric shells favored by the poets.

## Copernicus

Against this background, Nicholas Copernicus attempted a grand revision of calculating planetary positions (Figure 8.6).

Copernicus was born in Poland in 1473, studied astronomy at the University of Cracow, and earned degrees in law and medicine in Italy. He spent most of his life in Frauenburg (the modern town of Frombork, Poland) as a mathematician, doctor, and cleric.

Working with older collections of planetary positions and calculating schemes, he noticed that the orbital period of the Sun around Earth figured into all the calculations for the positions of the other planets. Dissatisfied with contemporary astronomers' geocentric view of the planets, Copernicus took it upon himself to read the literature of the ancients to see what other possibilities had been thought of. He found, among other theories, Heracleides' theory (mentioned in Chapter 1) that Venus and Mercury orbited the Sun; it was quoted in the writings of Martianus Capella, a fifth-century commentator himself quoted by John Eriugena. From this work, Copernicus began to consider whether a system with Earth itself circling the Sun might not be mathematically simpler and more elegant than the "almost endless multitude of circles" (to quote Copernicus) that the geocentric system needed.

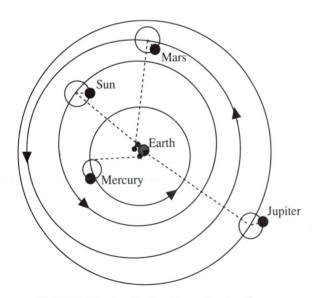

**FIGURE 8.5** In Ptolemy's epicycle theory, Earth did not move. Notice that the circles of the planets' orbits were not centered exactly at Earth nor did all the orbits share a common center. (This figure is not to scale.)

**FOR THE RECORD...**

# Precision and Accuracy

Tycho Brahe's observations had to be both precise and accurate. A **precise** measurement is one that has a small degree of measurement error. Nowadays it is represented by the number of decimal places reported in the measurement (for example, 1.245 is more precise than 1.2), and an honest scientist does not report more precision than her measuring instrument allows; if a watch measures time only to the nearest minute, the time cannot be reported to the nearest second. Brahe was the first astronomer to calculate quantitatively just how much measurement error was to be expected for each observation.

An **accurate** measurement is one that actually reflects the truth. This is quite different from precision; a watch with a second hand can be precise to the second, but if it runs slow it can be minutes away from giving an accurate time. To get accurate measurements, one must know how to anticipate and correct for systematic errors in the measurements. One of Brahe's advances was to account for the refraction of starlight near the horizon. Just as a stick in water appears to be bent, when seen at an angle, the light from stars near the horizon can be bent by the atmosphere, so the true position of the star or planet can be significantly different from its apparent position.

His manuscript outlining this theory was essentially complete by 1532, but fear of ridicule prevented him from publishing it. Indeed, as word leaked out he became an object of derision in Frauenburg; a local theater piece made fun of him and his ideas. He did write a short summary of his concept for private circulation, called *Commentariolus (Little Commentary)*. When Cardinal Nicholas von Schönberg saw this summary he offered to publish the whole work, at his expense, and present it to the pope. Still Copernicus declined. Finally, the German mathematician George Joachim Rheticus published a summary of Copernicus's ideas, *Narratio Prima (First Report)* and finally convinced Copernicus to let him publish the entire work. Dedicated to Pope Paul III, *De Revolutionibus Orbium Coelestium (On the Revolutions of the Heavenly Spheres)* finally appeared in 1543, the year Copernicus died.

Although it appeared to find favor in Rome, reaction to this theory in the Protestant world was decidedly unfavorable: It contradicted certain passages in the Psalms referring to a fixed Earth and rendered meaningless the story of Joshua commanding the Sun to stand still. When word of this system reached Martin Luther, his off-hand comment was that "the fool would overturn the whole world of astronomy." Ironically, nearly a century later the Catholic world sided against the Copernican view while the Protestants defended and advanced it.

Rheticus, himself a Lutheran, taught at the University of Wittenberg, the center of Lutheran thought. Out of fear

of local reaction, however, he arranged to have the work published in Nuremberg, not Wittenberg. Because Rheticus could not be present in Nuremberg while the manuscript was prepared, he turned the editing over to Andreas Osian-

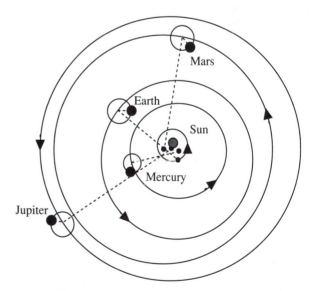

**FIGURE 8.6** The heliocentric theory of Copernicus took Ptolemaic ideas, but switched the positions of Earth and Sun. Copernicus still needed epicycles and off-centered orbits to be able to reproduce the motions of the planets accurately. (This figure is not to scale.)

**FOR THE RECORD...**

# Galileo

From the point of view of judging between the Copernican and Brahe theories, Galileo's contribution was important but not decisive. In 1610 he first trained a telescope onto the heavens and discovered that satellites orbited Jupiter, proving that objects in space can orbit a body other than Earth; he found that Venus and Mercury have phases, just as they would if they orbited the Sun; and eventually he found that Mars more than doubled in size as it travelled from behind the Sun to our part of the solar system. To his mind, these observations confirmed the Copernican system. However, as many other astronomers of his day pointed out, these observations were equally consistent with the Brahe system.

Galileo did something in the social realm, however, that altered everything. Where other Copernicans, such as William Gilbert, actively scorned popular philosophy, Galileo took it upon himself to preach the Copernican picture to the nonastronomically educated classes of Italy. He wrote his books in Italian, not Latin, and spent months at a time promoting his work at fashionable dinner parties and salons in Rome. He forced the world of poetry and literature, philosophy and theology, to confront that their world view was incorrect.

Of course, as we've seen, the world view of the general culture was never what the astronomers actually believed, but Galileo was the first to emphasize just how different these world views had become. His motives were mixed; partly driven by profit and self-glory, he also wanted to ensure in the heated theological debate raging through Europe at the time (about to break into the Thirty Years' War) that Catholicism would be on the side of scientific

truth. He wanted the Catholic church to adopt the Copernican world view officially. He had reason to believe that it would; the strict interpretation of the Bible that an earlier generation of Protestants had used to ridicule Copernicus was contrary to the tradition of Catholic thought.

Unfortunately, Galileo was talented at making enemies of the wrong people, and his books incited bitter heated debates in Italy itself. In 1616, urged on by Galileo's political enemies, Cardinal Bellarmine (the church's leading theologian at the time) instructed him that the Sacred Congregation of the Index had declared that "the doctrine attributed to Copernicus (that the Earth moves around the Sun and the Sun stands at the center of the world without moving from east to west) is contrary to Holy Scripture and therefore cannot be defended or held." In 1619, in the debate resulting from Galileo's writings, the Congregation of the Index decreed that all copies of Copernicus's work had to be "corrected" to show that his system was not a true description of the universe, but merely a useful mathematical tool to "save the appearances." However, Galileo's writings themselves were not censored at this time.

The Sacred Congregation of the Index was not the last word in the church; indeed, Cardinal Bellarmine himself had had books condemned by this congregation, which had later reversed itself. Bellarmine's own opinion was expressed in a letter to another cleric in 1615: "I say that if there were a true demonstration that the Sun is at the center of the world and the Earth in the third heaven, and that the Sun does not circle the Earth but the Earth circles the Sun, then one would have to proceed with great care

der. Just before it was printed, Osiander on his own initiative inserted an unsigned introduction explaining that the work was only intended as a mathematical device to "save the appearances." Other people reading this introduction assumed that it stated Copernicus's own intentions. If they had actually read the text of the book itself, they would have found passages such as "what appears to be a motion of the Sun is in truth a motion of Earth." (Book I, Ch. 10) that made it clear that Copernicus himself viewed his system as actually describing a real state of affairs.

This revolutionary idea still had serious scientific problems. To match the observations of planets, Coperni-

cus still had to resort to off-centered circles and epicycles (Figure 8.6). To explain why parallax was not seen, Copernicus had to postulate that the fixed stars were an immense distance away from Earth, compared to Earth–Sun distance. The other arguments against the heliocentric system, outlined above, remained unanswered. In spite of these difficulties, his system gradually was adopted by a sizable minority of scientists. (One estimate, based on numbers of seventeenth-century astronomy publications, suggests that one astronomer in 10 before Newton was a follower of Copernicus). Most notable among Copernicus's champions was William Gilbert, Queen Elizabeth's private in explaining the Scriptures that appear contrary, and say rather that we do not understand them than that what is demonstrated is false. But I will not believe that there is such a demonstration, until it is shown me."

In 1623 Galileo's friend and fellow philosopher, Cardinal Maffeo Barberini, was elected as Pope Urban VIII. The new pope was mostly preoccupied with the Thirty Years' War then raging in Europe; but on occasion he did continue his astronomical discussions with Galileo, defending the Tycho Brahe view. With his friend's tacit permission, in 1632 Galileo published, with official church permission, a discussion of the Copernican idea in the form of a dialogue between defenders of the heliocentric and geocentric views.

Unfortunately, the dialogue was extremely one-sided. The Copernican view was presented with sarcastic cleverness (even though Galileo's logic was sometimes faulty), whereas the pope's opinions were put in the mouth of a simpleton. Worse, the book appeared at a politically awkward time for Galileo's defenders. Spain, supporting the Catholic side in the war, had just suffered a serious defeat at the hands of Gustavus Aldolphus, the Swedish king leading the Protestant side. Catholic France was openly allied with the Protestants; it was rumored that the Grand Duke of Tuscany, Galileo's patron, and the pope himself secretly supported the Protestants as well. Both feared Spanish power in Italy, and the pope owed his election to France.

Partly to embarrass the grand duke and the pope, Galileo's enemies forced a trial. Ten cardinals examined the case; by a seven to three vote Galileo's book was "diligently examined and found to violate explicitly the above-mentioned injunction," noting that the Copernican system was "false and contrary to the divine and Holy Scriptures." Galileo was forced to "abjure, curse, and detest the above-mentioned errors and heresies" publicly, and he was confined to his estates outside Florence.

His house arrest did at least allow him the time to write the results of his lifetime's theorizing and experimentation about the nature of motion. His *Discourses and Mathematical Demonstrations on Two New Sciences Pertaining to Mechanics and Local Motion,* published in 1638, was without question his most important scientific work.

Neither the truth of his theory nor its compatibility with the faith was ever debated in Galileo's trial. Galileo's crime ultimately was disobedience, not disbelief; the issue constantly discussed at his trial was not if his science was correct or in conflict with the Bible, but whether or not he had violated his instructions of 1616. But the effect of the judgment against him was to tell scientists throughout Europe that the church condemned the Copernican system itself.

Galileo's agitation had provoked the exact opposite of what he had desired. However, his work opened the door for two great thinkers living in the Protestant north. They devised the laws that finally combined observation and theory into an elegant and complete picture of how the planets moved.

physician, who was most famous for his treatise on Earth's magnetism, the first description of a planetary dipole field (see Chapter 10).

## Brahe

One of the scientific motivations behind Copernicus's work was that tables of planetary positions based on the ancient Greek system did not accurately predict where those planets could be observed. The Danish astronomer Tycho Brahe, born 3 years after Copernicus's death, realized that better measurements of planetary positions were needed before any successful mathematical prediction scheme could be developed.

As the full-time astronomer to the king of Denmark, Brahe had the time and resources to build the best observing instruments of his day. The telescope had not yet been invented, so planetary positions were measured by naked-eye sighting along the diagonal of a quadrant, essentially a large protractor with angles marked around its circumference. Clearly, a larger quadrant can fit more divisions of an angle along its circumference, and so more precise measurements could be made; Brahe's largest quadrant was 19 feet in radius. With this he carefully plotted star positions, inscribing them on a celestial globe 5 feet in diameter. Thus on this globe he could follow the paths of the planets with unprecedented precision and accuracy.

Brahe was an admirer of Copernicus, calling him "a second Ptolemy...with wonderful intellectual acumen" who "restored the science of the heavenly motions in such a way that nobody before him reasoned more accurately about the movements of the heavenly bodies." However, for scientific reasons, Brahe rejected the idea of a moving Earth.

Denmark was strongly Protestant, which certainly made it impolitic to accept the Copernican view at this time; but there were important scientific arguments to consider as well. For one thing, as noted above, the lack of parallax meant that the stars must be immensely far away. However, reasoning from analogy with the Moon and planets, Brahe and most of the other astronomers of his era assumed that the stars also shone merely by reflected sunlight. Copernicus himself thought of the stars as objects embedded in a sphere of ice encasing the whole universe. For stars to be as far away as the Copernican system required, and still reflect as much light as they did, Brahe calculated that they would have to be unreasonably large. (His calculation was correct. Only his starting assumption was wrong.)

Instead, he modified the Copernican view in one essential point. Like Copernicus, he assumed that the planets orbited the Sun. However, he held that the Sun and all its satellites orbited a fixed Earth. This was essentially the same picture as John Eriugena's theory 600 years earlier but, unlike Eriugena, Brahe based his view on detailed observations and rational, geometric proofs.

Mathematically, the Copernican and Brahe models are identical, and no observation of planetary motion alone can discriminate between them. The same argument that said that Earth could as well be moving as standing still also said that it could as well be standing still as moving. And the lack of parallax argued convincingly for a stationary Earth. Fifty years later Galileo tried to use the existence of the tides on Earth as a proof of its motion; his arguments were fallacious. In 1729 the English astronomer James Bradley discovered **stellar aberration**, a slight change in the position of stars that could be attributed to the motion of Earth; however, not until 1838 was stellar parallax finally measured, and only Foucault's pendulum in 1851 finally demonstrated the spin of Earth without ambiguity.

## SUMMARY

The sixteenth century saw two major competing cosmologies, Ptolemy's stationary Earth surrounded by planets in epicentric orbits and Copernicus's Earth that itself orbited the Sun. Both faced severe observational problems. The Ptolemaic system, although somewhat capable of "preserving the appearances" of the planets' positions, was overly complex and as time went on less and less successful at predicting planetary positions accurately. The Copernican model, however, was no better; it could only work if a new physics of relative motion were developed, and it demanded that the fixed stars must be located a seemingly unreasonable distance from Earth. Tycho Brahe introduced a system where the planets orbited the Sun but the Sun orbited Earth, which seemed to get around the difficulties that the Copernican model faced. However, as we will see in the following sections, the seventeenth century brought new developments in both observations of planetary positions and the understanding of physics that validated the core idea of the Copernican system, the Sun as the center of the solar system.

## STUDY QUESTIONS

1. True or False: Before Copernicus, most astronomers thought the world was flat.

2. There were several reasonable scientific arguments against the Copernican view. Name one.

3. True or False: Even with his new system, Copernicus needed epicycles to predict correctly the positions of the planets.

4. Two scientists measure a widget with a ruler. One says that it's 5 cm in diameter, the other says that it's 5.24 cm in diameter. Is the second measurement an example of greater precision, or greater accuracy?

5. True or False: No observation available in the seventeenth century could distinguish between the Copernican view and the Brahe view of the solar system.

## 8.2   ORBITS

### Kepler's Laws

The motions of the planets were finally summarized some 70 years after Copernicus's work in three simple and elegant laws by Johannes Kepler, a student of Tycho Brahe. The first two were published in 1609 and the third in 1619. These three laws are stated in Table 8.1. Let's take a close look at each law.

**I. Each planet moves in a path shaped like an ellipse, with the Sun at a focus.** Kepler worked out his laws by trying to fit many different possible theories against actual records of the positions of the planets, measured to very high accuracy by his mentor, Tycho Brahe. Many theories involving perfect circles came close, but only ellipses

**TABLE 8.1**

Kepler's Laws and Newton's Laws

**Kepler's Laws (1609 and 1619)**

*I. Each planet moves in a path shaped like an ellipse, with the Sun at a focus.*

*II. The planets move faster when they are closer to the Sun and slower when they are farther away, so that the area of the segment of the ellipse swept out over a given period of time is constant.*

*III. The period of time it takes for a planet to complete an orbit, squared, is proportional to the mean distance between the planet and the Sun, cubed.*

**Newton's Laws (1687)**

*I. If a body is not accelerated, then its velocity stays constant.*

*II. The acceleration a body experiences is equal to the force acting on it, divided by the body's mass.*

*III. If one body exerts a force on another body, then the second body must also be exerting just as much force on the first body, but in the opposite direction.*

actually fit the data. The idea that the paths of the planets around the Sun are ellipses, and not circles, was the key advance. It was such a radical departure that even Galileo never accepted it.

What is an ellipse? To a mathematician, an ellipse is not just any flattened circle; it is a precisely defined curve. The size and shape of the ellipse are determined by the values for its **semimajor axis** $a$ (not to be confused with the acceleration **a**!) and **eccentricity** $e$. Given these quantities, the equation

$$r = \frac{a\left(1 - e^2\right)}{\left(1 + e\cos\eta\right)} \qquad (8.1)$$

is the mathematical statement of Kepler's first law. The position of the planet in its orbit is described by its distance from the Sun (the Sun is located at one focus, the other focus is empty), denoted by the radius $r$, and its angular distance around the orbit, $\eta$, called the **true anomaly**. For every angle $\eta$, the planet in orbit lies a distance $r$ away from the Sun as determined from Equation 8.1.

Note that the coordinate system used in this formula is not Cartesian $(x, y, z)$, but polar $(r, \theta, \phi)$. The focus of the ellipse is the center of the coordinate system, and the angle $\eta$ plays the role of $\theta$ in polar coordinates. The true anomaly is measured counterclockwise (looking down from the north) from the point on the ellipse closest to the Sun.

Using polar coordinates makes our formulas for velocity and acceleration more complex, but there is a powerful advantage to this choice. Kepler's first law, written in polar coordinates, says that the location of the planet can be described in terms of its position on an ellipse. The ellipse is defined by constants, such as $a$ and $e$, which don't change. If we stay in the plane of the ellipse, $\phi$ is constant. The only thing that changes is the planet's precise position

along that ellipse. If we know $\eta$, we can calculate $r$. So instead of having three independent position variables, $x$, $y$, and $z$, we now have only one, $\eta$, to worry about.

However, Equation 8.1 can be complicated to work with. Fortunately, most planets' orbits are very close to being circular. A circle is equivalent to an ellipse with an eccentricity of zero, or no flattening. In circular orbits (Figure 8.7), the semimajor axis distance $a$ becomes the same as the radius of the circle $r$, which is just the (constant) distance from the Sun to the planet. In circular orbits, the true anomaly $\eta$ is the only quantity that varies with time.

**II. The planets move faster when they are closer to the Sun and slower when they are farther away, so that the area of the segment of the ellipse swept out over a given period of time is constant.**

We want to define the motion of a planet, and to do so we need to worry about motion in two directions: the

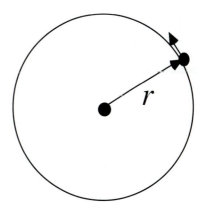

**FIGURE 8.7** Motion in a circle is simple compared with an ellipse. The radius, $r$, never changes as the planet orbits its star, and the motion of the planet is always at right angles to the radius line.

**FOR THE RECORD...**

# Ellipses

What is an ellipse and what do we mean by its focus? An **ellipse** is a specific kind of closed curve, similar to a flattened circle. The measure of how "flat" the circle is, we call the **eccentricity** of the ellipse.

One simple way to draw an ellipse is to stick two thumbtacks into a piece of paper and place a loop of string around the tacks. If you stick a pencil in the loop and pull it taut, then move the pencil all around the tacks, keeping the string taut, the pencil will draw an ellipse. The position of each thumbtack is a focus of the ellipse. Every ellipse has two foci.

The length of the ellipse (in the longest direction) is called the major axis; half the length of the major axis, the distance along the line from the center of the ellipse out to the edge, is called the **semimajor axis**. Knowing that an orbit is in the shape of an ellipse allows us to define the shape of that orbit simply by specifying the size of the semimajor axis (usually referred to as $a$) and the degree to which the path has been flattened, the **eccentricity**, or $e$.

Once the size and shape of the ellipse has been de-termined, any point on the ellipse can be uniquely deter-mined by specifying the angle from the semimajor axis to that point on the ellipse (see Figure 8.8). This angle, $\eta$, is called the **true anomaly**.

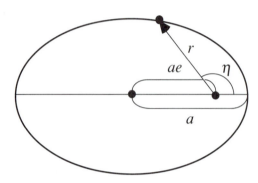

**FIGURE 8.8** An ellipse: $a$ is the semimajor axis, $e$ the eccentricity. The planet's position is denoted by distance $r$ and angle $\eta$, called the true anomaly.

speed that the planet moves around the Sun, $d\eta/dt$, and the speed at which it moves closer or farther from the Sun, $dr/dt$. Kepler's second law helps us describe these motions. However, it takes a bit of work to understand what this law is telling us.

First, what do we mean by an area "swept out" of an ellipse? Say we start at some point on the ellipse and draw a line from the Sun to the planet. A bit later on, as the planet moves to another point in its orbit, we draw another line from the Sun to the planet. The pie-shaped wedge enclosed by these two lines is the area swept out by the planet.

How do we measure this area? The infinitesimal area of a small segment of an ellipse is virtually the same as the area of a triangle. The base of our triangle is the ra-dius, $r$. To find the altitude of the triangle, recall that the distance along the orbit that a planet moves is equal to the radius times the change in true anomaly. For an infinitesi-mal change in $\eta$ this may be written as $r \cdot d\eta$:

$$dA = \tfrac{1}{2} \cdot r \cdot r \cdot d\eta \qquad (8.2)$$

as described in Figure 8.9.

Divide both sides by $dt$ and we have an equation for the amount of area $dA$ swept in a time $dt$: $dA/dt = \tfrac{1}{2}r^2 \, d\eta/dt$. (Remember that the symbol $dA$ means "a small change in $A$," so $dA/dt$ means "a small change in area $A$ that occurs during a small change in time $t$.")

The second law tells us that area swept by an orbit

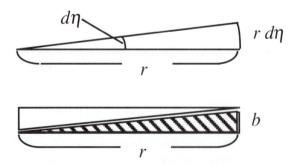

**FIGURE 8.9** A thin segment of an ellipse is equivalent to a thin triangle. The area of a tri-angle is $\tfrac{1}{2}$ (*base* × *height*); so the area of the segment of the ellipse is $\tfrac{1}{2}$ ($r \cdot rd\eta$).

over a given period of time is the same, everywhere in the orbit. Because this is a constant, we can equate $dA/dt$ for any tiny time $dt$ with the total area of an ellipse, $A$, divided by the total period of an orbit around the ellipse, $P$. The area of an ellipse is given by

$$A = \pi a^2 \sqrt{\left(1 - e^2\right)} \qquad (8.3)$$

Substituting and dividing through by $\frac{1}{2}r^2$, we find

$$\frac{d\eta}{dt} = \frac{2\pi a^2}{Pr^2} \sqrt{1 - e^2} \qquad (8.4)$$

Notice that for elliptical orbits, both $r$ and $\eta$ change with time, so this one equation alone is not enough to determine the motion of the planet in both directions. However, for orbits that are nearly circular $r$ hardly varies at all; it's practically equal to $a$. And the square-root term approaches 1 as $e$ goes to zero. Thus the speed of $\eta$ approaches $2\pi$ radians per period $P$ seconds. We call this speed, $n \equiv 2\pi/P$, the **mean motion** because it represents the average speed of the planet in its orbit. A planet in a perfectly circular orbit travels around the Sun at a constant angular speed equal to $n$.

**III. The period of time it takes for a planet to complete an orbit, squared, is proportional to the mean distance between the planet and the Sun, cubed**. Mathematically, we write $a^3 \propto P^2$.

This law tells us that the planets closer to the Sun move faster than the ones farther away. The rate follows a regular mathematical pattern: Take the distance from the Sun, cube it, and compare that with the time it takes to complete an orbit, squared. You should get the same result, regardless of what planet you choose. If you use a calculator and try it for all the planets, you obtain the results given in Table 8.2.

The results look quite impressive, but in fact Table 8.2 is something of a cheat. For instance, the planet Pluto was

discovered only 60 years ago; how did we manage to measure that its period is 248.5 Earth years long? We didn't; that period, and the periods of all the outer planets in that table, are in fact calculated from their distances from the Sun using Kepler's third law.

But the law does work for the inner planets, the ones that Kepler knew. It also works for the satellites orbiting each planet, including the thousands of artificial satellites that are orbiting Earth. So we have every reason to expect it should work for the outer planets as well.

There's one final step we want to take with this law, and that's to turn the proportionality into an equation. Recall how we defined the mean motion $n$ above. If it takes $P$ seconds to go through an orbit of 360°, or $2\pi$ radians, then $n$ is the portion of an orbit, on average, that the planet travels through in one second. Thus we state that $n = 2\pi/P$. Given this definition, we realize that $1/n^2 \propto P^2$. If we define a constant of proportionality $\mu$ we can write

$$n^2 a^3 = \mu \qquad (8.5)$$

This is Kepler's third law in equation form.

## Defining Keplerian Coordinates

So far, we have talked about how the position and the motion of a body in orbit can be described using an ellipse. Once we define the shape of the ellipse, by specifying its semimajor axis $a$ and eccentricity $e$, we can use the variable angle $\eta$, the true anomaly, to mark the position of our body on the ellipse. But this tells us even more than the position of the orbiting body; by knowing the size and shape of the ellipse, we know the direction that the body is travelling. In fact, Kepler's laws also tell us some things about not only where the body is travelling, but how fast it's moving.

Cartesian coordinates, the $xyz$-coordinate system familiar from high-school trigonometry, describe the position

**TABLE 8.2**

A Demonstration of Kepler's Third Law

| Planet | AU from the Sun | Period (Yrs) | Distance Cubed | Period Squared |
|--------|-----------------|--------------|----------------|----------------|
| Mercury | 0.3871 | 0.241 | 0.058 | 0.058 |
| Venus | 0.723 | 0.615 | 0.38 | 0.38 |
| Earth | 1. | 1. | 1. | 1. |
| Mars | 1.524 | 1.881 | 3.5 | 3.5 |
| Jupiter | 5.203 | 11.86 | 140. | 140. |
| Saturn | 9.539 | 29.46 | 870. | 870. |
| Uranus | 19.19 | 84.01 | 7,100. | 7,100. |
| Neptune | 30.1 | 164.1 | 27,000. | 27,000. |
| Pluto | 39.8 | 248.5 | 62,000. | 62,000. |

of a body in terms of its distance from a specified point in three different dimensions. We have three directions to consider, and so the position of the body is defined by three numbers, $x$, $y$, and $z$. The motion of the body, the speed and direction at which it is travelling at any given moment, also can be reduced to three variables. In Cartesian coordinates these variables are $v_x$, $v_y$, and $v_z$. Thus we need six terms to define position and velocity in Cartesian coordinates, three for position and three for velocity. In fact, we need six independent numbers to define the position and velocity of our body in any other coordinate system, as well.

But in Keplerian terms, we have defined our orbit only in terms of the size of the ellipse, $a$, its shape, $e$, and its position on the ellipse, $\eta$. That is only three coordinates. Where are the other three coordinates?

The missing coordinates are needed to orient our ellipse in space. An ellipse always lies on a plane. We need three coordinates, three constant angles, to show just how that plane is oriented relative to the three axes.

The **inclination**, $i$, defines how the plane of the ellipse is tipped relative to our original $xy$-plane. The **longitude of the ascending node**, $\Omega$, is the angle from the $x$-axis to the spot (called the ascending node) in the planet's orbit where it "ascends" up through the $xy$-plane. Finally, the **argument of pericenter**, $\omega$, is the angle from the ascending node to the pericenter of the ellipse, the spot where the planet comes closest to the Sun. These angles are all defined in Figure 8.10.

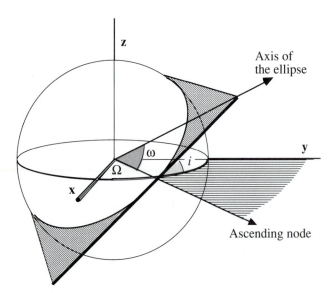

**FIGURE 8.10** Five constants define the path of an orbit: *a* and *e* define the shape of the ellipse, and *i*, *ω*, and *Ω* define the orientation of the ellipse in space.

Five constants define the shape of the orbit completely, and the sixth term, $\eta$, defines the precise location of the planet on that orbit. Notice the most important advantage of using the Keplerian coordinates. A body in a tipped, twisted orbit is always changing its $x$, $y$, and $z$ positions, and it's also constantly changing its speeds in the $x$-, $y$-, and $z$-directions. Thus to keep track of its motion, one has six complicated changing numbers to follow. However, of the six Keplerian coordinates, five of them are constants. Only $\eta$ changes with time.

In practice, the coordinate system used for the planets is based on Earth's orbit. The plane of Earth's orbit defines the $xy$-plane, with $z$ defined as positive in the northward direction. The center of the coordinate system is the Sun. The position of the $x$-axis is the line connecting the Sun to Earth at the moment when spring begins in the northern hemisphere.

When it was first defined, in ancient Greek times, this line pointed to the constellation Aries at a place in the celestial sphere traditionally called the **First Point in Aries**. As Earth wobbles in its spin, this line slowly moves (recall Chapter 6), and at present it has moved from Aries into the constellation Pisces. Eventually this point will enter the constellation Aquarius, beginning an important age for astrologers and fans of 1960s rock music.

## Newton's Laws

Fifty years after Kepler, Isaac Newton revolutionized physics, and the world, when he summarized the way bodies move with his three laws of motion, published in 1687. Even more astonishingly, he was able to apply these laws to celestial motions and show that the same force of gravity that caused an apple to fall from a tree also controlled the motion of the Moon and the periods of the planets. These laws are also stated in Table 8.1. Let's take a look at them.

**I. If a body is not accelerated, then its velocity stays constant**. Newton's first law defines acceleration as the change in the velocity of a body.

What exactly is meant mathematically by velocity and acceleration? **Velocity** is a measure of how fast something is moving (its speed) and in what direction it is moving. The direction is an essential part of velocity. Speed, on the other hand, is just how fast an object changes its position. If you drive a car at 40 miles per hour down a road going east, then your speed is 40 mph but your velocity is 40 mph east.

Mathematicians don't usually define directions in terms of north, east, south, and west, however, but with coordinate systems such as the Cartesian $(x, y, z)$ or polar $(r, \theta, z)$ systems. The movement of any object in three-dimensional space can be divided up into components of movement in these directions.

Because velocity is a measure of how fast something changes its position, it can be written as $v_x = dx/dt$,

$v_y = dy/dt$, and $v_z = dz/dt$. This way of writing velocity divides the movement into the $x$-direction, the $y$-direction, and the $z$-direction.

**Acceleration** is the rate of change of velocity with time. We can write acceleration just the same way we wrote velocity, as $a_x = dv_x/dt$, $a_y = dv_y/dt$, and $a_z = dv_z/dt$. This definition of acceleration is really Newton's first law. Basically, anything that speeds up a body (like the gas pedal of a car), slows it down (the brake), or makes it move in some path other than a straight line (the steering wheel) represents an acceleration.

Notice that our definitions for acceleration and velocity have been given for Cartesian coordinates. The formulas for other coordinate systems, including polar coordinates, can be much more complicated.

So, given these definitions, Newton's first law tells us that, in the absence of an acceleration, there is no change in velocity. From our definitions this seems obvious. But this law is hardly self-evident. The ancient Greeks, who were able to deduce many subtle principles of mathematics and physics, failed to see it at all, and for good reason. If you throw a stone through the air, or give it a push along the ground, it eventually stops; the velocity does not stay constant. Indeed, the only place where you see objects moving without rest is in the heavens; but as far as the ancient Greeks could tell, the stars and planets were travelling in circles and not straight lines.

One trouble was that Aristotle and the other ancient scientists didn't understand friction, an unseen force that constantly acts on bodies that slide across the ground or through air or water. Friction causes the bodies to slow down and come to a stop. In Newton's terminology, friction provides an acceleration to a moving body, to change its velocity until it drops down to zero.

**II. The acceleration that a body experiences is equal to the force acting on it, divided by the body's mass.**

Aristotle taught in *The Physics* that it takes a constant force applied to a body to result in the body moving at a constant speed; he thought that speed, not acceleration, was proportional to the force. Again, he didn't understand the role of friction.

Newton declared that the force acting on a body was directly related to the acceleration, not the speed, that the body felt. This can be expressed as $\mathbf{F} = m\mathbf{a}$, where $\mathbf{F}$ is the force, $m$ the mass of the body, and $\mathbf{a}$ the acceleration. The symbol $\mathbf{a}$ is used here to stand for all three components, $a_x$, $a_y$, and $a_z$; likewise, $\mathbf{F}$ also has components in each direction. A quantity that has more than one component is called a **vector**. The direction a body travels has no effect on its mass; $m$ is not a vector.

If we know the force $\mathbf{F}$ acting on a body, its mass $m$, and its position and velocity $x$, $y$, $z$, $v_x$, $v_y$, $v_z$ at some given time, this law states that we now have an equation, $\mathbf{F} = m\mathbf{a}$, which can be used to predict where the body will be, and how fast it will be moving (and in what direction), for any other time, past or future.

**III. If one body exerts a force on another body, then the second body must also be exerting just as much force on the first body, but in the opposite direction.**

Like the ancients, Newton realized that if a body feels a force there must be some other body that caused the force. However, Aristotle had reasoned that a body exerting a force on a second body must be moving along with that second body, not against it as Newton's third law asserts. After all, a person throwing a baseball has her arm moving in the same direction as the ball. What Aristotle didn't see, but what Newton understood, is that the ball thrower must push her legs back against Earth, like a baseball pitcher pushes off against the pitcher's mound, to project the ball forward. If she tries to throw the ball from a frictionless surface, like a rowboat, she'll feel the boat slide out from under her feet. The action of the ball, says Newton, results in an equal and opposite reaction.

Note that the forces, not the accelerations, are equal and opposite. In the case of the ball thrown from the rowboat, the ball has a much smaller mass so it will travel much faster than the rowboat travels in the opposite direction. If the pitcher throws the ball from the deck of the *Queen Mary*, the resulting reaction is too tiny to notice.

## Newton's Laws in Space

Friction complicates any attempt to find the underlying rules that describe how things move. In the motions of the planets, however, Newton was able to follow bodies that moved without friction. To get a taste of how this works, let's look at a simple example of a body in a circular orbit.

If you apply Newton's first and second laws to a planet in a circular orbit, you come up with an important result. Newton's second law says that a force in the direction that the body is moving results in the body being accelerated in that direction, speeding it up or slowing it down. But recall that a body orbiting in a circle moves at a constant speed; hence, there's *no* force pulling the planets in the direction of their motion.

On the other hand, Newton's first law states that bodies don't turn in their paths unless a sideways force acts on them. But, for a circle, the "sideways" direction is always pointing toward the center of the circle. So there *is* a force pushing the planet sideways and that force is directed straight at the Sun.

It can be shown that this result, that the only force moving the planets is a force directed at the Sun, turns out to be true for elliptical orbits, too. A force like this, always directed toward one point, is called a **central force**.

**FOR THE RECORD...**

# A Review of Calculus

One important thing about the laws of physics is that they can be directly translated into mathematical expressions. Once an object's motion is reduced to equations, you can use mathematics to predict what the body's motion is going to be in the future.

Up to the time of Newton (c. 1700), mathematics included most of the branches taught in high school today. The ancient Greeks, especially Euclid, had worked out the principles of plane (Euclidean) geometry. The medieval Arabs had invented algebra (the term comes from *al-jabr*, a word used to describe setting broken bones), which solves for unknown quantities by breaking apart and rearranging equations. A generation before Newton, René Descartes had combined these two into his "Cartesian" coordinates $(x, y, z)$ and showed how, with trigonometry, one could use equations to solve problems in geometry.

But all these techniques involved static systems, systems that never moved or changed. In many ways this reflected the preoccupation of medieval philosophers with matters eternal and unchanging. In studying the laws of motion, such mathematics is not enough. Thus, along with the laws of motion, Isaac Newton invented **calculus**.

Most upper-level science undergraduates are quite familiar with the workings of calculus, but some fields such as geology or chemistry depend on calculus less than physics does. For students in these fields, it is often useful at this juncture to review what actually is going on when the language of calculus is applied to physical problems.

Suppose that you were watching a car that constantly accelerated so that every second its speed was increased by 1 m/s. Starting from standing still at time zero, after 1 second its speed will be 1 m/s; after 2 seconds, its speed is 2 m/s; after 3 seconds, its speed is 3 m/s, and so on.

Now, after 3 seconds, how far has this car travelled? The distance travelled is the speed (m/s) times the time (s), but the speed is constantly changing. Which speed do you use? Furthermore, what if the acceleration isn't a constant 1 m/s every second? In real life, cars speed up and slow down at very irregular rates.

This is the sort of problem that can be solved with calculus. Arithmetic teaches us how to add numbers and how to undo this addition with subtraction, the opposite of addition. Multiplication is just repeated additions of numbers; its opposite is division. In algebra we learn to add and

subtract pieces of equations, **functions** made up of numbers and variables. In calculus we perform a new kind of operation on functions, called **integration**. It too has an opposite, **differentiation**. If you take a function, integrate it, and then differentiate it, you'll be left with the same function you had in the first place.

These two operations can be illustrated by pictures. Figure 8.11 shows a plot of the function $f(t) = t$. Let's look at the piece between $t = 0$ and $t = 4$. If we say that the line represents the function $f(t)$, then it turns out that the area under this line, the shaded area in the picture, is the **integral** of $f(t)$ from $t = 0$ to $t = 4$.

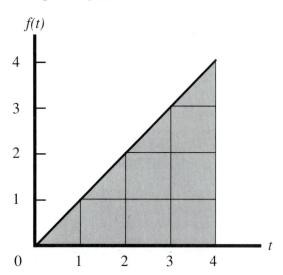

**FIGURE 8.11**   The line represents the speed of an object, which in this case is constantly increasing. The shaded area under the curve is total distance travelled.

What is the area of the shaded region? We can count up the squares, remembering that triangles have half the area of full squares. At 0, the area is 0; at 1, the area so far is $\frac{1}{2}$; at 2, the total area from 0 to 2 is 2; at 3, the area is 4.5; and at 4, the area is equal to 8 squares.

If we plot these numbers, in Figure 8.12, we see a curve that increases at an ever more rapid rate. In fact, the function $g(t) = \frac{1}{2}t^2$ produces a curve that fits these points.

By taking the area under the curve of the function $f(t)$, we say we have integrated the function $f(t)$ from 0 to 4; in this case, the integral of $f(t) = t$ is the function $g(t) = \frac{1}{2}t^2$.

Now let's look at the curve in Figure 8.12. At any point we can draw a tangent line that indicates the slope of the curve, which describes how rapidly the function is increasing at that point. The slope of such a line, tangent to

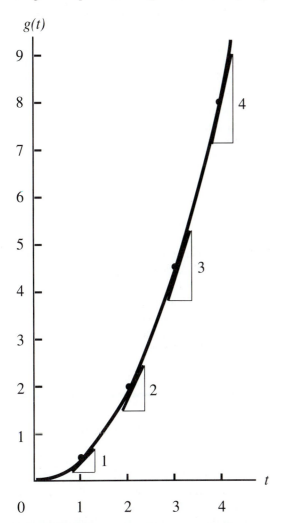

**FIGURE 8.12** Here, the curve represents the distance that the object has travelled as a function of time. The slope of a line tangent to the curve at any point is equal to the speed at which the object was travelling at that time.

the curve representing a function, is called the **derivative** of that function.

Notice that in our curve, at $t = 0$ the slope is horizontal; the slope is 0. Mathematically, we can say that

$$\left.\frac{dg}{dt}\right|_{t=0} = 0$$

or, in English, the change of the function $g(t)$ evaluated at the point where $t$ equals zero is zero. Notice further that at $t = 1$, the slope is 1; at $t = 2$, the slope is 2; at $t = 3$, it's 3, and at $t = 4$, it's 4. You can see that from the figure: The vertical leg of the little triangle drawn at point $t = 4$ is 4 times as long as the horizontal leg.

If we plot these numbers, we'll just get Figure 8.11 again. We say the derivative of the function $g(t) = \frac{1}{2}t^2$ is $f(t) = t$.

Let's look again at what we did with Figure 8.11. You could find those areas by taking the average value of $f(t)$ between two points (say, the value between 2 and 3, which is 2.5) and multiplying this value by the distance between the points (here, just 1, so 2.5 × 1 is 2.5), then adding this to the total sum we have found up to that point (between 0 and 2 the total area was 2, so we get the new sum of 4.5).

But what if we couldn't guess what the average value was? Well, if we'd taken the distance between 2 and 2.0000001 instead of between 2 and 3, the difference between the two values of $f(t)$ at these points would be so small that the average value would be very close to either of them. If you made the change in $t$ even smaller, the average value would be even closer, until the change in $t$—call it $dt$—was infinitesimally small. At that point, however, you'd have to add up an infinite number of thin slivers, rather than the four broad stripes that we started with.

The mathematical symbol for this infinite summation of the product of the function $f(t)$ times the infinitesimal width $dt$ is written $\int f(t)\,dt$, where the integral symbol $\int$ is simply an elongated $S$, standing for "sum."

It is impractical to do this infinite summation by repeated additions. (A computer can approximate such an infinite sum, of course.) Likewise, the values for the slopes we found in Figure 8.12 really aren't all that obvious to the naked eye. Just from looking at Figure 8.12, could you say

**FOR THE RECORD...**

# A Review of Calculus (continued)

for certain that the slope at $t = 2$ was 2 and not 2.1 or 1.9? But through the techniques of calculus these difficulties can be overcome.

Let's demonstrate calculus one more time. Looking again at Figure 8.11, we'll take the derivative of $f(t)$ this time. Because the function $f(t)$ is a straight line, the slope is the same at all the points. In this case we see that it is equal to 1. We plot this result, which we'll call $h(t)$, in Figure 8.13. Likewise, if we integrate $h(t)$ from Figure 8.13, by adding up the shaded blocks, we arrive at $f(t)$, just as we would expect.

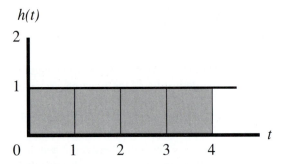

**FIGURE 8.13** The acceleration is constant. The area under this line represents the speed of the object. As time progresses to the right, the area increases at a constant rate, just as the speed is constantly increasing.

In physics, we know from Newton's laws that force is proportional to acceleration. But acceleration is the derivative of speed with respect to time, or $a = dv/dt$. And velocity is the derivative of position with respect to time, or $v = dx/dt$. Now let $h(t)$ represent the acceleration of our car, with the constant acceleration of 1 m/s. Then if we integrate acceleration, to find speed, we see that Figure 8.11 can represent the resulting plot of speed as a function of time. At $t = 4$ seconds, the car is moving at 4 m/s.

Doing it again, we integrate the speed $f(t)$ to produce Figure 8.12. This tells us that, at $t = 4$ seconds, the car has travelled 8 meters.

We can also work in the opposite direction. Let's say that you had a stopwatch while you were watching the car. You could record its position at each second and produce a curve like Figure 8.12. Then, through insight or trial and error, someone else may realize that the car's motions seem to obey the function $g(t) = \frac{1}{2}t^2$. Yet another scientist, who knew calculus, could then take the derivative of $g(t)$ to find that the velocity must obey the function $f(t) = t$ and take the derivative of $f(t)$ to see that the acceleration was a constant, equal to 1 m/s$^2$.

Multiply the acceleration of the car by the car's mass, $m$, and we have found the force $F = ma$ that is acting on the car, causing it to accelerate. Thus, just from observing the position of the car at different times, it is possible to deduce the nature of the force pushing the car. It is a constant force, in this case.

In exactly this way Brahe observed the positions of the planets, Kepler worked out the equation that describes those positions with time, and Newton was able to deduce the nature of the force that moves the planets in their orbits.

---

What more can Kepler's laws and Newton's laws tell us about this central force? That's the topic of the following section.

## SUMMARY

Kepler defined the paths of the planets in terms of an ellipse. Five constants define the shape and orientation of a planet's orbit. The shape is defined by $a$, the **semimajor axis** of the ellipse, which is related to how big the ellipse is, and $e$, the **eccentricity** of the ellipse, how "squashed" the ellipse appears.

The **inclination**, $i$, describes the tilt of the plane of the ellipse relative to some reference plane. The **longitude of the ascending node**, $\Omega$, is the angle describing how the plane of the orbit is twisted relative to some reference axis. The **argument of pericenter**, $\omega$, is the angle measured from the place where the ellipse crosses the reference plane, to the point in the orbit closest to the body being orbited.

The **true anomaly**, $\eta$, is the angle that defines the position of the planet on the ellipse at any given time. It is not a constant; it changes with time as the planet moves around in its orbit.

Newton outlined three basic laws of motion. His first law defines acceleration as a change in a body's velocity (either its speed or its direction of motion). The second law relates this acceleration to the force acting on the body and its mass. The third law notes that if one body exerts a force on a second body, then that second body exerts an equal and opposite force on the first one. Newton's laws of motion can be seen in action best when the force of friction can be ignored; the motions of the planets provide just such a case.

## STUDY QUESTIONS

*The following statements are all false. What could you change to make them true?*

1. "Each planet moves in a circle, with the Sun at the exact center of that circle."

2. "If a body does not experience a force, then regardless of how fast it starts out moving eventually it will slow down and stop."

3. "The force acting on a body is directly proportional to its mass and its velocity."

4. "Planets move slower as they approach the Sun and faster as they move away."

5. "An 'ellipse' is a mathematical term for any oval shape."

## 8.3 THE LAW OF GRAVITY

*In the 1970s, a famous author and astronomer was being interviewed on a college radio talk show. The subject matter had been drifting towards less orthodox views of the universe, bordering on pseudoscience, and the astronomer was beginning to sound distressed.*

*One of the interviewers, an earnest young man with a mystical lilt to his voice, then posed what he felt was the crucial question. "Professor, do you believe in a Cosmic Force that controls the universe? Like, a Force that formed the stars, a Force that holds the planets together and controls their motions?"*

*"Yes," replied the astronomer dryly. "It's called gravity."*

### Kepler's Third Law, Revisited

Newton's laws and Kepler's laws are not merely descriptive tools; rather, they are tools that we can use to predict the future motions of the solar system. But even more than that, by using these laws we can arrive at a fundamental principle of nature, the law of gravity.

We've already learned, from Newton's first two laws, that the force holding the planets in their orbits is a force always directed towards the Sun. Whatever that force is, the

Sun is its source. We've also hinted that the orbits of satellites going around bodies other than the Sun—the satellites of Jupiter, for instance, or our Moon—also obey Kepler's third law. Is this really true? Look at Table 8.3 and you may not think so; the numbers in the last two columns aren't the same, unlike Table 8.2.

And notice something interesting. Earth's Moon orbits Earth at a distance of just under 400,000 km, and it takes 27 days to complete an orbit. On the other hand, Io orbits Jupiter at a distance of a bit more than 400,000 km, but it completes an orbit in less than 2 days. What's going on here?

One reason everything worked out so well in Table 8.2 was that we used as our standard of measure 1 Earth year and 1 AU. For Earth (and so for all the other planets as well) these units are in a 1:1 ratio. Remember that Kepler's third law merely states that the cube of the period and the square of the distance are proportional, not equal; using years and AU in this special case made the constant of proportionality equal to 1.

But in the satellite table, we are comparing in kilometers and hours. The two columns are no longer equal. However, the ratio of the two columns—the value in the distance cubed column divided by the value in the period squared column—does stay the same. You can see that in Table 8.4.

The other thing about Table 8.2 is that every body in that table orbited the same object, the Sun. In Tables 8.3 and 8.4 we have some bodies that orbit Earth and others that orbit Jupiter. You'll notice in Table 8.4 that the ratio we get for all Jupiter's satellites is a different value from the ratio we get for all Earth's satellites.

Why the difference? Jupiter is much bigger than Earth. If the size of the body being orbited has something to do with the force pulling on the orbiting satellites, then this could explain the differences in the ratios.

What do we mean by "size"? If we just mean the physical dimensions of the planets, it doesn't work. We know how big Jupiter is and we know how big Earth is; the ratio of their sizes is not the same as the ratio of the two values in the first column of Table 8.4. But what Newton realized was that it wasn't the physical size of the body that mattered but rather the amount of material inside that body: the mass.

Find a moon orbiting a planet. Determine its distance from the planet; determine the period of its orbit. The distance cubed divided by the period squared is proportional to the mass of the planet being orbited.

Now we have a new insight into that force that keeps the planets in their orbits. The force is directed towards the body being orbited, and the strength of the force is directly related to the mass of the body being orbited.

We can even go further than that. We notice that the ratios from Kepler's third law may depend on the mass of

**TABLE 8.3**

A Demonstration of Kepler's Third Law: Satellites

| Moon | Distance (km) | Period (h) | Distance$^3$ | Period$^2$ |
|---|---|---|---|---|
| **Jupiter Satellites** | | | | |
| Io | 422,000 | 42.46 | $7.52 \times 10^{16}$ | 1,802.51 |
| Europa | 671,000 | 85.22 | $3.02 \times 10^{17}$ | 7,263.13 |
| Ganymede | 1,070,000 | 171.70 | $1.23 \times 10^{18}$ | 29,479.52 |
| Callisto | 1,883,000 | 400.56 | $6.68 \times 10^{18}$ | 160,448.31 |
| **Earth Satellites** | | | | |
| Space Shuttle | 6,678 | 1.50 | $2.98 \times 10^{11}$ | 2.25 |
| RCA SATCOM | 42,300 | 23.93 | $7.57 \times 10^{13}$ | 572.64 |
| The Moon | 384,400 | 655.72 | $5.68 \times 10^{16}$ | 429,968.72 |

**TABLE 8.4**

Masses of Earth and Jupiter from Kepler's Third Law

| Moon | Distance$^3$/Period$^2$ | Calculated Mass of Planet (kg) |
|---|---|---|
| **Jupiter Satellites** | | |
| Io | $4.2 \times 10^{13}$ | $1.90 \times 10^{27}$ |
| Europa | $4.2 \times 10^{13}$ | $1.90 \times 10^{27}$ |
| Ganymede | $4.2 \times 10^{13}$ | $1.90 \times 10^{27}$ |
| Callisto | $4.2 \times 10^{13}$ | $1.90 \times 10^{27}$ |
| **Earth Satellites** | | |
| Space Shuttle | $1.3 \times 10^{11}$ | $6.03 \times 10^{24}$ |
| RCA SATCOM | $1.3 \times 10^{11}$ | $6.03 \times 10^{24}$ |
| The Moon | $1.3 \times 10^{11}$ | $6.03 \times 10^{24}$ |

the planet we're orbiting, but those ratios do not depend on the mass of the satellite; the Moon and a little communication satellite are bent in exactly the same sort of orbit.

Now comes an especially tricky part: We've just said that the shape of the orbit is determined by how much the path of the orbiting body is turned inward, which is to say how much the direction of its velocity changes. A change in a velocity (even if it is only a change in direction, with the speed held constant) is what we mean by an acceleration. So the acceleration of an orbiting body does not depend on its mass.

But Newton's second law says that the force on the body equals its mass times it acceleration. Thus the force of gravity is the mass of the satellite times this constant acceleration.

To summarize our understanding of gravity: (1) The force of gravity (and the acceleration) is directed towards the body being orbited, (2) the force of gravity (and the acceleration) gets bigger as the mass of the body being orbited gets bigger, and (3) the force of gravity (*but not* the acceleration) gets bigger as the mass of the body in orbit gets bigger.

## The Inverse Square Law

Finally, let us examine how the force of gravity changes as the size of the orbit increases. As we noted earlier, most orbits in the solar system are nearly circular, and circular orbits are far easier to deal with mathematically than elliptical orbits. So for the rest of this section, let's make a simplifying assumption: Let's assume that the orbits are, in fact, circular. (In a following section we explore the more rigorous problem of elliptical orbits.)

From Kepler's third law, we saw that bodies farther from the Sun move more slowly, so it seems reasonable to assume that the force of gravity between two bodies decreases as the space between them gets bigger. But how much does it decrease? Does it ever fall to zero?

To answer this question, we'll first ask another one: How much force does it take to pull an object into a circular path?

Tie a string to a rock and twirl it around your head. Your hand feels the string pulling in it, that familiar tug we call **centrifugal force**. By Newton's third law, the force of the rock pulling on your hand is exactly as strong as the force with which the string, and your hand, is pulling on the

**FURTHER INFORMATION...**

# Acceleration in Uniform Circular Motion

Newton's first law states that if a body is not accelerated, it keeps a constant velocity. Velocity means both speed and direction; thus an unaccelerated body always moves in a straight line. This means that it takes acceleration to make a moving object travel in a circle.

Consider a body travelling around a circle at a constant speed. This is what we call **uniform circular motion**. A complete circuit carries this body through an angle of $360°$, or $2\pi$ radians. If it takes $P$ seconds to complete a circuit, we say that its angular speed is $n = 2\pi/P$ radians per second. For a period of time $\Delta t$, then, it travels through an arc of $n \cdot \Delta t$ radians. If its circle has a radius of $r$ meters, then it has a speed of $r \cdot n$ m/s, and the distance along the arc travelled during this time is $r \cdot n \cdot \Delta t$ m.

Because it is travelling in a circle, it must be feeling some sort of acceleration. Acceleration is defined as the change in velocity, over a given change in time. How much is this object accelerated, and what is the direction of the acceleration?

The direction is obvious: If the acceleration had any component in the direction of motion, the object would speed up or slow down. Because we are looking at an object whose speed is constant, the acceleration in the direction of motion, the direction of increasing angle $\eta$, $a_\eta$, must be zero. Thus the acceleration is totally in the $r$ direction, $a_r$, at right angles to the motion.

Look at Figure 8.14. After a time $\Delta t$, the velocity vector **A**, perpendicular to line $\overline{AC}$, has been twisted into the vector **B**. This vector is drawn again so that its point touches the point of the first vector; from the rules for vector subtraction, the difference between these vectors is simply the little vector **D**, the line $\overline{DA}$.

The triangle $\triangle ABD$ is an isosceles triangle, because the two vectors **A** and **B** are the same length. It can be shown that $\triangle ABC$, also an isosceles triangle, and $\triangle ABD$ are similar triangles; the tiny angle $\theta$ between the velocity vectors **A** and **B** is the same as the angle $\angle ACB$ between the two radius vectors.

The length of the tiny line $\overline{DA}$ is the magnitude of the difference between the velocity vectors. Bisect $\angle ABD$ to make two right triangles, each with a narrow angle $\frac{1}{2}\theta$,

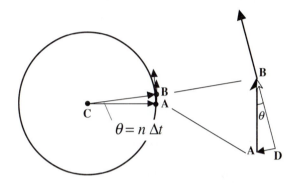

**FIGURE 8.14** After a short time $\Delta t$ in uniform circular motion, the velocity vector **A** has been twisted into the vector **B**. If we move the vector **B** until it touches the point of vector **A**, the difference between these vectors is the little vector **D**.

and then by trigonometry we can find this length:

$$\overline{DA} = 2rn \sin \tfrac{1}{2}\theta \qquad (8.6)$$

If the angle $\theta$ at point $B$ is very small (that is, if $\Delta t$ is very small and so the difference between these two vectors is very small), then $\sin \theta$ is approximately equal to $\theta$ and so we can simply write this as $\overline{DA} = rn\theta$.

On the other hand, $r \cdot \theta$ is simply the distance travelled by the object over a time $\Delta t$, $r \cdot n \cdot \Delta t$. Thus the change in velocity, $\overline{DA}$, is given by:

$$\overline{DA} = rn \left( \frac{rn\Delta t}{r} \right) = rn^2 \Delta t \qquad (8.7)$$

Recall that the acceleration is the change in velocity divided by the change in time, or $\overline{DA}/\Delta t$. Thus we find the acceleration in uniform circular motion

$$a_r = rn^2 \qquad (8.8)$$

and it is directed along the radial line back towards the center of the circle, exactly as we expected.

## FURTHER INFORMATION...

# Real Orbits, Real Gravity

Section 8.3 showed how we could derive the law of gravity for circular orbits, but circular orbits are an idealized case; no orbit is ever perfectly circular. We have mathematical expressions for an elliptical orbit described by Kepler's laws; can we deduce from them something about the force that controls this orbit? Can we use these laws to solve for the *real* law of gravity?

From Kepler's first law, we note that the orbit of each planet is an ellipse, and we observe that the elliptical orbit of each planet lies in a different plane, slightly tilted from all the other planes of all the other planets. However, all these planes intersect at exactly one point; this point is a focus of all the ellipses, and the Sun is located at this point. This strongly suggests that the Sun is the source of whatever force controls the motions of the planets.

Kepler's second law says that the area, $dA$, swept in a given period of time, $dt$, is equal to some constant. Thus we get

$$\frac{dA}{dt} = \frac{1}{2} \cdot r \cdot r \cdot \frac{d\eta}{dt} = \frac{1}{2}h \qquad (8.9)$$

or

$$dA = \frac{1}{2}r^2 d\eta = \frac{1}{2}h dt \qquad (8.10)$$

where we define our constant as $\frac{1}{2}h$. The reason we define the constant in this way is that $h$, which is called the **angular momentum per unit mass**, turns out to be a useful quantity. For instance, we can further use the rules of algebra to cancel the factors of $\frac{1}{2}$ and simplify this equation to

$$r^2 d\eta = h dt \qquad (8.11)$$

This is a mathematical statement of Kepler's second law.

It can be shown (with a bit of calculus) that in polar coordinates, the total acceleration can be broken into $r$ and $\eta$ components:

$$\mathbf{a} = \left[\frac{d^2r}{dt^2} - r\left(\frac{d\eta}{dt}\right)^2\right]\hat{r} + \left[\frac{d\left(r^2\frac{d\eta}{dt}\right)}{dt}\right]\frac{1}{r}\hat{\eta} \qquad (8.12)$$

Thus the acceleration in the $\eta$ direction alone is given by the second part of Equation 8.15:

$$a_\eta = \left[\frac{d\left(r^2\frac{d\eta}{dt}\right)}{dt}\right]\frac{1}{r} \qquad (8.13)$$

or

$$a_\eta = \left(\frac{dh}{dt}\right)\frac{1}{r} \qquad (8.14)$$

How does our constant $h$, the angular momentum per unit mass, change with time? It doesn't; it's a constant. Thus $dh/dt = 0$, and the acceleration in the $\eta$ direction, $a_\eta$, must also be zero.

This means that there is no force in the $\eta$-direction. The only possible acceleration is directed in the $r$-direction, directed along the line connecting the Sun and the planet.

Given enough calculus, it is possible to go a step farther and work out mathematically the formula that describes what the force must be. We know that

$$\mathbf{a} = \left[\frac{d^2r}{dt^2} - r\left(\frac{d\eta}{dt}\right)^2\right]\hat{r} + 0 \qquad (8.15)$$

rock. That's the force that is pulling the rock into a circular path. It works the same way for the force of gravity needed to pull a planet into a circular orbit.

How strong is this force? The rules of physics tell us that its strength depends on three factors.

The first is how fast the object is spinning: The faster the spin, the greater the tug you feel on the string, and the tug you feel is the same as the tug the rock feels. As a

matter of fact, this force varies as the square of the spin rate: doubling the spin rate causes the force not just to double, but to increase by a factor of four. Recalling our definition of the mean motion, the average spin rate $n = 2\pi/P$, we can say that the force depends on $n^2$.

You also need more force if you keep the rock spinning at the same rate but use a longer string. The longer the string, the more force it takes to keep the rock spinning

Using Equation 8.14 we find that

$$a_r = \frac{d^2r}{dt^2} - r\left(\frac{h}{r^2}\right)^2 = \frac{d^2r}{dt^2} - \frac{h^2}{r^3} \qquad (8.16)$$

Next, find $d^2r/dt^2$ by taking the derivative of Equation 8.1 with respect to time, twice. After a considerable amount of algebra, one arrives at

$$\frac{d^2r}{dt^2} = \frac{h^2 e \cos \eta}{r^3 (1 + e \cos \eta)} \qquad (8.17)$$

Inserting this into Equation 8.16:

$$a_r = \frac{h^2}{r^3}\left[\frac{e \cos \eta}{1 + e \cos \eta} - \frac{1 + e \cos \eta}{1 + e \cos \eta}\right]$$

$$= -\frac{h^2}{r^3}\left[\frac{1}{1 + e \cos \eta}\right] \qquad (8.18)$$

But again, using the relation of Equation 8.1, this becomes

$$a_r = -\left[\frac{h^2}{a\left(1 - e^2\right)}\right]\frac{1}{r^2} \qquad (8.19)$$

Notice that we have now derived the inverse square relation for the acceleration that we expected from the circular orbit case.

But we can go further. From Equation 8.4, Kepler's second law, we know that

$$\frac{dA}{dt} = \frac{h}{2} = \frac{\pi a^2}{P}\sqrt{1 - e^2} \qquad (8.20)$$

Thus we can solve for $h$:

$$h = \frac{2\pi a^2 \sqrt{1 - e^2}}{P} = na^2 \sqrt{1 - e^2} \qquad (8.21)$$

Substitute this into Equation 8.18 and use Kepler's third law, Equation 8.5, to get:

$$a_r = -\left[\frac{\left(na^2\sqrt{1 - e^2}\right)^2}{a\left(1 - e^2\right)}\right]\frac{1}{r^2}$$

$$= -n^2 a^3 \frac{1}{r^2}$$

$$= -\mu\frac{1}{r^2} \qquad (8.22)$$

The negative sign indicates that the acceleration is in the negative $r$-direction, that is, it is directed in towards the Sun, just as we'd expect. Notice that this acceleration does not depend on the mass of the planet. The way the planet moves depends only on its position, $r$, not on how large or small that planet is.

By Newton's second law, $\mathbf{F} = m\mathbf{a}$, and so we finally derive

$$\mathbf{F} = -\frac{m\mu}{r^2}\hat{r} \qquad (8.23)$$

which is the vector version of Equation 8.25, and now we see that this equation is valid for all orbits, not just circular ones.

at the same spin rate, the more force you'll feel tugging on the string, and the more force the rock will feel. The force depends on the length of the string, $r$.

If you hold the spin rate constant and the rope length constant, but tie a second identical rock to the rope, you'll double the amount of force in the rope. (Recall that force depends on mass, $F = ma$, according to Newton's second law; double the mass and you double the force.) Thus the force acting on the rock depends on the spin rate, squared, times the length of the rope, $r$, times the mass, $m$. So the magnitude of the force to hold an object onto a circular path is equal to $F = mrn^2$.

Finally, we note that this force has a direction; the rope on our hand pulls outwards, so the force on the rock must be equal and opposite by Newton's third law. The force on the rock is inwards. To indicate this, we add a negative sign:

$$F = -mrn^2 \qquad (8.24)$$

The length of the string, $r$, plays the same role as the distance between planet and moon. But for the case of a planet and a moon, Kepler's third law (Equation 8.5) states that $n^2 r^3 = \mu$ for circular orbits, where $r$ is always equal to the semimajor axis $a$.

Now substitute $\mu / r^3$ for $n^2$ in our force equation, Equation 8.24. We find that the force holding a planet in a circular orbit is

$$F = -\frac{mr \cdot \mu}{r^3} = -\frac{m\mu}{r^2} \qquad (8.25)$$

But the force holding the planets in their orbits is what we call the force of gravity.

By this calculation we determine that gravity varies as $1/r^2$. It is an example of an **inverse square law**. In other words, compared with the force of the Sun's gravity that is felt at 1 AU from the Sun, the force is one fourth as great at 2 AU, one ninth as great at 3 AU, and so forth. The force of gravity falls off rather rapidly with distance, but it never goes all the way to zero.

From all this we can complete our description of the force of gravity. Gravity is a force between two masses, directed from the one mass towards the other; and its strength is proportional to the mass $M$ of the body being orbited and the mass $m$ of the body in orbit and inversely proportional to the square of $r$, the distance between them. The constant of proportionality is usually given the symbol $G$; hence the formula

$$F = -\frac{GMm}{r^2} \qquad (8.26)$$

Comparing this formula with what we derived before, we see that the constant $\mu$ defined above must be equal to $GM$. Different satellites orbiting the same body use the same value of $\mu$. If we know $\mu$ and we know the universal value of $G$, we can solve to find the mass of the body being orbited. Thus we can explain the results of Table 8.4. The actual value of that constant $G$ (called the **Universal Gravitational Constant**) is, in truth, not all that well known. Best estimates put it at $6.672 \times 10^{-11}$ in SI units, $\text{N} \cdot \text{m}^2/\text{kg}^2$.

The final insight that Isaac Newton had was that this force that pulls the Moon around Earth is the same force that causes an apple to fall from a tree. In essence, every falling object is trying to move in an orbit around Earth. In the case of the apple, however, Earth just gets in the way before it can travel very far.

Newton's genius realized that the force of gravity controlled both apples and satellites and indeed is the force that ties the solar system together. The laws of physics are straightforward and logical; they can be understood by we mortals; and they are the same on Earth as they are in the heavens. That was a staggering insight into the nature of the universe.

## SUMMARY

Isaac Newton worked out the general rules for how any body responds to a force, and Johannes Kepler worked out three rules specifically describing how the planets move about the Sun. By comparing Kepler's rules with Newton's principles, it is possible to learn the nature of the force that holds the planets in their paths. The force of gravity on a planet due to the Sun is directed towards the Sun. It is proportional to the mass of the Sun and the mass of the planet and falls off as one over the square of the distance between them.

## STUDY QUESTIONS

*The following statements are all false. What could you change to make them true?*

1. "The force of gravity pulls planets around the Sun in the same direction as the planets move."

2. "There is no gravity in outer space, because things are weightless there."

3. "Heavier planets move more quickly around the Sun than lighter planets."

4. "The way a planet moves depends on the planet's position, on how big or small that planet is, and how big or small the Sun is."

5. "Planets experience a force that pulls them around the Sun and a second force that pulls them straight towards the Sun."

## 8.4   PROBLEMS

1. Tycho Brahe spotted stars by looking from a movable peg at one end of his 19-foot quadrant, past a peg at the center of the quadrant 9.5 feet away, and on to the star.
   (a) How far apart would two pegs be along the arc of this quadrant to represent an angle of 1°?
   (b) The human eye can resolve angles of a bit less than 1 arc-minute. (There are 60 arc-minutes to 1°.)

How far apart would this be, on Brahe's quadrant?

(c) Using common sense, estimate how precise Brahe's instrument probably was.

2. If a planet orbited in a circle exactly 4 AU from the Sun, how many years would it take to orbit the Sun?

3. Assume that a satellite is orbiting a planet in a circular orbit.

(a) Draw a sketch with arrows showing the direction the satellite is moving at any four arbitrary points in its orbit.

(b) Suddenly, by magic, the planet disappears. Draw a sketch describing the path the satellite will take once the planet disappears.

(c) Assume that the planet has a mass of $2 \times 10^{25}$ kg and that the satellite orbits at a mean distance of $5 \times 10^5$ km. When the planet disappears, in what direction will the satellite be moving? How fast will it be travelling?

4. (a) Mimas, a moon of Saturn, orbits Saturn at a distance of 185,520 km. Each orbit takes 22.6 h. Use this information to find the mass of Saturn. Assume that Mimas's orbit is circular and that the gravitational constant $G$ is equal to $6.67 \times 10^{-11} \mathrm{m}^3/\mathrm{kg} \cdot \mathrm{s}^2$. Compare this answer with the data given in the appendix.

(b) Tethys, another moon of Saturn, takes twice as long to orbit Saturn. If it has a circular orbit, how far from Saturn should Tethys be? Compare your answer with the data given in the appendix.

5. What is the first day of (northern hemisphere) spring on Earth? What's the first day of fall? Count the number of days between these dates: the number of days in fall and winter versus the number of days in spring and summer. Why aren't they exactly the same? In countries lying on the equator, which might be expected to have no seasons at all, the month of January is in fact warmer than the month of July. Why is this so?

6. When an asteroid is first discovered, it is simply a dot of light in the sky. Its position can be measured, but of course there is no way of telling exactly how far away it is. However, if its position can be determined at three different times, it is possible in theory to completely define its orbit, including its semimajor axis. Why are three observations needed?

7. The mass of the Sun is $2 \times 10^{30}$ kg; that of Earth is roughly $6 \times 10^{24}$ kg. One AU, the distance between the Sun and Earth, is $1.5 \times 10^{11}$ m. What is the acceleration (not the force!) that Earth feels due to the gravitational force of the Sun? How much is the Sun accelerated by Earth?

8. Derive Equation 8.17.

9. Both Copernicus and Rheticus were hindered by the religious climate in which they worked, and Galileo's difficulties with the church are well documented. Yet some historians of science maintain that the main reason science flourished in Europe, instead of the more advanced cultures of China or India, was the Judeo-Christian world view of the West. René Descartes in the seventeenth century argued that the notion of a Creator-God gave scientists the confidence to expect that the universe was run by rational, consistent, and harmonious laws. Does a scientist's philosophical or religious outlook affect the way that she does science? Should it?

10. According to this chapter, that "the laws of physics are straightforward and logical; they can be understood by us mortals; and they are the same on Earth as they are in the heavens" is a "staggering insight into the nature of the universe." Do you agree with this statement? Comment, with specific examples from your knowledge of history or literature.

## 8.5 FOR FURTHER READING

A. C. Crombie, *Augustine to Galileo* (London: Heinemann Educational Books, 1970), is perhaps still the best single-volume treatment of the science of the medieval and renaissance times; Edward Grant, *A Source Book in Medieval Science* (Cambridge, Mass.: Harvard University Press, 1974), is also excellent, if less inclusive.

Edward Rosen, *Copernicus and the Scientific Revolution* (Malabar, Fla.: Krieger Publishing Company, 1984), is a terse but exhaustive scientific biography of Copernicus.

Altogether too many books have been written about Galileo. The best place to start may well be Maurice A. Finocchiaro, *The Galileo Affair*, (Berkeley: University of California Press, 1989) who gives a brief and very even-handed overview of the facts of the case and a summary of the arguments that they have evoked over the years; best of all, the bulk of the book is simply a compilation of all the relevant letters, documents, memos, and minutes that surrounded all Galileo's legal problems. Read them yourself and draw your own conclusions.

Many standard physics texts cover the basics of gravitation, usually confined to circular orbits. A basic introduction to celestial mechanics can be found in Victor G. Szebehely, *Adventures in Celestial Mechanics: A First Course in the Theory of Orbits* (Austin: University of Texas Press, 1989). For a more advanced look, there are new editions of two widely used texts, John M. Danby's *Fundamentals of Celestial Mechanics* (Richmond: Willman-Bell, 1988) and A. E. Roy's *Orbital Motion* (Philadelphia: A. Hilger, 1988), which are classic, thorough, and complete introductions to the mathematics of celestial mechanics.

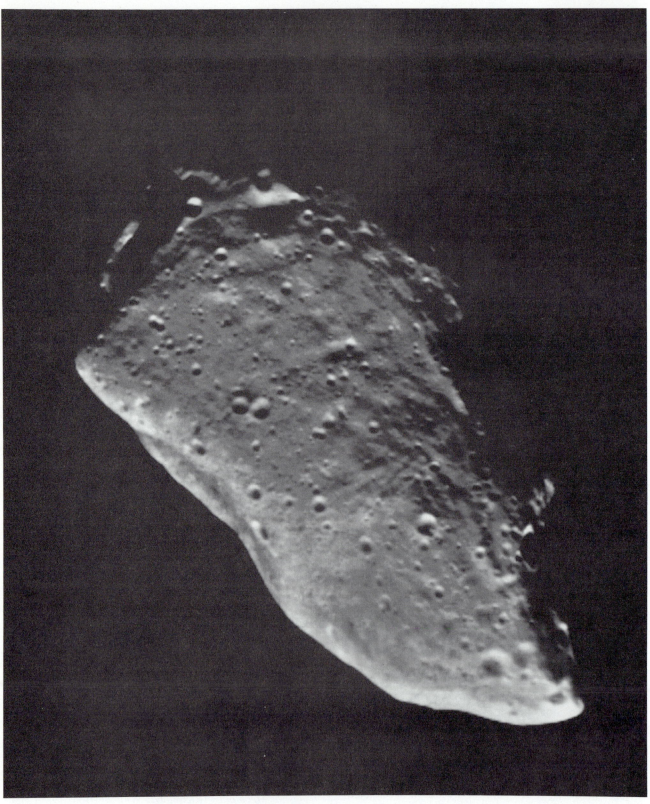

**FIGURE 9.1** The asteroid 951 Gaspra was photographed by the *Galileo* spacecraft during a flyby on October 29, 1991.

# CHAPTER **9**

# Meteorites and Asteroids

## 9.1 METEORITES

The eighteenth century is sometimes referred to as the Age of Enlightenment, a period of intellectual ferment when the great thinkers of the day viewed all traditional ideas with intense skepticism. Why should distances be measured in feet and miles? The metric system was invented. Why should nations be ruled by kings? The American and French revolutions followed. Why should we accept the biblical story of creation? The ideas of Newton and Laplace led to the vision of a relentless, clockwork universe, eternal and without change.

This skepticism was especially harsh on reported phenomena of nature that could not be readily observed. Stories of dragons and sea serpents were declared to be worthless superstitions, as were the strange tales, current among ignorant peasants (first reported by the ancient historians Pliny and Livy and so all the more disdained by eighteenth-century "modern thought") that now and again rocks fell from the sky.

Given the vagaries of weather and observing habits, roughly once a decade someone in Europe might observe a "fireball," a bright bolt of light that would streak across the night sky, changing night to day, sometimes accompanied by thunder or weird whistling sounds. These were just com-

mon enough that eighteenth-century science could admit they did occur. In 1794 the German physicist Ernst Chladni wrote a paper proposing that Pliny's mythical "rocks from the heavens" might actually be somehow associated with these fireballs. But for most scientists, the idea of rocks from the sky smacked too much of superstition. Thomas Jefferson, even while he was president, took an active interest in natural history; following the report of a meteorite fall in New England he is reputed to have said, "It is easier to believe that two Yankee professors would lie than that stones would fall from heaven."

Convincing proof of the existence of meteorites did not come until 1803, when a fall of stones was reported near the town of L'Aigle in France. A 29-year-old scientist, Jean-Baptiste Biot, was dispatched by the French Academy of Sciences to investigate. He collected testimony from several villages that confirmed the time and nature of the fall of stones. And, for the skeptical, he also collected actual samples of the stones themselves. They were all like each other, and none of them like any other rocks native to the area. They were found sitting on top of the ground, as if they were newly placed there. As Biot pointed out in his report to the Academy, there were no active volcanoes anywhere near this town that could have made these stones. His evidence was overwhelming and convincing: The stones had indeed

***TABLE 9.1***

Falls and Finds Up to 1985

| Meteorite Type | Falls | Percentage of Falls | Finds | Percentage of Finds | Total | Percentage of All |
|---|---|---|---|---|---|---|
| Irons | 42 | 5% | 681 | 27% | 723 | 22% |
| Stony-irons | 9 | 1% | 59 | 2% | 68 | 2% |
| Stones | 781 | 94% | 1,741 | 71% | 2,522 | 76% |
| Total | 832 | 100% | 2,481 | 100% | 3,313 | 100% |
| *Of the Stones:* | | | | | | |
| Chondrites | 712 | 91% | 1,667 | 96% | 2,379 | 94% |
| Achondrites | 69 | 9% | 74 | 4% | 143 | 6% |

fallen out of the sky. And so the study of **meteorites**—rocks from the heavens—was born.

(The evidence didn't completely convince everyone, of course. Jefferson took Biot's report to be typical French hyperbole. "The exuberant imagination of a Frenchman... runs away with his judgement," he wrote his friend Andrew Elliot in December of 1803. "It even creates facts for him which never happened, and he tells them with good faith... the evidence of nature, derived from experience, must be put into one scale, and in the other the testimony of man, his ignorance, the deception of his senses, his lying disposition.")

Early suggestions for the origin of these rocks included the idea that they might be condensations from the clouds, like large hailstones, or that they were formed by lightning strikes. But Biot's evidence was most consistent with the theory that they were indeed, in Chladni's words, "masses foreign to our planet."

They are samples of other worlds. We can measure their compositions, study the structures of the minerals, see how the different crystals grew together, measure trace elements, use radioactive age dating to determine their ages, and from all these studies learn what conditions are like now in space and what they must have been like back when these rocks were formed, back at the beginning of the solar system, some 4.6 billion years ago.

## Meteorite Classification

Meteorites come in three main varieties: **stones**, **irons**, and **stony-irons**. The names describe their compositions.

Meteorites that are discovered on the ground are called **finds**; those that were actually observed to fall are called **falls**. By far the greatest number of meteorites actually seen to fall (Table 9.1) are stones. But erosion by wind and water, and all the other processes that weather rocks on Earth, tends to break down and mix the constituents of these stones into Earth's soil. Thus stones are not easily found or rec-

ognized as being extraterrestrial after they've been on the surface of Earth for a few years.

Irons, on the other hand, are just big lumps of metallic iron and nickel; the material is something like stainless steel and does not weather much at all. Also, it is hard to mistake an iron meteorite for a terrestrial sample; iron does not normally appear on the surface of Earth as lumps of metal. The metallic iron first forged by humans in the iron age was actually metal from iron meteorites. Iron meteorites are sometimes turned up by farmers plowing fields.

Stony-irons, the least common type of meteorite, appear to be just mixtures of iron and rock. Like irons, they tend to be more commonly found than observed to fall.

Between falls and finds, scientists, until a few years ago, considered themselves lucky to obtain more than 10 or so new meteorites every year. However, a project started in the 1960s photographed a part of the sky over the central plains of North America every night to see how many meteorites could be seen falling through that part of the sky; from these results, it has been calculated that many thousands of meteorites ought to be hitting Earth every year. Most are simply never noticed, falling onto uninhabited land or into the oceans.

Some regions of Earth, where the surface has been relatively undisturbed for millions of years (parts of Australia, for instance), offer good hunting grounds for meteorites. But even regions such as these are not the most fruitful place to find them.

In 1969, Japanese scientists in Antarctica discovered that certain regions of the ice sheet are rich in meteorites. After a rock lands on the snowpack, it is likely just to be covered by more snow. But there are regions of the continent to which the ice flows and then evaporates away. When the ice evaporates, the rocks are left behind, while more ice flows in carrying its load of rocks frozen in with it. Thus, in these areas, the accumulation of thousands of years' worth of meteorites can be found. The dark masses are easy to pick out against the white ice. Searchers using helicopters

**TABLE 9.2**

A Summary of Meteorite Types

---

## Stones

### Chondrites

*Carbonaceous Chondrites*
*CI* are rich in water and organic material.
*CM* and *CO* contain some water and carbon.
*CV* may, like Allende, have inclusions
with extremely odd isotope patterns.
*Ordinary Chondrites* are the most abundant type, made mostly
of olivine and pyroxene.
*High Iron (H)*
*Low Iron (L)*
*Very Low Iron (LL)*
*Enstatite Chondrites (E)*

### Achondrites

*Basaltic Achondrites*
*Eucrites* are plagioclase–pyroxene basalts.
*Diogenites* are mostly pyroxene.
*Howardites* are a mix of Eucrites and Diogenites?
*Shergottites* may come from Mars.
*Nahklites* may come from Mars.
Others
*Enstatite Achondrites* contain very unusual minerals, made of
sulfides instead of oxides.
*Ureilites* contain diamonds.
Others

## Irons

### Nickel-Rich
### Nickel-Poor

## Stony-Irons

**Pallasites** are iron surrounding crystals of olivine.
**Mesosiderites** appear to be a mixture of irons and basaltic achondrites.

---

have recovered many hundreds of new meteorites in this way.

## Composition of Meteorites

The first step in making sense out of the thousands of meteorites in our collections is to organize them according to their composition and structure. A summary of this classification scheme is given in Table 9.2.

Stony meteorites come in two varieties, called **chondrites** and **achondrites,** depending on whether or not they have chondrules. **Chondrules** are little beads of rock (see Figure 9.3). Most meteoriticists assume that the chondrules were formed when grains of meteoritic material were melted

and then refrozen, but nobody really understands well why or how they were melted. Some chondrites look to be made of nothing but chondrules, clustered together like tiny grapes a millimeter across.

These chondrites can be further divided into three types. The most common, simply called **ordinary chondrites**, can be divided further into groups called **H, L,** and **LL**, representing high, low, and very low iron contents. They are primarily made of the minerals olivine and pyroxene with some small bits of metallic iron, and they tend to be rich in chondrules.

The minerals in meteorites often appear to have been recrystallized. The more reworked the minerals in the meteorite appear to be, the higher the meteorite's **petrographic**

**type** (determined by looking at an extremely thin piece of it—a **thin section**—in a microscope), which is designated by a number from **1** (no recrystallization, true only of *C*-type meteorites) to **6** (so recrystallized that there are few recognizable chondrules left). Thus a meteorite identified as an *H5* is a high-iron ordinary chondrite that has seen considerable reworking of its minerals; an *LL3* would be very low in iron oxide, with many well-defined chondrules and less recrystallization.

What formed the chondrules? And what recrystallized these meteorites? These questions are still hotly debated.

By contrast, a rarer class of stones called **enstatite chondrites** (designated *E3* to *E6*; no *E1*'s or *E2*'s have been found) are made almost entirely of the mineral enstatite, with metal and iron sulfide as minor constituents. These meteorites also tend to have some exotic minerals, such as titanium nitride, for instance, that are never seen in Earth rocks.

Perhaps the most intriguing of the chondrites are the **carbonaceous chondrites**. As the name implies, their minerals bear carbon, and they also have many minerals that are rich in hydroxides and may contain complex hydrocarbons. The most primitive of these meteorites, designated *CI*, can be made up of almost 20% water by weight. Minerals with so much water can be altered quite easily if they are heated to temperatures not much warmer than typical Earth conditions.

But age dating has determined that these rocks are 4.6 billion years old, the oldest materials known; thus it is clear that these rocks have been kept cold and relatively untouched since the very beginning of the solar system. However, looking in detail at the shapes and positions of the crystals in these meteorites, some scientists have suggested that they may have been in the presence of flowing, liquid water at some time in their distant past. Another type of carbonaceous chondrite, type *CV*, has only trace amounts of water and carbon.

Meteoriticists originally designated carbonaceous chondrites by the symbols *C1* through *C4*, but more detailed studies showed that this simple gradation in petrographic scale did not adequately describe all the differences between these meteorites. The designations are now based on comparisons with standard, well-studied samples. Thus *CI* (formerly *C1*) meteorites resemble the meteorite Ivuna; *CM* (formerly *C2*) meteorites resemble Murchison; other *C2* meteorites make a different class, *CR*, resembling the meteorite Renazzo. *CV* (roughly speaking, old type *C3*) meteorites resemble Vigarano; *CO* meteorites (once lumped in with the other *C3*'s) are those that resemble Ornans. (Individual meteorites are named for the place where they are found.) The essential point is that, while one could conceivably turn an *H3* meteorite into an *H6* meteorite by mechanically breaking up the chondrules, the chemical differences between a *CI* and a *CV* meteorite are so profound that one could not possibly be formed from the other.

Meteorites that do not have chondrules are called **achondrites**. Most of them are significantly different from the chondrites. Many appear to be basalts, and these are called **basaltic achondrites**. Two meteorites discovered in the Antarctic hoard look just like Moon rocks in all details and may have indeed come from the Moon; most of the rest of the basaltic achondrites may have come from a larger asteroid such as Vesta. It has been suggested that a few rare kinds of basaltic achondrites (called **SNCs** after the first three examples, **S**hergotty, **N**akhla, and **C**hassigny) might be samples from volcanoes on Mars, blasted off its surface by a giant impact.

Achondrites are generally quite rare, and they include some of the more bizarre kinds of minerals. Meteorites of one class, called **ureilites**, contain tiny diamonds possibly caused by an energetic collision of an ordinary meteorite with a carbon-rich asteroid. **Enstatite achondrites**, like the enstatite chondrites, have strange minerals that must have been formed where there was little oxygen and plenty of sulfur. A few achondrites are one of a kind, rocks different from any other known kind of rock including other meteorites.

**Iron meteorites** are mostly just that, metallic iron. They usually have about 6% nickel (one iron is known to be 60% nickel, but that's unusual). The nickel forms two crystal structures with iron, a nickel-rich crystal called **taenite** and a nickel-poor crystal called **kamacite**. When these two crystals grow together in a slab of iron, they make a very distinctive pattern called a **Widmanstätten pattern**, which can be seen when the metal is polished and etched with acid (Figure 9.2). To make this distinctive pattern, the metal must be rich in nickel, warmed to nearly melting, then allowed to cool slowly over a very long time.

**FIGURE 9.2** Widmanstätten patterns in the LaPorte iron meteorite.

The iron meteorites we see appear to have cooled over a period of millions of years. Because only meteorites have enough nickel to make this pattern, its presence is used as a test to differentiate between meteoritic and terrestrial iron.

The iron in stony-iron meteorites is the same as that in the irons. If the stony part is similar to certain basaltic achondrite classes, such meteorites are called **mesosiderites**. Other stony-irons, called **pallasites**, consist of meteoritic iron with big lumps of clear, green olivine, called **peridot**. However, there are certain one-of-a-kind varieties of stony-iron meteorites, too. In some cases, the iron seems to have been molten and surrounds rocky material that does not seem to have been heated. How did it happen?

## The Allende Meteorite

In February 1969 a large collection of stones, several tons in total weight, fell near the village of Allende in northwestern Mexico. Collectively, these stones are known as the Allende meteorite (Figure 9.3).

These stones happened to land at a most propitious time and place. The time was just before the first rocks were to come back from the Moon (*Apollo 11* landed on the Moon on July 20, 1969), and in preparation many laboratories around the world had put together elaborate equipment with trained personnel to study these first rocks from the Moon. Allende landed only a few hundred miles from the Johnson Space Center in Houston and several labs in southern California, including those at Cal Tech, UCLA, and the University of California at San Diego. Numerous

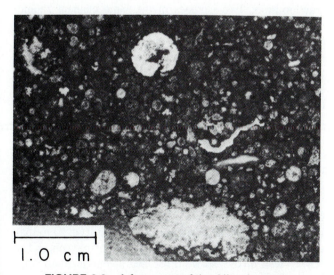

**FIGURE 9.3** A fragment of the Allende meteorite, showing round **chondrules** and irregular white inclusions. (The large object at the bottom is not a white inclusion; it is rich in olivine.)

samples were taken and analyzed as a sort of rehearsal for the return of the Moon rocks. In many ways, however, the Allende meteorite turned out to be as exciting as the Moon rocks themselves.

Allende is a *CV* carbonaceous chondrite. Along with the dark, fine-grained mass and the occasional chondrules, however, there are small white inclusions ranging in size from 1 mm to almost 2 cm across. These white blobs are made up of unusual minerals such as melilite, perovskite, and spinel, minerals not normally seen on Earth, but very similar to what some theorists had predicted the very first rocky material in the early solar system might have looked like. The arguments are still going on as to the origin of these white inclusions, however.

Further study of these white inclusions revealed an even greater surprise. Throughout the meteorite, the oxygen—the most abundant element in meteorites—was enriched in the isotope $^{16}O$. And in some, but not all, of the inclusions the isotope abundance patterns of several other elements appeared to be strangely different from the abundances of the isotopes as seen in Earth rocks.

What does this mean? We describe the whole problem in detail in the box, but to summarize some of the results, the observations are these. First, the oxygen in most of the inclusions appears to have an excess of the isotope $^{16}O$. This was so surprising that geochemists started to look for extra $^{16}O$ in other places and discovered that many other meteorite classes also seem to have smaller, but distinct, excesses or deficits in $^{16}O$ compared with Earth rocks. (Moon rocks show no such excess.) But there are few obvious processes, short of the sort of nuclear reactions that only go on inside large stars, that could produce pure $^{16}O$. So how did it get into the meteorites?

The problem doesn't stop with oxygen. In different inclusions of Allende—in some, but not all, of them—there is evidence that the magnesium present contains an excess of $^{26}Mg$. This excess is directly related to the aluminum content of the inclusion: the more aluminum there is, the more $^{26}Mg$ is seen. This can be explained by noting that there is a radioactive isotope of aluminum, $^{26}Al$, which decays to $^{26}Mg$. Obviously, there was some $^{26}Al$ in these inclusions when they were formed. The trouble is, $^{26}Al$ has a half-life of only 700,000 years, while many lines of evidence have suggested that at least 200 million years must have passed between nucleosynthesis and the time when the meteorites in the solar system were formed. Where did the $^{26}Al$ come from?

A very few of the white inclusions have been found that have bizarre isotope abundance patterns among the heavy elements Ba, Sr, Nd, and Sm. The easiest way so far to make sense out of these findings is if these elements saw r-process and s-process material in proportions that were different from the uniform pattern seen in every other meteorite and Earth rock. How could this occur?

**FURTHER INFORMATION...**

# Oxygen Isotope Anomalies

Recall the definitions of elements and isotopes. Each element is determined by the number of protons in the nuclei of its atoms; variations in the number of neutrons produce the various isotopes. The relative abundance of one isotope compared with another one was one of the tools we used in understanding the nucleosynthesis of the elements. On Earth, if you have a sample of oxygen, you expect the abundance of $^{18}O$ and $^{17}O$ to be much less than that of $^{16}O$. But will the ratio of $^{18}O$ to $^{16}O$ atoms be exactly the same in all Earth rocks? Not necessarily.

There are two ways that isotope abundances could change. One way is if there is some process making more of one of the isotopes. Outside of a star or a nuclear reactor, this does not occur very often. This process can occur if an isotope is hit by an energetic cosmic ray; this might result in a new isotope being formed, but such events are quite rare. A more common occurrence is that of a naturally radioactive isotope that decays. Then the abundance of that isotope always decreases, while the abundance of the isotope that is its daughter always increases. But this can be important only for those radioactive isotopes whose half-lives are long enough that there can be significant amounts of them available long after nucleosynthesis has stopped.

How long is it between the time a star goes supernova, spreading its newly formed atoms into space, and the time when those atoms get incorporated into a newly forming stellar system? Two lines of evidence give a roughly similar answer. The Milky Way, our galaxy, is a large disk filled with stars, but the stars in this disk tend to be clumped together into "arms." Stars move about within this disk as they orbit the center of the galaxy, and so they may move out of one arm and into another. Current measurements indicate that it takes about 200 million years to move from arm to arm. But because most stars are in arms, we expect most supernovas to take place there; and because most of the gas is in the arms, too, we expect most star formation to take place there. Present theories of star formation predict that material from a supernova in one arm will not form back into a star until it has reached another arm; thus there should be a 200-million-year lag between a supernova and the formation of lumps of rock from its debris.

Looking at the meteorites themselves, we notice that virtually none of them shows any variation in the abundance of an isotope that is the decay product of materials with a half-life of less than 10 million years. Because, after 10 half-lives, the amount of radioactive material remaining is less than one part in 1000 of the original abundance and we can usually detect variations on that scale, the majority of material in the solar system must have been synthesized, decayed, and uniformly mixed together at least some 100 million years before it could be trapped in rocks and show up as an isotope anomaly. On the other hand, anomalies are seen in stones that contain iodine. These meteorites tend to have an excess of $^{129}Xe$, which is the decay product of the radioactive isotope $^{129}I$. The half-life of $^{129}I$ is 16 million years. The implication is, therefore, that the time between formation of the elements and formation of the meteorites was not much more than 160 million years (using our 10-times the half-life rule).

Both lines of reasoning therefore lead us to the conclusion that isotopes with half-lives shorter than 16 million years should not survive in enough abundance to make any differences in the abundances of isotopes in meteorites.

Besides decay products, there is another way of altering the abundances of isotopes in a rock, called **chemical fractionation**. Two isotopes of the same element should have the same chemical behavior, because that chemical behavior is determined by the electrons of an atom, not its nucleus. However, the slightly heavier isotopes are just a little slower-moving when chemical reactions take place. Because they have more mass to drag around, the forces that move atoms into new chemical arrangements are just a little less effective on these isotopes. And so, in some processes, it is possible to enrich a chemical product in the lighter isotopes and leave the reactants with the heavier isotopes. But this process should behave in a regular way for each isotope of an element.

For example, look at the element oxygen. Most oxygen is $^{16}O$, so we compare the abundances of the other two isotopes with this one and contrast the ratio $^{17}O/^{16}O$ to that of $^{18}O/^{16}O$. Sea water on Earth has lots of oxygen and it is very well mixed, so we'll use that as our standard. The ratios turn out to vary by only a tenth of a percent or so from rock to rock on Earth, so to see this tiny change, we'll define two numbers $\partial_{17}$ and $\partial_{18}$.

Take a rock from Earth and measure the $^{17}O/^{16}O$ ratio. Divide this ratio by the standard $^{17}O/^{16}O$ ratio of sea water. The result is likely to be some number close to 1; call it, for our example, 1.0012. Subtract 1 from this number and then multiply what's left by 1000. This is then called the $\partial_{17}$, and it is equal to 1.2 in this case. Mathematically, we say

$$\partial_{17} \equiv \left\{ \frac{\left[^{17}O/^{16}O\right]_{rock}}{\left[^{17}O/^{16}O\right]_{seawater}} - 1 \right\} \times 1000 \qquad (9.1)$$

The ratio $\partial_{18}$ is defined in the same way.

When you plot $\partial_{17}$ versus $\partial_{18}$ for a number of Earth rocks, the points all line up, as can be seen in the solid line of Figure 9.4. This line has a slope of 1/2, which is exactly what we would expect, because $^{18}O$ is two neutrons heavier than $^{16}O$ while $^{17}O$ is only one neutron heavier. So the "differential difference" should be twice as great for the $^{18}O$ as for the $^{17}O$, and it is.

But if you plot where the oxygen from the Allende meteorite falls, it is completely different. First, it does not lie on the same line as Earth rocks. If you look at other meteorite types, you can see that their oxygen isotope abundances lie on different lines, too. But Allende's oxygen isn't even on a line with a slope of 1/2. It looks as though some process injected pure $^{16}O$ into the material that made up Allende, lowering both the $\partial_{17}$ and $\partial_{18}$ ratios at the same rate. Different inclusions seem to have gotten different amounts of $^{16}O$, so when you plot them all they fall on a line of slope 1. The change is the same for $^{17}O$ and $^{18}O$.

Likewise, it appears that the other meteorite classes also had varying amounts of $^{16}O$. (By contrast, the Moon and Earth appear to have identical $^{16}O$ contents.) Where is this pure $^{16}O$ in the meteorites coming from? Theorists are still arguing that one.

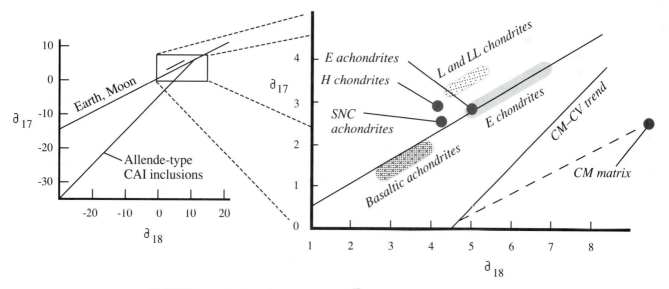

**FIGURE 9.4** A plot of the amount of $^{17}O$ in a rock versus the amount of $^{18}O$ (the $\partial$ terminology is explained in the text). For all Earth and Moon rocks, the abundances of these two isotopes of oxygen fall on a single line. However, various meteorite classes lie on different lines, suggesting that they came from different regions of the early solar system, which had distinctly different abundances of the isotopes.

These isotope anomalies in a meteorite that appears to date from the earliest days of the solar system have given rise to several possible explanations. One theory proposes that the early Sun was extremely energetic, pouring out intermittent bursts of energetic radiation that could have changed the isotope patterns in any dust grains it happened to encounter. Another suggests that a supernova went off near the Sun just as it was forming, and perhaps such a supernova was even instrumental in causing the Sun to form. Or it is possible that the white inclusions in Allende are not from our solar system at all, but are bits of dust that travelled to us directly from the supernova or star that formed their elements. If this is so, then meteorites may be made up of, quite literally, raw stardust.

Which theory is correct? Possibly all three mechanisms occurred. Further studies of these inclusions may inspire still other hypotheses concerning the region around our Sun when the planets were forming.

## SUMMARY

It is estimated that tens of thousands of rocks from space land on Earth every year. These meteorites are generally very old, and they are believed to be representative of material in the early solar system, at the time when the planets were forming.

Most meteorites seem to be stony meteorites, usually chondrites, containing small round beads of rock less than a millimeter across, called **chondrules**. A special type of chondrite, called a **carbonaceous chondrite**, seems to be very primitive because it is rich in carbon and water-bearing minerals. Enstatite chondrites seem to have been formed under very unusual chemical conditions, and they contain many unusual minerals. Stony meteorites without chondrules are called achondrites; most of these are basaltic achondrites, which appear to have been formed from lavas just as igneous rocks on Earth are formed.

Iron meteorites are noted for being rich in nickel as well as iron. When a smooth surface of the metal is etched with acid, the nickel-rich and nickel-poor crystals form a characteristic pattern called a Widmanstätten pattern. Some meteorites are a mixture of irons and stones, usually material similar to basaltic achondrites. These meteorites are called stony-irons.

Meteorites are material that has come to us relatively unchanged since the formation of the solar system. As such, they contain clues to many of the complex physical processes that went on at that time. Measurements of the white inclusions from a meteorite called Allende have prompted a new generation of theories about the formation of isotopes and the distribution of elements in the early solar system.

## STUDY QUESTIONS

1. Widmanstätten patterns are found in what type of meteorite?
2. Roughly how many meteorites hit Earth every year?
3. What type of meteorite contains water and complex organic material?
4. What are chondrules?
5. Some people think that the white inclusions of Allende contain bits of "raw stardust." Why?

## 9.2 ASTEROIDS

What exactly is a planet? Several moons of Jupiter are bigger than the planets Mercury or Pluto. Should Earth's Moon, which has had a complex history, be any less a planet than Pluto, just because it happens to be orbiting Earth instead of the Sun?

Traditionally, the answer is yes. To be called a planet, a body must orbit the Sun. Even with this definition we make a distinction between the nine traditional major planets and the thousands of smaller bodies that also populate the solar system.

Asteroids and comets are all considerably smaller than any planet. The largest asteroid, Ceres, is only 512 km in radius, comparable to the Saturnian moons Tethys and Dione. Comets are only a few kilometers in radius. We will discuss comets further in Chapter 13. An icy asteroid, Chiron, orbiting between Saturn and Uranus, might be considered a large comet by virtue of its composition; recently it has been seen to develop a gas cloud similar to a rudimentary tail.

That is not to say, however, that these **minor planets** are unimportant. For by virtue of being so small, they probably have come down to us relatively unaltered since the beginning of the solar system. They are the debris left over from the formation of the planets. As such, they can give us valuable clues as to what went on when the solar system was being formed.

Even more, the asteroids are almost certainly the source of most meteorites landing on Earth today. That means that the asteroids ought to be made up of the same material that we find in meteorites. But we know that the meteorites are rich in carbon, iron, nickel, and many rare (and precious) metals; even diamonds have been found in some of them. It seems reasonable that, in the twenty-first century, they will become an important commercial source of raw materials. So there are serious practical reasons to want to discover the sizes, compositions, and locations of the asteroids.

## Asteroid Sizes

Table 9.3 shows a listing of 18 asteroids with at least 125 km radius. Smaller than these, some 200 asteroids have radii greater than 50 km, and there are another 4000 smaller ones (down to about 5 km radius) that have been found, numbered, and named. Perhaps another 4000 or more may may lie undiscovered yet, orbiting in the asteroid belt.

**TABLE 9.3**

Asteroids with Radii ≥ 125 km

| | Asteroid | Radius (km) |
|-----|-----------|------------|
| 1 | Ceres | 512 |
| 4 | Vesta | 277 |
| 2 | Pallas | 269 |
| 10 | Hygiea | 221 |
| 704 | Interamnia | 169 |
| 511 | Davida | 167 |
| 65 | Cybele | 155 |
| 52 | Europa | 145 |
| 451 | Patientia | 140 |
| 31 | Euphrosyne | 135 |
| 15 | Eunomia | 130 |
| 324 | Bamberga | 128 |
| 107 | Camilla | 126 |
| 87 | Sylvia | 125 |
| 45 | Eugenia | 125 |
| 3 | Juno | 125 |
| 16 | Psyche | 125 |
| 24 | Themis | 125 |

(Asteroids are numbered in the order of their discovery. Unlike most astronomical objects, they can be named after anything, including living people. The discoverer gets to choose the name, subject to approval by a committee of the International Astronomical Union.)

Not surprisingly, most of these bodies are nothing but chunks of rock that, so far, have escaped being captured by a larger planet. Indeed, most of the smallest asteroids may be fragments of the larger ones. As these asteroids collide, they keep chipping pieces off each other, or they break each other up completely. Eventually all the larger asteroids will probably be broken down into small pieces of rubble.

Even the largest of these rocks are so small that they appear as no more than points of light in a telescope, too small for us to measure the apparent angular size of their disks. So how can we tell how large these asteroids are?

One way is to measure how bright they appear. The apparent brightness of the asteroid depends on three things.

1. How close is it? The closer the asteroid is to us, the brighter it appears.

2. How big is it? The bigger the asteroid, the more light it can reflect back towards Earth.

3. How dark is it? A large but dark asteroid can reflect as much light as a small but light one.

The **albedo** of a body is a measure of its lightness or darkness; the lighter its color, the more light it reflects. It is a number between 0 and 1, representing the fraction of the light hitting the surface of the body that gets reflected back into space.

A body with an albedo of 1 would be very bright, and very cold, because it reflects all the Sun's light and is warmed by none of it. (Enceladus, an icy moon of Saturn, has an albedo near 1.) At the other extreme, a body with an albedo of 0 would be warm, but pitch black and invisible. (Phobos, a moon of Mars, has an albedo of 0.06.) In nature, the blackest of materials has an albedo of about 0.03; coal dust is an example of such a substance. Most rocks have an albedo of about 0.1 to 0.2. Dirty ice has an albedo of 0.5, while very clean ice can approach an albedo of 1. The albedo can depend on the color of the light, of course. A red object is one whose albedo for red light is higher than its albedo for other colors (assuming that it is not red because it is very hot!). Measuring the albedo as a function of color is the same thing as measuring the reflectance spectrum of an object (see the box on reflectance spectra in Chapter 4).

Measuring the amount of light coming from the asteroid is the first step in solving the problem of finding an asteroid's radius. Next, by noting its positions at several different times, we can derive an asteroid's orbit and so deduce how far away it is from us and the Sun. But even given the brightness of an asteroid and its distance from the Sun and us, we still must choose: Is it smaller but lighter colored, or bigger but darker?

To find the answer, we need one more measurement. This time, instead of seeing the asteroid's brightness in visible reflected light, we measure it in radiated infrared light.

Recall that the wavelength at which a planet emits the most radiation depends on the temperature. But (as we discuss further in Chapter 10) the temperature depends on the albedo. It is a familiar experience even here on Earth that darker objects left out in the Sun get hotter than lighter objects; the same is true for asteroids in space. Therefore, measuring the infrared brightness and thus determining the temperature helps tell us the albedo of the asteroid (assuming that the visible and infrared albedos are the same, which may not be a good assumption).

In effect, the two formulas for visible brightness and infrared brightness as functions of the planet's size and albedo represent two equations with two unknowns, which we can solve simultaneously. Knowing the brightness and the temperature, we can then solve for the albedo and the radius of the planet.

## Composition Classes of Asteroids

By this technique, we've learned not only the size of the asteroids but also their albedos. It turns out that asteroids come in all sizes, but most of them can be divided into a few distinct albedo classes. By looking more carefully at the reflectance spectra of the asteroids (in essence, the color of the asteroid), we can further subdivide these groups.

Members of the most numerous group are quite dark, with albedos ranging from 0.02 to 0.06; that is, they absorb all but 2% to 6% of the sunlight they receive. This makes them almost pitch black, like coal dust. The spectra of some of the biggest dark asteroids appear to be similar to the class of meteorites called carbonaceous chondrites. Thus these asteroids are called **C**-type asteroids.

The next most common group are called **S**-type asteroids because in both albedo and spectra they appear to resemble stony or stony-iron meteorites. These albedos range from 0.07 to 0.23; that is, they reflect 7% to 23% of the Sun's light. Studies of hundreds of asteroids have revealed other classes as well (see Table 9.4) that often show similarities to various meteorite classes.

**TABLE 9.4**

Classes of Asteroids

| Type | Albedo | Interpretation |
|------|--------|----------------|
| S | 0.07–0.23 | Stony, metal |
| C | 0.02–0.06 | Carbonaceous |
| M | 0.07–0.2 | Metal |
| E | 0.2 + up | Enstatite |
| R | 0.16 + up | Iron oxide? |
| P | 0.02–0.07 | Dark metal? |
| D | 0.02–0.07 | Organic? |
| U | various | Unclassifiable |

## Location of the Asteroids

This similarity between asteroid spectra and meteorite spectra is one link implying that the meteorites are fragments chipped from asteroids. Another link is that the orbits of a few meteorites have actually been traced by photographs taken as they fell. In every case, the orbits extend back to the asteroid belt between Mars and Jupiter.

The vast majority of asteroids are seen to orbit somewhere between Mars and Jupiter. When one plots the number of discovered asteroids versus their mean distance from the Sun (as is done in Figure 9.5), one notices a strange pat-

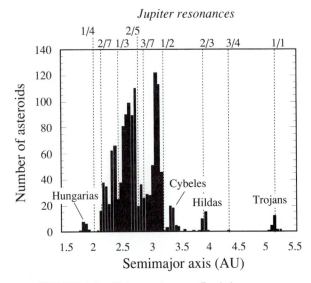

**FIGURE 9.5**   Kirkwood gaps. Each bar represents the number of asteroids whose orbits lie at a given distance from the Sun. The fractions (1/2 and so forth) indicate resonances with Jupiter. The locations of certain asteroid families are also marked.

tern. The distribution of asteroids is not smooth. There are distinct gaps at certain, well-defined places. These are called **Kirkwood gaps** after the American astronomer Daniel Kirkwood, who first noted them in 1867.

The gaps seem to occur when the orbital period of the asteroid is some simple fraction of Jupiter's year. The orbit of any asteroid in such a position is said to be in **resonance** with Jupiter. In such an orbit, the small gravitational perturbation of Jupiter on the asteroid can build up with time, until such asteroids are completely perturbed out of circular orbits and into orbits that are likely to cross the path of some other asteroid or planet (see Chapter 11). Asteroids in such orbits are likely to suffer collisions and be swept up by other asteroids, broken into smaller pieces, or at least knocked completely out of their original orbits.

It is interesting to note, however, that at a few of the resonances there is an extra number of asteroids instead of a deficit. As we will see in planetary ring systems, the effects of perturbations on a collection of small bodies can be very complex, and the results are not always simple or obvious. No one has yet come up with a simple explanation for these odd clusters.

Along with the gaps, other patterns can be seen in the distribution of asteroid orbits, shown in Figure 9.6. Although there is a lot of overlap, one can see that different classes of asteroids tend to be concentrated in different regions of the asteroid belt. The S-type asteroids tend to predominate in the inner part of the main belt, while the C-type asteroids tend to be found farther out. Other small classes such as E and M also have distinct neighborhoods.

**FURTHER INFORMATION…**

# The Potential Energy of a Body in Space

Newton's laws tell us that to accelerate a body of mass $m$ at a rate **a**, we need a force $\mathbf{F} = m\mathbf{a}$. If we know the force, we can calculate the acceleration and ultimately the eventual path of the body. However, a force is a vector quantity, and vector equations can be very cumbersome to work with. One of the more clever tricks in physics is to replace a force vector field with a potential energy field. Instead of defining the force at any given point, it is sometimes possible to assign an energy number to each point. Because energy is a scalar, the complexities of adding or manipulating vectors is avoided.

The amount of force in any given direction is simply the gradient of the energy field, that is, the change in the potential energy quantity from point to point divided by the distance between the points. The direction of the force is from higher to lower potential; think of a ball rolling downhill. The net force is always in the direction of the steepest gradient, just as water always finds the steepest and fastest route down a hill.

Because the force only depends on a difference in potential energy, it doesn't matter at what spot we define the point of zero potential energy. Say that point $A$ has a potential of 0 and point $B$ has a potential of $+10$. If we wished, we could make point $B$ our zero point and subtract 10 from the energy level at every other point; then point $A$'s energy would be $-10$. But in either case, the difference of $B$ from $A$ remains the same.

In space we are sometimes interested in moving from planet to planet or even star to star. Any system that assigns the zero energy point to some place on a particular planet will wind up being inconvenient whenever that planet is not involved in the calculation. The one spot that is the same for all stars and all planets is a spot infinitely far from them all. Conveniently, the force of gravity from all these planets also goes to zero at infinity. Thus we choose to set our zero potential point at infinity.

How do we derive the formula for the potential energy of a planet, given this choice of zero potential? Recall that the force **F** is the gradient of the potential energy $U$ in the direction of lower potential. Mathematically we write

$$\mathbf{F} = -\frac{dU}{dr}\hat{r} \tag{9.2}$$

where the negative sign indicates the gradient running downhill and $r$ is our measure of distance. Because the force of gravity only depends on $r$, we can drop the vector notation and write this as an integral equation:

$$\int F\,dr = -\int dU \tag{9.3}$$

The integration is taken from the zero point, infinity, to the point of interest, $R$. But we know what $F$ is, from Equation 8.26:

$$F = -\frac{GMm}{r^2} \tag{9.4}$$

where $M$ is the mass of the planet and $r$ is the distance from the center of the planet. So we can write

$$\int_{\infty}^{R} -\frac{GMm}{r^2}\,dr = -\int_{\infty}^{R} dU$$
$$= -U(R) \tag{9.5}$$

But evaluating the first integral yields:

$$\left(\frac{GMm}{R} + \frac{GMm}{\infty}\right) = \frac{GMm}{R} \tag{9.6}$$

We thus arrive at

$$U(R) = -\frac{GMm}{R} \tag{9.7}$$

Notice that, by this definition, $R$ is always a positive number and so $U$ will always be a negative number.

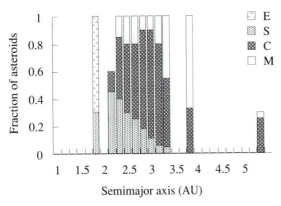

**FIGURE 9.6** The relative fractions of asteroid types are represented by different shaded bars, as a function of their distance from the Sun.

## Asteroid Groups and Families

Along with the general spread of small planetoids circling the Sun, there are several special groups and families of asteroids. A **group** of asteroids are those that are in similar orbits. Among the most important groups are the **Apollo** asteroids, which travel in paths that cross the orbit of Earth, and the **Eros** asteroids, which cross Mars's orbit but do not reach Earth. These asteroids appear to be very similar to ordinary chondrites in color and albedo, and it is believed that most meteorites that land on Earth originate from these asteroids.

Another famous group is the **Trojan** asteroids. They orbit not in the asteroid belt but in two special regions in Jupiter's orbit, 60° behind and in front of Jupiter. Thus they are at equal distances from both Jupiter and the Sun.

**Families** of asteroids have not just similar orbits but are also closely related in spectral class. These asteroids may have all originated from one larger asteroid that broke apart sometime in the last few million years. Like asteroid groups, families are generally named after the largest asteroid in the family.

One example is the **Themis** family. These are a collection of C-type asteroids in the outer part of the belt, with a small number of large objects and a cloud of smaller ones. Together they would have made a body 300 km in radius. Another asteroid, 171 Phelia, is an M-type asteroid made largely of metal, with the same orbital characteristics as the Themis family. It may represent the metal core of the original parent body. Other examples of asteroid families are marked in Figure 9.5.

## SUMMARY

Asteroids make up many tens of thousands of minor planets orbiting the Sun. Asteroids can be classified by their surface composition and appear to come in chemical types reminiscent of meteorite types. Each class of asteroid tends to be found in its own particular region of the asteroid belt between Mars and Jupiter, although these regions overlap considerably.

## STUDY QUESTIONS

1. Between which planets' orbits are the majority of the asteroids found?

2. Are the Apollo asteroids, all asteroids whose orbits cross Earth's orbit, an example of an asteroid group or an asteroid family?

3. When tabulating asteroid orbits, one notices that there are few asteroids orbiting at certain semimajor axes. What are these gaps called?

4. How big is the largest asteroid?

5. Give an example of an asteroid class and a meteorite type that may have come from that class of asteroid.

## 9.3 CALCULATING SIMPLE ORBITS

We have samples of meteorites in our labs, and we see asteroids that look in many ways like they could be the parent bodies of the meteorites. The obvious question thus arises: What does it take to get a meteorite from the asteroid belt to Earth? Another way of asking essentially the same question is, What would it take to get us from Earth out to where we might sample the asteroid belt, to collect rocks that would have considerable scientific interest or commercial value?

The best spaceship in the world, like the finest car or fastest plane, can't go anywhere without fuel. How much fuel does it take to get to the asteroid belt? But fuel is merely a way of storing energy; what we're really asking is, How much energy is involved in moving from one place in space to another?

The answer is to be found in the laws of orbiting bodies. In Chapter 8 we examined how the basic laws of physics and the techniques of calculus could be used to define the paths of planets orbiting the Sun and moons orbiting planets. Now that we know the general rules, how do we go about applying them? How do we set a spaceship from one orbit into a new orbit, or into an orbit that makes a path from one planet to another? In general, how do bodies behave when they are in orbit?

## Orbital Potential Energy

The first step in understanding this sort of celestial navigation is understanding the energy contained in any orbiting body.

We start by considering the most basic system. A small body of mass $m$, our spaceship, a meteorite, or an asteroid itself orbits a much larger body of mass $M$, the Sun, Earth, or some other planet. Our coordinate system is centered on the larger mass; $r$ is the distance from the center of that mass to our orbiting body.

The energy of a body is the sum of its kinetic and potential energies. The **kinetic energy** is the energy expressed in the motion of the body. For a body of mass $m$ moving at a speed $v$, it is always equal to

$$K = \tfrac{1}{2}mv^2 \qquad (9.8)$$

The **potential energy** is the "stored" gravitational energy of the system. For example, lifting an object against the pull of gravity stores energy in that object. Once we drop it, the object falls downwards, turning this stored potential energy into moving, kinetic energy.

Potential energy is measured relative to some arbitrary point that we define to have zero potential energy. (Recall our discussion of potential energy in Chapter 4.) In the case of lifting a rock from the surface of Earth, one might start by considering the surface to be the zero energy point. The higher the rock, the greater the potential energy.

In this case, the potential energy would be some positive number. But if the rock fell into a hole, lower than the surface, it would take a negative number to describe its potential energy. We could keep redefining our zero point to be the lowest spot of the deepest hole, but clearly this is not a practical solution. Furthermore, what would be the appropriate surface to consider when we look at objects off the surface of the planet altogether and out in space?

For orbiting bodies moving due to the force of gravity, it is usual to pick our zero-potential-energy surface to be located an infinite distance away from the central mass $M$. With this definition, it can be shown (see the box) that the potential energy is given by

$$U = -\frac{GMm}{R} \qquad (9.9)$$

where $G$ is the gravitational constant and $M$ the mass that is being orbited. But notice that because everywhere is "below" this surface at infinity, $U$ is always a negative number.

As a body moves closer to the central mass, $R$ gets smaller, $1/R$ gets bigger, and so $U$ becomes ever more negative; thus it gets smaller. But kinetic energy is equal to one half mass (always a positive number) times velocity squared (also always positive). So kinetic energy is always a positive number.

Energy is **conserved**; that is, the total energy of the system, the sum of the kinetic plus potential energies, must always stay constant. For the total energy to stay constant while $U$ becomes ever more negative; the kinetic energy must become a bigger positive number. The body must have more kinetic energy when it moves closer to the source of the gravitational pull. Thus when an orbiting body comes closer to the primary, it must move faster. This agrees with Kepler's laws.

## The Energy in an Orbit

Because total energy is constant for a body in orbit, every ellipse that represents a possible orbit about our big mass $M$ has a certain total energy associated with it. How can we find the value of that energy?

Consider an asteroid travelling in a circular orbit around the Sun. It is located a distance $R = a$ away from the Sun, so its potential energy must be given by

$$U(R) = -\frac{GMm}{a} \qquad (9.10)$$

To find its kinetic energy, we need to know the velocity of the body. From Kepler's third law, we recall that the distance cubed divided by the period squared was proportional to the mass of the object being orbited. We can write that in mathematical form: $a^3/P^2 = kM$. What is $k$, the constant of proportionality? It can be shown that $k = G/4\pi^2$. We can use algebra to rearrange this formula to solve for the period squared:

$$P^2 = \frac{4\pi^2 a^3}{GM} \qquad (9.11)$$

Now we ask, How fast does our orbiting body travel? It completes one orbit, a distance of $2\pi a$, in a time of $P$ seconds, so it must be travelling at a speed of

$$v = \frac{2\pi a}{P} \qquad (9.12)$$

The velocity squared, which is what we need for our kinetic energy formula, is $4\pi^2 a^2/P^2$. Substituting in Equation 9.11 for $P^2$, we find that the velocity squared is

$$v^2 = \frac{4\pi^2 a^2}{P^2} = \frac{GM}{a} \qquad (9.13)$$

Notice that the smaller the semimajor axis $a$ is, the bigger the velocity $v$ will be (for a given mass $M$). This agrees with Kepler's laws, too; Mercury does move faster than Pluto.

Substituting Equation 9.13 into the definition of kinetic energy, Equation 9.8, we find that the total kinetic energy, $\frac{1}{2}mv^2$, must now be

$$K = \frac{1}{2}\frac{GMm}{a} \qquad (9.14)$$

By adding this to the potential energy given in Equation 9.10, we derive the **total energy** of the body in orbit:

$$E = K + U$$
$$= \frac{1}{2}\frac{GMm}{a} - \frac{GMm}{a} \qquad (9.15)$$
$$= -\frac{GMm}{2a}$$

This derivation assumed a circular orbit, but it turns out (see the box) that the energy of an elliptical orbit sums to the same amount.

Notice that the total energy is a negative number. The total energy for all closed orbits is negative. This is a result of defining the potential energy at infinity to be zero. It means that the potential energy always dominates over the kinetic energy. For circular orbits, as we saw above, the absolute magnitude of the potential energy is twice that of the kinetic energy. In addition, the magnitude of the kinetic energy is equal to the magnitude of the total energy, but is opposite in sign.

If we substitute the more general form for the kinetic and potential energies, valid for all closed orbits (not just circular), we arrive at the formula

$$\frac{mv^2}{2} - \frac{GMm}{r} = -\frac{GMm}{2a} \qquad (9.16)$$

This formula is often called the *vis viva* **equation**. Notice that the mass of the orbiting body, $m$, appears in each term. It can be cancelled from them all. The energy per unit mass of an orbiting body does not depend on the total mass of that body, only on its position $r$ and the mass $M$ of the big body it is orbiting. It also does not depend on the shape of the orbit, only on its size (the semimajor axis, $a$). Different orbits with different eccentricities or different inclinations have the same total energy, if they have the same semimajor axis (Figure 9.7).

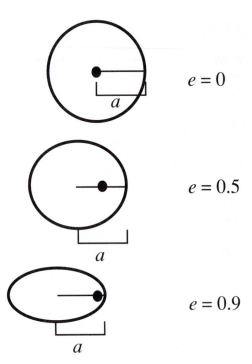

**FIGURE 9.7** Three orbits, with the same semimajor axis but different eccentricities, have the same amount of orbital energy.

## Escape Velocity

A body in a closed orbit never leaves that orbit. It never escapes the planet or star it circles. But say that our orbiting body has an engine that can speed up the satellite, increasing its kinetic energy. The orbit will change. As energy gets transferred from the rocket engine to the orbit, the orbit's energy will increase; that is, it will become less and less negative and eventually approach zero.

The orbit becomes a parabola instead of an ellipse when $E = 0$; adding still more energy turns it into a hyperbola. These are no longer closed curves; a body orbiting along such a path does not return, but "escapes" to infinity (Figure 9.8). In this case, the *vis viva* equation tells us that

$$v^2 = \frac{2GM}{r} \quad \text{(parabolic orbits)} \qquad (9.26)$$

This means that, for any point in space a distance $r$ away from the center of our mass $M$, if the total speed is equal to (or greater than) $\sqrt{(2GM/r)}$, then the satellite has enough energy to carry it away from the central mass out to infinity. It will escape the system. Thus we say that

$$v_e = \sqrt{\frac{2GM}{r}} \qquad (9.27)$$

is the **escape velocity**.

**FURTHER INFORMATION...**

# The Energy of Elliptical Orbits

Consider a body at the pericenter of its orbit. By definition, the angle $\eta = 0$ at this point, and so $r = a(1 - e)$. Thus we know that the potential energy must be

$$U = -\frac{GMm}{a(1-e)} \qquad (9.19)$$

To find the kinetic energy, we need to know the velocity at this point. We can split the velocity into its $r$ and $\eta$ components; the speed the body moves around the planet is the $\eta$ velocity, while the rate it moves closer to, or further away from, the planet is the $r$ velocity. The velocity at pericenter is completely in the $\eta$-direction and the $r$ velocity is zero, because at closest approach to the planet the body has just finished moving in and is not yet moving away again. From Chapter 8 we know that

$$v_\eta = r\frac{d\eta}{dt} \qquad (9.20)$$

and that

$$r^2\frac{d\eta}{dt} = h \qquad (9.21)$$

so at pericenter we find that

$$v = \frac{h}{r} = \frac{h}{a(1-e)} \qquad (9.22)$$

Thus the total energy at this point, the sum of the kinetic and potential energies, is

$$E = m\frac{h^2}{2a^2(1-e)^2} - \frac{GMm}{a(1-e)} \qquad (9.23)$$

But it was shown in the Chapter 8 that

$$GM = \frac{h^2}{a\left(1-e^2\right)} \qquad (9.24)$$

So with a little algebra, we can solve for $h$, substitute this into the first part of the equation, and find that the whole equation simplifies into

$$E = -\frac{GMm}{2a} \qquad (9.25)$$

This is the total energy at the pericenter. But because energy is conserved, this must also be the energy everywhere else in the orbit. Notice that this energy depends only on $GM$ and $a$, not on $e$.

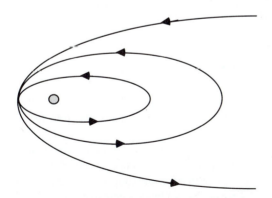

**FIGURE 9.8** Closed orbits are in the shape of ellipses; as the energy increases, the orbit stretches out towards infinity until the orbit is a parabola and the body escapes.

Consider a body on the surface of Earth. It is 6378 km away from the center of Earth, so $r = 6378 \times 10^3$ m. The mass of Earth is $5.97 \times 10^{24}$ kg, so $GM_{\text{Earth}} = 3.99 \times 10^{14}$. (**G**, the **universal gravitational constant**, is equal to $6.672 \times 10^{-11}$ N m$^2$/ kg$^2$.) We can thus calculate that the escape velocity from Earth is 11.2 km/s, or about 25,000 mph.

If we launch a rocket with this velocity, it will escape Earth's gravity and never fall towards Earth. But the direction of motion does not matter. If we had a race car moving on a road on the surface of Earth and somehow managed to get its speed up to 25,000 mph, it too would escape the pull of gravity! (See Figure 9.9.) Remember, Earth is round; anything travelling on the surface of Earth is going around a round object, so the race car is actually travelling in a circle around the center of Earth. When the race car reaches

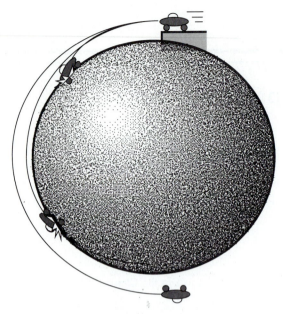

**FIGURE 9.9** Any object that reaches escape velocity is no longer bound to Earth by gravity. The direction of motion doesn't matter.

escape speed, you could think of it as being whipped off Earth by its own centrifugal force.

The point is that the direction of the velocity doesn't matter. Only the total speed determines whether it escapes or not. Of course, if a rocket is pointed down instead of up it probably won't go very far, but only because its velocity drops precipitously from 11.2 km/s to 0 km/s when it smashes into the surface of Earth. A ghostly rocket that could pass through stone would be able to come out the other side of Earth and still escape into space.

Of course it's not only Earth that has an escape velocity; for all planets, including asteroids, there is a minimum speed that needs to be attained before a body can leave the surface and break entirely free of the force of gravity holding it to that planet. Even the Sun has an escape velocity. For a body starting at Earth, orbiting the Sun in a circle of radius $r = 1\,\text{AU} = 1.5 \times 10^{11}$ m, the speed needed to travel completely out of the solar system is

$$\sqrt{\frac{2GM_{\text{Sun}}}{r}} = 42,000 \text{ m/s} \qquad (9.26)$$

## Synchronous Satellites

Note that the escape velocity is just $\sqrt{2}$ times the velocity of a body in a circular orbit the same distance from the planet. (Compare Equations 9.13 and 9.27.) In a circular orbit $r = a$, and the *vis viva* equation reduces to $v_c = \sqrt{GM/r}$. This simple formula can be used to calculate some interesting properties of satellites.

Consider a typical low-Earth orbit like that of the space shuttle. It generally travels some 300 km above the surface of Earth, in an orbit that is close to circular. In this case, then, $r = 300 + 6378 = 6678$ km, and so its orbital velocity is 7.7 km/s. How long does it take to complete one orbit? The circumference of a circle of radius $r = 6678$ is $2\pi r = 41,959$ km, and so it takes 41,959/7.7 or 5449 s, about 90 minutes.

How far away must a satellite be for it to take a whole day to complete one orbit? The Earth spins once every 24 hours, or 86,400 s. That must be the period, $P$, of our satellite. Recall that we wrote Kepler's third law in terms of a quantity called the **mean motion**, $n = 2\pi/P$. This means that we're looking for an orbit with a mean motion $n = 7.3 \times 10^{-5}$ s$^{-1}$. By Kepler's third law $n^2 a^3 = GM$; so $a = 42,000$ km from the center of Earth, or about 35,600 km (22,000 miles) above Earth's surface.

If a satellite were put into an orbit with a semimajor axis of 42,000 km, it would take exactly 1 day to complete one orbit. If the orbit were circular, poised right over the equator, and going from west to east, then the satellite would always be travelling with exactly the same angular motion as the ground underneath it. From the vantage point of the spinning Earth, it would appear that the satellite was hanging motionless in space. Such an orbit is called **geosynchronous** (Figure 9.10).

The advantage of such a satellite is obvious. A simple fixed antenna can send a signal to it, which it can relay to any other simple antenna underneath the satellite. Such satellites are obviously very popular with telephone companies, especially for transoceanic calls, or in countries such as Indonesia or Australia where geographical conditions make conventional wiring an expensive or difficult proposition. Weather satellites at these positions provide a continuous record of storms, droughts, and their patterns of change. Geosynchronous satellites are also used by TV networks and pay-TV channels, whose signals can be picked up by simple home satellite dishes that can stay aimed at one point in the sky.

## Moving Asteroids to Earth

Earth orbits the Sun at 1 AU; most asteroids are located at around 2.5 AU. How do meteoroids get from the asteroid belt to Earth?

For a meteoroid to enter a new orbit that crosses Earth's path, it must lose energy. Assume that it started with a circular orbit of semimajor axis 2.5 AU. Each kilogram of asteroid has a kinetic energy of

$$\frac{1}{2}v^2 = \frac{1}{2}\frac{GM_{\text{Sun}}}{2.5\,\text{AU}} = 1.78 \times 10^8 \text{J/kg} \qquad (9.27)$$

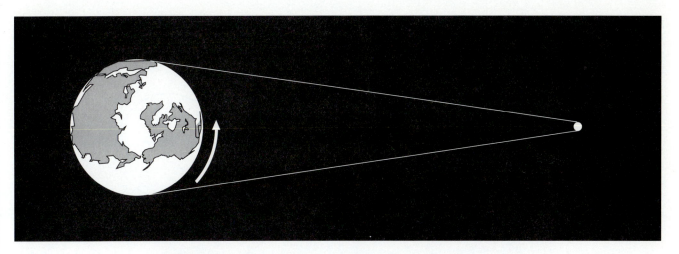

**FIGURE 9.10** Geosynchronous satellites orbit Earth at exactly the same rate as Earth spins. As a result, they appear to be stationary to any observer on Earth. This makes them ideal stations to relay TV or telephone messages around the globe. (The picture *is* to scale, except for the size of the satellite.)

and a potential energy of

$$-\frac{GM_{\text{Sun}}}{2.5 \text{ AU}} = -3.56 \times 10^8 \text{J/kg} \qquad (9.28)$$

The potential energy is fixed by its position. The only energy that can be changed is its kinetic energy; the asteroid can be speeded up or slowed down. If we remove all its kinetic energy, all we are left with is the potential energy. In this case its total net energy equals $-3.56 \times 10^8$ J/kg $= GM_{\text{Sun}}/2a$. Such an orbit would have a semimajor axis of $a = 1.25$ AU. With all kinetic energy removed, the meteoroid would stop orbiting and start falling directly into the Sun. This is equivalent to having an extremely eccentric orbit with a semimajor axis of half the original semimajor axis. Removing $1.78 \times 10^8$ J/kg (the entire kinetic energy of the asteroid) allows this meteoroid to reach any of the inner planets or to hit the Sun.

What if we only want to just barely reach Earth? In this case we need not remove so much kinetic energy. An asteroid whose orbit has an **aphelion** (the point farthest from the Sun) of 2.5 AU and a **perihelion** (the point closest to the Sun) of 1 AU, travels from the asteroid belt to the orbit of Earth but no further inward. The semimajor axis of this new elliptical orbit is just the average of the perihelion and aphelion, or 1.75 AU.

How much energy does each kilogram of asteroid travelling in this orbit contain?

$$E = -\frac{GM_{\text{Sun}}}{2 \times 1.75 \text{ AU}} = -2.54 \times 10^8 \text{ J/kg} \qquad (9.29)$$

At the asteroid belt, this body still has the same amount of potential energy as it did in the circular case, $-3.56 \times 10^8$

J/kg, so for energy to be conserved, the new orbit must have $1.02 \times 10^8$ J/kg of kinetic energy. This is about two thirds the kinetic energy that 1 kg in a circular orbit at 2.5 AU would contain. Thus, to move 1 kg of material from a 2.5 AU circular orbit into an orbit that crosses Earth's orbit, one must remove at least $0.76 \times 10^8$ J/kg of energy from its orbit.

A 1-megaton nuclear device explodes with $4.2 \times 10^{15}$ J of energy. If applied in the right direction, this would be enough energy to move about 10 million kg, or 10,000 tons, of material from the asteroid belt to Earth. This is roughly equivalent to an iron asteroid just under 10 m in radius.

For the purpose of moving asteroids into Earth-crossing orbits, if we can place nuclear bombs in just the right places, then the energies involved are not wildly impossible. But the Apollo asteroids are in Earth-crossing orbits already. Some of them are considerably larger than 10 m in radius. And one does not expect that there were any 1-megaton nuclear bombs in the asteroid belt in any earlier epoch! So where did these asteroids get enough energy on their own to leave the asteroid belt and enter Earth-crossing orbits? The question remains important, and unsolved.

## Hohmann Transfer Orbits

Another issue is raised by the example given above. How much energy would it take to travel from low Earth orbit out to the asteroid belt? Clearly boosting a spacecraft into any elliptical orbit that crosses both Earth's orbit and the asteroids' orbit does the job. The spacecraft is then given additional energy, an increase in velocity, to stay in the larger orbit.

The most efficient transfer orbit, the one that needs the least energy to perform the task, is one that has its pericenter distance exactly equal to the orbital distance of Earth and its apocenter—the furthest point of its orbit—at the asteroid's distance. Such an orbit is called a **Hohmann transfer orbit** (Figure 9.11). Walter Hohmann, a pioneer of spaceflight engineering, published his seminal work *Die Erreichbarkeit der Himmelskorper (The Attainment of Heavenly Bodies)* in Munich in 1925.

We can calculate from the laws of ellipses that the semimajor axis of a Hohmann transfer orbit from Earth to the asteroid point is 1.75 AU (just the average of Earth's and the asteroid's semimajor axes).

How much energy would it take? First we need to escape the gravitational pull of Earth. From ground level ($r = 6374$ km, measured from the center of Earth) it takes

$$\frac{GM_{Earth}}{r} = 6.24 \times 10^7 \text{ J/kg} \qquad (9.30)$$

of energy. (As before, all calculations are per kilogram of spaceship.)

This puts us away from Earth, but still in a more-or-less circular orbit around the Sun, still at 1 AU, travelling at a speed of

$$v = \sqrt{\frac{GM_{Sun}}{1 \text{ AU}}} = 29.8 \text{ km/s} \qquad (9.31)$$

(the orbital velocity of Earth around the Sun). The total energy of this orbit about the Sun is

$$-\frac{GM_{Sun}}{2 \times 1\text{AU}} = -4.45 \times 10^8 \text{ J/kg} \qquad (9.32)$$

The net orbital energy of the transfer orbit is

$$-\frac{GM_{Sun}}{2 \times 1.75 \text{ AU}} = -2.54 \times 10^8 \text{ J/kg} \qquad (9.33)$$

Thus we must add $1.91 \times 10^8$ J/kg of energy to the spacecraft. Adding this to the original kinetic energy of the orbit,

$$\tfrac{1}{2} \cdot (29.8 \text{ km/s})^2 = 4.5 \times 10^8 \text{ J/kg} \qquad (9.34)$$

yields $6.36 \times 10^8$ J/kg; setting this equal to $\frac{1}{2}v^2$, we calculate that the speed of the spacecraft $v$ must be increased to 35.7 km/s.

Half an orbit later, the satellite will finally be 2.5 AU from the Sun, at the same orbital distance as our asteroid. Substituting 2.5 AU ($= 3.74 \times 10^{11}$ m) for $r$ into the *vis viva* equation (Equation 9.16), we see that the velocity of our satellite will have slowed down to 14.3 km/s. (Notice that $a$ in Equation 9.16 for the transfer orbit is 1.75 AU.)

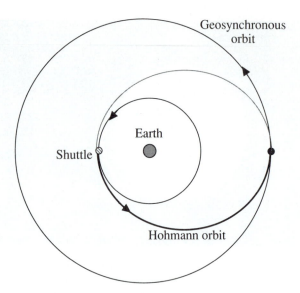

**FIGURE 9.11** The Hohmann (minimum-energy) transfer orbit. (This figure is not to scale.)

We calculated above that the asteroid's orbit has a total orbital energy of $-1.78 \times 10^8$ J/kg. This means that to stay in a circular orbit at this radius, the satellite must be speeded up to 18.9 km/s. This takes another $0.76 \times 10^8$ J/kg of energy (the same amount of energy it took to move an asteroid out of a circular orbit and into an Earth-crossing one.)

Ignoring the energy it took to leave Earth, we see that we added energy twice to move from 1 AU to 2.5 AU. We added $1.91 \times 10^8$ J/kg to enter the Hohmann transfer orbit and then another $0.76 \times 10^8$ J/kg to move from the transfer orbit to the asteroid orbit. The net result is that it takes $2.67 \times 10^8$ J to move a 1-kg satellite from Earth to the asteroid's orbit. This is the same as the differences in orbital energies between Earth's orbit and the asteroid's orbit. Thus in this two-step transfer no energy is wasted.

## Rendezvous in Orbit

Notice what happened in the Hohmann transfer orbit calculation above. At the lowest point in the orbit, we added energy to our satellite; but still, by the time it reached the asteroid's orbit, it was going slower than when it started. Looking at the *vis viva* equation, we see that if you add energy to an orbit you increase $a$; but the farther away you are from the body you're orbiting, the slower you travel.

This leads to some odd-seeming behavior of satellites in orbit. Consider two spaceships in the same orbit, one just ahead of the other (Figure 9.12a). If the astronaut in the trailing one wants to catch up to the leading one, she might naively think to fire her rockets to push herself forward. If she does so, however, she's adding energy to her orbit. This makes her move "up" (away from the body being orbited)

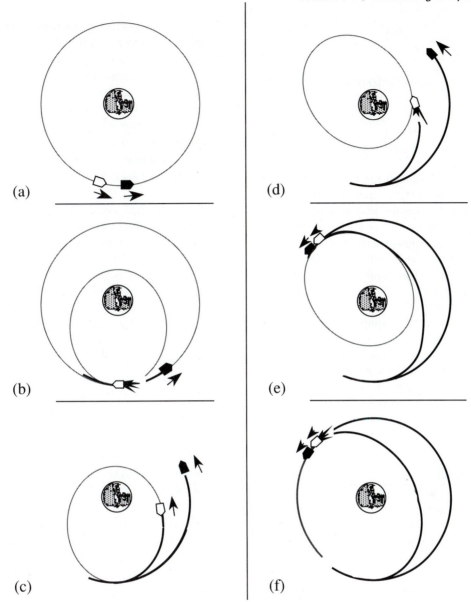

**FIGURE 9.12** How to match orbits with another spacecraft. See the text for details.

into a bigger orbit, where her velocity is slower than when she started. Gradually she'll drop farther behind the lead spaceship.

If she fires her thrusters backwards, however, her orbit will lose energy (Figure 9.12b). She will drop "down" into an orbit closer to the planet (Figure 9.12c). Doing so, she will speed up until she passes underneath the first spaceship. Once in front, all she has to do is fire her rockets to push herself forwards (Figure 9.12d). This will bring her back up into the old orbit and slow her down (Figure 9.12e). If timed right, she'll be right next to her neighbor. One last firing returns her to a circular orbit (Figure 9.12f).

## SUMMARY

The total energy of a body in orbit is constant. Every orbit therefore has a unique energy associated with it. This energy depends only on the semimajor axis of the orbit. A circular orbit and an extremely eccentric orbit with the same semimajor axis must have the same energy. The *vis viva* equation describes the speed of a body in orbit at any point in that orbit, in terms of the semimajor axis of the orbit and the mass of the body being orbited.

From this equation, the speed sufficient to escape the gravity of the body can be derived:

$$v_e = \sqrt{\frac{2GM}{r}} \qquad (9.35)$$

In a similar fashion, the speed of a body in a circular orbit is found to be

$$v_c = \sqrt{\frac{GM}{r}} \qquad (9.36)$$

A Hohmann transfer orbit is the most efficient way to move satellites from one orbit to another. If the old and new orbits are circular, the transfer orbit must have a pericenter equal to the semimajor axis of the lower orbit and an apocenter equal to the semimajor axis of the upper orbit. The semimajor axis of the transfer orbit is just the average of the semimajor axes of the two circular orbits. It will take exactly half a Hohmann transfer orbit for a satellite to move from one circular orbit to another.

The pilot of a spacecraft attempting to rendezvous with another orbiting spacecraft must keep in mind that the speed of an orbit increases as its energy decreases. Thus, to move ahead of another satellite, it must fire thrusters against its motion and lose energy; to move behind another satellite, it must increase its energy by firing thrusters to increase its forward motion.

## STUDY QUESTIONS

*True or False?*

1. "Synchronous satellites always orbit over the equator."

2. "A race car travelling on the surface of Earth at a speed faster than the orbital velocity will fly off Earth and into Earth orbit."

3. "According to our conventions, the total energy of a satellite in orbit around Earth is a negative number."

4. "Two satellites are in the same orbit, one following the other. For the trailing satellite to catch up to the leader, it must first turn around and fire its rockets as if it were trying to move away from the leader."

5. "The most efficient way to move from low orbit to high orbit is via an eccentric orbit whose pericenter is at the low orbit and whose apocenter is at the high orbit."

## 9.4   PROBLEMS

1. Compare how scientists before 1800 studied meteorites with the attitude that scientists have today toward UFOs. How are UFOs different from meteorites? How are they similar? Is the current skepticism about UFOs among scientists today justified? What would it take to convince scientists that UFOs exist?

2. Notice in Table 9.1 how the relative abundances of stones versus irons is different between falls and finds. Which set of numbers do you think represents the true abundance of irons versus stones out in space? Which set do you expect is closer to the abundances seen in the Antarctic meteorites?

3. Iron-age artifacts in museums have been identified as having been made from iron from meteorites. How can we tell?

4. NASA has equipped a U2 spy plane to fly in the stratosphere to collect dust in the air. Some of this dust has the chemical composition of a CI meteorite. Speculate on the origin of this dust.

5. What class of meteorite should have a composition closest to cosmic abundances? Why?

6. A farmer brings you a lump of metal that she found one day while plowing her field. What would you look for to determine whether or not it was a meteorite?

7. A theory of the nineteenth century speculated that the asteroids came from an "exploded" planet. Using the data from Table 9.3, find the volume of each of the largest asteroids and add these volumes together to find the volume (and, from this, the radius) of a possible parent planet. Add in another 2000 asteroids of average radius 10 km. How big would such a parent planet be? In light of your answer, and what we know of asteroids, discuss this theory.

8. Most asteroids are C types, but most meteorites are not carbonaceous chondrites. Why might this be?

9. It has been proposed that some basaltic achondrite meteorites, the eucrites, howardites, and diogenites, come from the asteroid Vesta. Another theory maintains that rarer basaltic meteorites, the shergottites, nakhlites, and chassignites, may come from Mars.

   (a) What sort of volcanic features would you expect to see on Vesta if this theory were true? Given that there is no evidence the eucrites were ever exposed to water, what sort of volcanic feature would you not expect to see?

   (b) The presumptive Vesta meteorites are much more numerous than the supposedly Martian ones. Which takes more energy: to move a 1-kg rock from the surface of Mars into Earth-crossing orbit or to remove 1 kg of rock from Vesta to Earth-crossing orbit? Vesta orbits at 2.36 AU; to calculate its mass, assume that it is spherical and has a rocklike density of 3.0 g/cm$^3$. Does this explain why eucrites are more common than shergottites, or must we look elsewhere for an explanation?

10. At one point in the section on orbits we state that "when an orbiting body comes closer to the primary, it must move faster"; several paragraphs later, we say that "the smaller the semimajor axis $a$ is, the bigger the velocity $v$ will be." What's the difference between these two statements?

11. Using the data from the appendix, calculate the semimajor axes of synchronous satellites about Mars, Saturn, and Pluto. In relation to the moons or rings of these planets, as listed in Chapter 11, where would such synchronous satellites lie?

12. Inspired by the Jules Verne novel *Around the World in Eighty Days*, Rachel Ramjet wants to build a space capsule that will orbit her around the world in 80 minutes. What would the semimajor axis of her orbit have to be? What problem is she likely to encounter?

13. Virtually all satellites launched from Earth—not just synchronous satellites—travel in orbits moving from west to east. Why?

14. Over 100 years ago, Jules Verne (in his novel *From the Earth to the Moon*) set a fictional American space-launching facility on the east coast of Florida, near the actual location of the Kennedy Space Center on Cape Canaveral. Was this just a lucky guess? What geographical advantage does the United States have over Europe or Russia when it comes to launching satellites? (Hint: Consider your answer to Problem 13; also consider the advantage of launching a rocket over uninhabited areas such as the ocean.)

15. Assume that the space shuttle orbits in a circular orbit 300 km above the surface of Earth. Describe the various steps necessary to send a satellite from the shuttle to a synchronous orbit, via a Hohmann transfer orbit. If the satellite weighs 500 kg, how much energy is needed for the transfer?

16. Following the example given in the chapter, how long would it take for a spacecraft to travel from Earth orbit to asteroid orbit (at 2.5 AU) via a Hohmann transfer orbit?

17. A new asteroid, "Darb," has been discovered with an orbital period of exactly 8 Earth years; for simplicity, assume both Earth and asteroid have circular orbits.

(a) What is the semimajor axis of Darb's orbit, in terms of AU?

(b) Now assume that an interplanetary probe is to be launched from Earth to Darb. What will be the semimajor axis of the Hohmann transfer orbit?

(c) How long will it take for this probe to get to its destination?

(d) Draw a sketch showing the approximate position Darb would have to be in, relative to the position of Earth, at time of launch, and indicate the positions of the two planets when the probe reaches Darb.

## 9.5   FOR FURTHER READING

The standard reference text for meteoritics is John T. Wasson's *Meteorites: Their Record of Early Solar System History* (New York: Freeman, 1985). An older classic, and in many ways still the most readable introduction to the modern study of meteorites, is Brian Mason's book, *Meteorites* (New York: Wiley, 1962).

The University of Arizona Press has produced three excellent anthologies of meteorite and asteroid review papers: *Asteroids*, edited by Tom Gehrels and Mildred Shapley Matthews (Tucson: University of Arizona Press, 1979); *Meteorites and the Early Solar System*, edited by John Kerridge and Mildred Shapley Matthews (Tucson: University of Arizona Press, 1988); and *Asteroids II*, edited by Rick Binzel, Tom Gehrels, and Mildred Shapley Matthews (Tucson: University of Arizona Press, 1989). Asteroids *2873 Binzel*, *1777 Gehrels*, and *878 Mildred* were named for these editors.

A less technical introduction to meteorites and their history can be found in John G. Burke, *Cosmic Debris: Meteorites in History* (Berkeley: University of California Press, 1986).

Reference works for celestial mechanics topics were discussed in Chapter 8.

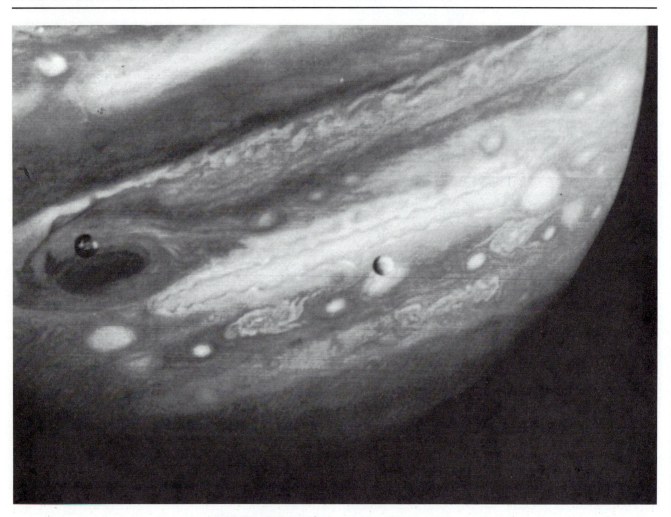

**FIGURE 10.1** Jupiter and two of its moons.

# CHAPTER 10

# Jupiter

## 10.1 ON JUPITER

Imagine that you are falling directly into Jupiter.

As you approach the planet, you can make out the bands of clouds, alternately white and shades of orange brown, which swirl around the planet. Towards the poles, the broad bands break up into a jumbled mix of small swirls. But along the equator, the colored clouds gently blend into each other, making patterns that seem almost like marbled paper or abstract art. Getting closer, you pick out dark spots between the white zones and the dark belts. These seem to be holes in the clouds. There are also red ovals, the most spectacular of which is the Great Red Spot sitting in the southern hemisphere, as wide as three planet Earths, with white clouds and dark spots swirling around it.

You've fallen past the moons, past the thin ring of dust as fine as flour, and into the thin upper atmosphere of the planet. The flat "surface" of Jupiter reveals itself to be three-dimensional. The Red Spot stands out like a mountainous, immensely wide plume of up-bubbling gas, spilling fine red dust into the clear upper air.

Next you plunge through a zone of white clouds (which smell distinctly of ammonia). Below them you see darker brown and orange clouds. Where sunlight can hit them, they are darkly colored, but beneath the white am-

| Jupiter's Vital Statistics | |
|---|---|
| Radius, equatorial | 71,492 km |
| Radius, polar | 66,854 km |
| Surface area | $6.25 \times 10^{10}$ km$^2$ |
| Mass | $1.899 \times 10^{27}$ kg |
| Density | 1.33 g/cm$^3$ |
| Local gravity | 24.8 m/s$^2$ |
| Escape velocity | 60 km/s |
| Albedo | 0.34 |
| Atmospheric temperature (0.5 bar) | 125 K |
| Length of year | 11.86 yr |
| Length of day | 9 h 55 min |
| Distance from the Sun | 5.2 AU |

monia clouds the colors are less intense. Passing through them you smell ammonia and rotten eggs: hydrogen sulfide. (A slight taste of bitter almonds is also in the air; there's hydrogen cyanide in the atmosphere, too. Of course, you'd be seriously ill from even the slightest whiff of Jupiter's atmosphere, but we'll ignore that on this imaginary trip.)

Below the dark-colored clouds, sunlight pours through only the holes that you saw as dark spots from above. These holes seem to be downdrafts of cold, dry air.

At the cloudtops, the air pressure is half that of Earth and the temperature a frigid 125 K (nearly −300° F). The air pressure through the brown clouds is about 1 Earth atmosphere (about $10^5$ Pa in SI units), and the temperature is around 200 K, still colder than anywhere on Earth. But as you go deeper, the pressure and temperature keep rising.

The next cloud layer below the ammonia–hydrogen sulfide clouds is the ammonia–water cloud. Plunging deeper and deeper, you may pass through still other cloud layers and dust layers. But through it all, these clouds represent nothing more than a fraction of a percent of the mass of the basic atmosphere, which is three quarters hydrogen gas and nearly one quarter helium.

As you go deeper, the gas becomes denser. At 250 km into the planet, the pressure is greater than 100 atmospheres and the temperature is now 1000 K. At 1500 km down, the pressure is 10,000 atmospheres and the temperature is 3000 K. Finally, 21,000 km below the cloudtops, nearly a third of the way into the planet, the pressure exceeds 1 million atmospheres and you notice a distinct change in the planet. It's no longer made of a thick, sluggish gas but a liquid, and one with strange properties.

When you shine a light on it, this liquid looks something like quicksilver (the element mercury), shiny and viscous. But what is it? The pressure here is so high that even hydrogen atoms get crushed; there is no room for each proton of a hydrogen nucleus to hold an electron around itself. Instead, the protons get packed together and the electrons flow as they can among them, hopping from proton to proton. This is how electrons flow in a metal; the hydrogen itself deep inside Jupiter has become a liquid metal.

Mixed with this metal are the other heavy elements, perhaps sorted out into a rocklike core some 10 times the mass of Earth, tiny compared to the bulk of Jupiter. The temperature here at the center of Jupiter is 20,000 K and the pressure is 10 million times the pressure of Earth's atmosphere.

## The Composition of Jupiter

The scene above may sound like the setting for a science fiction story, but it is based on our best understanding of the interior of Jupiter. Indeed, it describes what we expect to find when the entry probe from the *Galileo* orbiter enters Jupiter's atmosphere in December 1995. Yet how can anyone make such specific predictions about a place that no human or spacecraft has visited yet?

One key that allows us to predict what the interior of Jupiter must be like is our understanding of the behavior of hydrogen gas at high temperature and pressure. It is not a bad approximation to assume that the behavior of hydrogen reflects the state of the interior of Jupiter as a whole, because Jupiter is 75% hydrogen.

How do we know that Jupiter is 75% hydrogen? Recall from Chapter 4 that the bulk composition of a planet can be found in several ways. The most basic is to look at its mean density.

If one had a ball of pure, cold hydrogen, how would the radius of such a ball change as more and more hydrogen were added to it? Rupert Wildt asked this question in the late 1930s. His answer was quite surprising. One would expect that adding more gas to our cold ball would simply make it bigger, and for a while this is true. However, Wildt knew that a gas like hydrogen becomes denser as it is subjected to higher and higher pressures. And he realized that the more mass that was added to this hypothetical ball of hydrogen, the higher the pressures in the interior would be. Thus as more and more mass is added, the interior pressures start squeezing the gas so that eventually the planet stops growing so fast. In fact, he found that once the mass of the ball exceeded about a tenth of our Sun's mass, adding more hydrogen would squeeze the interior so much that the ball would actually get smaller!

It actually doesn't happen quite this way. After a gas ball's mass is greater than about one tenth of a solar mass, the high pressure inside causes the hydrogen nuclei to fuse, which heats up the ball and turns it into a star with a very hot interior. Wildt was assuming his ball of hydrogen was cold.

But one other interesting result came from this work. It showed that the densities of the giant planets, especially Jupiter and Saturn, were well matched by this very simple model (see Figure 10.2). Observations of Jupiter's upper atmosphere, both from Earth-based telescopes and the passing *Pioneer* and *Voyager* probes, have since confirmed that its composition is 99% very light gases, 75% hydrogen and 24% helium by mass. There's good reason to believe that this upper-atmosphere composition holds true throughout the entire planet.

Helium is denser than hydrogen, so the density of Jupiter or Saturn should be raised slightly off the "cold pure hydrogen" line, and this is in fact observed in Figure 10.2. That Jupiter is fairly warm inside tends to make the whole planet less dense than this model predicted, but this effect is not enough to cancel out the extra density due to helium. Modern, more detailed models take these and other effects into account.

## Hydrogen under Pressure

We concluded above that Jupiter is made primarily of hydrogen. We can study the behavior of hydrogen in a laboratory under a variety of temperatures and pressures. Measurements at pressures as high as 1 million atmospheres can be made by placing a container of hydrogen next to a nuclear bomb and watching the hydrogen as the shock wave from the explosion passes through the container. Also, be-

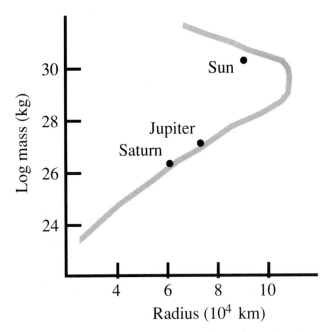

**FIGURE 10.2** A ball of cold hydrogen grows as mass is added until it reaches a critical mass, then it starts to shrink as the weight of the hydrogen compresses the gas deep within the planet.

cause hydrogen is such a simple element—just one proton and one electron—it is possible to use theories of solid-state physics to predict how it should behave under conditions of extreme temperature and pressure.

From these measurements and theories, the phase diagram of hydrogen shown in Figure 10.3 has been deduced. The horizontal axis shows the pressure, and the vertical axis shows the temperature.

If you take a gas and cool it enough or squeeze it hard enough, it will condense into drops of liquid. With further cooling or pressure, it will freeze into a solid. Solid, liquid, and gas are the **phases** of the substance. However, at a high enough temperature, the difference between liquid and gas disappears. The precise values of pressure and temperature where this difference stops are called the critical values. This point on the phase diagram is called the **critical point**.

The dotted line in Figure 10.3 shows the plot of temperature versus pressure inside Jupiter today. This curve is called an **adiabat** (we'll discuss below how this curve was derived). As you can see, the line goes above the critical point for molecular hydrogen. If Jupiter were colder at the top of its atmosphere, this line would be moved down lower in the diagram and it could have cut the boundary between liquid and gaseous hydrogen. In that case, at a certain depth inside Jupiter we would expect the atmosphere to stop and a liquid ocean of hydrogen to start. But this apparently doesn't happen in Jupiter. Instead, the atmosphere just keeps getting thicker as we get deeper into the planet.

However, it does cross the boundary where hydrogen becomes a metallic liquid, at a pressure of about $10^{11}$ Pa, equal to 1 million Earth atmospheres. Recall that the SI unit of pressure, the **Pascal (Pa)**, is defined as 1 newton of force per square meter.

Pressure is often thought of as a "force per area"; for instance, the pressure of the atmosphere at the surface of Earth is listed as 14.7 pounds per square inch, or $10^5$ newtons per square meter ($N/m^2$). However, recall from basic physics definitions that force times distance is energy and that area times distance is volume. So if you multiply both force and area by distance, then force per area can also be thought of as energy per volume. In other words, the pressure of a gas is in some sense a measure of its "energy density."

If we take the energy needed to ionize a hydrogen atom (that is, remove its electron) and divide this energy by the volume of the atom, we get an energy density equivalent to $10^{11}$ Pa. Thus a gas with a pressure of $10^{11}$ Pa has enough energy packed into the volume of one hydrogen atom to ionize the atom, allowing the electrons to flow freely from atom to atom as electrons do in a metal. Like a metal, hydrogen in this state is an excellent conductor of electricity. Thus we conclude that a strong magnetic field could be generated in the core of Jupiter as this liquid flows and convects.

## The Clouds of Jupiter

The multicolored cloud layers described in the beginning of this chapter have long been a topic of fascination and speculation, and even modern space probes to Jupiter have not completely explained them. From Earth, alternating light **zones** and dark **belts** appear to change slowly over a period of 10 years or so. Because of their relative stability, names have been given to each zone and belt on the basis of its position on the planet ("south tropical zone," "north equatorial belt," and so on, as illustrated in Figure 10.4). However, at any given moment not all of these cloud bands may exist or be visible (Figure 10.5).

Detailed study from Earth has now been supplemented by many series of images from the four spacecraft that have encountered Jupiter to date (*Pioneer 10* and *11*, and *Voyager 1* and *2*) and most recently by photographs from the Hubble Space Telescope. In these observations both the change in the zones and belts and the motions of spots on these areas can be traced. Each band seems to travel around the planet at a different rate, with high-speed streams in the clouds showing evidence of strong winds.

Along with the bands of clouds, one feature that has been visible for hundreds of years is the **Great Red Spot**. Around this spot, turbulent swirls show the cyclonic motions of the winds.

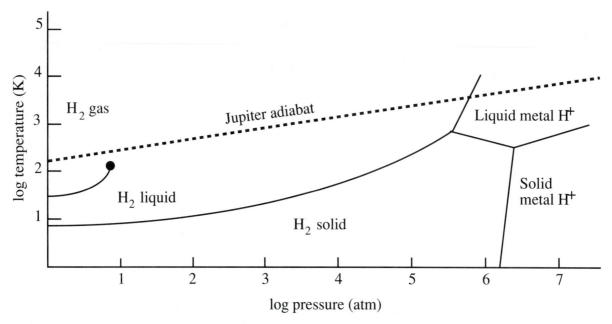

FIGURE 10.3 The phase diagram of hydrogen. The line marked Jupiter adiabat marks the temperature and pressure conditions thought to exist inside the planet. The boundary into the H⁺ liquid region marks the beginning of a conductive liquid core; the exact location of this boundary is still uncertain.

The structure of the clouds and of the spots that move in them is a result of the rapid spinning rate of Jupiter. At high latitudes, this rapid spin produces the multiple swirls seen in Figure 10.6. At lower latitudes, it confines the clouds to layers of constant latitude. Much of the detail in the clouds is due to pressure waves in the atmosphere. The Red Spot may be an example of a solitary standing wave in the atmosphere.

The chemistry of the clouds as described in the opening section is based on estimates of what species would be present in the Jovian atmosphere if it had the same proportions of all the elements as is seen in the Sun. The highest clouds are expected to be ammonia, which needs the coldest temperature to freeze. The next layer may be an ammonia–

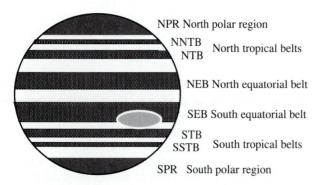

FIGURE 10.4 Names of the various belts and zones on an idealized map of Jupiter. The bands constantly change their sizes and colors, so it is unlikely that all bands will be visible at any given time.

FIGURE 10.5 Photographs of Jupiter, taken on January 24, 1979 (by *Voyager 1*, left), and May 9, 1979 (by *Voyager 2*, right). Notice how the different cloud belts have changed in four months. For instance, the white belt below the Red Spot has dissipated, a white oval originally southwest of the Red Spot has moved 60° to the east, and spots in the northern hemisphere have disappeared.

**FIGURE 10.6** A view of Jupiter's south polar region, photographed by *Voyager 2*. Notice the swirling eddies, different from the horizontal bands seen in the equatorial region.

water solution with ammonium sulfide, $NH_4SH$, a combination of ammonia ($NH_3$) and hydrogen sulfide ($H_2S$). In the warmer regions below these, water clouds should be seen, with ammonia dissolved in the water.

The difficulty with this theoretical scheme is that all these clouds would be white and colorless; but in fact only the highest clouds are white. Many explanations have been proposed for the colors of the clouds. The most likely is that solar ultraviolet (UV) light breaks up the hydrogen sulfide, and the free sulfur atoms recombine in a number of ways to form compounds whose colors can range from red to brown to yellow. Likewise, the Red Spot's color may be due to sulfur or to red phosphorus from the breakup of the compound phosphine ($PH_4$), whose absorption bands have been seen in the Jupiter spectrum. Or these colors may be due to organic materials, as was discussed in Chapter 6.

## SUMMARY

Jupiter, the largest of the planets, is a ball of gas, 75% hydrogen and 24% helium. The other elements, the last 1%, are probably also present in the same proportions as in the Sun. Because it is made up almost entirely of gas, it does not have a surface, but merely gets thicker as one goes deeper into the planet. When the pressure in the planet exceeds 1 million Earth atmospheres, the hydrogen gas becomes ionized and acts like a liquid metal. The clouds of Jupiter consist of several layers, including white bands of ammonia in the upper atmosphere, and lower clouds rich in sulfur and sulfur compounds. Solar UV light probably produces the coloring agents that make these clouds orange, brown, or red.

## STUDY QUESTIONS

1. The clouds of Jupiter are made of many different chemicals. Name one.
2. How deep into the atmosphere of Jupiter must you travel before the atmosphere stops and a surface is seen?
3. What are the dark-colored clouds that make the dark stripes across Jupiter called, belts or zones?
4. What element makes up 75% of Jupiter's mass?
5. Does Jupiter spin faster or slower than Earth?

## 10.2 ATMOSPHERIC PHYSICS

Some planets, such as Jupiter and Saturn, are almost all atmosphere; they're just big balls of gas. Other planets, such as Mercury, have no atmosphere at all. Mars, Earth, and Venus are surrounded by relatively thin blankets of gas. But even for planets with a thin atmosphere, the effects of an atmosphere can be significant, far out of proportion to the atmosphere's mass. For Jupiter, to understand the atmosphere is to understand the planet itself.

The atmosphere above the clouds of Jupiter traps some wavelengths of light, scatters others, and in general determines the temperature of the clouds. On the terrestrial planets, it weathers and erodes the surface of the planet, hiding scars of the past and producing new features. From a more subjective viewpoint, the atmosphere contains the "climate," the feel of a planet. It's what determines what it would feel like to sit on the surface of Venus or to plunge through the clouds of Jupiter.

Atmospheres are collections of gases. As we see in Jupiter, these gases, simple chemical substances, condense to form clouds. The gases themselves react with each other, with particles in clouds, and with ultraviolet radiation from the Sun. On Venus, Mars, and Earth the atmosphere also reacts with the material on the surface of the planet. But to determine how these reactions will proceed, we have to know the temperature and pressure of the gas.

Once some new chemical species is made, be it bits of complex sulfur chemicals in the upper clouds of Venus or drops of methane rain on Titan, it will move; gases are wonderfully mobile. They are able to react with clouds or rocks at one point on the planet and carry the products away to another point; that makes an atmosphere a powerful tool to shape the surface of a planet. For a gas giant planet like Jupiter, understanding the physics of the atmosphere is the same thing as understanding the physics of the whole planet.

What makes gases move? What heats them, and what cools them? How do they maintain their temperatures and pressures?

## Heat Input

All physics starts with energy. The prime source of energy at the surface of a planet is sunlight. How much does the Sun heat up a planet? That depends on three factors:

1.  How much energy does the Sun put out?
2.  How far from the Sun is the planet?
3.  How efficient is the planet in absorbing the Sun's energy?

We can measure the **energy flux** from the Sun, the amount of energy 1 m² of area on Earth, facing the Sun flat-on, receives every second. This number is called the **solar constant**, $F$. Its value is 1367 J/(s · m²).

If 1 m² at Earth receives this much energy per second, how much energy does the whole Universe receive from the Sun? Assume that the Sun radiates energy in all directions equally. Imagine a huge sphere, with a radius as big as Earth's orbit, centered on the Sun. Our value for the solar flux gives us the amount of energy falling on 1 m² of that sphere's inner surface; multiply this number by the whole area of the sphere ($4\pi a^2$, where $a$ is 1 AU). The net energy output of the Sun is $4 \times 10^{26}$ J/s.

Now place another imaginary sphere right at the surface of the Sun. Recall that the area of the surface of this sphere can be found by the formula $4\pi R^2$, where $R$ is the Sun's radius. The Sun's surface has an area smaller than that of the sphere out at Earth's orbit, by a factor of $(R/a)^2$. The energy passing through the sphere at the Sun is still $4 \times 10^{26}$ J/s; only, at the Sun's surface, it is concentrated onto a much smaller area. The energy flux at the Sun's surface is thus quite intense. But by the time that energy reaches Earth, the amount of heat that any square meter receives has dropped by $(R/a)^2$; at any other distance $r$ from the Sun, the energy is $(R/r)^2$ times the energy at the Sun, or $(a/r)^2$ times the energy at Earth. Thus we say that

the energy flux follows what is called an **inverse square law**; it falls off as 1 over the distance from the energy source, squared (Figure 10.7).

So we know how much energy the Sun is putting out, and we know how the distance from the Sun affects how much energy a planet's surface sees. But how efficient are planets at trapping this energy?

When sunlight reaches an asteroid or a planet, it has one of two fates. Either it is reflected from the planet (which makes the planets appear to shine, so we can see them) or it is absorbed, heating up the planet. Recall that the fraction of light that is reflected is called the **albedo** of the planet (see Chapter 9).

Finally, how many square meters of area does a planet have to capture sunlight? A planet is lit on one hemisphere and dark on the other. Furthermore, as we saw in Chapter 6, the part of a planet where the sunlight does not strike flat-on (like the poles, for most planets) is less efficient in collecting sunlight than the part of the planet where the sun does strike flat-on (like the equator at high noon). In effect, the total effective area a planet presents to collect sunlight is its **cross-sectional area**: the area of a disk, $\pi b^2$, where $b$ is the radius of the planet.

Put this all together. How much energy does a planet absorb? The total energy input into a planet starts with the solar constant, $F$ (the energy from the Sun hitting 1 m² each second). Multiply this times the square of the distance from the Sun to Earth, $a^2$ (where the solar constant was measured), divided by the square of the distance from the Sun to the planet we are interested in, $r^2$. This takes care of the inverse square effect. Multiply this times $(1 - A)$; if the albedo $A$ measures how much light is reflected, then

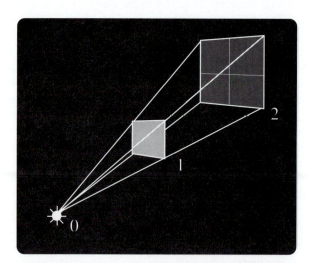

**FIGURE 10.7** How hot a planet gets depends on how near the Sun it is. The Sun's rays, as they travel away from the Sun, have to fill more and more of space; as a result, they tend to appear weaker and weaker.

(1 − A) measures how much light gets absorbed. Only light that is absorbed can heat the planet. Multiply this times the total number of square meters that are collecting sunlight, the cross-sectional area $\pi b^2$, where $b$ is the radius of the planet. Thus

$$\text{Heat in} = F\left(\frac{a}{r}\right)^2 (1 - A)(\pi b^2) \qquad (10.1)$$

What happens to this heat? The planet can't keep absorbing it forever; it would just keep getting hotter and hotter. It must be radiating heat away at the same rate that it absorbs it.

It should not be surprising that the hotter a substance is, the more energy it radiates. In fact, it turns out that energy radiated varies not simply with the temperature, but as the fourth power of the temperature. A body at 200 K radiates 16 times as much heat as a body at 100 K. Mathematically, we can write this rate (energy per square meter of surface area per second) as $\sigma T^4$, where $\sigma$ is the constant of proportionality called the **Stefan–Boltzmann** constant.

If the planet spins rapidly, all sides of it get hot eventually and so all sides can radiate away heat. If it doesn't spin (or if it always keeps one face towards the Sun), the nighttime side never gets hot and thus it can't radiate anything away. In either case, the total heat it radiates is simply equal to the surface area of the planet, times the energy radiated per square meter per second. For a rapid spinner,

$$\text{Heat out} = \sigma T^4 (4\pi b^2) \qquad (10.2)$$

But if only one side of the planet sees the Sun for a long period of time,

$$\text{Heat out} = \sigma T^4 (2\pi b^2) \qquad (10.3)$$

For the temperature of a planet to stay in balance, all the heat that is absorbed must eventually get radiated. If it didn't, the planet would heat up to a larger temperature $T$, thus increasing the "heat out" term until eventually it equals the "heat in" term. Mathematically we account for this by setting the "heat in" terms equal to the "heat out" terms. Using algebra, we then solve for the temperature $T$:

$$T = \left[\frac{F(1 - A)a^2}{4\sigma r^2}\right]^{1/4} \qquad (10.4)$$

If we measure $T$ in kelvins and $r$ in AU, the formula reduces to

$$T = 280\left[\frac{\sqrt[4]{(1 - A)}}{\sqrt{r}}\right] \qquad (10.5)$$

If the planet doesn't spin, the daytime side is hotter by a factor of $\sqrt[4]{2}$.

Notice that the radius of the planet, $b$, divides out: The temperature of a planet depends only on the planet's albedo and how far it is from the Sun, not on how big it is!

The results of these calculations are shown in Table 10.1. They seem to work well for the rocky planets, but they give numbers that are substantially low for planets with atmospheres. One reason for this, the greenhouse effect, was described in Chapter 5. But there are other reasons operating in gas giant planets such as Jupiter, as we shall see.

## Pressures and Temperatures

Is there any way that we can predict how the temperature and pressure of an atmosphere should change as we move up and down within it? To do this would allow us to calculate the conditions anywhere inside Jupiter.

It can be done by using two laws that describe the behavior of gases. One is a simple law that relates the pressure, temperature, and density of gases; the other describes how a gas behaves under the pull of gravity.

Gases in most planetary atmospheres (including cloud layers of Jupiter, but not the gases deep within its interior) are quite closely approximated by what physical chemists call an **ideal gas**. We know that real gases are made up of molecules of a given finite size and are subject to chemical reactions. An ideal gas is a hypothetical substance made up of point masses (hence having no size themselves) that collide and bounce off each other without ever sticking

**TABLE 10.1**

Calculated Equilibrium Temperatures of the Planets

| Planet | Albedo | $T_{calc}$ | $T_{obs}$ |
|--------|--------|--------|--------|
| Mercury | 0.12 | *620 K* | 700 K |
| Venus | 0.75 | *325 K* | 740 K |
| Earth | 0.36 | 250 K | 290 K |
| Moon | 0.07 | *385 K* | 400 K |
| Mars | 0.25 | 210 K | 240 K |
| Jupiter | 0.34 | 110 K | 125 K |
| Saturn | 0.34 | 80 K | 95 K |
| Uranus | 0.30 | 60 K | 60 K |
| Neptune | 0.29 | 45 K | 60 K |
| Pluto | 0.30 | 40 K | ~ 40 K |

Note: Numbers in italics were calculated assuming that the planet spins slowly.

together, chemically reacting, or interacting in any of the other complicated ways that real gases do. This idealization, however, provides simple results that are highly accurate over a wide range of circumstances.

Suppose that you had such a gas trapped inside a container. The pressure felt on the walls of that container arises from the impact of all those little molecules hitting the walls as they bounce around inside. You can increase the gas's pressure either by increasing the temperature—that makes the molecules move faster—or by increasing the number of molecules in the box, thus increasing the number of molecules hitting any square meter of the box's walls. (Shrinking the box accomplishes the same thing. Either way, the density of the gas is increased.) All this can be related mathematically by the formula called the **ideal gas law**:

$$P = R \left( \frac{\rho T}{\mu} \right) \qquad (10.6)$$

As you can see, $P$ goes up if $T$ goes up or $\rho$ goes up. Here, $P$ is the pressure, $\rho$ the density, and $T$ the temperature of the gas. $R$ is the constant of proportionality, the **gas constant**, that keeps the units straight. It is equal in SI units to 8.31 J/mole · K.

The term $\mu$ represents the molecular weight of the gas, the number of grams in 1 mole of gas. (A mole is **Avogadro's number** of molecules, roughly $6 \times 10^{23}$ molecules.) Hydrogen is the lightest gas, with a molecular weight of 1 g/mole, which translates into $10^{-3}$ kg/mole in SI units. By moles, Jupiter's atmosphere is roughly 65% hydrogen and 34% helium (with a molecular weight 4 g/mole), its average molecular weight is $2.70 \times 10^{-3}$ kg/mole. By contrast, the nitrogen–oxygen atmosphere of Earth has a molecular weight of $29 \times 10^{-3}$ kg/mole.

Next, we place this ideal gas on a planet and allow the planet's gravity to compress the gas. If a gas is sitting stable in a gravity field, it obeys the law of **hydrostatic equilibrium**. This law says that as you go deeper into an atmosphere, the pressure of the gas must increase, because it has the weight of all the atmosphere above it pushing down on it. In mathematical terms, this can be written

$$\frac{dP}{dz} = -g\rho \qquad (10.7)$$

Here, $dP/dz$ stands for the change in the pressure, $P$, as the height above the surface of the planet, $z$, changes; $g$ is the acceleration of gravity; and as before, $\rho$ is the density of the gas in the atmosphere. Notice that this law applies anywhere inside an atmosphere. You don't have to assume that the gas is ideal. Indeed, it doesn't even have to be a gas; this law can even be applied to rocks to calculate

the pressure inside a terrestrial planet. Of course, we must be careful to keep track of the changes in density, $\rho$, and gravity, $g$, as we travel deep inside a planet.

The pressure can range over many orders of magnitude as one moves from the near-vacuum conditions at the top of the atmosphere to a pressure equal to the weight of $10^5$ Pa at the cloudtops of Jupiter to $10^{11}$ Pa deep in its interior. On the other hand, the temperature may change by only a factor of two or so, from 125 K at the top of Jupiter's atmosphere to about 250 K in its clouds. Even deep into the atmosphere of Jupiter and the other gas giant planets we find that the temperature does not change by more than a factor of 10. Certainly this holds for the part of the atmosphere where the gas is close to "ideal." So for the moment, to make our mathematics easier, let us assume that the temperature $T$ stays constant throughout the whole atmosphere.

We can use algebra to rewrite the ideal gas law and solve for $\rho$ in terms of $T$ and $P$: $\rho = \mu P / RT$. Substituting this into the hydrostatic law, we get

$$\frac{dP}{dz} = -g\frac{\mu P}{RT} = -\left[ \frac{g\mu}{RT} \right] P \qquad (10.8)$$

The term in the brackets has the dimensions of 1/distance. Its reciprocal is called the **scale height**, $H$:

$$H \equiv \frac{RT}{g\mu} \qquad (10.9)$$

Notice the factors that go into the scale height: temperature, which for most planets is usually on the order of 100 to 300 K; the molecular weight of the gas, which is about 3 for the gas giant planets but about 30 for atmospheres of oxygen or nitrogen (like Earth or Saturn's moon Titan) to 44 for carbon dioxide atmospheres; and the acceleration of gravity, $g$. But $g$ on Jupiter is only about three times the value for Earth or Venus. Thus the scale height really doesn't vary tremendously from planet to planet. For most planets, it is on the order of 10 km.

What is the physical meaning of this scale height? Notice that our equation can be solved, using calculus, to find the pressure $P$ at any point $z$:

$$P = P_0 \exp\left( -\frac{z}{H} \right) \qquad (10.10)$$

where $P_0$ is the pressure at the point in the atmosphere where we define $z$ to be zero (for instance, at the rocky surface of a terrestrial planet or the cloudtops of Jupiter). According to this equation, the scale height $H$ is the distance over which the pressure drops by a factor of $e$ (recall that $e$, the exponential, is equal to 2.71828...). In other words, when rising up a scale height in an atmosphere, the pressure drops down to about 37% of its original value.

The scale height is a useful number. The atmosphere can be assumed to be more or less constant in temperature, pressure, and density over distances that are small compared to a scale height. On the other hand, if you are comparing conditions of regions in the atmosphere separated by several scale heights, you know that both pressure and density must be changing significantly.

## Layers of the Atmosphere

In deriving a scale height, we assumed that $T$ was constant. But is the temperature constant throughout an atmosphere? Not really.

Consider how sunlight heats a planet and its atmosphere. The very top of the atmosphere receives much ultraviolet (UV) and infrared (IR) radiation, which most gases absorb, and so the very thin upper atmosphere is very hot. On the other hand, the solar radiation that gets through this layer, which includes most of the visible light, is not absorbed at all until it reaches the surface (or the dark cloudtops) of the planet.

Once it is heated up, this surface radiates heat back as infrared light. But infrared radiation is absorbed by water or carbon dioxide in the atmosphere; thus the air just above the clouds or surface is heated up. When a parcel of air is heated, the ideal gas law demands that either the pressure gets higher or the density lower. But the pressure is already fixed by the hydrostatic law. Thus the density must drop.

As the air becomes less dense, it becomes buoyant. The parcel of hot air floats upwards through the surrounding cold air until it reaches a level in the atmosphere where the amount of water or carbon dioxide above it is low enough that it can radiate its heat away into space. As it cools off, it becomes more dense than the surrounding atmosphere and sinks back down to the surface. This motion of air parcels is called **convection** (Figure 10.8). It leads to a widespread circulation of air in the lower atmosphere.

While the parcel is moving up or down through the atmosphere, it is surrounded by enough infrared-absorbing gas that no heat escapes from the parcel of air. Such a process is an **adiabatic** process.

When the atmosphere gets thin enough, the rising warm parcels of air can start to lose heat by radiating to space. Once the parcels are able to lose heat by radiation faster than the convection can carry the heat away, their convection stops. Because radiation cools all the levels of this region equally, the temperature in this region is the same everywhere. The atmosphere above this level is no longer well mixed. The air is stratified; that is, each layer of air in this region is stable and doesn't mix with layers above and below it. This part of the atmosphere is called the **stratosphere**.

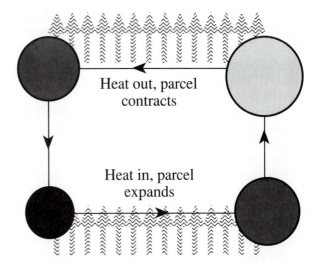

**FIGURE 10.8** Convection. When a parcel of air becomes warmer than its surroundings, it expands and floats upwards to the top of the atmosphere. There it can radiate its heat into space, cool down, and sink back to the bottom of the atmosphere.

The other regions of the atmosphere have names, too. The well-mixed convecting region below the stratosphere is called the **troposphere**. The Greek word *tropos* means "turn"; the convecting air in this region is quite turbulent, always "turning" in on itself. The boundary between the troposphere and the stratosphere is called the **tropopause**.

At the top of the atmosphere, ultraviolet light from the Sun hits the molecules; as they absorb its energy, they become quite hot. This thin, hot air is called the **thermosphere**; the boundary between it and the stratosphere is called the **mesopause**. All planets with atmospheres can be expected to have these layers; thus, Earth and Venus, like Jupiter, have troposphere and stratospheres.

The thermosphere absorbs the most energetic of the UV radiation; however, slightly less energetic UV photons (although still energetic enough to cause damage to living things) are able to penetrate most gases and reach deep into Jupiter or all the way to the surface of Mars. It is probably responsible for the chemical reactions that form the colors of Jupiter's clouds and coats the dust of Mars with a thin layer of free radicals and hydrogen peroxide. Earth's stratosphere, however, contains a molecular variant of oxygen, $O_3$ (ozone), which absorbs this light, except where manmade pollution has destroyed the ozone layer.

## Hadley Cells

The eighteenth-century meteorologist George Hadley realized that along with the vertical mixing due to convection there should also be considerable mixing of air from the poles to the equator of a planet. He reasoned that the hottest parts of a planet are the equator, and so air there should rise most readily, while the poles are the coolest, which is where the air should descend. Winds between these two regions should thus blow polewards in the upper atmosphere and equatorwards along the planet's surface.

This simple picture however, is not an accurate description of the actual weather on Jupiter or Earth for two major reasons.

First, these planets spin quite rapidly. For instance, air at the equator of Earth is moving around the center of Earth at the same speed as Earth is turning, or about 2000 m/s (about 1000 mph) eastward. Jupiter is 10 times bigger in radius than Earth and spins nearly three times as fast, so the gas at its equator travels at over 50,000 m/s!

But at the poles, the speed of the spin is zero.

Newton's first law of motion says that in the absence of a force, a body tends to maintain its speed and direction. Thus air moving from the poles toward the equator has the ground below it beginning to travel in an eastward direction relative to its own motion. However, relative to the ground, the parcel of air seems to be moving westward. Likewise, air moving from the equator to the pole appears (from the ground) to be moving eastward, because it has more eastward motion than the underlying ground.

Now things get complicated. Start with a parcel of air near the pole, moving south. The faster the air moves south, the faster the speed of the ground below it increases, and so the faster it appears to be moving westward, relative to the ground. But any westward motion gives the air less centrifugal force, relative to the ground, because it appears that the air is moving in the opposite direction as the spin of the planet. Less centrifugal force means the air is being pushed in, towards the poles, and away from the equator. But this air, now moving north, appears to have an eastward motion; air going east has more centrifugal force, and more centrifugal force has the opposite effect: The air is pulled south. Thus the parcel that started moving southward first turns westward, and this westward motion causes the parcel to move northward; a northward parcel moves eastward, and an eastward parcel moves south, starting the cycle all over again (Figure 10.9).

The end result is that the air masses are broken up into turbulent, swirling pieces. The force which seems (from our vantage point) to be spinning the air into cyclones is called the **Coriolis force**, named after the nineteenth-century French scientist Gaspard Gustave de Coriolis. It's an example of an "apparent" force. There's no actual force pushing the air sideways; rather, it's Earth that is spinning around underneath the air. (Centrifugal force is another example of such an "apparent" force.)

On Earth this effect is strongest in the middle latitudes, between latitudes of about 30° and 55° north and south of the equator. Closer to the equator, wind motions

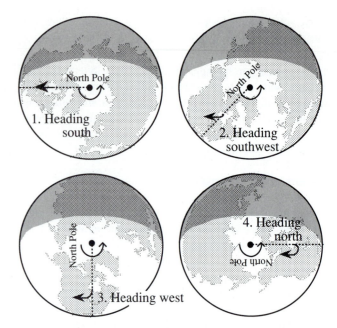

**FIGURE 10.9** The Coriolis force. A parcel of air travelling in a straight path from the poles appears to head first south, then southwest, then west, and eventually north as Earth turns underneath it.

carry air mostly parallel to Earth's axis so little Coriolis effect occurs. Closer to the poles, the air is so little heated by the Sun that strong winds can't arise to carry the air parcels long distances away from the spin axis. These are the regions that typically have cyclonic weather patterns. This includes most of North America and Europe. Winds near the poles and the **trade winds** near the equator tend to be more constant.

With Jupiter's rapid spin, the regions of trade winds extend to much higher latitudes on the planet. The result is the band and zone structure of the atmosphere. Only near the poles (as imaged by *Voyager 2*, seen in Figure 10.6) do we see the characteristic turbulence of cyclonic weather patterns.

Another effect on Earth is that surface features break up the global wind patterns. Some features, such as mountain ranges, physically stand in the way of the winds. Cities or dark mountains and fields standing in sunlight create hot spots, while evaporating ponds and lakes cool the air and create cool spots; these set up their own local convection patterns. The oceans on Earth play a dominant role in its climate. The water is able to store heat and transport it across the planet in currents such as the Gulf Stream of the North Atlantic. A gas planet such as Jupiter has no mountains or oceans, but light- and dark-colored clouds might create local convection patterns.

The result of the rapid spin and the local hot and cool spots is that the nice, orderly convection predicted by Hadley is overwhelmed by other effects on Jupiter and Earth. However, recall that the planet Venus does not spin very quickly at all (its day is several hundred Earth days long). And it has neither oceans nor quite the extensive mountain ranges of Earth. The result is that the sort of convection first predicted by Hadley 250 years ago for Earth is actually seen on Venus.

## Temperature of Jupiter's Interior

Recall that the equilibrium temperature of a planet heated by the Sun depends only on how close it is to the Sun and how much light it absorbs. Recall too, as we saw in Chapter 2, that the infrared spectrum of a planet indicates its temperature. Jupiter, located 5 AU from the Sun with an albedo of 0.343, ought to have an equilibrium temperature of 110 K at its cloudtops. Instead, its observed temperature is 125 K. We saw above that the energy radiated by a planet is proportional to the fourth power of its temperature. But the ratio $(125/110)^4$ is equal to 1.67. In other words, Jupiter is radiating more than half again as much energy as it is receiving from the Sun!

Where is this energy coming from? Of all the ways a planet can heat itself, the most likely heat source for Jupiter is the heat of its own gravitational collapse. We think that Jupiter grew from a large cloud of gas and dust (perhaps a part of the same cloud from which the Sun itself formed). We saw above that the addition of mass to a growing gaseous planet causes its interior to become more and more compressed. This compression heats up the interior; but even more importantly, as the compression occurs the outer layers of the planet start to collapse inward towards the center, as the center itself shrinks. Gravitational energy is released and heats the gas, just as the gravitational energy from forming an iron core heated the interior of Mercury (described in Chapter 4).

The heat created by the collapse has to be transported out of the center to the surface, where it can be radiated away to space. The only efficient way to carry so much heat is for the gas that makes up Jupiter to convect. Convection leads to an adiabatic temperature profile in the gas, which means that we can use the theory of adiabatic gases to estimate the temperature anywhere inside Jupiter. We know the pressure and temperature of the gas at the cloudtops; using the equations of thermodynamics, we can find the temperature at any other pressure and the pressure at any depth.

Convection also tends to keep the gases in the atmosphere well mixed. Thus we can conclude that the composition of gases in the upper atmosphere, except for those components that are condensed into clouds before the gases reach the upper atmosphere, reflects the composition of the whole planet.

As convection cools the interior, the planet contracts further. This contraction leads to more shrinking of the planet; more energy from this infall of gas gets released; and thus more convection is needed to carry away this heat. Over the entire age of the solar system, Jupiter must have been slowly shrinking and radiating away energy as it shrank.

When it was first forming, still adding mass, its rate of shrinking must have been greatest. We conclude that the young Jupiter must have been a large, low-density disk of gas, collapsing rapidly and getting very hot as it shrank. Out of this disk of gas, it is very likely that some solid particles were left behind when the gas collapsed into Jupiter. The moons of Jupiter may well have been formed from this material and the compositions of these moons determined by the heat of the hot, collapsing Jupiter.

## SUMMARY

Sunlight powers the motions of atmospheres. The temperature of a planet is determined by how near the Sun it is and how much sunlight it absorbs.

The sunlight heats the atmosphere in two ways. Energetic photons are absorbed at the very top of an atmosphere, creating a hot but very tenuous region called the thermosphere. And sunlight heats the surface of a planet, which then transmits this heat into the air immediately above it.

Hot air from the surface of a planet rises, gradually dropping in pressure and temperature, until the air above it is thin enough to allow infrared radiation to escape into space. This cools the gas, preventing further convection upwards. Between this convecting troposphere and the thin, hot thermosphere is a stable, isothermal layer of gas called the stratosphere.

The convecting motions of the troposphere are the basis for horizontal motions of the gas across the surface of a planet. If the planet spins and has a varied topography, like Earth, then these motions can become quite complex. The result of these complex motions is that the air at the bottom, densest, part of the atmosphere is always well mixed. On Jupiter, the action of its spin on these lateral motions results in large horizontal bands of clouds from the equator to relatively high latitudes; near the poles, the atmosphere breaks into a chaotic pattern of cyclones.

The formation of Jupiter released considerable heat, which is still being carried from the interior of Jupiter to its cloudtops (by convection) and radiated to space. As a result, Jupiter radiates 1.67 times as much heat as it receives from sunlight.

**FURTHER INFORMATION...**

# Adiabatic Gases

How can we predict precisely how the pressure, temperature, and density of a mole of gas changes as it expands adiabatically?

The first law of thermodynamics says that a change in the energy that a parcel of gas contains is equal to the change in the heat put into the parcel, $dQ$, minus however much work is done when the parcel expands due to its internal pressure. Mathematically, we can write:

$$dE = dQ - P \, dV \tag{10.11}$$

where $P$ is the pressure and $dV$ is a small change in the volume of the gas.

In an adiabatic system, $dQ$ is zero. No heat enters or exits from the parcel of gas. So in this case

$$dE = -P \, dV \tag{10.12}$$

The energy in a parcel of gas is related to its temperature; we can write an equation

$$dE = C \, dT \tag{10.13}$$

Here $C$ is a value called the **heat capacity**. (More specifically, this is the heat capacity at constant volume). So it follows that we can equate the two formulas for $dE$ and find that

$$C \, dT = -P \, dV \tag{10.14}$$

If we write

$$V = \frac{\mu}{\rho} \tag{10.15}$$

we can use the ideal gas law (Equation 10.6) to substitute for $P$:

$$C \, dT = -\left(\frac{RT}{V}\right) dV \tag{10.16}$$

then divide through by $T$, and integrate:

$$C \int \frac{dT}{T} = -R \int \frac{dV}{V} \tag{10.17}$$

$$C \ln \frac{T}{T_0} = -R \ln \frac{V}{V_0} \tag{10.18}$$

or finally,

$$\left(\frac{T}{T_0}\right)^{C/R} = \left(\frac{V_0}{V}\right) = \left(\frac{\rho}{\rho_0}\right) \tag{10.19}$$

We define

$$\alpha \equiv \frac{C}{R} \tag{10.20}$$

It's related to the number of ways that a molecule of gas can absorb energy. The more complicated the gas molecule, the more ways the atoms there can absorb energy by spinning, vibrating, and so forth. Thus for monatomic gases, $\alpha$ is usually a small number, close to 1.5, but this number grows to 3.5 or greater for complicated molecules.

It is also sometimes useful to define

$$\gamma \equiv 1 + \frac{R}{C} = 1 + \frac{1}{\alpha} \tag{10.21}$$

When $\alpha$ varies from 1.5 to 3.5, $\gamma$ drops from 1.67 to less than 1.3.

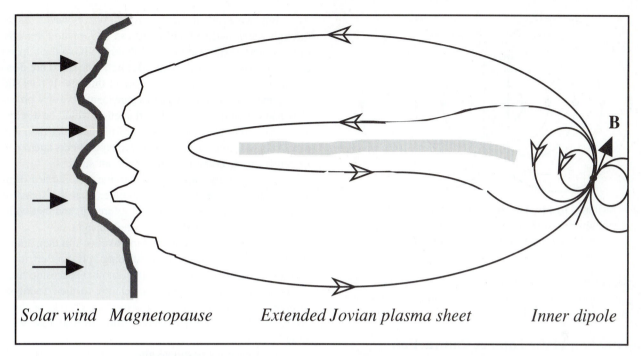

*Solar wind   Magnetopause        Extended Jovian plasma sheet            Inner dipole*

**FIGURE 10.10** Jupiter's magnetic field fills a region of space nearly 100 times the radius of Jupiter itself. Notice that it looks like a dipole near Jupiter, but far from Jupiter it becomes quite distorted.

## STUDY QUESTIONS

1. The intensity of sunlight reaching Earth is 1367 $J / s \cdot m^2$. What is the intensity of sunlight 5 AU from the Sun?

2. Does the atmosphere in a stratosphere convect?

3. If you squeeze a gas *adiabatically* to increase the pressure, does the temperature of the gas get hotter, cooler, or stay the same?

4. Hadley cells do a nice job of describing the cloud patterns on Venus. Why don't they work for Jupiter?

5. As one rises up in an atmosphere the distance of one scale height, roughly how much does the pressure drop?

## 10.3 MAGNETIC FIELDS IN THE SOLAR SYSTEM

Huge as Jupiter is, its size is dwarfed by the volume of space surrounding it that is filled with charged particles and controlled by Jupiter's magnetic field (Figure 10.10). Indeed, appropriate for its size, the strength of Jupiter's magnetic field is second only to that of the Sun. Where does this strong magnetic field come from? For that matter, just what is a magnetic field?

Any child who has played with a bar magnet has been fascinated by the way some materials, such as iron, magically jump onto a magnet and stick there when the magnet is brought up close to them. Most other materials seem inert. Even more bizarre is the way that the end of one magnet can repel one end of a second magnet and attract the other end.

These simple examples illustrate three important points about magnetism. First, magnetic fields can affect the motions of solid matter, but only certain kinds of matter under certain circumstances.

Second, the effective range of these fields drops off very quickly. A bar magnet has an inverse cube field rather than an inverse square field like gravity or light; doubling your distance from a magnet means you experience only one eighth the field. Inverse cube fields are typical of **dipoles**. Recall Chapter 2, where we discussed the electric field arising from a positive and negative charge placed slightly apart from each other. Magnetic fields look as if they were caused by a positive and a negative magnetic pole, always paired together (Figure 10.11). Because of the way these poles act in a compass on Earth, we call them "north-seeking" and "south-seeking," or **north** and **south poles**. But unlike electric fields, which are caused by distinct protons and electrons, no one has ever been able to detect or create a magnetic particle with only north or only south orientation.

The third point also arises from this dipole nature. Opposite poles of a magnet attract, but magnets repel one another when the like-signed poles are brought close together.

All this describes the behavior of a magnetic field, but what exactly is a magnetic field? Ultimately, it's a

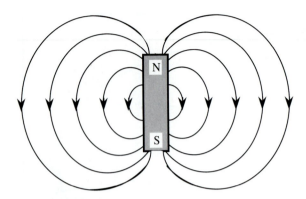

**FIGURE 10.11** The shape of a dipole field. In the shaded box could be a magnet or a loop of wire carrying a current; both give the same shape of field.

consequence of the motions of electrons and protons. The magnetic force field turns out to be a peculiar manifestation of the electric force field.

## Relativity and the Lorentz Contraction

At the beginning of the nineteenth century, scientists believed that electricity and magnetism were two distinct kinds of forces. However, experiments of the British physicist Michael Faraday showed that a changing magnetic field produced a current of electricity, and his countryman James Maxwell showed that a changing electric field produced a magnetic field. Obviously there had to be some connection between the two. With the development of Einstein's theory of relativity it was possible to show how the magnetic force was the consequence of the motions of electrons or protons.

Many of the seemingly bizarre consequences of relativity occur when objects are moving at speeds close to the speed of light. Recall that the three basic measures used by physicists to describe the universe are the units of distance, mass, and time. Relativity predicts (and experiments have confirmed) that the measure of all three units changes when the object being measured moves past the measurer, especially at speeds approaching $c$, the speed of light.

This result follows from the observation that nothing can move faster than the speed of light. Imagine that you have a spaceship moving at 60% of the speed of light, $0.6c$. To everyone on the spaceship, life seems quite normal. In fact, as long as their speed is constant, there is no way to tell they're moving at all, except that they may see planets that appear to whiz past them at $0.6c$.

In other words, there's no preferred frame of reference; planets that seem to be moving in the spaceship's frame appear to stand still in the planet's frame of reference. Einstein called this principle, that no frame of reference is any better than any other frame, **relativity**.

But what about light waves? Water waves travel through water, and so their speed can be measured relative to the water; sound waves travel through air, so their speed can be measured relative to air. But light waves do not need a "medium" like water or air to travel through. If they did have a medium (hypothesized by nineteenth-century physicists and called the **ether**), then that ether would be a reference frame that was somehow more fundamental than any other frame. However, several clever experiments attempted to detect this hypothetical ether, without success.

To get around this preferred frame of reference, Einstein postulated that there was no ether, that the speed of light was the same in all frames of reference, and that nothing could travel faster than the speed of light.

Now, consider our spaceship travellers. In their frame of reference, they appear to be standing still. So they think that it is perfectly possible to launch another spaceship from theirs and have it travel at $0.6c$ relative to their frame of reference. But won't an outside observer, on one of those whizzing planets, see that second ship travel at $1.2c$, 20% faster than light?

By Einstein's postulate, this could not be so. He realized what must happen instead: For an observer not to see speeds greater than light speed, the very units of space and time, meters and seconds, must look different in different frames of reference.

It turns out that, to an observer sitting on a stationary planet, three effects will happen

1.  The original spaceship moving at $0.6c$ must appear to be much shorter than the length it is when standing still.
2.  Time, as measured by a clock on the spaceship, must appear to move more slowly on the spaceship than it does on the planet.
3.  The ship must appear to be more massive than its original mass (a point which we needn't worry about at the moment).

(These effects are not noticeable in our everyday life because we usually deal with velocities much lower than $c$. Nevertheless, relativity is a fantastically successful explanation of a multitude of widely varying laboratory and astronomical experiments involving high velocities, and therefore relativity is a confident cornerstone of modern physics.)

With distances shorter, and time longer, the apparent speed of any object on that spaceship must always appear slower to the observer on the planet than it does to the people on the spaceship. The precise amounts that these quantities change can be calculated; they depend solely on the relative speed between the object being measured and the person being measured. It turns out that, in our example, the measure of length and time must be altered such

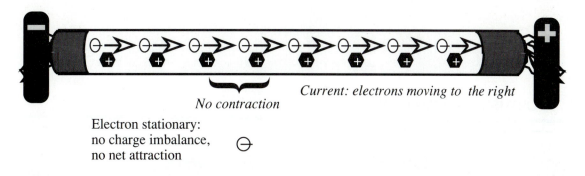

*No contraction*

*Current: electrons moving to the right*

Electron stationary:
no charge imbalance,
no net attraction

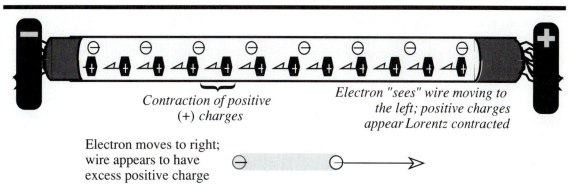

*Contraction of positive
(+) charges*

*Electron "sees" wire moving to
the left; positive charges
appear Lorentz contracted*

Electron moves to right;
wire appears to have
excess positive charge

**FIGURE 10.12**   A stationary electron sees no net force from a wire carrying a current. But if it moves, the Lorentz contraction of the wire and the Lorentz expansion of the electrons mean that the moving electron sees a net imbalance in the charge in the wire and is attracted to it.

that the little ship launched by the first one appears to an observer on the planet to be travelling at only $0.88c$.

Effect 1 is called Lorentz contraction. (H. A. Lorentz proposed it even before Einstein explained it.) It is this effect that gives rise to magnetism.

## Magnetic Field of Current in a Wire

Consider a wire carrying a current. The current is nothing more than a line of electrons carrying a negative charge, moving along a string of positively charged nuclei, the nuclei of the atoms that make up the wire. As one electron leaves the wire, another electron is added (by a battery, for instance) at the other end of the wire. Thus the total number of electrons in the wire stays constant. A "test charge" sitting somewhere far away from this wire (call it a positive charge, a proton, to make the math simpler) does not feel any net attraction or repulsion because the positive charges of the nuclei in the wire are exactly balanced by the negative charge of the moving electrons.

But if the outside proton is moving in the same direction as the flowing electrons in the wire, then the speed of the single moving proton relative to the flowing electrons is less than the relative speed between the moving electrons and the proton when it was stationary. The proton's

speed relative to the positive nuclei in the wire, however, is greater than the speed (zero!) it had when it was stationary. As a result, as far as the moving proton is concerned, the flowing electrons appear less Lorentz contracted than before and the positive nuclei in the wire appear more contracted.

The nuclei appear to be packed together while the electrons are spread out, so now the outside charge sees a *net positive charge* in the wire. The moving proton must be repelled from the wire.

Likewise, if we replace the proton with a loose electron, it feels an attraction (Figure 10.12). If that outside electron is actually one of a current of electrons in some second wire, flowing in the same direction as the current in the first wire, we'll see the whole wire being pulled towards the first wire. This can actually be seen in the lab. Two wires, each carrying 10 milliamps of current, separated by 1 cm, feel a force between them of 0.01 N, a force equal to the weight of 1 g (on Earth). This attraction is the force we call **magnetism**.

Notice what we can learn from this example. First, a current of moving charges sets up a magnetic field. Second, a charged particle reacts to this field only if it is moving in a certain direction; reverse the direction that the charged particle moves and you reverse the direction of the force. Third, the faster it moves, the stronger the force it feels. Finally, notice that the direction of the force is perpendicular

**FURTHER INFORMATION...**

# Dimensionless Numbers

Much of the difficulty in understanding convection in fluids or plasma/field interactions is that the equations of motion for fluids, magnetic fields, and plasmas are all nonlinear. In a linear equation, a small change in one parameter produces effects that are proportionate in size with the change. But for nonlinear equations, small changes can have unpredictably large effects someplace else. Nonlinear equations (except for certain special cases) are notoriously difficult to solve.

Fortunately, it is often sufficient to simply look at an equation and solve only the part of it that is the most important in our circumstances, neglecting the other pieces. But how do we judge which part is most important?

One way is to look at the "characteristic times" for a system to evolve under two different situations. We form a **dimensionless number** from the ratio of the two characteristic times. (It is dimensionless because it's the ratio of two terms with identical dimensions.) If this dimensionless number is much greater (or much less) than unity, then we know that one term dominates and the other can be ignored. Another method is to take the ratio of two different pieces of the nonlinear equation; again we obtain a dimensionless number. Both techniques are illustrated in the following examples.

### *Ra*, the Rayleigh Number

Under what circumstances will a fluid convect? Consider the equation of Newton's second law, $F = ma$, as applied to a fluid. First, we divide both sides of the equation by the volume of a unit of fluid, so on the left side of the equation we have $f$, the force acting on a unit volume of fluid, while on the right side we have a density times an acceleration. Then we assume a steady-state situation so that the net acceleration of our fluid is zero. Thus the sum of all the forces acting on the fluid must be zero.

Recall the equation for hydrostatic equilibrium, Equation 10.7. Under static conditions, the force of gravity, $-\rho g$, is balanced by the increase in fluid pressure with depth, $-dP/dz$. (The minus signs indicate that the direction of these forces is downwards, opposite the direction of increasing $z$.)

However, other forces might also be acting. If the fluid is viscous and moving relative to some outside object, then its viscosity can carry force into (or out of) the fluid. For instance, a paddle moving through fluid can be a source of force acting on the fluid. (A paddler in a canoe who simply lets her paddle drift with the current doesn't move the canoe forward at all.) The velocity appears to diffuse away from the point where the velocity is highest (just like heat diffuses away from the hottest spot) and so we use a diffusion equation to describe this force: $f_{\text{viscosity}} = \eta \, d^2v/dz^2$, where $\eta$, the diffusion constant, is called the **diffusive viscosity**. (We assume constant viscosity; inside a gas giant planet such as Jupiter, however, the extreme changes in pressure lead to changes in $\eta$, making the math much more difficult to solve.) Adding this term to the equation, we arrive at

$$-\frac{dP}{dz} - \rho g + \eta \frac{d^2v}{dz^2} = 0 \qquad (10.22)$$

In hydrostatic equilibrium, with no fluid motion, the third term is zero and the other two terms are balanced so that the net force is zero. Now, what happens if the temperature in this parcel of fluid increases by a factor $\Delta T$? A rise in temperature causes the fluid to expand; if $\alpha$ is the thermal expansion coefficient of the fluid, then the density increases by a factor $\Delta \rho = \rho_o \alpha \Delta T$.

When the temperature increases, two possible changes might result. In the one case the warmer fluid might cool off by conducting its heat away, as represented by the heat diffusion equation

$$\frac{\partial T}{\partial t} = \kappa \frac{\partial^2 T}{\partial z^2} \tag{10.23}$$

where $\kappa$ is the thermal diffusion coefficient, equal to the thermal conductivity divided by the heat capacity and the density of the fluid.

The other thing that could happen is that the fluid, now having a different density from the hydrostatic case, begins to move with a velocity $v$; the third term in Equation 10.22 is no longer zero.

What is more likely to happen: Will the parcel cool off, or will the fluid rise? Let's look at characteristic times.

Equation 10.23 gives us an equation for the change in temperature with time. Let's approximate this equation by replacing derivatives with deltas: $\partial t$ becomes $\Delta t$, the characteristic time, $\partial z$ becomes $\Delta z$, the characteristic distance of the system, $\partial T$ becomes $\Delta T$, the change in temperature with time, and $\partial^2 T$ becomes $\Delta^2 T$, the change in $\Delta T$ with time. Also, $\Delta T \approx \Delta^2 T$, because $\Delta T$ was zero before we started heating the fluid. The characteristic distance is whatever distance controls the physics in the problem; for a paddle in a stream, it might be the width of the paddle. In our convection case, it's the distance over which the temperature changes by our factor $\Delta T$. Our equation becomes

$$\frac{\Delta T}{\Delta t} = \kappa \frac{\Delta^2 T}{(\Delta z)^2} \tag{10.24}$$

$$\frac{1}{\Delta t} = \kappa \frac{1}{(\Delta z)^2} \tag{10.25}$$

$$\Delta t_{\text{conduction}} = \frac{(\Delta z)^2}{\kappa} \tag{10.26}$$

How far does our fluid move during this time? If it moves much less than our characteristic distance, then all the heat will have flowed out of the parcel of fluid before convection can start and convection will not occur. On the other hand, if it moves a distance much larger than our characteristic distance, then the heat will be transported primarily by the motion of the fluid itself and conduction can be ignored.

The speed of the fluid $v$ is equal to the distance travelled per unit time. In our case $v = \Delta z / \Delta t$, where $\Delta t$ is now the characteristic time for convection. We find $v$ by solving Equation 10.22.

$$-\frac{dP}{dz} - \Delta \rho g - \rho_0 g + \eta \frac{d^2 v}{dz^2} = 0 \tag{10.27}$$

But we know the first and third terms sum to zero, so

$$\Delta \rho g = \rho_0 \alpha \Delta T g = \eta \frac{d^2 v}{dz^2} \tag{10.28}$$

$$\rho_0 \alpha \Delta T g = \eta \frac{v}{(\Delta z)^2} \tag{10.29}$$

($v$ and $\Delta v$ are the same because $v$ started out equal to zero),

$$\rho_0 \alpha \Delta T g = \eta \frac{\Delta z / \Delta t}{(\Delta z)^2} \tag{10.30}$$

$$\Delta t_{\text{convection}} = \frac{\eta}{\rho_0 g \alpha \Delta T \Delta z} \tag{10.31}$$

Taking the ratio of the conductive characteristic time (Equation 10.26) to the convective characteristic time (Equation 10.31) yields our definition of the **Rayleigh number**:

$$Ra = \frac{\rho_0 g \alpha \Delta T (\Delta z)^3}{\eta \kappa} \tag{10.32}$$

# Dimensionless Numbers (continued)

Notice the strong dependence on the length scale $\Delta z$. A thin layer of fluid will not convect. That's why insulating materials are made of lots of little air pockets instead of one large airspace. On the other hand, planetary atmospheres have characteristic lengths of a scale height, and so convection is inevitable.

(A slightly different formulation of the Rayleigh number, appropriate for cases where heat is generated within the convecting region, is given in Chapter 12.)

## $R_M$, the Magnetic Reynolds Number

Magnetic fields travelling through a plasma obey the **plasma equation**

$$\frac{\partial \mathbf{B}}{\partial t} = \frac{\partial}{\partial z} \times (\mathbf{v} \times \mathbf{B}) + \frac{10^7}{4\pi\sigma} \frac{\partial^2}{\partial z^2} \mathbf{B} \qquad (10.33)$$

The first term describes how the magnetic field $\mathbf{B}$ can be changed by dragging it in or out of a region with a plasma moving at a velocity $\mathbf{v}$; the second term describes how the field can diffuse through the plasma, leaking out of a region without the plasma itself moving. The electrical conductivity of the plasma is represented by $\sigma$.

If the conductivity of the plasma is very high, the second term goes to zero and we say the magnetic field is "frozen" into the plasma. As the plasma moves, the magnetic field follows. On the other hand, if the conductivity is low, the second term becomes large and we note that the magnetic field can diffuse through the material quite easily.

These terms can be simplified in terms of characteristic lengths, as before, yielding

$$\frac{\Delta B}{\Delta t} = \frac{\Delta(vB)}{\Delta z} + \frac{10^7 \Delta^2 B}{4\pi\sigma(\Delta z)^2} \qquad (10.34)$$

$$\frac{1}{\Delta t} = \frac{v}{\Delta z} + \frac{10^7}{4\pi\sigma(\Delta z)^2} \qquad (10.35)$$

Taking the ratio of the first term over the second term gives us the **magnetic Reynolds number**:

$$R_M = \frac{4\pi\sigma \Delta z(v)}{10^7} \qquad (10.36)$$

If the conductivity is sufficiently high, the velocity of the plasma sufficiently large, or the length scale sufficiently large, transport of the magnetic field lines by the plasma will be more important than diffusion of field lines. None of this is surprising; a high conductivity freezes in the field lines. Even if field lines are only loosely tied to the plasma, if the plasma moves fast enough that it can still warp the lines (just as a fast jet of water can pull on our "rubber bands" as effectively as slow molasses), and if the scale of the system is big enough, it will take so long for field lines to diffuse that distance that they might as well be considered to be part of the plasma. The advantage of the magnetic Reynolds number, however, is that it takes these very vague, hand-waving ideas and allows us to put some numbers to them. How much conductivity is high enough? How fast is fast enough? The magnetic Reynolds number allows us to make specific quantitative predictions.

Another result comes from this analysis. Consider the case where the magnetic Reynolds number is much less than 1 and diffusion dominates. The solution to the diffusion equation is well known; if the magnetic field obeys an equation of the form

$$\frac{\partial \mathbf{B}}{\partial t} = \frac{10^7}{4\pi\sigma} \frac{\partial^2 \mathbf{B}}{\partial z^2} \qquad (10.37)$$

then we know that the magnetic field decays exponentially, with a decay time equal to

$$\tau = \frac{4\pi\sigma \, \Delta z^2}{10^7} \qquad (10.38)$$

Consider field lines generated in Earth. We can estimate values for the conductivity of Earth's core; the radius of the core is the characteristic length for this system. For Earth, we find that the field should decay away in a span of 10,000 years. But in fact remanent magnetism has shown that Earth's field is as old as Earth itself, and even field reversals only occur on a time scale of millions of years. This demonstrates that the field inside Earth must be actively generated and is not some remnant of an externally imposed field. The Sun is the opposite situation; there we know that the field reverses sign with an 11-year period, but the decay time for a field in the Sun is on the order of $10^9$ years. Once again we conclude that active field generation must be occurring inside the Sun.

to the direction of the current and the direction in which the charged particle is moving.

We can draw magnetic **field lines** just like we drew electric field lines in Chapter 2. These field lines, in the case of current flowing through a wire, by convention look like rings surrounding the wire. We choose this direction for the field lines so that we can use the vector mathematics of cross products to describe the force on the proton. If a proton is moving at a velocity **v** through a magnetic field of strength and direction **B** (a vector whose direction is represented by these field lines), we can say that the force is proportional to $\mathbf{v} \times \mathbf{B}$. The cross product notation indicates that the force is perpendicular to both the direction that the particle is moving and the rings of the magnetic field.

If the wire carrying the current is twisted into a loop (as in an electromagnet), then the magnetic field rings look just like the field lines of a dipole, as if there were positive and negative magnetic particles inside the ring. If we have two separate loops of current, the electrons and protons in one are moving relative to the protons and electrons in the other, and thus the rings can be attracted or repulsed, depending on which way the currents circulate.

## The Dynamo

Running a current through a wire creates a magnetic field. Increasing the current increases the field. But it turns out that the inverse is also true: If we have a simple loop of wire, with no current running through it, and we bring a bar magnet over to the loop, the act of introducing a magnetic field near the loop induces a current to start flowing in the loop! Changing the amount of magnetic field near a loop of wire changes the current in the loop, just as changing the current in the loop changes the strength of the magnetic field near it. This is called **Faraday's law**.

Say there's a current flowing through one wire. If we move another wire in towards the first wire, that means we are moving that second wire through the magnetic field produced by the current in the first wire. As we move it, we induce a current to flow in the second wire. This current sets up its own magnetic field, that, in turn, changes the current in the first wire, changing its magnetic field, which changes the current in the second wire, and so on.

In fact, it is actually possible to set up a self-sustaining dynamo. With the right geometry, one can move a wire (or anything that conducts electricity) through a "seed" magnetic field and set up a current in that conductor. That induced current then generates its own magnetic field. At this point, we can take away the seed field; if we bend the wire around properly, the wire's own field can be strong enough and directed in the right direction to keep the current going inside the wire and keep producing its own magnetic field, as long as we keep the conductor moving through this field. The field produced by this process can even be

stronger than the original seed field. As long as one keeps the conductor moving in the right direction, moving through its own magnetic field, its motion creates a current that can keep creating the field.

At first this might seem like a perpetual motion machine, but in fact it is not. The energy of the electrical current, and of the magnetic field, comes from the mechanical energy needed to keep the conductor moving. Anyone who has ridden a bicycle with a light-and-generator attachment (the generator is a simple dynamo) knows that it takes extra work pedalling to keep the generator turning and the bicycle lamp lit. The dynamo converts this mechanical energy into electromagnetic energy.

## Generating Magnetic Fields in Planets

How does all this apply to planets? We know that Mercury, Earth, Jupiter, Saturn, Uranus, and Neptune all have magnetic fields (Table 10.2). So does the Sun. What is the source of these fields?

To get a magnetic field, we need to start with a seed field. We need to move a conductor through this field to start a current flow; that means we also need some energy source to keep the conductor moving. And the motions of the conductor must have the right orientation for a self-sustaining dynamo to result.

With the planets' magnetic fields, the seed field was probably the magnetic field of the Sun, itself produced by a dynamo seeded from the interstellar magnetic field. The liquid metallic cores of the planets (probably iron sulfide and iron in Earth and metallic hydrogen in the gas giant planets) are the conducting materials. Convection in these cores provides the motion of the conductor.

Simple, symmetric convection patterns do not lead to a self-sustaining dynamo, however. For the complex motions needed to make a dynamo, the planet must be spinning. The Coriolis force that results from spin, the same force that produces complex weather patterns in planetary

***TABLE 10.2***

Magnetic fields

| Planet | Field Strength |
| --- | --- |
| Mercury | 0.02 |
| Venus | $\approx 0$ |
| Earth | 3.1 |
| Mars | $\approx 0$ |
| Jupiter | 42.5 |
| Saturn | 2.1 |
| Uranus | 2.3 |
| Neptune | 1.4 |

Note: Strength at the surfaces or cloudtops of the planets in $10^{-5}$ teslas.

atmospheres, must also set up turbulent eddies in the core, and some of those eddies must have the proper orientation to produce a self-sustaining magnetic field (Figure 10.13).

At least, that's the present theory. One of the biggest failures of this theory, however, is the magnetic field observed around the planet Mercury. Mercury spins very slowly, so that turbulent motions in its core would not be very likely. Some theories of planet composition (see Chapter 14) predict that Mercury should not have an iron sulfide core but rather a core of pure iron, which is much harder to melt. Even if the core had iron sulfide, it isn't clear that it would be warm enough inside Mercury today to keep that core molten and convecting, because Mercury is such a small planet. Nonetheless, Mercury does have a small magnetic field. Its origin is still not known.

## SUMMARY

Magnetic fields are the relativistic consequence of electric currents moving in a conductor. A current in a loop, or a conductor carrying charge in any circular path, produces a dipole magnetic field. Magnetic fields can be produced by the motions of a conductor, such as a convecting metallic liquid; such liquids (liquid hydrogen metal in the gas giants, molten iron and iron sulfide in Earth) probably exist in the cores of the planets that have magnetic fields.

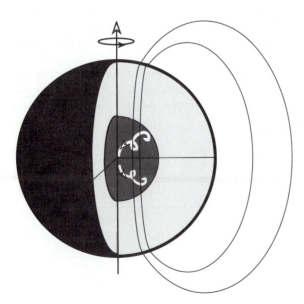

**FIGURE 10.13** Convective motions in the core, twisted by the Coriolis force as the core spins, may produce eddies in such a way that a dipole magnetic field can be generated.

## STUDY QUESTIONS

1. Many other planets besides Jupiter have magnetic fields. Name one planet that does not.

2. Which theory of twentieth-century physics explains how moving electric charges create magnetic fields: relativity, quantum field theory, or blackbody radiation theory?

3. The measure of three basic quantities of an object changes when an object travels at close to the speed of light. Name those quantities.

4. What must the core of a planet do (according to present theory) to produce a magnetic field? All the following except (choose one):
   (a) convect
   (b) conduct electricity

5. A self-sustaining dynamo produces electricity by moving wires through a magnetic field and then uses this electricity to produce the magnetic field itself. Where do we add outside energy to the dynamo to make this whole process work?

## 10.4   PROBLEMS

1. In our travelogue of Jupiter at the beginning of this chapter, we predicted that the colors of the clouds underneath the zones of ammonia clouds would not be as brilliant as the colors that were exposed to direct sunlight. Why?

2. Go through the travelogue that started this chapter and list 10 specific aspects of Jupiter stated there. Of these 10, which ones are observed facts, which ones are the result of theoretical calculations, and which ones are speculations based on the observations and the theories?

3. Jupiter is the largest of the gas giant planets and Earth is the largest of the terrestrial (rocky) planets. Name five other ways in which they are similar.

4. When a meteorite hits the Moon, it forms a crater. What is likely to happen when a meteorite hits Jupiter?

5. Which planet collects the most total energy from the Sun: Mercury, Earth, or Saturn? Be sure to consider both the size of the planets and their distances from the Sun, but ignore albedo differences.

6. There is concern that a fleet of supersonic airliners, flying regularly in the stratosphere, would lead to serious pollution problems. Why should pollution in the stratosphere be more of a problem than pollution in the troposphere?

7. The cities of Tucson and Phoenix are only 100 miles apart; both are in the desert, in similar surroundings. But Tucson is at an altitude of roughly 900 m (3000

feet), while Phoenix is set in a valley at 600 m (2000 feet). Assume that the atmosphere is diatomic, adiabatic, and ideal. The scale height on Earth is 8.4 km.
(a) If the pressure at sea level is $10^5$ Pa (1 bar), what is the pressure in Phoenix? In Tucson?
(b) If the temperature in Tucson is 300 K (80° F), predict the temperature in Phoenix.

8. Will every planet that has an atmosphere have the same layers as Earth: a troposphere, stratosphere, and thermosphere? What layers could be missing if the atmosphere is thin?

9. The planet Uranus spins in a most peculiar way. It is tipped over nearly on its side. As a result, for part of its year only the northern hemisphere receives any sunlight; then, for the next season, the entire planet has days and nights. Following that, only the southern hemisphere receives sunlight, and so on. Describe, in a general way, what the temperatures of the poles and equator should be like during each season. Describe what wind patterns you might expect during a season when the pole faces the Sun.

10. In Chapter 2 we said that every electric wave made by an oscillating electron also produced a magnetic wave. Why is this so?

11. The electric field lines of a point charge are all radial; they all point straight towards, or away from, the point charge. However, as can be seen in Figure 10.11, the field lines of a dipole are not all radial. Define distance $r$ as the distance from the center of the dipole and angle $\theta$ as the angle from the dipole axis. At what values of $\theta$ is the magnetic field entirely in the $r$-direction? At what value of $\theta$ is it entirely in the $\theta$-direction?

12. Jupiter's surface magnetic field at the equator is $4.25 \times 10^{-4}$ tesla. (**Tesla** is the SI measure of magnetic field strength.) The field at the surface of Earth, by comparison, is $0.3 \times 10^{-4}$ tesla. Assuming that both Earth and Jupiter have dipole fields, how far from Jupiter must you travel before its field is as strong as Earth's? How strong is Earth's field one Jupiter radius away from the center of Earth?

13. Why is it no surprise that Venus lacks a magnetic field? Would you expect that Pluto has a magnetic field?

14. How does a compass work? If you were standing right over the north magnetic pole, what direction would

your compass point in? (Hint: Consider your answer to Problem 11.)

15. *Computer Spreadsheet Problem.* The amount of energy in a magnetic field is proportional to the square of the magnetic field strength times the volume of space occupied by that field. Make a rough estimate of the energy necessary to generate each of the magnetic fields listed in Table 10.2 by squaring those field strengths and multiplying each by the volume of the planet where they are generated. How does this energy compare with the total amount of energy in the sunlight falling on the surface of each planet? Do you think it's possible that the energy of the magnetic fields could come from sunlight? Give your reasons.

## 10.5   FOR FURTHER READING

Although it is now rather old, the best single source for scientific papers on Jupiter is still the anthology *Jupiter* edited by Tom Gehrels (Tucson: University of Arizona Press, 1976). For current research on atmospheres, consult the University of Arizona book *Origin and Evolution of Planetary and Satellite Atmospheres*, edited by S. K. Atreya, J. B. Pollack, and Mildred Shapley Matthews (1989).

The physics of atmospheres is explained at an introductory level in Richard Goody and James Walker, *Atmospheres* (Englewood Cliffs, N.J.: Prentice-Hall, 1972). A more advanced look, including discussions in depth about Jupiter, can be found in John Lewis and Ronald Prinn's *Planets and Their Atmospheres* (New York: Academic Press, 1984). An exhaustive advanced discussion of planetary atmospheres is found in Joseph W. Chamberlain and Donald M. Hunten, *Theory of Planetary Atmospheres* (New York: Academic Press, 1987).

For an introduction to relativity and its role in magnetic fields, Edward Purcell's undergraduate textbook *Electricity and Magnetism* (New York: McGraw-Hill, 1985) is recommended. For more information about the generation of magnetic fields in planets, consult the appropriate chapters in an advanced geophysics text such as Frank D. Stacy, *Physics of the Earth* (New York: Wiley, 1977), or Donald Turcotte and Gerald Schubert, *Geodynamics* (New York: Wiley, 1982).

**FIGURE 11.1** Saturn.

# CHAPTER **11**

# The Outer Planets

## 11.1 SATURN, URANUS, NEPTUNE, AND PLUTO

Saturn, Uranus, and Neptune are three large planets, giants compared with Earth (but smaller than Jupiter) that lie in the outer reaches of the solar system, three balls of gas like Jupiter (although in fact it might be more proper to refer to them as balls of liquid, with very thick atmospheres). These three outer planets lie at the outer reaches, not only of the solar system, but of our knowledge of the solar system.

And out beyond them all, for most of its orbit, lies the little ice ball Pluto.

### An Overview of the Outer Planets

*Saturn*

Forget about Saturn's rings, just for a moment. The second largest planet in the solar system sits inside those rings. Saturn the planet is a large yellow ball, about 60,000 km in radius, with a mean density around 0.7 g/cm$^3$. It is lower in density than water; if you could drop it in a very large tub of water, it would float. It is the least dense of the planets.

Saturn is tilted on its axis by 27°, giving it seasons much like Earth or Mars. Although its composition is sim-

| *Saturn's Vital Statistics* | |
|---|---:|
| Radius, equatorial | 60,268 km |
| Radius, polar | 54,364 km |
| Surface area | $4.38 \times 10^{10}$ km$^2$ |
| Mass | $5.69 \times 10^{26}$ kg |
| Density | 0.69 g/cm$^3$ |
| Local gravity | 10.4 m/s$^2$ |
| Escape velocity | 36 km/s |
| Albedo | 0.34 |
| Atmospheric temperature (0.5 bar) | 95 K |
| Length of year | 29.46 yr |
| Length of day | 10 h 39 min |
| Distance from the Sun | 9.53 AU |

ilar to Jupiter's, its clouds are not nearly as colorful; rather than a sharp contrast between white zones and dark belts, Saturn's clouds are a more subdued yellowish brown. The composition of the clouds is probably not that different from Jupiter's; of the compounds seen so far in Saturn's atmosphere, all the same compounds have been seen in Jupiter. However, the abundance of helium in Saturn's upper atmosphere is only two thirds that seen in Jupiter.

| Uranus's Vital Statistics | |
|---|---|
| Radius, equatorial | 25,559 km |
| Radius, polar | 24,973 km |
| Surface area | $8.13 \times 10^9$ km$^2$ |
| Mass | $8.66 \times 10^{25}$ kg |
| Density | 1.27 g/cm$^3$ |
| Local gravity | 8.8 m/s$^2$ |
| Escape velocity | 21 km/s |
| Albedo | 0.30 |
| Atmospheric temperature | |
| (0.5 bar) | 60 K |
| Length of year | 84.01 yr |
| Length of day | 17 h 14 min |
| Distance from the Sun | 19.2 AU |

Like Jupiter, Saturn has a strong magnetic field. Furthermore, like Jupiter it spins rapidly: A day on Saturn is only 10.5 hours long.

## Uranus and Neptune

Uranus and Neptune are almost twin planets. They are considerably smaller than Saturn and only about 5% as big as Jupiter. Uranus is the larger of the two, but Neptune is slightly more massive. Even though Neptune is almost half again as far away from the Sun as Uranus, both planets are observed (by measuring their infrared emission) to have the same temperature, 60 K, near the tops of their atmospheres. Both planets appear dark blue-green in color. Uranus is unusual in its spin; it is tilted completely over on its side, with the direction of its spin axis oriented more than 90° from the plane of its orbit.

Distinct clouds are difficult to make out on Uranus, but clouds on Neptune (including a Great Dark Spot reminiscent in its shape of Jupiter's Great Red Spot) were discovered by the *Voyager* spacecraft during the 1989 flyby (Figure 11.2). By tracking the clouds from picture to picture, it was possible to estimate the wind speed in the up-

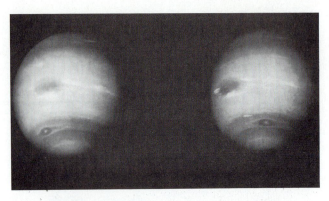

**FIGURE 11.2** Neptune as seen by *Voyager*. Notice the differential rotation of the spots.

per atmosphere of Neptune. It turns out to be a remarkably windy place; in some cases wind speeds exceed 600 m/s (over 1300 mph).

## Pluto

The smallest and most recently discovered of the planets (Pluto was discovered in 1930 and its moon, Charon, in 1978), Pluto is quite different from the other outer planets. It is not a ball of gas, but rather a ball of ice similar in many respects to the icy moons of its larger neighbors. Its surface is 97% covered with frozen nitrogen. Methane and carbon dioxide ices cover another percent or two; other ices (such as water) are also presumably present. Some researchers are convinced that Pluto is in fact an escaped moon of Neptune; other scientists have argued strongly against this hypothesis.

Pluto and Charon make up a small double-planet system. Pluto itself is about two thirds the radius of Earth's Moon, with a density close to that of Jupiter's icy moon Ganymede; its companion, Charon, is a bit smaller than the Saturn moon Iapetus. Although too small and too far away to be seen as more than a dot of light in a telescope (and, alone of the planets, Pluto has not yet been the target of any spacecraft missions), one can carefully measure the change in brightness of the combined Charon–Pluto system that

| Neptune's Vital Statistics | |
|---|---|
| Radius, equatorial | 24,765 km |
| Radius, polar | 24,340 km |
| Surface area | $7.65 \times 10^9$ km$^2$ |
| Mass | $1.03 \times 10^{26}$ kg |
| Density | 1.64 g/cm$^3$ |
| Local gravity | 11.2 m/s$^2$ |
| Escape velocity | 24 km/s |
| Albedo | 0.29 |
| Atmospheric temperature | |
| (0.5 bar) | 60 K |
| Length of year | 164.1 yr |
| Length of day | 16 h 6 min |
| Distance from the Sun | 30.1 AU |

| Pluto's Vital Statistics | |
|---|---|
| Radius | 1152 km |
| Surface area | $1.67 \times 10^7$ km$^2$ |
| Mass | $1.28 \times 10^{22}$ kg |
| Density | 2.0 g/cm$^3$ |
| Local gravity | 0.6 m/s$^2$ |
| Escape velocity | 1.2 km/s |
| Albedo | 0.3 |
| Surface temperature | ~ 80 K |
| Length of year | 248.5 yr |
| Length of day | 6 d 9 h 18 min |
| Distance from the Sun | 39.8 AU |

occurs when one body gradually covers the surface of its twin, as it orbits in front of the other; this information then can be used to tell us something about lighter and darker regions of the surface that were hidden during the transit. From such studies we can conclude that both bodies have bright polar caps; there appears to be a dark band around Pluto's equator, while most of Charon is dark.

Pluto is also special for having the most eccentric and inclined orbit of any planet. Although the semimajor axis of its orbit is nearly 30% larger than Neptune's, the eccentricity of its orbit is large enough to bring Pluto closer to the Sun than Neptune for a substantial part of its orbit. In fact, on January 21, 1979, Pluto did cross the orbit of Neptune, and it will stay closer to the Sun until March 14, 1999. In addition, Pluto and Charon are in mutually synchronous orbits; Pluto always keeps the same face towards Charon and Charon towards Pluto. Pluto's spin axis and the axis of Charon's orbit about Pluto are tilted perpendicular to the plane of Pluto's orbit about the Sun, just like Uranus and its moons.

## A Day on Uranus and Neptune

Saturn is close enough and has enough distinct clouds visible from Earth that it has long been known to have a rotation period of a little over 10 hours. What about Uranus and Neptune? How long is a day on these two planets? That's been a controversial question for many years.

One problem in determining a spin rate for Uranus from Earth is that the spin axis is tilted 98° from the ecliptic plane, the plane of the planet's orbit. We knew the direction of Uranus's spin even before *Voyager*, because the planet has rings and moons that all orbit in a plane tilted 98° from the plane of Uranus's orbit. Uranus is like a top on its side, "rolling" as it orbits the Sun. This means that for part of its year (which lasts for 84 Earth years) it has only its north pole pointing at the Sun and Earth; half a Uranus year later, it's the south pole that faces us (as it did during the *Voyager* encounter in 1986). Only for a few years out of every 84 does the side of the planet face the Sun and Earth so that Earth-based observers could see its appearance change as it spins. Thus most of the Earth-based observations discussed below were performed in the 1960s and early 1970s, when Uranus could be viewed side-on.

Several Earth-based techniques were tried before *Voyager* to determine Uranus's spin rate. One method was to look for spots on the surface of the planet and see how fast they moved. This proved to be virtually impossible for planets as far away as Neptune; even the best telescopes of the day could not make out markings smaller than a quarter the size of Neptune itself.

A variant on this approach was to measure the brightness of the whole planet to see if it changed at some regular rate. The data (especially for Uranus) were not very good,

however. One of the major troubles with observing the spin rate of Uranus in this manner is that its atmosphere is much too uniform to make the brightness vary as it spins. Neptune, by contrast, has methane clouds that come and go in its atmosphere. A problem with Neptune, however, is that different clouds at different levels in its atmosphere could (and, in fact, do) move at different rates, much as the clouds of Venus circle the planet at a completely different rate from that at which the planet itself spins. These observations for Neptune suggested a period of about 18 hours; for Uranus, no consistent result was reached by any of the different observers attempting this measurement.

Another method, equally difficult to accomplish, was to measure the precise wavelengths of some well-understood absorption lines in the spectrum of either planet. Light absorbed by gas on the side of the planet spinning towards us is slightly **Doppler-shifted** to a shorter wavelength and that from the receding side is shifted to a longer wavelength (much as the siren of a fire engine on a freeway seems to be high-pitched when approaching and lower when moving away). Thus, by measuring how much these absorption lines were shifted from their laboratory value, one could estimate how fast the limbs (that is, the edges) of the planet were moving towards or away from us as the planet spun. These measurements appeared to give a spin rate in agreement with the brightness curve for Neptune, but they gave a much shorter period for Uranus than most of the brightness curve determinations. Neither measurement was very precise, however.

In fact, before *Voyager*, the best Earth-bound observations of Uranus, using all these methods, gave published rotation rates of 10.8 hours, 12 hours, 15.6 hours, and 23 hours!

About 10 years after these disappointing results were obtained, yet another technique was attempted. This new method used the size of Uranus's equatorial bulge to estimate how fast it was spinning. In March 1977 Uranus passed between Earth and the faint star SAO 158687. Several observers in different locations on Earth carefully timed how long Uranus covered the star. Their observations produced several chords across the disk of the planet so that a precise value for the size and shape of Uranus could be determined, reminiscent of how the size of the Sun was determined by transits (as discussed in Chapter 1). This was the same occultation that revealed the rings around Uranus and suggested that one of Uranus's rings is eccentric. By seeing how this eccentric ring **precesses** (that is, how its orbit drifts around Uranus with time), it was possible to determine the second harmonic of Uranus's gravity field. (Harmonics were discussed in Chapter 4.) An equation relates this harmonic, called $J_2$, to how much a planet's poles are flattened and its equator bulges while it spins, assuming that the planet is perfectly fluid. If one could get $J_2$ from the behavior of the eccentric ring and then measure how flattened Uranus appears to be, it should be possible to solve for how fast Uranus is spinning. The answer from this technique for the spin rate was 13 hours, in disagreement with all the other measurements!

Finally, in January 1986 the *Voyager 2* spacecraft passed by Uranus and took pictures of the planet as it spun. It also recorded the variation in the magnetic field of Uranus as the interior of the planet, where the field is generated, was spinning. The unambiguous result from *Voyager* is that Uranus spins once every 17.24 hours.

Although nobody predicted that particular number, it is unlikely that any scientist was particularly surprised by it. There are times when data are hard to get and one must resort to trying to interpret a pile of numbers that represent both real information from the planet (**signal**) and random fluctuations within the instrument taking the measurements (**noise**). In the case of the Uranus spin rate, the signal-to-noise ratio for most of the spin measurements was very low, and so it was difficult to make any precise statements about the spin rate. Those who tried to get around the difficulty by a clever application of gravitational theory were equally misled. The theory failed to account for many other effects that could change the orbit of the ring (such as the presence of small moons, pulling it in a way different from the equatorial bulge of Uranus). It also had to assume a perfectly fluid Uranus.

If all these techniques had given the same rotation rate, then a different *Voyager* result would have been more of a surprise. But even in such a case a good planetary scientist would always hold a cautious, and perhaps skeptical, attitude towards data from inherently difficult experiments.

## Clouds

The atmospheres of Saturn, Uranus, and Neptune are all rich in hydrogen, like Jupiter's. Like Jupiter, one might expect cosmic abundances of the other elements to exist there as well. Why don't these planets look like Jupiter, with alternating dark belts and white zones and great red spots? There are at least three reasons.

First, these planets are all at least twice as far from the Sun as Jupiter and so they receive much less ultraviolet (UV) radiation than Jupiter does. It is this UV light that takes essentially colorless clouds and gases such as methane and hydrogen sulfide and breaks them apart, freeing sulfur and combining hydrocarbons to make the colors of Jupiter. With less UV light, less coloring material is made.

Being farther from the Sun, these planets are also much colder. On Saturn, it is more likely that more ammonia clouds are formed than on Jupiter, because the deeper parts of the atmosphere, which have more ammonia, are colder on Saturn than on Jupiter. Ammonia clouds are white, even on Jupiter; furthermore, they also cut down on the UV

light reaching the sulfur-rich clouds below them. It's also cold enough for white methane haze to form at the poles of Saturn and in the cloudtops of Uranus and Neptune.

Finally, observations have shown that the compositions of these three planets' atmospheres are different in certain important ways. Saturn's upper atmosphere seems to be depleted of helium, which might indicate of other nonsolar abundances in the region of the clouds. Additionally, no ammonia has been detected in the atmospheres of Uranus or Neptune. On the other hand, in Uranus and Neptune methane is superabundant, a factor of 20 times more abundant than cosmic composition. There appear to be methane haze droplets throughout the upper atmosphere of Uranus, and methane clouds are seen on Neptune. The dark greenish color of Uranus and Neptune results from their lack of white ammonia clouds and the presence of methane.

The only other place in the solar system known to have methane haze is Titan, the largest moon of Saturn. An interesting but still unexplained phenomenon occurred during the late part of the 1970s. Over a period of 4 years, all the places with methane haze—Uranus, Neptune, and Titan—grew much brighter. The best explanation seems to be that the climate of all three places changed enough to allow more methane clouds to form. This brightening coincided with a peak in solar activity. Weather patterns on Earth have also been seen to coincide with the sunspot cycle on the Sun. But the connection, why sunspots should affect clouds and weather on all these planets, remains a mystery.

## Internal Structure and Excess Heat

As noted above, Neptune and Uranus have similar infrared temperatures, even though Neptune is half again as far away from the Sun as Uranus. Recall that the temperature of Jupiter is seen to be much higher than an equilibrium with sunlight would predict; instead, Jupiter is radiating twice the energy it is receiving from the Sun. This excess energy is believed to come from the slow collapsing of the Jovian gas as Jupiter continues to compress itself by its own gravity.

Saturn also has a much higher temperature than equilibrium would predict (Table 11.1); so does Neptune. But

**TABLE 11.1**

Calculated Temperatures of the Outer Planets at the 0.5-bar Level

| Planet | $T_{calc}$ | $T_{obs}$ | $E_{obs}/E_{calc}$ |
|--------|-----------|----------|-------------------|
| Jupiter | 110 | 125 | 1.7 |
| Saturn | 80 | 95 | 2.0 |
| Uranus | 60 | 60 | 1.0 |
| Neptune | 45 | 60 | 3.6 |

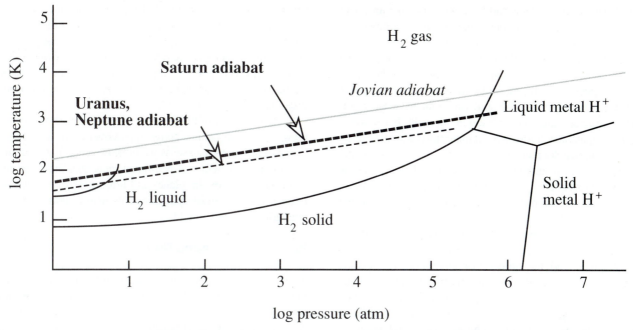

**FIGURE 11.3** The phase diagram of hydrogen, with lines showing the temperatures and pressures likely to be seen inside Saturn, Uranus, and Neptune (assuming an adiabatic temperature gradient). Because these planets have colder temperatures than Jupiter, it is possible that an ocean of liquid hydrogen could occur inside these planets.

Uranus does not; Uranus is quite close to the temperature one would expect for a slowly spinning body in equilibrium with sunlight. Why do the other planets have excess heat? Why doesn't Uranus?

The simple explanation for Jupiter's excess heat won't work for Saturn. Saturn is sufficiently smaller than Jupiter that one would expect much less heat to remain from gravitational collapse. Certainly Neptune is too small for so much extra heat to be formed in this way.

However, notice Figure 11.3, the phase diagram of hydrogen. Recall that the temperatures and pressures inside Jupiter are greater than the **critical temperature** where gas and liquid are no longer separate phases. On Saturn, Uranus, and Neptune, however, the temperature at the cloudtops is lower, and so the temperatures of the atmosphere at any given pressure inside these planets are likewise lower than the temperatures inside Jupiter. As can be seen in Figure 11.3, it is likely that in all three planets there is a definite transition between a hydrogen-rich atmosphere and a thick ocean of liquid hydrogen appearing at several tens of bars inside the planet.

Experiments on hydrogen–helium mixtures have indicated that under certain temperatures and pressures, liquid hydrogen and liquid helium are **immiscible**; like oil and water, they do not mix together, but rather the helium tends

to separate from the hydrogen and clump together. These little droplets of liquid helium, being more dense than hydrogen, can rain out into the center of the planet. They release energy as they fall, thus heating the interior just as the formation of an iron core heated the interior of Mercury. This would also explain the depletion of helium in the upper atmosphere of Saturn. The mean density of Saturn tells us that helium must be present somewhere; we infer that it is present as an immiscible, liquid core.

A further complication occurs in Uranus and Neptune. From their mean densities, it is clear that these planets do not have cosmic abundances of the elements. Where a water-to-hydrogen ratio of 1/320 prevails on Jupiter and Saturn, on Uranus and Neptune water and hydrogen are 1/1, equally abundant. This means that the simple phase diagram of hydrogen does not tell the whole story of what the interiors of these planets look like.

Experiments on the high-pressure behavior of water show that at about $1.5 \times 10^{10}$ Pa of pressure (about 150,000 atmospheres) the electrical conductivity of the water suddenly changes. Under these conditions, water behaves as if it were a molten **salt** made of $H_3O^+$ and $OH^-$ ions, similar to the way table salt is made of $Na^+$ and $Cl^-$ ions. This ionized water may be abundant enough in Uranus and Neptune to rain out of the upper atmosphere and form a large, thick **ionic ocean**. Uranus and Neptune may have an ocean of ionic water 10,000 km deep underneath an atmosphere 6000 km thick of mostly dry hydrogen.

Such an ocean would have an important effect on the other substances in the atmosphere of Uranus and Neptune. Water (especially in this ionized state) is a very strong **polar** solvent. It tends to dissolve materials that have an asymmetric structure and a distinct dipole moment, but it won't dissolve symmetric molecules. Ammonia is a nonsymmetric, polar molecule; methane is the opposite, symmetric and nonpolar. As a result, we would expect all the ammonia in these atmospheres to be dissolved into the thick water oceans while all the methane and helium would be rejected into the atmosphere.

This is consistent with the observed superabundance of methane and helium in the upper atmospheres of Uranus and Neptune and the absence of ammonia or ammonia clouds. This ocean could also have dissolved whatever rocky elements exist in the planet. The present challenge is to build detailed models of this thick ocean that are also consistent with the observed gravitational harmonics $J_2$ and $J_4$.

Finally, why is Neptune as warm as Uranus? One suggested solution looks at the methane clouds on Neptune. Because Uranus is closer to the Sun than Neptune, methane clouds are less likely to form in its upper atmosphere; the temperature is too warm for them. However, in Neptune, methane clouds easily form at its cold equilibrium temperature. Like water clouds on Earth and carbon dioxide on Venus, methane haze absorbs infrared light. Thus these methane clouds could trap absorbed visible light or heat from Neptune's formation, raising its temperature to that of Uranus. Once the temperature reaches that of Uranus, the clouds might start to dissipate; but then heat can finally leak out, lowering the temperature and allowing the clouds to form again. Thus the clouds may act like a thermostat, regulating the temperature of Neptune.

## Magnetic Fields

The magnetic fields of Saturn, Uranus, and Neptune were first measured by the Voyager spacecraft. All three planets turn out to have strong but odd fields.

Saturn's field is about 10% as strong as Jupiter's, and there is nothing odd about that. But its orientation is quite unexpected. Unlike Jupiter and Earth, where the north and south magnetic poles are some tens of degrees in latitude away from the geographic poles, the magnetic field of Saturn is lined up to within 1° of the geographic poles. This is disturbing to theorists, because present theories for how a planet should generate a magnetic field (see Chapter 10) predict that the field ought to be more askew, like Earth's or Jupiter's.

The Uranian field is about as strong as Saturn's; its oddness, however, is the opposite of Saturn's. Instead of

being formed in the core of the planet and aligned with its spin, the field seems to be centered a third of the way out from the center of Uranus and its poles are tipped 60° away from the geographic poles of the planet! However, the location of this field does support the idea that the mantle of Uranus is a deep, ionized (and thus electrically conducting) water layer.

Does the tilt of the Uranian magnetic field have something to do with Uranus being tipped over on its side? This is apparently not the case. *Voyager* discovered that Neptune has a magnetic field that is aligned more than halfway out from the center of the planet and tilted nearly as much as Uranus's, 47° from its spin axis. This field is about half the strength of Uranus's field.

## Is Pluto an Escaped Moon?

Two lines of argument suggest that the Pluto–Charon double planet system may have been formed near Neptune and later driven into its current eccentric orbit about the Sun.

First, the Pluto–Charon system orbits the Sun in a very unplanetlike orbit. It is inclined nearly 20° from Earth's orbit; Mercury's, the next most inclined planetary orbit, is only 7° from Earth's orbit, while all the other planets lie within $3\frac{1}{2}°$. Furthermore, its orbit is so eccentric that it actually crosses the orbit of Neptune.

Although at first it might seem extremely unlikely, it is possible that something such as a passing star or protoplanet could have added enough energy to Pluto and Charon to cause them to escape from an orbit that initially circled Neptune. The amount of energy that would have to be added to Pluto's orbit is similar to the amount needed to move Triton from a prograde to a retrograde orbit. The strange orbits that these three bodies have today could thus all be attributed to a single unusual catastrophic event early in the solar system's history.

Second, the chemistry of Pluto may also be more consistent with it being a moon than a planet. Pluto's radius is 1152 km; Charon's is 615 km. By observing their orbits about each other and assuming that they are both made of the same material, one can infer a total mass and from that a mean density of 2 g/cm$^3$, which implies that they are made up primarily of ices. The observed infrared colors of the Pluto–Charon system are consistent with the presence of methane on their surfaces.

Methane is found on several other moons, most notably Triton and Titan. This is not surprising, because the chemistry of a thick atmosphere surrounding a forming gas giant planet should be rich in methane. However, out in the far reaches of the early solar system, the gas pressure is so low that it is doubtful that the chemical reactions to form methane could occur. Instead, one would expect carbon to be present as carbon monoxide ice instead of methane ice. Evidence for carbon monoxide is seen in comets, which we assume did form in the outer reaches of the solar system, far from planets. Because we see methane on Pluto instead of carbon monoxide suggests that it was formed close to a major planet, presumably as a satellite of that planet.

There are some holes in the chemical argument, however. For one thing, a moon rich in methane would be expected to have a much lower density, about half that actually observed. Furthermore, the observation of methane ice does not prove that it is the dominant carbon-bearing ice present. Methane is easier to detect than carbon monoxide; carbon monoxide ice could be present in much greater abundance than the methane and we still might not see it. It is also denser than methane ice. Thus the high density of Pluto suggests that carbon monoxide, not methane, is really the dominant form of carbon in Pluto. If so, our chemical argument could be reversed and used to prove that Pluto was *not* an escaped moon! Instead, it might be thought of as an extremely large comet. Only improved observations, capable of measuring methane and carbon monoxide abundances, will be able to resolve this dilemma.

## SUMMARY

Saturn, Uranus, and Neptune are all large, gas-rich planets, so far from Earth that they are difficult to study except with spacecraft. Because of their distance from the Sun, they are colder worlds than Jupiter and have less active atmospheric chemistry and less colorful clouds. Each planet may have a large liquid interior resulting from the sorting out of various chemical species: helium in the case of Saturn, water and ammonia for Uranus and Neptune. Methane haze is seen in Uranus, and methane clouds are abundant on Neptune. The clouds may trap Neptune's internal heat and allow it to be as warm as Uranus. Saturn, Uranus, and Neptune have magnetic fields that have been studied by *Voyager 2*. The structure of these fields is at odds with some of our ideas about how magnetic fields are generated inside planets, but they are consistent with the presence of ionic liquid interiors.

Unlike the other outer planets, Pluto is a small icy body. Its unusual orbit, and the presence of methane on its surface, suggests it might be an escaped moon of Neptune; however, its relatively high density is consistent with a composition more like a very large comet, a primordial piece of the early outer solar system.

## STUDY QUESTIONS

Of the four planets described in this chapter, name one that:

1. . . . has a liquid hydrogen core that dissolves helium.
2. . . . has a large blue spot.
3. . . . spins at an unusual angle, tipped over 98° from the plane of its orbit.
4. . . . has a moon that was discovered in 1978.
5. . . . is yellow in color, probably due to UV reactions with sulfur compounds in its clouds.

## 11.2  RINGS

Seen from a distance, rings around planets are disks of beautiful simplicity. Understanding their detailed structure, however, is a task of breathtaking complexity. In the motions of the ring particles that orbit Jupiter, Saturn, Uranus, and Neptune are illustrated nearly every principle of celestial mechanics, combined and blended to a rich but at times mind-boggling effect.

What are the rings? Why do they appear where they do, and why do they look like they do?

### A Gallery of Rings

We'll start by describing each set of rings observed in the solar system so far (Table 11.2) and then try to understand some of the reasons for the way they look.

#### Saturn's Rings

Saturn's rings, the most brilliant, are visible through any small telescope. Galileo saw them in 1610, although he didn't know quite what he was looking at; he called them "ears" and described them as "handles" on the side of the planet. (The Latin word for handles, *ansae*, is still used when referring to rings.) It wasn't until 1656 that the Dutch astronomer Christiaan Huygens, with a better telescope, realized that the "ears" were a disk of material surrounding Saturn at its equator. Over the next 300 years further improvements in telescopes led to further refinements in our understanding of the structure of this ring system.

The ring system is over 20,000 km wide but less than 2 km thick. It is bright and icy: The infrared spectrum of the rings shows the distinct colors of water ice. It is separated into several distinct bands of varying brightness, named by letters. Rings A through D are labelled in Figure 11.4. The E ring is a thin dusty ring centered on Enceladus (which is probably the source of the dust), but it stretches from Mimas out almost to Rhea, from 3 to 8 Saturn radii. The F ring is a thin, braided ring just outside the A ring; just beyond it lies the diffuse, dusty G ring.

The rings are not solid objects. Stars can be seen to shine through them, and the disk of Saturn itself is also easily seen through the rings. They are made up of millions of moonlets orbiting together. In Saturn's rings, the moonlets

**TABLE 11.2**

The Rings of the Outer Planets

| Ring Name | Position (km) | $R/R_{planet}$ |
|---|---|---|
| **Jupiter** | | |
| Inner halo? | 95,000 ? | 1.35 ? |
| End of halo | 122,090 | 1.71 |
| End of main ring | 129,130 | 1.81 |
| Gossamer edge | 210,000 | 2.94 |
| **Saturn** | | |
| D inner edge | 66,970 | 1.11 |
| C/D boundary | 74,510 | 1.225 |
| B/C boundary | 92,000 | 1.525 |
| B outer edge | 117,580 | 1.949 |
| *Cassini division between B And A rings* | | |
| A inner edge | 122,170 | 2.025 |
| *Encke division* | 133,570 | 2.214 |
| A outer edge | 136,780 | 2.267 |
| F ring | 140,180 | 2.324 |
| G ring | 170,000 | 2.82 |
| E ring | 240,000 | 4(3–8) |
| **Uranus** | | |
| U2R inner edge | 37,000 | 1.43 |
| U2R outer edge | 39,500 | 1.5 |
| 6 | 41,850 | 1.616 |
| 5 | 42,240 | 1.631 |
| 4 | 42,580 | 1.644 |
| Alpha | 44,730 | 1.727 |
| Beta | 45,670 | 1.763 |
| Eta | 47,180 | 1.822 |
| Gamma | 47,630 | 1.839 |
| Delta | 48,310 | 1.865 |
| *Dust ring from Delta to Epsilon* | | |
| Lambda | 50,040 | 1.932 |
| Epsilon | 51,160 | 1.975 |
| **Neptune** | | |
| 1989N3R inner edge | 38,000 | 1.5 |
| 1989N3R max | 41,900 | 1.69 |
| 1989N3R outer edge | 49,000 | 2.0 |
| 1989N2R | 53,200 | 2.15 |
| 1989N4R inner edge | 53,200 | 2.15 |
| 1989N5R | 57,500 | 2.32 |
| 1989N4R outer edge | 59,000 | 2.4 |
| 1989N1R | 62,900 | 2.54 |

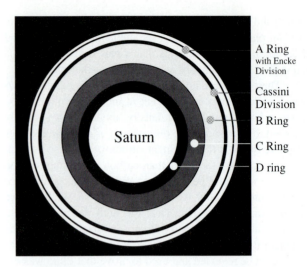

**FIGURE 11.4** A sketch of Saturn's rings, as viewed looking down from above Saturn's pole. The thin black line in the A ring represents the Encke division.

the ring particles, but not so close that the ring would really appear to be solid when viewed close up.

*Voyagers 1* and *2* gave the first good close pictures of the rings and showed that each band was made up of thousands of individual ringlets. The darker bands appear fainter because they have less material in them (notice in Figure 11.1 how Saturn can be seen through the faint C ring).

Voyager also found other oddities about Saturn's rings. The ringlets within a band do not keep themselves in the exact same position as time goes on. A number of small ringlets were discovered outside the main rings, which are "kinked" and "braided" (Figure 11.5), and dark spokes often appear in parts of the main ring system.

These rings raise several new questions. Why are there different bands? Why are the bands broken into ringlets? Why do the ringlets move? Why are there gaps? What kinks the braided rings? What causes the spokes?

### Uranus's Rings

Uranus has a set of rings that at first seem to present quite a different picture from Saturn's system. The rings are narrow, not broad, and very dark, like carbon-contaminated rock, not bright and icy. They were discovered by accident when astronomers observing the occultation of a star by Uranus saw that star "blink" several times before it reached the planet and then blink again at the same respective positions on the other side of the planet. A number of such observations confirmed the presence of opaque material at nine distinct positions around Uranus. Following the discovery, a careful study of previous photographs of Uranus found a thin dark swath across the disk of the planet made

are typically a few meters across—roughly the size of an automobile—and composed of water ice. The average ring chunk probably orbits tens of meters away from its nearest neighbor, close enough to lead to a continual jostling among

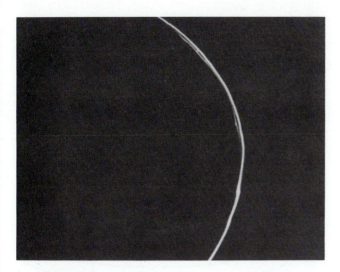

**FIGURE 11.5**   The braided F ring of Saturn.

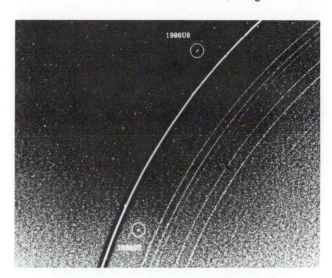

**FIGURE 11.6**   The rings of Uranus, as seen by *Voyager*, with two of the shepherd satellites on either side of the epsilon ring.

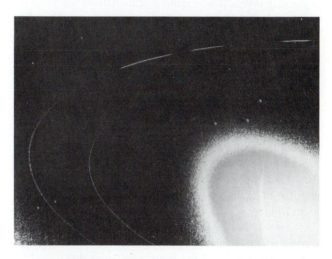

**FIGURE 11.7**   Neptune's rings and ring arcs, as seen by *Voyager*.

by these coal black rings. Two sets of astronomers viewing this occultation came up with ring discoveries; one group gave their rings Greek letter names, while the other merely numbered their rings.

In addition to the rings discovered by this occultation, *Voyager* found two new narrow rings and two broader rings of fine dust (thought to be similar in composition and structure to the dust rings of Jupiter discussed below). These dust rings are located inside the 6 ring between the delta and epsilon rings. The epsilon ring itself, however, is remarkably free of dust.

These rings are not uniform as they circle the planet but vary in thickness and position. Data from *Voyager* enabled scientists to conclude that the particles that make up these rings are on average about 20 cm across, and they are distributed through the rings a few meters apart from each other. The outermost ring, epsilon, is eccentric and has a width of nearly 100 km when farthest from the planet while being much narrower when it passes closer to the planet (Figure 11.6). The other rings are much narrower, only a few kilometers wide.

### Neptune's Rings

Neptune has three narrow rings somewhat similar to those around Uranus, but one ring has an unusual feature. Along with material spread evenly around the planet, the outermost ring has much of its matter concentrated into three distinct **ring arcs** (Figure 11.7).

Like Uranus, Neptune also has dusty rings. One dust ring starts from the innermost narrow ring and extends 6000 km out towards the outer ring with the ring arcs. This dusty ring itself encloses a faint narrow ring. The second dust ring starts 4000 km inside the inner narrow ring and extends about 11,000 km down towards the planet. This ring is estimated to be about 2000 km thick.

### Jupiter's Ring

Jupiter's ring is quite different from the other rings (Figure 11.8). The existence of this faint, wispy band of dust was first suggested by detectors on *Pioneer 11* that saw a drop in the number of energetic electrons in the plasma close to Jupiter. Photographs by *Voyager* confirmed that a faint ring of dust, mostly 3 microns or less in diameter (motes as fine as flour), was absorbing charged particles less than 1 Jupiter radius above the cloudtops. This thin ring is less than 30 km thick.

In addition, a halo of even smaller dust grains, perhaps only a few tenths of a micron in diameter, is seen above and below the disk. An even fainter "gossamer" ring extends outwards for perhaps 100,000 km from the main ring.

**FIGURE 11.8**    The dusty ring of Jupiter.

Such small dust can be pushed away from the ring by many different forces. Friction with the plasma around Jupiter, interactions with the magnetic field, even the force of sunlight falling on this dust is enough to sweep it out of the ring in less than a million years. Unlike the rings of boulders or iceballs around Saturn, Uranus, and Neptune, it is clearly impossible for Jupiter's ring to be left over from the time Jupiter was formed. This introduces a new question: Where did this ring come from?

## Basic Structure of Rings

All rings have certain aspects in common, but there are peculiarities unique to each system as well that must be considered separately.

First, we know that the rings cannot be made of a single solid disk. Rather, they must be a collection of millions of small moons, each in its own orbit about the planet.

James Clerk Maxwell first pointed this out from a completely theoretical point of view in 1857. He considered what would happen if the rings were solid disks, and he realized that such solid rings would eventually fall into the planet they circled.

If solid rings were positioned so that the planet were exactly in the center of the ring, he reasoned, then the attraction of its gravitational field on the ring would be the same everywhere, and so it might appear that such a situation were physically reasonable. But if the rings were nudged slightly out of position (by the gravitational pull of an orbiting moon, for instance), then the part of this ring slightly closer to the planet would feel a stronger gravitational pull while the part opposite would feel less force. The closer part would fall even closer to the planet, and as it fell it would push the far side of the ring even farther away. Thus the whole disk would keep accelerating until it crashed into the planet (Figure 11.9).

Putting spin on a disk, to make it imitate a satellite in orbit, doesn't help. Moons in orbit react to the chang-

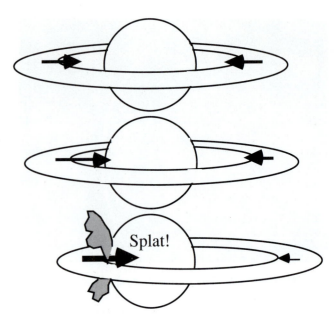

**FIGURE 11.9**    A solid ring is not stable; a slight perturbation draws the near part into the planet.

ing strength of the gravity field as they move nearer to and farther from a planet by moving faster when they are nearer and slower when farther away. But a solid ring is constrained to spin always at the same rate everywhere. Thus the part of the ring closer to the planet is moving too slowly to stay in orbit while the far side is moving too fast. The result is the same. The near side falls into the planet, and the far side is pushed away.

If, however, the rings are not solid but rather broken into individual moonlets, this problem is solved. Each moonlet can adjust its own individual speed to keep itself in orbit, without being tied to the speeds of the other moons. In fact, observations of the rings of Saturn have shown that the inner parts of its ring do indeed move faster than the outer parts, just as we would expect if each part were in a Keplerian orbit.

## Roche's Limit

With millions of little moonlets orbiting the same planet in the same region, we might expect that they would run into each other fairly often. Why don't the little moons eventually stick together and grow into one large moon?

As we discuss in Chapter 12, moons and the planets they orbit raise tides on each other. A tide is raised on a moon because the force of the planet's gravity on the side of the moon nearest the planet is slightly stronger than the planet's gravity pulling on the far side of the Moon. From the moon's point of view, this difference in forces looks like a force trying to rip the moon apart.

The closer a body is to the tide raiser, the greater the difference in gravitational attraction from one side to the other. Assuming that the planet and moon are made of similar materials, it can be shown that once a moon is within 2.45 planetary radii of the planet, the tides become stronger than the moon's own gravity holding itself together (Figure 11.10). (A more precise analysis of the strengths of tides and the strengths of materials slightly changes the exact point where solid moons can survive, but as a general rule 2.45 planetary radii is a good estimate of the critical distance. It can be seen from Table 11.2 that most planetary rings lie inside this limit.)

This radius is called the **Roche limit**, after its discoverer, the nineteenth-century French astronomer Édouard Roche. Small bodies, like rocks or artificial satellites, are held together by molecular bonds; their own internal strength allows them to survive these tidal stresses. But inside the Roche radius a collection of rocks held together only by gravity is pulled apart by tides.

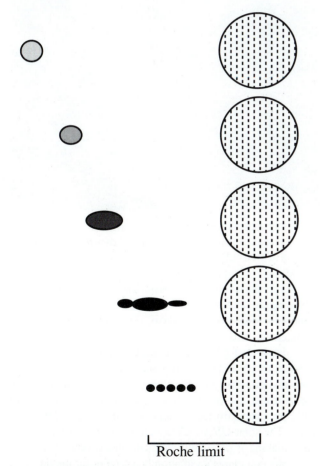

Roche limit

**FIGURE 11.10** The Roche limit. As a moon approaches a planet, there will come a point where the pull of the planet's tides is greater than the gravity holding the moon together.

Thus ring particles cannot gather together under the force of their own gravity to form a moon because the rings lie too close to the planet: The tides raised by the planet pull apart any moon that would try to form. The rings could be made up of planetesimals that never were able to accrete into moons, unlike their luckier counterparts outside the Roche limit. Or it is possible that the rings might actually be the debris of a comet or small moon perturbed to within the Roche limit and pulled apart by tides. In either case, the rings remain a collection of individual rocks or ice balls, jostling each other in space.

## Why Are Rings Flat?

The effect of this jostling on the rings is quite straightforward. Each time two objects in space hit and bound off each other, they share their kinetic energy and move in new directions that are the average of the original directions. A cue ball hitting another ball standing still on a pool table acts the same way; the cue ball slows down as the other ball starts to move.

All these collisions in the ring system eventually serve to average out the velocities of all the particles. And because the particles can collide anywhere in their orbits, each moon tends to orbit at a constant speed, meaning that each moon tends towards a circular orbit.

Collisions also affect the plane of this orbit. Spinning planets with bulges at their equators cause the plane of a moon's orbit to precess. The drifting of these orbits ensures than any particle in an inclined or eccentric orbit eventually crosses the path of some other moonlet, leading to further collisions. The effect of these collisions is to damp down the tilt of the orbit as well as to make it circular.

By "collisions" we do not necessarily mean that the ring particles actually have to touch each other physically (although they can, and do). When two ring particles pass by each other close enough to perturb each other's orbit, they have a similar (although smaller) effect. After many such encounters, the orbits of the ring particles become more circular and tend to lie closer to the same plane.

## Perturbations of Rings

Because of these multiple collisions and close encounters, it is not surprising that the rings of Saturn appear circular, very flat, and aligned with Saturn's equatorial plane. The epsilon ring of Uranus, however, looks very different. Even though it is the largest of the Uranian rings, it is quite elliptical in shape. The individual ringlets in Saturn's rings also are not exactly circular. The braided F ring of Saturn in this context is quite bizarre. What forces are warping the shapes of these rings?

The best explanation so far for these bizarre rings is that small satellites near the ring are perturbing the orbits of the ring particles. In several cases (but not for all rings) the responsible moons have actually been discovered in *Voyager* photographs. How might these moons act on the rings? Two ways have been suggested.

One theory notes that many of these the small satellites are observed to lie just inside or outside the ring. A ring particle orbiting just inside a shepherd moon moves faster than the shepherd. As it approaches, the ring particle sees a forward acceleration due to the gravity of the shepherd. This initial positive acceleration gives it a slightly larger orbit as the particle passes the shepherd. Thus it is closer to the shepherd after it passes than it was as it approached. It also moves more slowly now, because of its larger orbit, so it spends more time near the shepherd after passing it than it did approaching it. The result is that the backward acceleration it feels from the shepherd moon after passage is stronger, and lasts longer, than the forward acceleration it felt on approach. The net result is to rob energy from the ring moonlet, forcing it into a smaller orbit.

A shepherd moon just inside the ring has exactly the opposite effect, pushing the ring moonlets into larger orbits. In this way the shepherd moons keep the ring moonlets confined to a narrow ring.

Shepherd moons can also set up waves in a narrow ring, like the kinks seen in Saturn's F ring. If the shepherd is actually embedded in the ring itself, the approaching ring moonlet with a slightly lower orbit than the shepherd is so strongly accelerated by the shepherd that it actually moves outside the orbit of the shepherd before it has a chance to pass the moon. Once outside, it orbits more slowly and, from the point of view of the shepherd, appears to fall back away from the moon. Eventually the shepherd "laps" the slower-moving moonlet as they circle the planet. When it catches up to the ring moonlet, the shepherd robs energy from the moonlet's orbit and pulls it back down into a lower but faster orbit, where the process reverses itself. Such orbits are called **horseshoe** orbits, from the horseshoe shape of their paths as seen in a frame of reference rotating with the shepherd moon.

These processes help explain how shepherd moons maintain narrow, kinked rings. What is still not clear, however, is how the narrow ring and shepherd moon system got formed in the first place. Why do these shepherd moons exist? How were they formed?

The shepherd satellites cannot themselves be too large; after all, being near the rings, they too must be within the Roche limit. However, close to the Roche limit such moons are more likely to survive the tidal stresses that rip apart larger moons that are held together by their own gravity. Thus it is not surprising that these odd rings are among the outermost rings of their planets.

## Dusty Rings and Electromagnetic Forces

The concept of shepherd or embedded moons has been invoked to explain most of the peculiar features of the rings of Saturn and Uranus. In addition, such moons may be responsible for the dust that makes **spokes** in the Saturn rings (Figure 11.11), and they may be the source of the dusty ring systems, like Jupiter's rings. Both spokes and dusty rings seem to be made up of very small (micron- or submicron-radius) dust grains; these grains may be fragments knocked off of larger parent moons (Figure 11.12). (These parents are sometimes called "mooms," a whimsical shortening of "mother moons.")

**FIGURE 11.11** The dark streaks cutting across Saturn's rings in this *Voyager* image are examples of **spokes**.

The dust could get spread into a ring, and eventually dragged into the planet, by a number of forces. One important force acting on dusty ring particles is the electromagnetic force.

Each dust grain, and indeed everything in space, is exposed to UV light from the Sun and to high-energy electrons and protons from the solar wind or the radiation belts that surround the planets. The result of this exposure is to build up an electrostatic charge on the surface of the dust grain. The solar UV knocks electrons off the surface of the dust, making it positively charged; on the other hand, hot electrons in a planet's radiation belt tend to hit the dust and make it negatively charged. Which effect wins depends on how much light and how much hot electron gas the dust is exposed to.

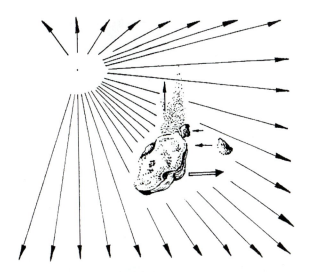

**FIGURE 11.12** Boulder-sized moons collide with larger "mooms," knocking off dust. This dust, charged by radiation belts or solar UV, is lifted out of the plane of the ring by the planet's magnetic field.

Electrostatic charging may explain the spokes of Saturn's rings. When dust comes out of Saturn's shadow and sees sunlight, it gets charged up. The electrostatic repulsion may move dust particles off the surfaces of larger ring moonlets, thus changing the way the rings scatter light. This repulsion may even move the dust out of the ring plane altogether, where it then could partly obscure the rings from our view. Either process would appear to form dark **spokes** on the rings.

Once charged up, the dust also interacts with the charged radiation belts surrounding the planet (see Chapter 13). The radiation belts move at the same speed as the planet spins, but the dust particles (inside the synchronous point) orbit faster than the belts spin. Thus the charged dust appears to be running into a rain of protons, electrons, and other ions in the radiation belts. These collisions rob energy from the dust particles and so they are dragged down into the planet. The ions do not actually have to touch the dust physically for this effect to occur, because the dust itself has an electric charge.

The dust can also be moved around by the magnetic field of the planet. The magnetic field has several effects: It can push the dust closer to, or farther from, the planet; and it can also perturb the dust up or down, out of the original plane of its orbit.

Small dust particles from Jupiter's ring get spread out into a thin flat ring by the magnetic forces acting on them. The ring fades slowly into Jupiter's cloudtops; this is the result of dust being dragged down by the radiation belts out of the bright, outer part of the ring. And some dust gets perturbed up out of the plane of the ring by magnetic fields,

forming the thin halo seen above and below Jupiter's dust ring.

The mass of a spherical dust particle grows as the cube of its radius, but the amount of charge on the particle of a certain voltage grows only linearly with radius. Thus the net acceleration due to the electromagnetic force varies inversely as the square of the dust-grain radius. In other words, dust particles twice as big as the 2 micron grains in Jupiter's ring see a force only a quarter as strong. Because the ring moonlets in the major rings of Saturn, Uranus, and Neptune are at least 10,000 times larger than those in Jupiter's dust ring, it is assumed that electromagnetic forces can be neglected in these rings.

## SUMMARY

Rings are formed by millions of small moonlets orbiting inside the Roche limit of a planet, the distance where the tidal forces of the planet are strong enough to pull a big moon apart. Small moons near or within these rings can be responsible for warping the shapes of the rings, confining them to narrow ringlets, or even setting up kinks in the rings. Dust knocked off these moons can be responsible for spokes and haloes above and below the thin disks of the rings.

## STUDY QUESTIONS

1.  Are the rings of Saturn solid disks or millions of individual fragments?

2.  A planet's gravity will pull a small moon apart and scatter the pieces into a ring, if the moon is inside a certain distance from the planet. What is this limiting distance called?

3.  What do we call small moons that perturb ring particles and confine them into a narrow ring?

4.  What force produces the spokes on Saturn's rings and shapes the dust ring of Jupiter?

5.  Which planet has "ring arcs"?

## 11.3   ORBITAL PERTURBATIONS AND THE THREE-BODY PROBLEM

The rings orbiting the outer planets are made up of individual moonlets, each with its own individual orbit. Start with one such moonlet in a simple, elliptical orbit. What happens if it is hit by a meteorite? What if another large satellite passes nearby, tugging at it with its gravity during the encounter? What if the planet it orbits is not a smooth, uniform sphere but a real planet, with a bulge at the equator, mountains on its surface, and lumps of mass distributed in an irregular fashion throughout its interior? What if that moonlet is, in fact, a space probe that can fire rockets to change its course?

What happens is that the moonlet changes its orbit.

Our interest in studying these perturbations is twofold. First, it helps us understand how rings, satellites, and moons can move out of their original orbits. This can help explain the shapes of the rings or the flux of meteorites from the asteroid belt to Earth. When we examined how shepherd satellites controlled narrow rings, we were using the techniques of orbital perturbation theory.

Second, and more intriguing, once we understand these perturbations we can follow the orbit of a moon very closely, see how it changes, and deduce from these changes the nature of the accelerations that must be acting on it where it sits in space. For example, by observing the behavior of a satellite's orbit we can learn how the planet it orbits is put together and find if there is a concentration of mass in its core or buried at a shallow depth just below its surface, or deduce the presence or the masses of other nearby satellites. This was the idea behind using the precession of the Uranus epsilon ring to calculate the spin rate of Uranus.

## Osculating Orbital Elements

In perturbation theory, we assume that at any given moment the satellite being perturbed is in an elliptical Keplerian orbit. Recall how, by knowing the exact location and velocity of a body in orbit at any given time and the acceleration of gravity acting on that body, we could deduce the six Keplerian coordinates of the orbit (Figure 11.13). Even a body that is about to be perturbed has, at any given instant, a hypothetical orbit that can be calculated based on the body's current location and velocity. This calculated orbit is called the **osculating orbit**, and its elements are the **osculating orbital elements**. Perturbations may change the real orbit; but for the moment, this calculated orbit—the one that the satellite would follow if it were not perturbed—is just touching the real orbit.

Let us review what the six Keplerian coordinates tell us. Two of the coordinates define the shape of the ellipse: $a$, the semimajor axis, tells how big the orbit is, and $e$, the eccentricity, describes how elongated it is. Three of the coordinates tell us how this ellipse is oriented in space. The inclination, $i$, tells how much the plane of the orbit is tilted from some reference plane (such as the equator of the planet being orbited). The longitude of the ascending node, $\Omega$, tells where the orbiter's path crosses the reference plane (that is to say, how the plane of the orbit is oriented in space). The argument of pericenter, $\omega$, is the angle between

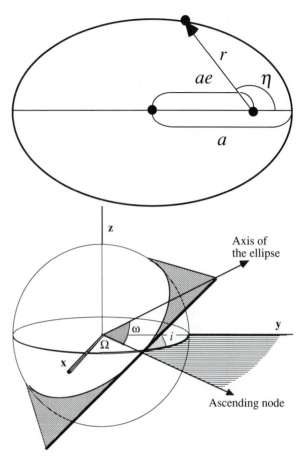

**FIGURE 11.13** A review of the orbital elements. The upper figure shows how $a$, $e$, and $\eta$ are determined within the ellipse; the lower figure shows how the terms $\omega$, $\Omega$, and $i$ orient the plane of the ellipse itself in space.

the ascending node and the location of pericenter. It tells how the ellipse is oriented within its own plane. Finally, the last Keplerian coordinate, the true anomaly, $\eta$, specifies where on the ellipse the satellite happens to be located at the moment.

How is a perturbing acceleration going to act? Some accelerations provide short little jolts: for instance, a rocket firing briefly or a meteorite hitting the satellite. Others may provide a small but constant change that builds up over a long time, such as those caused by slight variations in a planet's gravity field due to some extra mass buried below its surface. We will consider each type of perturbation separately.

## Instantaneous Accelerations

There are three components to any acceleration that suddenly hits a satellite. Part of the acceleration can be perpendicular to the plane of the orbit; this is the **normal acceleration**, denoted in Table 11.3 as $N$. Part of the ac-

**TABLE 11.3**

Equations for the Change in Orbital
Elements due to Instantaneous Perturbations,
where $p = a\left(1 - e^2\right)$ and $h^2 = GMp$

$$\frac{da}{dt} = \frac{2a^2}{h}\left(eR\sin\eta + T\frac{p}{r}\right)$$

$$\frac{de}{dt} = \frac{p}{h}\left(R\sin\eta + T\left[\left(1 + \frac{r}{p}\right)\cos\eta + \frac{re}{p}\right]\right)$$

$$\frac{d\eta}{dt} = \frac{h}{r^2} + \frac{1}{eh}\left[Rp\cos\eta - T(p+r)\sin\eta\right]$$

$$\frac{di}{dt} = \frac{r}{h}N\cos(\omega + \eta)$$

$$\frac{d\Omega}{dt} = \frac{r}{h\sin i}N\sin(\omega + \eta)$$

$$\frac{d\omega}{dt} = -\cos i\frac{d\Omega}{dt} - \frac{d\eta}{dt} + \frac{h}{r^2}$$

celeration may push it towards or away from the primary body that it's orbiting; this is the **radial acceleration**, $R$. A third acceleration direction, at right angles to the other two, is called the **transverse acceleration**, $T$. If the satellite is in a circular orbit, this direction is the direction in which the body is moving. But for eccentric orbits, the satellite moves in both transverse and radial directions as it goes about the planet.

To understand how each part of an acceleration will affect the planet's orbit, one principle of orbital physics is important to remember: The energy of a body in orbit depends on its position and its speed. Given a body in a certain position, the only way to change its energy is to change its speed. This can only be done by an acceleration, positive or negative, in the direction in which the body is moving already. An acceleration at right angles to the motion of the body will not change its speed or its energy, but it changes the body's direction; a constant push sideways to the motion of a body does not speed it up, but it makes it go in circles.

The exact equations showing how the Keplerian coordinates change with each type of acceleration are given in Table 11.3. With practice, it is possible to see intuitively how a certain acceleration changes the shape or orientation of an orbit.

## Normal Acceleration

A normal acceleration by definition is not in the direction of motion, and so it will not change the size or the shape of the orbit. Instead, it changes the tilt of the orbit, the inclination; it also has another effect.

If you've ever tried turning the direction of a spinning wheel, such as a bicycle wheel, you may have felt its spinning plus your turning motion combine to make the wheel twist in the direction perpendicular to both motions. For example, turning the front wheel of a moving bicycle causes it to lean over, and likewise, leaning over on a bike causes it to turn. This complicated motion, which allows us to ride and steer the bicycle, is a result of the principle of conservation of angular momentum.

A spinning top shows the same effect. As gravity tries to pull the top over, it wobbles around instead of falling over. This process works on planets in orbits just like it works on bicycle wheels and tops (Figure 11.14).

When the normal acceleration makes the inclination of the orbit change, it will also change the angles $\Omega$ and $\omega$. The normal acceleration thus tips and twists the whole orbit, without affecting its shape. This change in $\Omega$ or $\omega$ is called **precession**. These changes are illustrated in Figure 11.15.

## Transverse Acceleration

An acceleration in the direction of the orbit, transverse acceleration, changes the orbit's energy. Because the energy depends only on the semimajor axis, this acceleration must change the value of $a$.

The eccentricity must also change. Consider a sudden acceleration that acts on a circular orbit. Both old and new orbits are simple, closed orbits; and the point where the perturbation took place is a point in both old and new orbits.

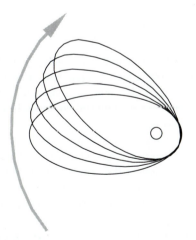

**FIGURE 11.14** The wobble of a top as it slows down its spin is an example of precession. Orbits can precess, too; when it feels a normal acceleration, an eccentric orbit slowly changes the orientation of its pericenter.

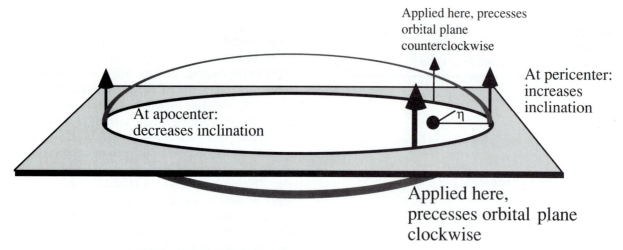

**FIGURE 11.15** Perturbations of an orbit by normal accelerations. The assumption is made that longitude of pericenter, $\omega$, is zero. However, as the orbit precesses, this will no longer be true; if the orbit precesses 180° from this case, then the normal acceleration has an effect exactly the opposite of what is illustrated here.

(This is a useful principle of perturbation theory: In the absence of further perturbations, a satellite always returns to the spot where it was last perturbed.) This new orbit has a new semimajor axis, but it must still intersect the old, circular orbit. This is possible only if the new orbit is not circular. Therefore, we can see its eccentricity must be changed (Figure 11.16).

A transverse acceleration adds a torque to the motion of the satellite. Not only is the energy changed, but the total angular momentum per unit mass, $h$, is also changed (at the rate $dh = rT\,dt$). Thus the angular speed of the orbit, $d\eta/dt$, also changes. These changes are illustrated in Figure 11.17.

### Radial Acceleration

A radial acceleration acting on a circular orbit does not change $a$; but from a similar argument as above, we see that it must change $e$. If a body in a circular orbit picks up a radial velocity, then its orbit can no longer be circular; its eccentricity must have changed.

A radial acceleration acting on a body in an eccentric orbit, however, can partly speed up or slow down the body. It depends on where in the orbit the perturbation occurs. An outward acceleration on a body moving away from the primary planet speeds it up, while that same outward acceleration slows it down when the body is on the return portion of its orbit. In either case, however, it changes the semimajor axis, $a$. Likewise, depending on where in its orbit the perturbation occurs, the radial acceleration can speed up or slow down the body, thus changing the value of angular velocity, $d\eta/dt$. The effects of this acceleration are summarized in Figure 11.18.

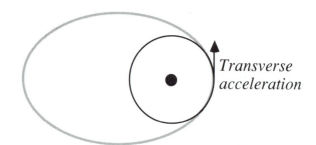

**FIGURE 11.16** A new orbit must always cross the old one at the point where the perturbation took place. Thus a transverse acceleration turns a circular orbit into an elliptical one.

### Disturbing Potential

The derivation of Keplerian orbits assumes implicitly that both the primary body and the satellite are points of mass and that they're the only points of mass in the universe. Neither assumption is true, of course.

The gravity field of a real planet can be described in terms of its gravitational potential energy. As we noted in Chapter 4, this potential can be expanded into **spherical harmonics**, an infinite series of terms. The first term is simply the point-mass approximation we've used up to now. (It is sometimes called the **zeroth harmonic**.) It leads to elliptical, Keplerian orbits. All the other terms are much smaller than this, and they lead to perturbations on Keplerian orbits. These perturbations are described in Table 11.4.

The first harmonic can be set to zero by defining the center of mass to be the center of our coordinate system.

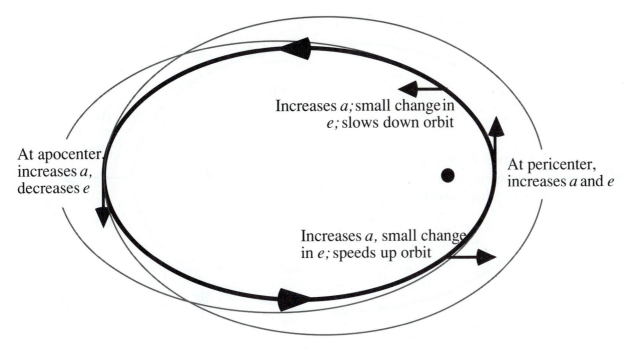

At apocenter,
increases $a$,
decreases $e$

Increases $a$; small change in
$e$; slows down orbit

At pericenter,
increases $a$ and $e$

Increases $a$, small change
in $e$; speeds up orbit

**FIGURE 11.17** Perturbations of an orbit by transverse accelerations.

**TABLE 11.4**

Equations for the Change in Orbital Elements
due to a Distributed Mass Giving Rise to a
Harmonic $J_2$ in the Gravity Potential Energy,
where $p = a(1 - e^2)$, $n$ is the Mean Motion,
and $t_0 =$ Time of Pericenter

$$\frac{d\omega}{dt} = -\frac{3}{2}J_2\frac{n}{p^2}\left(\frac{5}{2}\sin^2 i - 2\right)$$

$$\frac{d\Omega}{dt} = -\frac{3}{2}J_2\frac{n}{p^2}\cos i$$

$$(t - t_0)\frac{dn}{dt} = -\frac{3}{2}J_2\frac{n\sqrt{1 - e^2}}{p^2}\left(\frac{3}{2}\sin^2 i - 1\right)$$

The next harmonic is due to the bulge that spinning planets develop at their equators due to centrifugal force (Figure 11.19). This **second zonal harmonic** describes the little extra pull of gravity that a satellite feels from this bulge, through a mildly complicated collection of terms: sines and cosines of the latitude and longitude of the satellite, multiplied by $1/r^3$ instead of $1/r$ like the point-mass potential energy. This means that this harmonic and all the higher harmonics are strongest close to the planet.

Each $n$th harmonic is multiplied by a constant called $J_n$. Consider $J_2$, which represents the strength of the second harmonic. Every planet has its own value of $J_2$.

$J_2$ is a constant whose value depends on how mass is distributed inside our particular planet. It is related to

the moments of inertia of the planet. From the theory of potentials one finds that a symmetric spinning planet has a value of $J_2$:

$$J_2 = \frac{C - A}{Mr^2} \tag{11.1}$$

where $C$ is the moment of inertia measured from the spin axis and $A$ is the moment of inertia measured from any axis going through the equator, perpendicular to the spin axis. As before, $M$ is the mass of the planet and $r$ is its radius at the equator. Thus $J_2$ is not quite the same as the moment of inertia; rather, it's the difference between two moments of inertia measured in perpendicular directions. The bigger the bulge, the greater the difference between how material is distributed away from the spin axis and how it's distributed along that axis.

The bulge at the equator described by $J_2$ acts as a little extra pull of gravity. It acts as a radial perturbing force on the satellite. But because it is constantly acting, its effect on $a$ and $e$ during the first half of the orbit is undone by its effect on the second half of the orbit, and so it has no net effect on these terms. What does change, though, are the angles $\Omega$ and $\omega$. By observing how quickly a satellite's orbit precesses, one can directly measure a value for $J_2$ and thus learn how mass is distributed inside the planet being observed. From $J_2$ one can also derive a "reference spheroid" (in the case of Earth known as the **geoid**) to use in mapping the gravity field of the planet. This technique was discussed in Chapter 4.

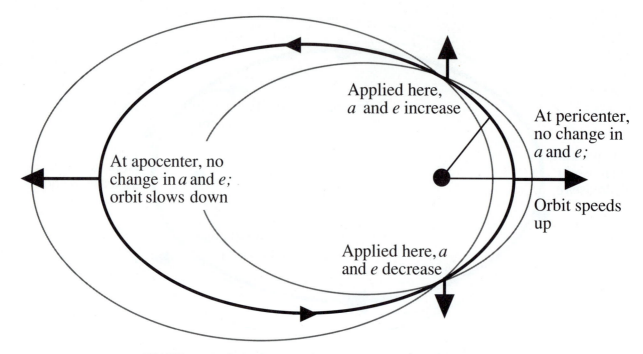

**FIGURE 11.18**   Perturbations of an orbit by radial accelerations.

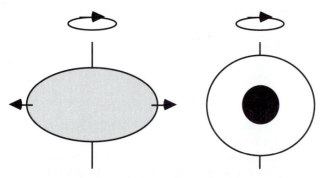

**FIGURE 11.19**   Once a planet starts to spin, its equator bulges out into the shape of an oblate spheroid. Its gravity field has higher harmonics, the largest one being the $J_2$ term.

However, recall that $J_2$ is not itself a direct measure of the moment of inertia of a planet but merely the difference between polar and equatorial moments of inertia. To actually find the moment of inertia we need to make an additional measurement.

A spinning planet, like Earth, obeys the same laws of physics as a spinning gyroscope. Because the equatorial bulge of Earth is subjected to gravitational tugs from both the Moon and the Sun, the spin of Earth tends to "wobble." Over a period of 26,000 years, the position of the spin axis changes so that at one time the north pole may point to the pole star Polaris (as it does today), where a few thousand years ago (during the beginnings of Babylonian astronomy) it pointed to another star, Thuban. This precession of the spin axis is what produces the **precession of the equinoxes.**

It turns out that the rate of this precession depends on a factor called the **dynamic ellipticity,** $H = (C - A)/C$, where $A$ and $C$ are the moments of inertia as defined above.

By measuring the rate that a satellite's orbit precesses around Earth, we can measure $J_2$; by measuring the rate that the spin of Earth itself precesses, we can measure $H$. Dividing $J_2$ by $H$ yields

$$\frac{J_2}{H} = \frac{C}{Ma^2} \tag{11.2}$$

which is the **moment of inertia factor** about the spin axis of the planet, introduced in Chapter 4.

## Many-Body Perturbations

What about the effects of other satellites? They too add slightly to the gravity field of the primary planet and so alter the paths of bodies in orbit around it.

Because the other satellites are much smaller than the primary, their effects are small. We only need to worry about the pull of these moons when they are passing closest to our satellite.

Recall that the effect some perturbations have depends on where in the orbit the perturbation takes place. But other perturbations have the same effect no matter where they occur. If our two satellites meet randomly, anywhere in their orbits, then we can ignore the first sort of perturbation. Whatever effect the one moon had on the other the last time they met will likely as not be undone the next time they meet. (These perturbations cannot be ignored, of course, if the satellites' meetings are not random. This case is discussed in Chapter 12.)

The second sort of perturbation, called **secular perturbations**, are the ones that build up with time, and so they are important. Because it doesn't matter where the encounters take place when calculating secular perturbations, we can treat the perturbing satellite as if it were a ring of mass, instead of a satellite. (The low density of matter resulting from spreading the mass of a satellite into a ring is exactly analogous to the low probability that the satellite is near our body at any given time.) And we can treat this ring of mass as if it were part of the mass of the primary. In other words, it just contributes to the $J_2$ and higher harmonics of the planet's gravity field. This reduces the problem back to the one just discussed.

These tricks work, of course, not only for satellites orbiting planets, but also for planets orbiting the Sun. By contrasting the $J_2$ of the Sun that each planet's orbital precession indicates, we can sort out which part of that $J_2$ is due to the bulge of the Sun itself and which is due to the masses of the other planets. In this way we can solve for the mass of the moonless planets Mercury and Venus. This trick has been called "throwing the planets into the Sun."

Systems of many, many particles, such as all the asteroids or the millions of particles in a planet's rings, provide an especially difficult challenge to celestial mechanicians. In this case, there are so many different bodies present whose gravities can significantly perturb each other that many of the simplifying assumptions described above are no longer valid. In the last 10 years, a new branch of mathematics called **chaos theory** has been applied successfully to modelling systems like these. This theory concludes that the orbit of any particular asteroid or ring particle may ultimately be unpredictable over a long time.

## The Three-Body Problem

A small moonlet near a larger moon is subject to the gravity of that larger moon and the gravity of the planet they both orbit. Likewise, small bodies or spacecraft (Figure 11.20) orbiting the Sun can be perturbed as they pass near other planets. The interactions of the coorbital moons of Saturn, the asteroids near Jupiter, or the celestial navigation of a spacecraft are all examples of a **three-body problem**.

The perturbation theory discussed above is adequate if the perturbations make only small changes to the orbit of our body. It's also the best we can do; it is mathematically impossible to derive an exact general formula for the motion of a body subjected to the gravitational field of more than one large neighbor. Even iterative computer solutions can

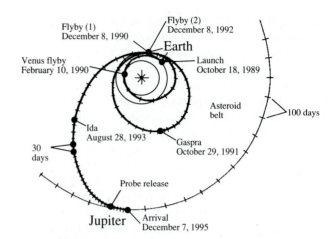

**FIGURE 11.20** The *Galileo* spacecraft has an unusually complicated orbital path, called VEEGA, for **V**enus **E**arth **E**arth **G**ravity **A**ssist.

only give us the path for one particular object under one particular set of starting conditions; they can't give us an overall picture of how bodies in general are likely to behave around two or more large masses.

However, we can put some restrictions on the motion of a very small object pulled by two masses orbiting in circles about their center of mass. Most planets have nearly circular orbits about the Sun, and most large moons are in nearly circular orbits around their planets, while comets, asteroids, moonlets, and spacecraft are indeed much smaller than these moons or planets. Thus this special case has widespread applicability throughout the solar system. This problem is called the **restricted three-body problem**.

To illustrate the restricted three-body problem, we use a very simple system, illustrated in Figure 11.21. Call $M$ the smaller mass and $M^\star$ the larger mass (for a planet going around a star, $M^\star$ is the mass of the star). If we place the center of the coordinate system at the center of mass of the

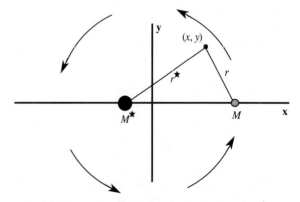

**FIGURE 11.21** A rotating reference frame centered on two orbiting masses sets the stage for the restricted three-body problem.

system and if the two masses are set on the $x$-axis, then $x$ represents the position of the smaller body and $x^\star$ the position of the larger one (the star).

Because the two bodies are both in circular orbits, they always stay this same distance apart from each other. One can think of an imaginary rod connecting the two, and picture this rod tumbling in space, spinning about the center of mass. This suggests a special, new way of looking at the system: To simplify our picture, let us spin our coordinates at the same rate as the imaginary rod tumbles, as if we were painting $x$- and $y$-axes on an old vinyl record and then placing the record on a turntable and letting it spin. This is called a **rotating frame of reference**. The advantage of this rotating frame is that the positions of the two planets appear to stand still, and we can concentrate more closely on the motions of the small third body relative to them.

The rate at which the record spins, the period of the tumbling "rod" connecting the planets, is the orbital period $P$ of the system. Recall that the mean motion, $n$, equals $2\pi/P$; the entire frame of reference rotates at an angular speed of $n$ radians per second.

Now let's examine the energy in the orbit of some third mass $m$ that is much smaller than $M$ and $M^\star$. The total energy $E$ of this particle is the sum of the potential energies due to its gravitational attractions to the other two bodies and its kinetic energy.

The potential energy is easy. We just add together the potential energy due to each of the two planets:

$$U = -m\frac{GM}{r} \qquad (11.3)$$

$$U^\star = -m\frac{GM^\star}{r^\star} \qquad (11.4)$$

where the terms $r$ and $r^\star$ are the distances from our little mass $m$ to $M$ and $M^\star$, respectively. As before, we define the potential to be zero at infinity, so it must always be negative everywhere else.

The kinetic energy is trickier. We would normally write $\frac{1}{2}mv^2$, but we have to be careful now because the velocity we measure in our rotating system is measured against a coordinate system that is also moving.

We don't want to include the energy of the coordinate frame. Consider an object at position $(x, y)$ that is stationary in the inertial frame of reference. In our rotating frame it would appear to be moving "backwards" at the speed that the frame rotates. The speed of the coordinate frame is equal to the distance from the center of the coordinate system, $\sqrt{x^2 + y^2}$, times $n$, the mean motion of the frame. Our stationary object thus would appear to have a kinetic energy of $\frac{1}{2}mn^2\left(x^2 + y^2\right)$. In fact, its real kinetic energy is zero, because (in the inertial frame) it is not moving at all.

Thus we must subtract this energy from its apparent kinetic energy as measured in the frame.

Like potential energy, the term $-\frac{1}{2}mn^2(x^2 + y^2)$ depends only on position; it is sometimes called the **rotational potential energy**, and the gradient of this term yields the familiar expression for centrifugal force. Adding this and the regular kinetic energy to Equation 11.4, we get

$$E = \frac{1}{2}mv^2 - \frac{1}{2}mn^2(x^2 + y^2) - m\frac{GM}{r} - m\frac{GM^\star}{r^\star} \qquad (11.5)$$

Because the total energy $E$ stays constant and $m$ (the mass of the third body, which also stays constant) appears in each term, we can lump these constants together into a new term $C = -2E/m$, and solve for $v$:

$$v^2 = n^2x^2 + n^2y^2 + 2\frac{GM}{r} + 2\frac{GM^\star}{r^\star} - C \qquad (11.6)$$

(Why did we include the minus sign when we defined $C$? Recall that the total energy $E$ of a closed orbit is negative. Thus by our definition $C$ is always positive. It's convenient.)

Notice what this equation does. If we know the position ($x$ and $y$, $r$ and $r^\star$) of a body and its velocity $v$ at any time, we can solve for the constant $C$. Then at any other position, we can calculate $v$ and find how fast the body must be moving. But what we don't necessarily know, and can never solve for exactly, is the direction the body is moving.

The constant $C$ is always positive, as are all the other terms in the equation for $v^2$. What would happen if $C$ were greater than the sum of the other terms? This would mean that $v^2$ would have to be negative, which is impossible. Thus the sum of the other terms must always be larger than $C$. But these terms are determined by the position of the small body $m$. Thus, given a value of $C$, we can see that there are some places where the small body can never go.

Let's say that $C$ is a fairly large number. Then the only other places our body can travel to are places either far from the center of mass so that $x$ and $y$ are very large, or else close to either $M$ or $M^\star$ so that $r$ or $r^\star$ is very small. If $C$ is smaller, these restrictions start to relax. Figure 11.22 shows what the areas of restricted motion look like for various values of $C$. Energy is proportional to $-C$; so the smaller $C$ is, the greater is the energy of the orbit. It makes intuitive sense that the more energy an orbit has, the less restricted it is.

## Lagrange Points

Certain interesting points in Figure 11.22 are the **Lagrange points**, those labelled *L1* through *L5*. The *L* stands for the French mathematician Joseph-Louis Lagrange, who first de-

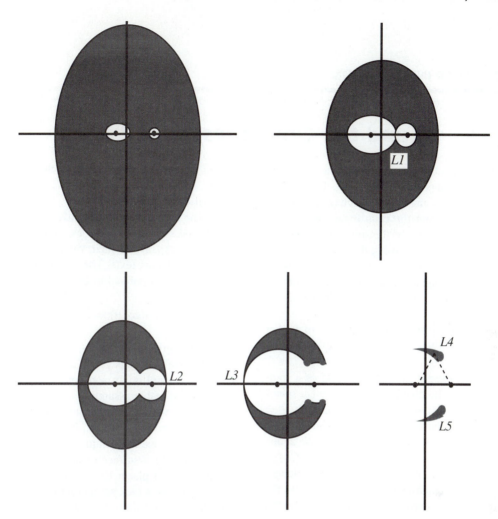

**FIGURE 11.22** Areas of restricted motion for five (decreasing) values of *C*, a constant related to the energy of the third body. The shaded regions mark the regions of space, in our rotating coordinate system, that the third body cannot reach without changing its energy (and thus the value of *C*). The boundaries are those places where the third body's velocity is zero. The five points marked *L1* through *L5* are the Lagrangian points of equilibrium, where the third body may be stationary. Only *L4* and *L5* can be stable.

scribed these points in 1772. (Since that time, the convention for numbering points *L1* through *L3* has varied from author to author.)

Recall that the boundaries between the shaded and unshaded regions are the locations where the velocity *v* is zero. Thus these points represent spots where the third body would be standing still, relative to the two main bodies. Is it possible that any of these points are stable resting spots for our third body? Consider each in turn.

*L2* and *L3* are points where the centrifugal force of the spinning frame is exactly matched by the gravitational attraction of the main masses. (Notice that *L3* is the point on the far side of the larger body sharing the same orbital distance as the smaller body.) If a third body at one of these points were nudged slightly inward, then the force of gravity would get stronger while that of centrifugal force would get smaller. The body would continue to plunge inward, away from its initial spot. Likewise, if nudged outward, centrifugal force would take over, spinning it ever further outward. Thus *L2* and *L3* are not stable.

*L1* is likewise not stable. It's the balance point between the two gravity fields (with a little centrifugal force

## FURTHER INFORMATION...

# J₂ and the Oblateness of a Planet

As first discussed in Chapter 4, a spinning planet bulges at its equator and its poles appear to be flattened. The amount of bulge depends on how mass is distributed inside the planet. The bulge arises from the centrifugal force acting on the mass in the outer layers of the spinning planet; the less mass there is in these outer layers, the weaker the centrifugal force and so the smaller the bulge. Thus if mass is distributed uniformly throughout a planet, there is a bigger bulge; if much of the mass is concentrated into a dense core, the bulge is smaller.

How can we quantify this insight and use it to predict the size of the bulge as a function of the planet's structure and spin rate?

We begin by looking at the potential field of a spinning planet. Recall that the potential energy $U$ of a spherically symmetric mass $M$ is given by Equation 9.9: $U = -GMm/r$, where $G$ is the gravitational constant and $r$ is the distance from the mass $M$ to our test mass, $m$. We define a slightly different quantity, $V$, as $U/m$. Thus $V = -GM/r$ for a spherically symmetric mass. Without $m$ it no longer has the correct units to be an energy; $V$ is technically the "potential energy per unit mass" or the "specific potential energy," but most commonly it's simply referred to as the **potential**.

For a planet with a bulge, we must consider the higher harmonics of the gravity field in calculating its potential. The potential in this case is represented by an infinite series of terms, or **harmonics**. The simplest case, involving only the equatorial bulge due to spin, can be represented by the first nonzero term in this series:

$$V(r) = -\frac{GM}{r}\left[1 - \left(\frac{a}{r}\right)^2 J_2 P_2(\cos\theta)\right] \quad (11.7)$$

where $a$ is the equatorial radius of the planet, $J_2$ is a constant describing how the mass is distributed inside the planet, and $P_2(\cos\theta)$ is the second Legendre polynomial. The angle $\theta$ is the colatitude angle, 0° at the north pole and 90° at the equator. The subscript 2 in $J_2$ and $P_2(\cos\theta)$ denotes that these terms are in fact the *second* gravity harmonic; however, for a symmetric planet with a coordinate system centered on the center of the planet, the first harmonic, which would have terms such as $J_1$ and $P_1(\cos\theta)$, is equal to zero.

The **Legendre polynomials** are a series of polynomials that take the place of sines and cosines in our gravity harmonics expansion; their values can be found in many advanced calculus texts. The polynomial we are interested in, $P_2(\cos\theta)$, is defined to be

$$P_2(\cos\theta) = \tfrac{1}{2}(3\cos^2\theta - 1) \quad (11.8)$$

For a given value of $J_2$, how large a bulge should we expect to see? Let us begin by assuming that our planet is in hydrostatic equilibrium and completely fluid. This is generally a reasonable assumption if the planet is not a small lumpy rock with relatively large mountains and valleys or a body warped out of shape by vigorous internal convection. The surface of a fluid planet in hydrostatic equilibrium should be a surface of constant potential. If any part of such a fluid planet had a higher potential than its surroundings (like a rock sitting on the top of a high hill), then the planet would flow (the rock, and the hill itself, would slump down) until its potential were the same as that of all the surrounding region.

What is the potential on the surface of such a planet? One might be tempted to simply apply Equation 11.7, which gives the potential due to the gravity of a planet with a bulge. But there's an additional term that must be added to take account of the material on the surface of the planet itself spinning around the planet's axis. If the planet spins at an angular rate of $\omega$ radians per second, then an object sitting on the equator has a speed of $r\omega$ m/s; if it's located at any other latitude than the equator, then this speed becomes $r\omega\sin\theta$. Thus it has a kinetic energy per unit mass of $\frac{1}{2}r^2\omega^2\sin^2\theta$. This is the energy supporting the bulge.

Compare this spinning planet with a planet of identical shape that did not spin. This nonspinning planet is obviously not in equilibrium; the material in the bulge is raised to a higher potential than the material at the pole. But in the spinning planet, which is in equilibrium, the material on the bulge and that at the poles are at equal potentials.

The poles of both planets (where the apparent speed of the spinning planet drops to zero) have identical potentials. Thus the material on the bulge of the spinning planet has a net potential less than the material on a static planet, less by an amount equal to the kinetic energy per unit mass of the spin.

Therefore, to find the net potential at the surface of a spinning planet, the kinetic energy per unit mass due to its spin must be subtracted from the potential given in Equation 11.7, to yield

$$V(r) = -\frac{GM}{r}\left[1 - \left(\frac{a}{r}\right)^2 J_2 P_2(\cos\theta)\right]$$
$$- \frac{1}{2}r^2\omega^2\sin^2\theta \tag{11.9}$$

Take this equation and set $r$ to $r_s$, the radial distance from the center of the planet to the surface. This distance $r_s$ is not a constant; it is smaller at the poles than at the equator. We can assume that it follows an equation of the form

$$r_s = a\left(1 - f\cos^2\theta\right)$$
$$= a\left[1 - \tfrac{1}{3}f - \tfrac{2}{3}fP_2(\cos\theta)\right] \tag{11.10}$$

where $f$ is a fraction, sometimes called the **optical eccentricity** of the planet, representing the proportional difference between polar and equatorial radii. If $c$ is the polar radius, then

$$f \equiv \frac{a - c}{a} \tag{11.11}$$

Notice how in Equation 11.10 the $\cos^2\theta$ term was rewritten in terms of the Legendre polynomial $P_2(\cos\theta)$. In the same way, using the laws of trigonometry, one can rewrite the term $\sin^2\theta$ in Equation 11.9 as $\sin^2\theta = \tfrac{2}{3}[1 - P_2(\cos\theta)]$.

Notice also that both $f$ and $J_2$ are small numbers. If we substitute our value of $r_s$ for $r$ in Equation 11.10, expand using the binomial theorem, and drop any terms of order $f^2$, $(J_2)^2$, $f\cdot J_2$, or higher, we can eventually rearrange this equation into two pieces:

$$V(r_s) = \left[-\frac{GM}{a}\left(1 - \frac{1}{3}f\right) - \frac{1}{3}a^2\omega^2\right]$$
$$+ \left[\left(J_2 - \frac{2}{3}f\right)\frac{GM}{a} + \frac{1}{3}a^2\omega^2\right] \tag{11.12}$$
$$\times\ P_2(\cos\theta)$$

The first piece in square brackets is independent of $\theta$ while the second piece is multiplied by $P_2(\cos\theta)$ and thus would appear to have different values at different latitudes. But if our surface is equipotential, it cannot have any dependence on latitude. Therefore, we can conclude that the piece in the second set of square brackets must go to zero, so that

$$f = \frac{3}{2}J_2 + \frac{a^3\omega^2}{2GM} \tag{11.13}$$

Notice the power of this equation. Assume that we already know the size and mass of a planet. If we know $J_2$ and the spin rate, we can predict how flattened the poles should be and how big a bulge we should see at the equator. Turning it around, by actually observing the shape of a spinning planet and thus measuring $f$ directly, we should be able to predict either $J_2$ if we know the spin rate (a technique often applied to the icy satellites) or the spin rate $\omega$ if we know $J_2$ (a technique that, as we saw in the text, gave a somewhat erroneous value for the spin rate for Uranus).

For many planets, as for instance Uranus after *Voyager*'s flyby, we have independent ways of finding both the spin rate and $J_2$. By calculating a value of $f$ and then comparing this value to the observed oblateness of the planet, we have a test to see just how close this planet is to being in hydrostatic equilibrium. From this test, most of the outer planets (even Uranus, where the true value of $J_2$ was not well known before *Voyager*) are not too far out of equilibrium. (Vigorous convection inside these planets can warp the shape slightly out of equilibrium.)

thrown in to help the smaller of the fields). Again, a slight motion towards either mass leads to an ever greater acceleration towards that mass, away from the initial resting point.

However, *L4* and *L5* are much more complicated. These points are located equidistant from both *M* and *M\**. In the frame of reference of the larger mass *M\**, they appear to be in the same orbit as *M*, orbiting 60° in front of and behind *M*. Here it's the Coriolis force (see Chapter 10) that is joining with the centrifugal force to counter the attraction of the two masses. It can be shown that if *M*, the smaller mass, is less than about 4% of the total mass, or about 25 times smaller than the larger mass, then these points are stable. Particles perturbed from these points tend to circle around the points rather than drift away from them.

The Jupiter–Sun system meets this mass requirement. Jupiter is 1000 times smaller than the Sun, and 60° before and behind Jupiter in its orbit about the Sun are two clusters of asteroids. The largest of these asteroids (and thus the first to be discovered) was named Hector. As others were found they also were given names out of the *Iliad* and *Odyssey* of heroes from the Trojan wars. As a result, these asteroids are called the **Trojan asteroids**, and the *L4* and *L5* positions of orbits in general are often called the **Trojan points**. Other notable bodies to be discovered in *L4* and *L5* positions are several small moons of Saturn that orbit in Trojan points before and behind the larger moons Tethys and Dione.

According to this criterion, there should also be stable *L4* and *L5* points in the Earth–Moon system. It has been proposed (and championed by a group originally called the "*L5* Society") that large space stations could be put at these points. These space colonies could house thousands of people and factories that would take advantage of the constant sunlight and low gravity of space. However, careful search has failed to find any small asteroids at these points, although there may be some slight concentrations of interplanetary dust there. The Earth–Moon–colony system is actually part of the larger Sun–Earth–Moon–colony system, and nobody's been able to solve the restricted four-body problem.

## SUMMARY

The paths of most real bodies in space can be considered in terms of simple Keplerian orbits that are subjected to small perturbations. Some perturbations act over a brief period of time, producing instantaneous accelerations. Others are small perturbations that act continuously on a body over an extended time. Often the latter can be described in terms of the harmonics of the gravity field in which the satellite orbits.

Perturbations perpendicular to the path of a body change the shape or orientation of the body's orbit. Perturbations along the direction of the orbit change the orbital energy, and so the semimajor axis of the orbit, as well as changing its shape.

Although the motions of bodies affected by more than one large mass can be very difficult to determine, certain limits can be set for the specific case of the possible orbits of a small body moving near two large masses, which are in circular orbits about their common center of mass. From the analysis of this **restricted three-body problem**, a constant *C* is derived that indicates where the total velocity of the third mass drops to zero and that describes boundaries that this third body can never cross. Five special points in space, called Lagrange points, are the places where all the forces acting on the third body balance out to zero. Three of these points are always unstable. The other two can be stable if the larger mass is at least 25 times more massive than the second mass.

## STUDY QUESTIONS

1. Of the following list, which does *not* perturb the orbit of a satellite?
   (a) the bulge of the equator of the planet it orbits
   (b) the masses of other satellites
   (c) the impact of a meteorite on a satellite
   (d) a change in the total mass of the satellite

2. How does the $J_2$ of a planet affect a satellite's orbit?

3. Which of the Lagrangian points (*L1* through *L5*) can be stable?

4. What is "restricted" in our definition of the restricted three-body problem?

5. What is the name of the class of asteroids that orbits 60° in front of or behind Jupiter?

## 11.4 PROBLEMS

1. Uranus and Neptune were the first planets to be discovered in modern times. Look up the history of their discovery in an encyclopedia. Have we observed their positions for a full orbit of either planet?

2. From Figure 10.2 look up what the mass of a Uranus- or Neptune-sized ball of cold hydrogen should be. What is the mean density of such a ball? Compare this with the observed densities of Uranus and Neptune.

3. What might the weather on Earth be like if it were as tilted as Uranus?

4. Even though Saturn, Uranus, and Neptune are all many times bigger than Earth, the acceleration of gravity at their cloudtops is not much different from that of Earth. How can you explain this odd fact?

5. Is the Roche limit of Saturn inside or outside of its synchronous point? Find a planet where the opposite is true.

6. Everything in space subjected to a hot electron gas and a given amount of solar UV light tends to have the same electric potential on its surface. Because this electric potential is the same for both big moons and small dust, why should small dust be more strongly perturbed by electrostatic force than big moons?

7. How many kilometers from Earth's surface is its Roche limit? What would happen if all the water in Earth's oceans were pulled away from Earth, out into space, to a position outside the Roche limit? What would happen to this water as it was brought back inside the Roche limit?

8. A communication satellite is being lifted from a circular, noninclined orbit at the altitude of the space shuttle to a synchronous orbit, via a Hohmann transfer orbit. Each time a rocket is fired, the orbit of this satellite is changed. How many rocket firings does it take to complete the transfer to a circular, synchronous orbit? In what directions are these rocket firings; are they normal, radial, or transverse?

9. At the end of the Hohmann transfer orbit in Problem 8, just before the rockets are fired to make the orbit circular it is found that the position of the satellite is in error. It is at the correct altitude, but it is sitting over the wrong part of Earth. It is decided to keep the satellite in the transfer orbit, but to cause this orbit to precess until it drifts into the correct position. What direction should the thruster be fired at apocenter?

10. Look up the word *osculating* in a dictionary. Why do you suppose this term is used to describe an orbit that momentarily touches the real orbit?

11. Consider a satellite orbiting a spinning planet with a significant (positive) value for $J_2$. Is the change in $\omega$ positive or negative when the inclination of the orbit is 0°? What if the inclination is 70°? At what inclination angle is there no net precession?

12. The technique of "throwing the planets into the Sun" has worked quite well for finding the masses of the planets Venus and Mercury, but fails miserably at finding the mass of Pluto. Why?

13. A small satellite orbits two bodies under the conditions of the restricted three-body problem. When it fires its thruster to increase its speed, what happens to the value of $C$?

14. Given the five different pictures in Figure 11.22 for five different values of $C$, which one illustrates a third body with the greatest energy? The least energy?

15. Are the *L4* and *L5* points in the Pluto–Charon system stable?

16. A number of science fiction stories have been based on the premise of an "alternate Earth" orbiting the Sun at a position exactly opposite the position of Earth and thus hidden from our view. Discuss the stability of such an orbit.

17. Pictures showing the path of the Apollo spacecraft from Earth to the Moon resemble a figure eight, because the Apollo spacecraft went around the Moon in the opposite direction from the way it went around Earth. Considering Earth, Moon, and spacecraft as a restricted three-body problem, why was such a path chosen? What is the significance of the point where the two loops of the figure eight meet?

## 11.5   FOR FURTHER READING

The University of Arizona Press, Tucson, has a number of anthologies that make excellent starting places to learn more about current scientific research on the outer planets and their rings. They include *Saturn*, edited by Tom Gehrels and Mildred Shapley Matthews (1984); *Uranus*, edited by Jay T. Bergstralh, Ellis D. Miner, and Mildred Shapley Matthews (1991); and *Planetary Rings*, edited by Richard Greenberg and André Brahic (1984). Another book, on Neptune, is in the works.

A more popular introduction to the outer planets can be found in Mark Littmann's book *Planets Beyond: Discovering the Outer Solar System* (New York: Wiley, 1988); it includes scientific background, a review of the history of their discoveries, and a first-person account of the *Voyager* flyby missions. A readable history of the discovery of the asteroids, the discovery of Neptune, and a general review of nineteenth-century astronomy can be found in *The Discovery of Neptune* by Morton Grosser (Cambridge, Mass.: Harvard University Press, 1962).

A much more complete mathematical discussion of the perturbation of orbits and the restricted three-body problem can be found in John M. Danby, *Fundamentals of Celestial Mechanics* (Richmond: Willman-Bell, 1988) or A. E. Roy, *Orbital Motion* (Philadelphia: A. Hilger, 1988). The relationship between the oblateness of a planet and its gravity field is discussed in A. H. Cook, *Interiors of the Planets* (Cambridge, England: Cambridge University Press, 1980).

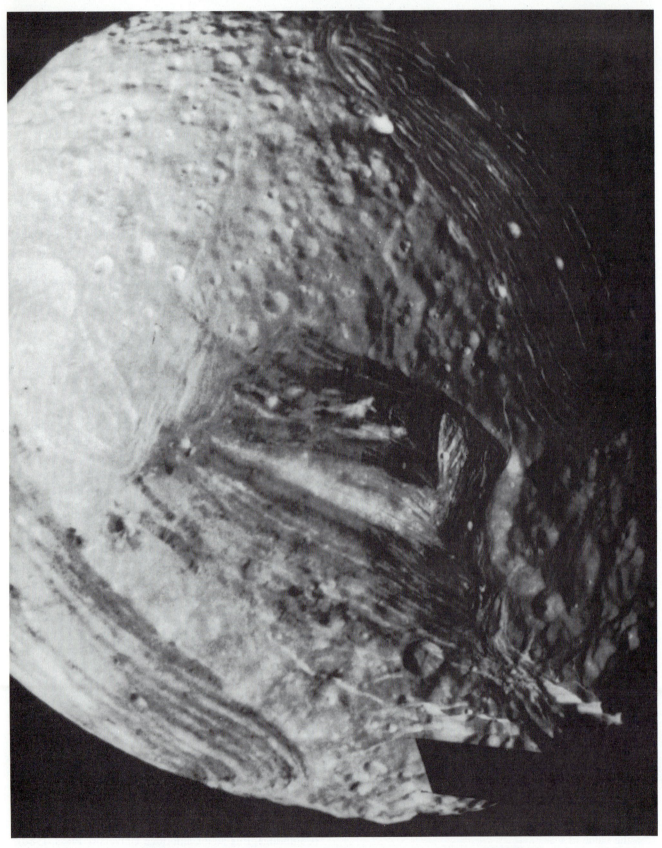

**FIGURE 12.1** Miranda has a complex history, which is written on its face.

# CHAPTER 12

# Planetary Satellites

## 12.1 A GALLERY OF MOONS

Every planet except Mercury and Venus has at least one natural moon (and even Venus has artificial moons because several spacecraft, starting with *Pioneer Venus*, have been put in orbit about it). Earth has one moon, Mars two, Jupiter 16, and Saturn 17 discovered so far, not counting their ring particles. There are now 15 known moons about Uranus (again, ignoring ring particles); Neptune has two moons visible from Earth, with another six discovered by *Voyager*, along with its rings; and even Pluto has a moon. There are doubtless other small moons around these outer planets that have not been discovered yet.

Some of these moons are little more than irregular chunks of rock a few kilometers across. Others are worlds larger than the planet Mercury. In this short chapter, we hardly have time to do justice to them all; the most we can hope for is to catalogue them, listing their most outstanding features.

### Earth's Moon

Earth's **Moon** is one of the moons of planetary dimensions, 1738 km in radius, placing it sixth in size among all satellites. It is discussed in detail in Chapter 3. In addition, many thousands of items of human manufacture, ranging from the manned Mir Space Station to complex communication satellites to spent booster rockets, orbit Earth.

### Martian Moons

**Phobos**, the inner moon, and **Deimos**, the outer one, are both small, odd, potato-shaped chunks of rock (figure 12.2). Phobos is 27 km long and roughly 20 km wide; Deimos is a little smaller, 15 km by 12 km in length. These moons are dark in color, almost as dark as carbonaceous chondrites; it has been suggested that they may be two C-type asteroids captured from the asteroid belt by Mars early in its history.

The surfaces, as photographed by *Mariner 9* and the *Viking* orbiters, are heavily cratered. It's possible that these moons are just fragments of initially larger bodies that were shattered by impacts in the final stages of planetary accretion, when Mars and the other planets were gathering the last bits of rock and dust from which they were formed. One odd feature of Phobos, which may be related to the stress of some of these large impacts, are the grooves seen on its surface.

Phobos orbits inside the synchronous point of Mars, and its orbit is gradually decaying; it is slowly spiralling inward and it may crash into Mars in about 30 million

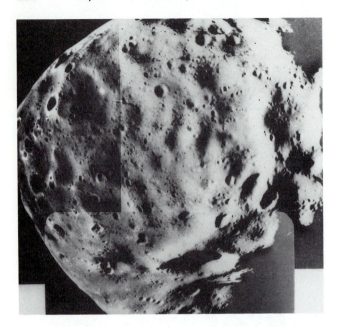

**FIGURE 12.2** Phobos is the inner of two asteroidlike moons of Mars. It is 27 km long and roughly 20 km wide.

**FIGURE 12.3** Europa, an ice-covered moon orbiting Jupiter, may have liquid oceans beneath its ice crust.

years. Because it moves about Mars faster than Mars spins, from the surface of Mars it appears to rise in the west and set in the east. Deimos, on the other hand, orbits outside the synchronous point and is slowly spiralling away from Mars. It is conceivable that these two moons started out as one larger object close to the synchronous point. A violent impact could have split them in two and sent them into their different orbits.

## Jovian Moons

The moons of Jupiter can be divided into four classes: the Galilean satellites, the inner satellites, the regular outer satellites, and the irregular outer satellites.

The four moons discovered by Galileo in 1610—Io, Europa, Ganymede, and Callisto—are all large, planet-sized bodies in nearly circular orbits around the equator of Jupiter. **Io** (diameter 3630 km, density 3.5 g/cm$^3$) is a sulfurous volcanic satellite slightly larger than Earth's Moon. It will be discussed at length in Chapter 14.

**Europa** (diameter 3138 km, density 2.9 g/cm$^3$) is a bit smaller than our Moon (Figure 12.3). Its density is typical of rocky material, but the surface we see is virtually all water ice. There are large systems of cracks that run across the surface but few craters, implying that this surface has been formed or reformed relatively recently in its history, in contrast to the heavily cratered primeval topography typical of the oldest regions of the Moon or Mercury.

The best guess today is that Europa was formed of rocks rich in hydrous minerals such as serpentine, minerals that had considerable amounts of water incorporated in their crystal structure. As Europa warmed up, the water was driven out of the rocks in the interior and was frozen on the surface of the moon. The cracks on the surface may be evidence of the stress caused by tides raised on Europa by Jupiter.

**Ganymede** is the largest moon in the solar system (Figure 12.4); with a diameter of 5262 km it is bigger than Mercury or Pluto. Its very low density (1.9 g/cm$^3$) can be explained by a composition that is half rock and half water ice. The surface reflects this composition. There are dark, heavily cratered regions that look like carbon-rich rock overlaying ice, and there are younger regions that are very bright and icy. It appears as though the surface was all dark and rock-covered at one time, but internal melting and convection of the water ice seems to have broken up this old surface. The icy regions show many cracks and grooves, evidence of materials upwelling between pieces of the old crust.

**Callisto**, on the other hand, is only a bit smaller than Ganymede (4800 km in diameter) and a little less dense (1.8 g/cm$^3$), but its surface is very old, untouched by tectonics or upwelling ice (Figure 12.5). It is heavily cratered, with no evidence that it has ever been resurfaced. The dark surface of both Callisto and the oldest parts of Ganymede are richer in rock than the average composition, as derived from their densities. Perhaps the "rock" is dust from space that collected on the surface, covering the ice. Or possibly the icy part of the surface has evaporated or been driven away

**FIGURE 12.4** Ganymede, the largest moon in the solar system (bigger than Mercury), has an icy crust that shows evidence of extensive tectonism.

by solar ultraviolet (UV) radiation or high-energy particles from Jupiter's magnetosphere, leaving behind only denser rocky materials.

The known **inner satellites** are four moons found inside the orbit of Io. **Amalthea** is an irregular chunk of rock quite unlike the grander Galilean satellites; at 270 km long and roughly 150 km wide, it is the largest of the non-Galilean moons. It was discovered in 1892. Its surface is distinctly reddish, perhaps coated with sulfur sputtered off

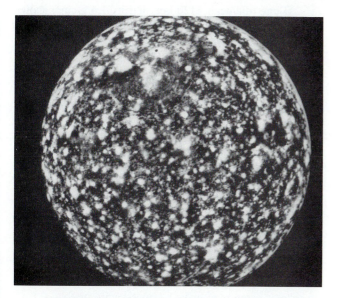

**FIGURE 12.5** Callisto, another large icy moon of Jupiter, has a surface that appears to be very primitive, unaltered by internal evolution.

Io; because of its color and shape, it has been compared to a very large strawberry.

Between Amalthea and Io is the recently discovered moon **Thebe** (110 km long and 90 km wide), about which little is known. **Adrastea** and **Metis** are both tiny fragments a few kilometers across discovered by *Voyager* near the thin dust ring of Jupiter.

The **regular outer satellites** are four moons orbiting roughly six times Callisto's distance from Jupiter. **Himalia**, with a diameter of 185 km, is the largest and first discovered (in 1904). Just inside it, the tiny (16 km in diameter) rock named **Leda** was found in 1974; **Lysithea** (36 km across) and **Elara** (76 km in diameter) were discovered just outside the orbit of Himalia in 1938 and 1905, respectively. These satellites are all dark chunks of rock. They bear the name "regular" because, although far from Jupiter, they travel in stable **prograde orbits**: They go around Jupiter in the same direction as the Galilean moons, the same direction as Jupiter spins, in fact in the same sense as Jupiter and all the planets go around the Sun. (Looking down from a point over Jupiter's north pole, the prograde direction is counterclockwise.) All their orbits are inclined about 28° from the plane of Jupiter's equator, and they all have mildly eccentric orbits; the paths are ellipses that vary a bit, but not too much, from perfect circles.

This regularity is in contrast to the **irregular outer satellites**, which include the moons **Ananke, Carme, Pasiphae,** and **Sinope**. (Note that the irregular moons have names ending in *e*; the regular ones end in *a*.) All four are found in very eccentric orbits, and they are twice as far from Jupiter as the regular outer moons. These small moons, all about 10 km in diameter, lie in orbits so inclined that they are close to perpendicular to Jupiter's equator; indeed they are inclined *more* than 90° and are technically in **retrograde orbits** (their orbits are in the opposite sense from all the other moons, looking clockwise from Jupiter's north pole). With these large, highly inclined orbits, the perturbing forces of Jupiter's equatorial bulge (see Chapter 11) and the Sun's gravity prevent them from following the same path two orbits in a row. All these peculiarities are consistent with the idea that the irregular moons are captured asteroids.

Information about the moons of Earth, Mars, and Jupiter is summarized in Table 12.1.

## Saturnian Moons

Saturn has a system of rings, small moons, and nine major moons (with some odd companions). The only one comparable in size to the Galilean moons, Titan, was discovered by Christiaan Huygens in 1655. Tethys, Dione, Rhea, and Iapetus are all bright enough to be seen in a fairly small telescope, however, and they were also discovered in the late seventeenth century by Giovanni Domenico Cassini

**TABLE 12.1**

Meet the Satellites: Earth, Mars, Jupiter

| Satellite | Radius (km) | Comments |
|---|---|---|
| **Earth** | | |
| Moon | 1738 | Only extraterrestrial body visited by humans |
| **Mars** | | |
| Phobos | 13.5 × 10.8 × 9.4 | Grooves on surface; captured asteroid? |
| Deimos | 7.5 × 6.1 × 5.5 | Captured asteroid? |
| **Jupiter** | | |
| Metis | 20 | Orbits in Jupiter's ring; source of ring dust? |
| Adrastea | 12.5 × 10 × 7.5 | Orbits at edge of ring; source of ring dust? |
| Amalthea | 135 × 83 × 75 | Reddish; coated with dust from Io? |
| Thebe | 55 × 45 | Little known |
| Io | 1815 | Spectacular yellow-orange volcanic plumes |
| Europa | 1569 | Smooth, heavily cracked icy surface |
| Ganymede | 2631 | Largest moon in solar system |
| Callisto | 2400 | Heavily cratered dark surface |
| Leda | 8 | Tiny regular moon near Himalia |
| Himalia | 93 | Main fragment of broken moon? |
| Lysithea | 18 | Associated with Himalia |
| Elara | 38 | Another piece of Himalia? |
| Ananke | 15 | Irregular moons; All have: |
| Carme | 20 | Retrograde orbits |
| Pasiphae | 25 | Highly inclined orbits |
| Sinope | 18 | And may be captured asteroids. |

(who also discovered a gap in Saturn's rings now called the **Cassini division**). Mimas and Enceladus were discovered by William Herschel in 1789; Hyperion and Phoebe were discovered in the nineteenth century.

**Mimas** (diameter 392 km, density 1.4 g/cm$^3$) is observed to be heavily cratered, with little evidence that the surface has been reworked by volcanism or tectonism since its formation. It has one very large crater, Herschel, on its surface that extends an eighth of the distance around the circumference of the moon. (See Figure 12.11, in Section 12.3.)

**Enceladus** (diameter 500 km, density 1.1 g/cm$^3$), by contrast, shows few craters. Instead, long, ropy flow features can be seen covering much of its surface. It also has one of the highest albedos in the solar system, close to 1; it reflects nearly all the sunlight it receives. This suggests that it has been resurfaced in geologically recent times. Tectonism and volcanism may be occurring even today.

**Tethys** (diameter 1060 km, density 1.2 g/cm$^3$) also shows a heavily cratered surface, like Mimas, including one large impact crater whose diameter is about an eighth the circumference of the whole moon. In addition, it has a large groove or crack on the opposite side of the moon, probably caused by the impact that made the large crater. The coincidence in size of the large craters on Mimas and Tethys, plus theoretical calculations on impact energies, has led some scientists to suspect that "one eighth the circumfer-

ence" may be an upper limit on the possible size of craters. Any larger impact, rather than making a larger crater, would probably have had enough energy to break the moon up into pieces instead.

**Dione** (diameter 1120 km, density 1.4 g/cm$^3$) shows both heavily cratered regions and smoother plains with extensive wispy features that may be chasms and rifts made by the moon expanding and splitting its surface. It appears to have had a complicated geological history.

**Rhea** (diameter 1530 km, density 1.3 g/cm$^3$) also shows a hemisphere of bright, wispy material on a dark background, while on the other side of the moon some regions appear to be more heavily cratered than others, again indicating some significant geological evolution.

All the above moons are predominantly icy, with low densities and water-ice surfaces.

**Titan** (diameter 5150 km, density 1.9 g/cm$^3$) is nearly as large as Ganymede. It has a dense atmosphere of nitrogen, with 1.5 times the pressure of Earth's atmosphere and a temperature of 95 K at its surface. The surface itself is obscured from direct view, however, by a thick haze of methane droplets. Methane makes up about 3% of the atmosphere; argon may make up another 5% or so.

Solar UV light hitting the methane haze breaks up the CH$_4$ molecules, allowing them to reform as more complex hydrocarbons and polymers, much like smog is formed from auto emissions. This organic "gunk" is responsible for the orange color of the haze (whereas methane itself would have colored the atmosphere blue, as we saw in Uranus and Neptune).

Because the surface is obscured, we can only speculate at what might be found there. Liquid nitrogen or liquid methane may rain out of the sky, like water rains on Earth; the methane would probably evaporate again before it hit the ground, but the liquid nitrogen might form lakes and rivers on the surface. Organic compounds formed in the methane haze have been proposed as precursors to life, although the very low temperature of the atmosphere argues against this suggestion. More likely flakes of carbon-rich compounds fall like a dark snow, covering the surface. To go along with the smog-filled atmosphere, the ground may be covered with material chemically similar to asphalt!

**Hyperion** is an irregularly shaped fragment, over 400 km long and roughly 250 km thick. Its surface is somewhat dark in color compared with the other icy moons. Unlike the other moons, it does not rotate on an axis, but rather it tumbles chaotically as Titan and Saturn alternately put a torque on its spin.

**Iapetus** (diameter 1460 km, density 1.15 $g/cm^3$) shows two remarkably different faces. As it orbits Saturn, one side always facing Saturn, the leading hemisphere is extremely dark, seven times darker than the trailing hemisphere, which is icy and heavily cratered. Some of the craters on the icy side have very dark floors, however. The origin of the dark material remains a subject of intense speculation. One theory suggested that dust from Phoebe spiralling in towards Saturn by the Poynting–Robertson effect (see Chapter 13) coats the leading side of the Iapetus; however, the spectral colors of Phoebe are noticeably different from the dark side of Iapetus. Another possibility suggests that impacts (which would be more energetic on the leading side) could induce chemical reactions in methane-bearing ices on the surface, leading to the formation of dark, carbon-rich organic compounds.

**Phoebe** (diameter 220 km) travels in an irregular retrograde orbit about Saturn, like the irregular moons of Jupiter. It is believed to be a captured asteroid.

**Tethys** and **Dione** are observed to have companion satellites, small moons that orbit Saturn at the same distance as the larger satellites but positioned on average 60° before or behind the positions of the larger moons, just like the Trojan asteroids are positioned relative to Jupiter. (These points are the stable Lagrangian points, as described in Chapter 11.) Dione has one known companion, **Helene**; Tethys has two, **Calypso** and **Telesto**. All three appear to be irregular lumps of ice and rock, roughly 35 km long and 30 km wide.

Their unusual shapes, along with the shape of Hyperion, suggest strongly that these bodies are fragments of some larger satellite or satellites. The large craters seen on

Mimas and Tethys also indicate that violent impacts were occurring early in the history of the Saturn system. It seems likely that the moons of Saturn may have been accreted from smaller bodies, broken up by large impacts, and then reaccreted, with the entire procedure repeated possibly several times. Another indication of this breakup and accretion may be the mean densities of these moons. A body with ice and rock in cosmic proportions would have a density of about 1.5 $g/cm^3$; the densities of the Saturnian moons vary randomly from this predicted average, which might be evidence that chunks of ice or rock from broken-up moons were redistributed among the satellite system.

Information about the Saturnian moons is summarized in Table 12.2.

## Uranian Moons

The moons of Uranus include five that were observed from Earth plus 10 new ones photographed by *Voyager 2* in January 1986. Titania and Oberon were found in 1787 by William Herschel, the discoverer of Uranus; Ariel and Umbriel were found by the English brewer and amateur astronomer William Lassell in 1851. Because Uranus and its largest moons were discovered by Englishmen, the names of all the moons are taken from English literature and mythology. (Miranda was found by a Dutch-American, Gerard Kuiper, in 1948.)

The newly discovered moons all lie in nearly circular orbits between Uranus and Miranda, the innermost of the five major moons. All are quite dark in color and range in size from 170 km in diameter down to 26 km in diameter. Most are associated in some way with the Uranian rings.

**Miranda** (diameter 480 km, density 1.3 $g/cm^3$) has an extremely complex surface, especially considering its small size. Most remarkable are sets of interlaced grooves in the shapes of trapezoids and ovals, resembling "racetracks" at times, and a distinct set of grooves in the shape of a chevron (like a sergeant's stripes). Valleys in the moon may be as much as 15 km deep, the deepest known chasms in the solar system. A number of ideas have been suggested to explain these features. Perhaps Miranda was completely broken apart by a giant impact and then reassembled, with chunks of a partly layered and differentiated interior now turned over onto its surface.

**Ariel** (diameter 1058 km, density 1.6 $g/cm^3$) has a very bright and apparently very young surface. Craters larger than 50 km don't exist on its surface, and there appear to be icy flows showing that material inside the moon has evolved out onto its surface. Furthermore the surface has several deep chasms and is crisscrossed with a network of cracks that may be evidence of tidal distortions earlier in its history.

**Umbriel** (diameter 1172 km, density 1.4 $g/cm^3$) appears to have a very old surface. There are many large

**TABLE 12.2**

Meet the Satellites: Saturn

| Satellite | Radius (km) | Comments |
|---|---|---|
| **Saturn** | | |
| Atlas | $20 \times 10$ | Near A ring |
| Prometheus | $70 \times 50 \times 50$ | Inner F ring shepherd satellite |
| Pandora | $55 \times 45 \times 35$ | Outer F ring shepherd satellite |
| Epimetheus | $70 \times 60 \times 50$ | Twin of Janus |
| Janus | $110 \times 100 \times 80$ | Co-orbits with Epimetheus |
| Mimas | 196 | Large crater Herschel |
| Enceladus | 250 | Very reflective, extensively resurfaced |
| Tethys | 530 | Large, long Ithaca Chasma |
| Calypso | $17 \times 11 \times 11$ | Co-orbits with Telesto and Tethys |
| Telesto | $17 \times 14 \times 13$ | Co-orbits with Calypso and Tethys |
| Dione | 560 | Wispy pattern on surface |
| Helene | $18 \times 16 \times 15$ | Co-orbits with Dione |
| Rhea | 765 | Relatively large, somewhat complex surface |
| Titan | 2575 | Large moon with thick nitrogen atmosphere |
| Hyperion | $205 \times 130 \times 110$ | Chaotic tumbler |
| Iapetus | 730 | Two faces, one icy and one dark |
| Phoebe | 110 | Retrograde orbit; captured asteroid? |

impact craters, and the surface appears to be quite dark compared with the other Uranian moons.

**Titania** (diameter 1480 km, density 1.7 g/cm$^3$) is similar to Ariel, covered with smaller craters, but none larger than 50 km in radius, and an extensive series of chasms, some 5 km deep, where fresh ice appears to be visible.

**Oberon** (diameter 1524 km, density 1.5 g/cm$^3$) is heavily cratered like Umbriel, but not quite as dark. Some of the craters on its surface have dark patches, whose origin is still unclear, in their floors.

## Neptunian Moons

Two moons of Neptune were known before *Voyager 2* arrived in 1989; another six have been found so far in the *Voyager* images.

**Triton** (Figure 12.6) is a large moon; its radius, 1350 km, is three quarters that of Io or Earth's Moon. It was discovered by Lassell in 1846. Before *Voyager*, the most unusual thing known about Triton was its retrograde orbit about Neptune. Unlike every other major moon in the solar system, Triton orbits its planet in the direction opposite to the way the planet spins. It is very difficult to understand how such a large body could have formed near Neptune yet move in the direction contrary to all the other material in the Neptune system.

The reflectance spectrum of the surface of Triton, as observed from Earth, showed a hint of a thin methane atmosphere; this was confirmed by the *Voyager* flyby in August 1989. Although covered with something bright, like ice, the characteristic spectral features of water ice are not visible.

Instead, the ice covering the surface may be frozen nitrogen, the same element that, as a gas, makes up 80% of Earth's atmosphere.

The surface of Triton as seen by *Voyager* is quite unusual. Parts of it are pocked with round features that might at first glance be craters; however, unlike craters seen on other moons, these round features all seem to be roughly the same size and shape. Instead of being formed by mate-

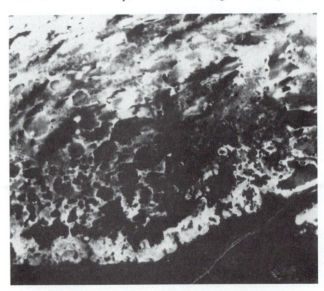

**FIGURE 12.6** Dark plumes on Triton (the dark streaks in the top of the photo) occur when sunlight heats material below the frozen nitrogen surface, causing it to vaporize and burst through the ice.

**TABLE 12.3**

Meet the Satellites: Uranus, Neptune, Pluto

| Satellite | Radius, km | Comments |
|---|---|---|
| **Uranus** | | |
| Cordelia | 13 | Inner epsilon ring shepherd |
| Ophelia | 15 | Outer epsilon ring shepherd |
| Bianca | 21 | Near resonance with ring 5 |
| Cressida | 31 | Near resonance with ring $\eta$ |
| Desdemona | 27 | Near resonance with ring $\alpha$ |
| Juliet | 42 | Near resonance with rings $\alpha$ and $\beta$ |
| Portia | 54 | Discovered by *Voyager 2* |
| Rosalind | 27 | Near resonance with ring $\delta$ |
| Belinda | 33 | Resonant with gap between rings $\eta$ and $\gamma$ |
| Puck | 77 | First Uranian moon discovered by *Voyager 2* |
| Miranda | 240 | Chevron pattern valleys on surface |
| Ariel | 579 | Extensive cracks from tidal stress |
| Umbriel | 586 | Dark, heavily cratered surface |
| Titania | 790 | Considerable tectonic activity |
| Oberon | 762 | Dark, cratered surface |
| **Neptune** | | |
| Naiad | (29) | Orbits near rings |
| Thalassa | (40) | Found by *Voyager 2* |
| Despina | 75 | Found by *Voyager 2* |
| Galatea | 80 | In resonance with Neptune ring arcs |
| Larissa | $104 \times 89$ | Found by *Voyager 2* |
| Proteus | $218 \times 208 \times 201$ | Found by *Voyager 2* |
| Triton | 1350 | Large moon with thin atmosphere, black plumes, retrograde orbit |
| Nereid | 170 | Large, eccentric orbit; tumbling spin? |
| **Pluto** | | |
| Charon | 615 | Pluto's sister planet |

rial outside Triton hitting it at random, they may instead be formed by something, not yet understood, going on inside the moon. Likewise, other parts of Triton are covered with elaborate grooves untouched by craters, again implying that the surface is relatively young, untouched by impacts, and stretched by some internal geologic processes.

One startling discovery *Voyager* made was of dark plumes (see Figure 12.6), geysers in the southern part of the moon that erupt several hundred kilometers into the thin Triton atmosphere until the dark material is caught and carried downwind by winds in the upper atmosphere. These plumes appear to be formed when pockets of material just below the surface of the nitrogen ice are warmed by sunlight and burst out through the ice.

The next-discovered moon of Neptune, **Nereid**, was found by Kuiper in 1949. It is in a prograde orbit, but one that is very eccentric and inclined. Until recently, very little was known about this moon. However, recent observations have revealed that not only is its orbit very peculiar, but it also appears to have a very odd shape or surface. Attempts to determine its spin rate have come up with several contra-

dictory results. Depending on when it is observed, several Earth-based observing teams have noted that its brightness may stay nearly constant or that it may vary by as much as a factor of four. On the other hand, the *Voyager* cameras (which unfortunately never came close enough to Nereid to get a clear picture of its surface) did not see any brightness variation at all. Analogy with the Saturnian moons suggests two possibilities: Either Nereid has a dark spot like Iapetus, but one that can change in size or darkness over the course of several years; or else it is irregularly shaped and tumbling chaotically, like Hyperion.

The peculiar orbits of both Triton and Nereid and that Pluto and its moon Charon are bodies similar in size to Triton and lie in an orbit that crosses Neptune's orbit all give credence to the speculation that (presumably early in its history) the Neptune system suffered a colossal catastrophe. According to this notion, Neptune could have started with several sizable moons: Pluto, Charon, Nereid, and perhaps Triton. However, as the material that makes up the outer solar system was still being accreted, some large planetesimal —one version suggests Triton itself for this role—passed close by this system. (We know that such large planetesimals did exist in the outer solar system; note the giant impacts and fragments of broken moons seen in the Saturnian moons and Uranus being tipped over onto its side).

This passing planetesimal could have strongly disrupted the orbits of the Neptune satellites, turning the orbit of Triton around, sending Nereid well out into a very eccentric orbit, and giving Pluto and Charon enough energy to escape Neptune's gravity altogether. Numerical simulations of the orbits of moons around Neptune subjected to this sort of cosmic billiards game show that such an outcome, while not trivial, is not impossible.

However, a number of points still need to be settled before such a scenario can become widely accepted. Is the chemical composition of Pluto and Charon, as discussed in Chapter 11, really compatible with their being formed near a planet? Or are they more cometlike, suggesting that they were formed directly in the solar nebula itself? Does Nereid look more like a regular moon or like a captured comet? Can another, more reasonable explanation for Triton's orbit be devised? These questions will almost certainly be intriguing planetary scientists well into the twenty-first century.

Several other moons of Neptune were all discovered by the *Voyager* spacecraft. As yet, little is known about any of them. Likewise, little is known about Pluto's moon **Charon** except its orbit, size and mean density.

Information about the moons of Uranus, Neptune, and Pluto is summarized in Table 12.3.

## SUMMARY

At least 60 moons are known to orbit the nine major planets. The majority are small, dark chunks of rock, many of which look like, and may be, captured asteroids. Six of Saturn's moons, five of Uranus's, one of Neptune's, and Pluto's moon Charon all appear to be substantial-sized worlds, on the order of 1000 km in diameter, made mostly of ice. Seven of the moons—Earth's Moon, the Galilean satellites of Jupiter, plus Titan and Triton—are comparable in size to Mercury and can be considered planetary bodies in their own right.

## STUDY QUESTIONS

1. What is "irregular" about the irregular moons of Jupiter?
2. What planets do not have icy moons?
3. What is the largest nonspherical moon?
4. What moons have active volcanoes or plumes on their surfaces today?
5. What moons have atmospheres?

## 12.2   TIDES AND RESONANCES

In Chapter 11 we explored how distortions of the gravity field of a planet could perturb the orbits of its moons. Now let's examine how the gravity of a moon can, in turn, distort the gravity field of the primary. There are two very effective ways a moon can do this. First, it can raise a tidal bulge on that planet; second, it can orbit in resonance with the spin of the planet or another moon.

### Origin of Tides

The connection between the ocean's tides on Earth and the phases of the Moon has been known since antiquity; the Chinese had realized this connection by the second century B.C., and the Stoic philosopher Posidonius wrote about it around 100 B.C. In medieval times, Robert Grosseteste and Roger Bacon discussed how the "virtue" of the Moon, connected in some way with moonlight (the highest tides occur at new and full Moon), raised a mist from the sea floor that pulled the oceans up. Galileo, on the other hand, attempted to devise a theory of the tides based only on Earth's motion, ignoring the Moon.

With Newton's laws it became possible to explain the tides in terms of the force of gravity. It also became clear that all moons, not just Earth's Moon, give rise to tides; all planet–moon systems are subjected to tidal interactions.

A tide-raising body can distort the shape of another body because that second body is not a point but an extended mass. Some part of that mass is going to be closer to the tide-raiser than other parts, and so that part sees a slightly stronger gravity field and thus is pulled towards the tide-raiser more than the rest of the body.

In turn, the distorted shape has an effect on the orbit of the tide-raising body. It can also perturb the orbits of other satellites (which might be too small or too far away to raise noticeable tides themselves). It also changes how fast the planet spins.

Physically, what happens is that the part of the planet closest to the moon feels a little bit stronger pull from the moon's gravity than the center of the planet feels. Likewise, the center feels a pull slightly stronger than the far side of the planet.

Recall that gravity is an inverse square force. If two bodies are a distance $a$ apart from each other and the planet with the tides has a radius $r$, then the force on the near side is proportional to $1/(a-r)^2$, the force on the center varies as $1/a^2$, and the force on the far side of the planet varies as $1/(a + r)^2$. So the difference in the forces between the near side and the center is

$$\frac{1}{(a-r)^2} - \frac{1}{a^2} = \frac{a^2 - a^2 + 2ar - r^2}{a^2(a^2 - 2ar + r^2)}$$

$$= \frac{2r - \left(\frac{r^2}{a}\right)}{a^3 - 2a^2r + ar^2} \qquad (12.1)$$

Because $r$ is usually very small compared with $a$, the first terms in both numerator and denominator are the the largest and the rest of the terms can be ignored. We conclude that the size of the tidal bulge varies as $2r/a^3$. It depends on the diameter of the planet, $2r$: Big planets have larger tides than small planets. And it depends on the cube of the distance between the moon and planet, $1/a^3$, so that closer moons raise much larger tides than moons farther out. From these facts, one can see that the net tide-raising force acting on one side of a planet has several interesting features.

First, the planet bulges out on the side facing the moon, just as one might expect. However, one also finds that the side of the planet facing *away* from the moon also appears to bulge out (Figure 12.7). You can think of this as the center of the planet itself being pulled towards the moon and that the material on the far side of the planet is, in a sense, "left behind."

Second, the size of the tidal force, and of the resulting bulge, is smaller if the moon is farther from the planet. This again is not surprising, because we recall that the force of gravity between two bodies drops away as the square of the distance between them. What the mathematics of tides tells us, however, is that the size of the tide falls as the *cube* of the distance. A small change in distance can have a very large effect on the size of the tidal bulge, which can be very noticeable if the moon has an eccentric orbit around its planet.

Finally, remember that the tidal bulge arises because the gravitational pull of a moon at the surface of a planet is different from that felt at the planet's center. Thus it is not surprising that the larger the planet (and thus the greater the distance from its center to its edge), the larger the tidal force is.

## Tidal Evolution of Orbits

Consider a tide raised on a rapidly spinning planet. The tide raises a small bit of extra mass on the surface of the primary planet, and this extra mass has an extra gravitational pull on the moon. The tidal bulge ought to lie exactly underneath the moon raising the tide, but the spin of the planet tends to carry the bulge a little bit away from this position. (This effect is familiar to anyone used to ocean tides on Earth. If there were no lag, then high tide during a full moon would

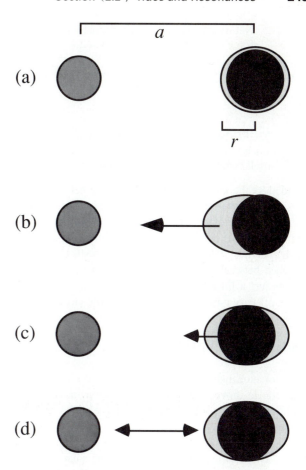

**FIGURE 12.7** (a) The moon on the left pulls with its gravity on all parts of the planet to the right, but those parts nearest to the Moon get pulled the most. The material on the near side of the planet gets pulled more than the center of the planet (b), but the planet center itself gets pulled more than the material on the far side (c). As a result, the tidal bulge is symmetric on both sides of the body (d).

occur at noon and midnight, whereas in fact true high tide can lag as much as six hours behind this time, depending on local geography.) This means that there is a component of the force, from the gravitational attraction of the tidal bulge, that is in the direction of the moon's motion. This is a *transverse* perturbing force.

How strong a pull is it? The strength of the attraction follows the general law of gravity and varies as the square of the distance between moon and planet, $1/a^2$, times the size of the extra mass, which varies as the strength of the tide, or as $1/a^3$. So the force due to the tidal bulge varies as $1/a^5$, the fifth power of the distance between planet and moon.

But not all this extra force acts in the direction of the moon's orbit (Figure 12.8). The fraction of the tidal force that acts in the transverse direction is just the ratio of how far away the spin of the planet moves the bulge

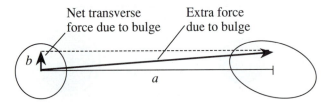

**FIGURE 12.8** Because the tidal bulge is carried in front of a planet by its spin, there is small transverse force on the tide-raiser.

divided by the moon–planet distance. The net force adding or subtracting energy to the moon's orbit, in the direction of $b$, is equal to the total extra force of the bulge times the fraction $b/a$. Thus the transverse force is proportional to $b/a \cdot 1/a^5$, or $b/a^6$. It varies as the sixth power of the distance between the planet and the moon! As the moon moves just a little farther away from the planet, the force drops off very quickly. As it moves a little closer, the force goes up rapidly.

## Outcomes of Tidal Evolution

This force changes the energy of the moon's orbit, causing the orbit to move in or out. Which way it moves depends on how fast the moon is moving relative to the spin of the planet (Figure 12.9).

*Case 1: Prograde Orbits Outside the Synchronous Point.* The most common situation in the solar system is a satellite moving in the same direction as the planet's spin and more slowly than the planet spins. Earth's Moon is an example of this situation. In such a case the tidal bulge is moved ahead of the moon, in the direction of the moon's orbit. Thus the bulge's attraction adds a little energy to the moon's orbit. As the energy increases, the orbit gets bigger. The moon slowly recedes away from the planet. Earth's Moon is currently receding from Earth at the rate of a few centimeters per year.

*Case 2: Prograde Orbits Inside the Synchronous Point.* Low Earth satellites such as the space shuttle travel in the same direction Earth spins but complete an orbit in less than a day. Such satellites are too tiny to raise appreciable tides, however. An example of a bigger moon orbiting inside the planet's synchronous point is Phobos. An observer on Mars would see the far moon, Deimos, rise and set like our own Moon, while Phobos would rise in the west and cross to set in the east before the night was over. Both moons actually travel in the same direction; it's just that Deimos travels slower than Mars spins, while Phobos travels faster.

In the case of Phobos, the satellite outruns the tidal bulge it creates. The bulge always lags behind the satellite, and so it robs energy from the moon's orbit. The moon slowly spirals in until it hits the planet, leaving a sizable elongated crater. (This is an exception to the usual "round crater" rule discussed in Chapter 3 because the impact velocity is at a very shallow angle and is relatively slow.) Some researchers are looking for such craters on Mars to see if it once had other moons in orbits like Phobos's.

*Case 3: Retrograde Orbits.* It doesn't matter how fast the moon is going in this case. If it is orbiting the planet in the opposite direction to the planet's spin, the tidal bulge is always carried behind the moon, robbing energy from its orbit.

The prime example of this case is the moon Triton, one of the largest moons in the solar system, which orbits retrograde about the planet Neptune. It is rapidly losing energy from its orbit. Theoretical calculations predict that it will be pulled into the cloudtops of Neptune in a matter of 100 million years, a relatively short time in the age of the solar system. However, because the tidal pull depends strongly on the distance between moon and planet, most of the change in Triton's orbit won't occur until the last million years or so.

## Tidal Spindown

If tides can put energy into a satellite's orbit, that energy has to come from some place. The energy transfer occurs because the planets are spinning. This spin, then, is the source of the energy. Case 1, the most common case, has energy being transferred from spin to moon's orbit. As the moon moves out, the planet also slows down. (This also happens in case 3. In case 2, the moon's orbit tends to make the planet spin slightly faster; the orbit of the moon becomes the source of energy, and the planet's spin is the sink of energy.)

Fossils of corals and nautiloids, sea creatures whose shells show seasonal variations as well as daily and monthly growth cycles, demonstrate that in the past Earth was spinning much faster and the Moon was much closer to Earth than today. The fossil corals tell us that over the past several hundred million years the days have grown longer at a rate of 25 seconds per million years. Fossil nautiloids from 79 million years ago show 22 days to the month (each day being 32 minutes shorter than our present day), and fossils from 400 million years ago indicate that there were only 10 days in the lunar month. The Moon must be twice as far from Earth now as it was back then. This effect can also be seen in finely laminated sediments called **varves** that formed in tidal plains (Figure 12.10).

This points to another important feature about tides and how they change spin rates and orbits. The energy transfer occurs because the spin of the planet pushes the bulge ahead (or behind) the moon. If a planet is perfectly fluid, the bulge is not dragged by the spin of the planet; if the planet is perfectly rigid, there is no bulge. The greatest interaction happens with a planet like Earth, that has oceans that react strongly to tides but that also has continents to

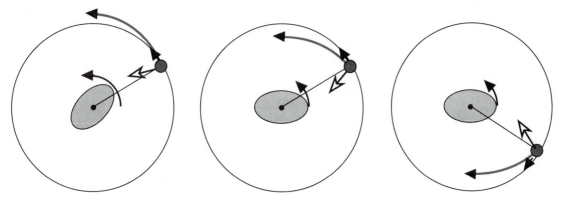

**FIGURE 12.9** Outcomes of tidal evolutions. Left (case 1): The moon is outside the synchronous point; the planet spins faster than the moon moves, and so the bulge adds energy to the moon's orbit, causing it to spiral out away from the planet. Middle (case 2): The moon is closer than the synchronous point and thus orbits faster than the planet spins; the planet's bulge robs energy from the moon and pulls it towards the planet. Right (case 3): The planet spins in the opposite direction of the moon's orbit; the moon is pulled into the planet.

**FIGURE 12.10** This photograph of varves shows how the length of the month has varied.

slow down the response of the oceans and push them away from the point just below the Moon.

The amount of tidal interaction between Earth and the Moon depends on the shape of the continents. Most of the interaction occurs in only a few shallow sea passages (such as the Bering Strait and the English Channel). Because of the current orientation of continents, tidal spin down presently is larger than the average rate over the history of Earth. That most of the evolution of the Moon's orbit has occurred in the past half billion years indicates that only recently have the continents had their present arrangement.

What will happen in the future? Eventually, as the Moon recedes, Earth will continue to slow down until one side of Earth always faces the Moon. When this happens, tidal evolution will cease, as the spin will no longer carry the tidal bulge in front of the Moon, and Earth and the Moon will continue forever locked, face to face. Pluto and its moon Charon are in this configuration already.

Earth raises tides on the Moon, too. One result of these tides has been to slow down the initial spin of the Moon and to pull the denser side, where the mare-filled basins are, towards Earth. This same effect has occurred with most of the large moons of the outer solar system. Stress features on the surface of these moons are silent testimony to the force of those early tides, when the moons were much closer to their parent planets.

## Resonances

In Chapter 11 we examined the effect of perturbing forces on the orbit of a satellite or planet, listing the equations for how the various Keplerian constants of the orbit could be changed by small accelerations. The terms in those equations can be divided into two types. Some perturbations act in the same direction, no matter where in the planet's orbit they occur. Others have a $\sin \eta$ or $\cos \eta$ term, where $\eta$ is the angle indicating how far around the orbit from pericenter the body has travelled; for these terms, the effect of the perturbation changes sign depending on the value of $\eta$ when the perturbation occurs. Because two moons may generally encounter one another anywhere in their orbits, it seems reasonable to assume that these terms will have a negligible long-term effect if the encounters can happen at any point in the orbit. Each encounter has the same chance of having a negative effect as a positive effect, depending on the sign of the sine or cosine term. The pushes and pulls, in the long run, would tend to average out.

But what happens if two moons always encounter each other in exactly the same alignment? Consider the moons Io, Europa, and Ganymede orbiting Jupiter. The period of Io is half that of Europa, and the period of Europa is half that of Ganymede. These moons are arranged such that every time Io and Europa are lined up, Ganymede is 180° away from them on the other side of Jupiter. Obviously, every time Io and Europa, or Europa and Ganymede, are closest together (and so have the strongest perturbing effect), the value of $\eta$ for each moon is always the same. Thus the values of $\sin\eta$ and $\cos\eta$ are the same, so the perturbing forces always have the same magnitude and are always in the same direction. Rather than undoing each other, the small perturbations build up with time until they become quite significant.

Such an alignment of orbiting bodies is called a **resonance**. Two moons are said to be in resonance if the ratio of their orbital periods can be expressed as some fraction of small integers. For example, the period of Ganymede is twice that of Europa, and the period of Europa is twice that of Io; each case is an example of a 2/1 resonance. In a resonance, the moons always meet in conjunction at the same spots in their orbits.

Resonances are quite common in the solar system. Besides the three Galilean satellites, the Saturnian moons Enceladus and Dione, and Titan and Hyperion, are each resonant pairs. The planets Neptune and Pluto also have an orbital resonance.

### Resonances and Rings

Given the orbit of any moon, there are a certain set of locations in resonance with that orbit. If there is a ring of moonlets spread inside (or outside) the orbit, then some of those moonlets may be in a position of resonance with the moon. What happens to those moonlets? The perturbations of the orbiting moon act on the semimajor axes of the orbits of the moonlets, changing them until their orbits are no longer in resonance with the moon.

One would expect that these resonance positions ought to be free of any moonlets and, in fact, this is seen. The Cassini division between the two brightest Saturnian rings is close to a resonance position of the moon Mimas; the edge of the outermost ring of Uranus, the epsilon ring, lies near a resonance with Miranda. The Kirkwood gaps (see Chapter 9) also illustrate this phenomenon; these are the orbits in the asteroid belt where asteroids would be in resonance with the orbit of Jupiter.

However, a careful analysis of the orbits of ring moonlets has shown that these resonance positions are not always exact. For instance, the Cassini division is only *close* to a resonance. Why isn't it exactly in resonance? The me-

chanics of ring particles are complex; along with resonant moons, the ring moonlets themselves also perturb one anothers' orbits. The complex structure of Saturn's rings as seen by *Voyager* show just how complicated these rings can become. The celestial mechanics governing rings is a subject still in its infancy, not well understood by anyone yet.

## SUMMARY

Tides and resonances are two important ways that moons orbiting a planet can affect its gravity field and thus change their own orbits and the orbits of any other moons about that planet.

A tide is raised on a planet because planets have substantial size and are not point masses. The parts of a planet closest to its moon feel a slightly stronger gravitational force than the parts farther away. The size of the tidal bulge depends on the size of the planet, but it varies inversely as the cube of the distance between the planet and its moon. (The same relations hold when planets raise tides on their moons.)

This tidal bulge raised by a moon changes the orbit of the moon. If the planet spins at a rate different from the orbital period of the moon, then the position of the bulge is carried away from a spot directly below the moon. In such a situation, energy is exchanged between the spin of the planet and the orbit of the moon, changing both. Eventually, either the moon will crash into the planet or both moon and planet will be locked face to face, the period of the moon's orbit equalling the period of each body's spin.

One effect of this perturbing force is to remove small bodies from regions of resonance. In this way, Jupiter clears out the Kirkwood gaps of the asteroid belt and Mimas causes divisions in the rings of Saturn.

## STUDY QUESTIONS

1. Are tidal forces pulling Earth's Moon to spiral in towards Earth, or causing it to spiral out away from Earth?

2. Which moons raise stronger tides: moons closer to a planet or moons farther from a planet?

3. Give an example of two bodies in the solar system whose orbits are in resonance.

4. Give an example where resonances have cleared debris out of a region of space, leaving "gaps."

5. Triton orbits Neptune in a retrograde orbit. What seems to be Triton's inevitable fate?

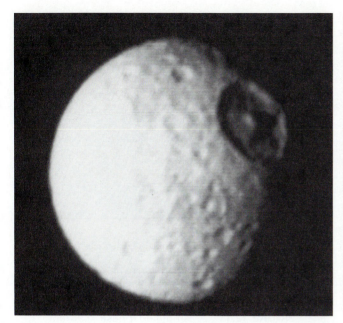
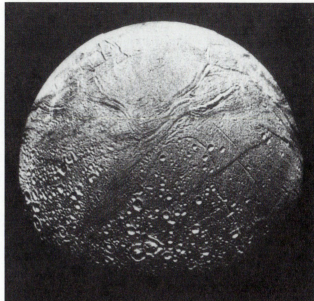

**FIGURE 12.11** Mimas, the moon of Saturn shown on the left, and Enceladus, on the right, are roughly the same size and were almost certainly formed under similar conditions. Yet Mimas is heavily cratered, its surface little affected by internal evolution, while many of the craters on Enceladus have been covered by material that flowed up from its interior.

## 12.3 THERMAL EVOLUTION

Planets change and evolve. Heat generated inside a planet flows to its surface where it is radiated to space. As the interior of a planet heats up, some of the minerals inside may begin to react chemically with one another. If the interior starts to melt, then denser materials are free to fall towards the center of the body and lighter constituents can float to the top. If the heat sources die away, the planet may freeze again. As it melts and freezes, the material inside the planet can expand or contract, or even start convecting, leading to tectonism—from growing cracks to moving plates—and forming valleys and mountains on the surface of the planet.

In principle, we know all these processes occur. We see tectonism and we see evidence of chemical reactions; once-molten lava flows cover the surfaces of many planets. From the moment of inertia of a planet we can discover to what degree it has differentiated itself into a light crust and a dense core. When we sound out the planets today with seismic detectors we find that some have molten interiors and some do not.

Is it possible to predict for any body how much melting has occurred and when that melting must have taken place? How can we tell that Io has a different source of heat than the Moon? Can we make sense out of the complex evolution of a terrestrial planet?

The larger a planet is, the more complicated (and difficult to unravel) its evolution. To introduce the principles of planetary evolution, therefore, let us concentrate on the simplest cases: the smallest bodies. For this reason, we examine here the thermal evolution of outer solar system moons.

### Evolution of the Moons

Although moons are generally smaller than most planets, they are nevertheless places that have each had unique histories. The complex geology of Ganymede, Miranda, and some of Saturn's moons and the absence of such history on other moons of comparable size (Figure 12.11) is sufficient evidence that these moons indeed deserve to be treated like planets.

First, consider the pressure inside a moon. The central pressure inside any spherical body (assuming that its density $\rho$ is constant) can be estimated by the formula

$$P_c = \tfrac{2}{3}\pi G \rho^2 R^2 \tag{12.2}$$

where $G$ is the universal gravitational constant and $R$ is the total radius of the moon. According to this equation, the central pressure of the larger moons can range from $10^7$ Pa (100 Earth atmospheres) for a body a few hundred kilometers in radius such as Enceladus to over $3 \times 10^9$ Pa (30,000 atm) at the centers of Ganymede and Titan.

Water ice, the principle constituent of many of these moons, is transformed into several new phases, in a complex way, at pressures greater than $10^8$ Pa (1000 atm). The simple ice that we are familiar with, Ice I, completely re-orders its crystal structure, changing many of its properties such as its density and its melting point. One high-pressure phase of ice, Ice VII, has a melting point higher than the boiling point of water under normal conditions! Figure 12.12 shows the phase diagram for water ice; superimposed are the internal temperature and pressure conditions predicted for the Jupiter moon Callisto.

Second, consider the temperature inside a moon. How might the interiors of these moons heat up? The energy of formation may have melted the ice in these moons, allowing the rocky portions to settle into a core. Rocky material in these moons contain radioactive isotopes such as $^{40}K$, $^{232}Th$, $^{235}U$, and $^{238}U$ that give off heat as they decay. Recurring tides in a moon can also be an important source of heat; the materials in the moon pulled up into a tidal bulge can rub together, producing a considerable amount of frictional heating.

As the temperature and pressure change, the density changes of the material inside the moon can cause it to expand or contract. Melting and other phase changes in ice produce even greater changes in density. The moon shrinks or grows to accommodate these changes in density, producing tectonic features such as cracks and grooves.

Molten water (or a eutectic melt involving water and ammonia) has a lower density than the ice and rock mixture of a moon's crust; this density gradient forces the water upward until it produces flooding on the surface of the moon. The water and ammonia mixture (shown in Figure 12.13)

is especially interesting in this regard, because it melts at a mere 173 K, a full 100 K colder than ice melts; any moon that gets warm enough to produce it is likely to have a surface covered with watery "lava flows."

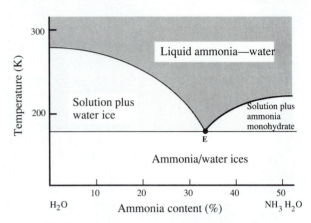

**FIGURE 12.13**   A simplified phase diagram for the ammonia–water system. (Phases near the eutectic point *E* are actually more complex than shown here.)

Because ice flows relatively easily (it has a lower viscosity than rock, for instance), the solid ice itself can convect (see the box). Such convection not only cools off the interiors of these moons; it may also lead to stresses on the surface visible as tectonic features.

As tectonic processes and craters raise topographic relief on the surfaces of these bodies, this relief is subject to relaxation. Icy mountains, with their low viscosity, tend to slump back into plains. The features also may become isostatically compensated, just like terrestrial mountains (see Chapter 4); the floors of some icy craters appear to be bowed upwards as a result of isostasy.

As can be seen, the evolution of these moons mirrors the processes seen in the larger terrestrial planets. In addition, there are certain effects on icy moons that have no terrestrial analogue.

For example, most of these moons have no atmosphere, and thus the ices on their surfaces are able to evaporate slowly into space. Furthermore, high-energy particles and solar UV radiation chip off ices or cause them to react chemically, much as gases in the atmospheres of planets are affected.

## Simple Thermal Models

To attempt to understand in detail how a specific moon evolved, we construct a mathematical **model**. The idea of keeping track of heat inside a moon or planet is, in essence, quite similar to the sort of bookkeeping that an accountant does.

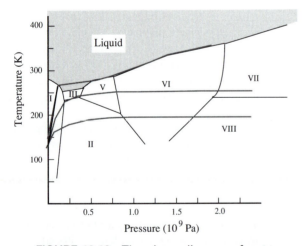

**FIGURE 12.12**   The phase diagram of water, as a function of pressure. As a large moon grows, increasing pressure deep inside can change ordinary Ice I into Ice II, Ice III, Ice V, and other ice phases. Dashed lines represent model pressure and temperature conditions at three different times in Callisto's early history.

We know the size and density of a moon; using the techniques discussed in Chapter 4, we can make a reasonable guess at its bulk composition. Given its size and composition, we try a reasonable starting state and keep track of the energy (heat) acting inside the moon. We note where the heat is coming from and how it flows from one part to another, at a series of points inside the body, for a progression of times in its history. When heat is generated at a given point inside the moon, or when more heat is conducted to that point than can be carried away, then we conclude that the temperature rises at that point. When the temperature at that point reaches the melting point of one of the body's ingredients, then we assume that the heat goes into melting that substance. When the heat is carried away faster than it arrives, the process is reversed.

With a computer we can keep track of hundreds of different spots inside the body and see how they warm up or cool off as we step through 100,000 increments of time in the body's history. If the final result predicted by our model looks vastly different from the real body, we go back and try to figure out why. Did we neglect some heat source? Were the initial conditions we assumed incorrect? Then we make the appropriate changes and try the model again.

What specifically goes into such models? We need the size and composition of a moon or planet, we must determine where the heat comes from, and we must calculate how it flows inside the body.

## Sources of Heat

We discussed accretional heating and decay of radioactive isotopes in Chapter 5. How do these processes heat up moons?

Consider accretional heating. Recall that we found asteroids and moons to be too small for this heat source to be important. Our argument was that, even falling from infinity, material would be so little attracted by the weak gravity fields of these small bodies that little overall heating could result. However, real moons (and planets) don't accrete by themselves from infinity. Rather, they form when a collection of planetesimals, all in orbit around some planet or star, coalesce. They will collide at velocities determined by the gravity of the planet they are orbiting, not the gravity of the accreting moon. Thus it is possible that accretional heating may have been important after all.

Radioactive heating occurs on icy moons, just as it does in rocky planets. But now we must be careful to remember that the bulk of these moons are ices, which rarely have radioactive isotopes. (One exception may be ammonia-bearing ices, where occasionally a $^{40}K$ ion can substitute for an ammonium ion.) The amount of heat produced by this process will be directly related to the amount of rock present in the moon. The more ice that is present, the less the radioactive-bearing rocky material is available to warm the moon's interior.

## Tidal Heating

A third process that can heat moons is tidal stresses. This is not an important heat source for most planets, because most planets are so much larger than their moons that the tidal energy dissipated inside them is negligible. (An exception may be Pluto, whose moon Charon is half its radius.) However, the same energy applied to a relatively small moon can result in significant heating.

A moon causes a tide to be raised on the planet it orbits, and likewise the planet raises a tide on the moon. Recall that the strength of the tidal force varies as $1/r^3$, inversely as the cube of the distance between the two bodies, so a small change in the distance between the two bodies (which happens if the moon has an eccentric orbit) quickly alters the size of the tidal bulge. Furthermore, the bulge moves around on the surface of the moon if the moon is not tidally locked with one side always facing the planet. The motion of this bulge, in and out or from point to point on the surface, means that some material in the moon is alternately being squeezed and released. This dissipates energy, as heat, inside the moon.

The actual mathematics describing how much energy is dissipated is remarkably complicated. The most important factors are rate and extent of the change in tidal force and the physical strength of the body being flexed. A simplified formula to estimate the heat input into a moon is

$$\frac{dE}{dt} \approx \frac{2\rho^2 R^7 n^5 e^2}{\mu q} \qquad (12.3)$$

where $\rho$ is the density and $R$ the radius of the moon; $n$ is the mean motion and $e$ the eccentricity of its orbit; and $\mu$ is the rigidity and $q$ the quality factor of the material that the moon is made of. Typical values for a solid moon (like Io) put $\mu$ at $6.5 \times 10^{11}$ N/m$^2$ and $q = 100$. Once the moon's interior melts, the amount of heat produced by tidal heating increases; when the rigid top of the moon is only 5% of the radius of the whole moon, the heating reaches a maximum, more than 10 times the original heat input. Thus once tidal heating starts to weaken the inside of a moon, it increases its power and causes even further melting to occur until all but the top 5% is melted. This process is sometimes called "runaway melting."

The energy heating the moon has to come from some place; generally it comes from the kinetic energy of the moon's orbit. But robbing energy from an orbit tends to make the orbit circular and noninclined, until the tidal strength no longer changes and so the heating stops. Only if the moon feels another force, like the gravitational attraction of another moon in resonance with the first moon, is there a significant, long-term heating process.

Also, the heating depends on how rigid the moon is. Once it starts to melt it loses rigidity, and thus the nature of the heating changes. Molten moons tend to flex more, but only the solid parts feel enough internal friction to get heated by this flexing. Thus a partially molten moon sees increased heating in its solid crust but decreased heating in its molten interior.

Finally, it turns out that the strength of the heating is a complicated function of the tide strength. The heating dies away as $1/r^{7.5}$, where $r$ is the distance from the moon to the planet. Doubling the distance from the planet causes the heat to drop by $2^{7.5}$, or a factor of 180, so the effect becomes very small quite quickly as one moves away from the planet.

Tidal heating is most important on Io, a moon close to Jupiter and in resonance with Europa, and to a lesser degree on Enceladus, a moon of Saturn in resonance with Dione. It may also be important on Europa itself (which is in resonance with Ganymede), and some studies suggest that Ariel might have been in resonance with Titania, 3 to 4 billion years ago.

## Modelling Heat Flow

Heat can be generated (by radioactive decay, for instance) and spent (as it is radiated from a planet's surface), and it flows in and out of each region inside a planet, just as money may flow in or out of a bank account. This sort of model is sometimes called a **heat budget**.

Once heat is produced, it flows according to the laws of conduction. Heat flows only from hotter regions to cooler ones. The greater the difference in temperature, the faster the heat flows. Likewise, the greater the conductivity of the rock between the two regions, the faster heat flows between them. These ideas can be expressed mathematically; once this is done, a computer model can be used to predict the temperature at any point inside a planet at any given time in the planet's history.

The general formula for heat flow is

$$\frac{\partial T}{\partial t} = \kappa \frac{1}{r} \frac{\partial^2 (rT)}{\partial r^2} + A \tag{12.4}$$

The term $\partial T/\partial t$ is the change of temperature with time; $\partial^2 (rT)/\partial r^2$ is the rate at which the change in temperature with distance changes; $\kappa$ is the **thermal diffusion coefficient**, the rate at which heat can diffuse through material (it depends on both the material's conductivity and its heat capacity); and $A$ is the rate at which heat is generated in the system.

In principle, this equation is very difficult to solve without a computer. A special case, which is much simpler to solve and is often a very good approximation, is to assume that the heat flow is in a **steady state**: that heat is being carried to the surface and radiated away at exactly the same rate as it is being produced.

Under this assumption, the temperature inside the planet at any given place would stay the same for all time. All planets are trying to evolve to this steady state; if heat sources were constant with time, then after a sufficiently long period the interiors would settle down to such a steady state. Once they've finished accreting, the main heat source in most planets is radioactive decay, which is always decreasing with time; but the change is so slow that the approximation of steady state is usually quite good.

If the temperature does not change with time, then we can set the term $\partial T/\partial t = 0$ in the equation above and solve for $T$ as a function of radius. A simpler way to see what's going on, however, is to write a new equation, setting the heat flow out of the planet equal to the heat generated inside the planet.

If radioactive decay generates heat at a rate of $Q$ J / kg $\cdot$ s (that is, every kilogram of material generates $Q$ joules of energy each second), then the total amount of heat generated in a uniform spherical planet is

$$\text{Heat in} = Q \cdot \tfrac{4}{3}\pi \rho r^3 \tag{12.5}$$

A value for the $Q$ of meteoritic material can be derived from the data in Table 5.3 and the cosmic abundances given in Chapter 2. The density term, $\rho$, tells us the number of kilograms there are per cubic meter, and the other terms give us the total number of cubic meters there are in the planet. Thus this combination of terms gives us the total number of joules produced inside the planet each second.

Next, we look at the rate in which heat flows out of a spherically symmetric shell of thickness $dr$ inside the planet. Heat always flows from hotter areas to cooler areas and so the bigger the difference in temperature, the faster the heat flows. The temperature gradient $dT/dr$ is a measure of how many degrees the temperature changes across this shell. (Note that because $T$ is now only a function of position, $r$, and not of time, we no longer need to use the $\partial$ symbol for partial derivatives.) The **conductivity** $K$ measures how many joules of energy can pass through a square meter of surface each second for each degree of temperature that the body is hotter on one side of the surface than on the other. Finally, we multiply all these terms by the surface area of a sphere, $4\pi r^2$, which is how much area we have for heat to pass through. Thus we arrive at

$$\text{Heat out} = 4\pi r^2 K \frac{dT}{dr} \tag{12.6}$$

The heat flows out of the planet and radiates into space at the planet's surface. Calculating the precise rate at which heat is lost from the surface is tricky, because the radiation rate itself depends on the temperature of the surface to the fourth power (as noted in Chapter 10). However, for interior thermal models of planets such a detailed calculation is not really necessary. The radiation rate is orders of magnitude more efficient than the rate at which heat flows to the surface. Except for the gas giant planets as noted above, the temperature of the surface is really determined by the amount of sunlight it absorbs. Thus we can simply fix the surface temperature, $T_s$, to be the equilibrium temperature (as determined in Equation 10.4).

With this assumption, we can use the equations above to calculate the temperature anywhere else inside the moon. Our steady-state postulate says that there can be no accumulation of heat; the heat generated in a region must be balanced by the heat flowing out of the region. Thus, setting Equations 12.5 and 12.6 equal, we arrive at the equation

$$\frac{4}{3}\pi\rho Q r^3 = 4\pi r^2 K \frac{dT}{dr} \tag{12.7}$$

The term $4\pi r^2$ can be divided from each side, resulting in our final equation:

$$\frac{dT}{dr} = \frac{\rho Q r}{3K} \tag{12.8}$$

This can be integrated easily, if we assume that $Q$, $K$, and $\rho$ are constant. (That's not a good assumption for major planets, but acceptable for small bodies.) The temperature at any point a distance $r$ from the center of this body becomes

$$T = T_s + \frac{\rho Q}{6K}(R^2 - r^2) \tag{12.9}$$

where $T_s$ is the temperature at the surface of this body and $R$ is its total radius.

Notice one important result in this equation. The temperature inside a body depends on the square of its radius. Large bodies are less efficient at losing heat than small bodies. The bigger the moon or planet, the more likely it is to melt.

As the rocks or ices inside a moon warm up, they eventually may get hot enough to melt. It takes heat to melt a rock, so heat that would otherwise go to making the body hot must be absorbed instead by the rocks that are melting. (The same principle keeps drinks cool in the summer. By putting ice in a drink, you keep it cool because all the heat that might go into warming the drink is absorbed by the melting ice instead. Once the ice is melted, though, the drink doesn't stay

cool very long.) As the various minerals melt, or chemically react with one another, heat may be absorbed or added to the body. This all contributes to the heat budget, of which a good detailed model must keep track.

Obviously, our steady-state model ignores all these effects. It is a very simple model. It's useful for making guesses for the thermal structure of small moons and asteroids; it's clearly inadequate if the body in question is likely to undergo any melting, internal chemical reactions, or external heating. But if this simple model predicts that the interior of the moon or asteroid is not likely to ever get hot enough to melt or react, then this model is probably just as good as any more complicated computer program.

## Results of Thermal Models

Given all these factors, what can thermal models tell us about the history and evolution of the moons and planets in our solar system today? A full evolutionary history of each body is beyond the scope of this book, although as an example we end Chapter 14 with a sketch history of Io. But some general comments can be made.

Most moons and asteroids have radii of less than 1000 km. For bodies this small, it appears to be unlikely that either accretional heating or long-lived radioactive elements in the rock can produce enough heat to cause the ice or rock to extensively melt and evolve.

Nonetheless, a few asteroids and small moons do show evidence of some sort of melting. Vesta, for instance, is an asteroid whose surface resembles basaltic meteorites; Miranda, Enceladus, and Ariel are all small icy moons that show clear evidence of extensive melting and tectonic activity. Thus other heat sources must be postulated for these bodies. Such heat sources include the dissipation of tidal energy within these bodies, which may be quite significant for Io, Europa, Enceladus, and possibly Miranda and Ariel, explaining the unusual amount of geological activity found on these particular moons (Figure 12.14). This is not a likely heat source for Vesta, however.

Potassium, uranium, and thorium are all known to exist in rocks today. They produce enough heat to melt rock inside terrestrial-sized planets but not asteroids. According to detailed time-dependent computer thermal models, melting from these heat sources should take place about a billion years after the body was formed. As confirmation of this prediction we note that mare basalts on the Moon, for example, were formed from molten lavas inside the Moon about a billion years after the Moon was formed. The heat from the decay of long-lived radioisotopes should lead to complicated thermal histories for the larger moons, such as Ganymede, Callisto, Titan, and Triton. That Ganymede shows such evolution, while Callisto does not, is still not well understood.

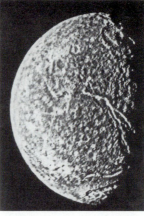

**FIGURE 12.14**   Ariel (left) and Titania (right) both show signs of complex thermal evolution.

Very early in the solar system, $^{26}$Al and $^{60}$Fe may have been very important heat sources, and evidence for these radionuclides has been reported in some meteorites. These isotopes would have completely melted even asteroid-sized bodies, but such bodies would soon freeze again as the isotopes quickly decayed away. Vesta may have been heated by short-lived radioisotopes; another more exotic heat source for Vesta might have been electric currents generated by super-strong solar magnetic fields that may have existed early in the Sun's history.

The thermal evolution of the terrestrial planets has also been modelled. From these calculations we derive a picture of Mercury as a body that underwent early episodes of heating and expansion, followed by cooling and contraction, to explain the scarps described in Chapter 4. Mars, a planet known to contain numerous ancient volcanoes, was likewise probably thermally active in the first 2 billion years of its existence, before it had radiated away most of its heat. Both Mars and Mercury are expected to be thermally inactive today.

Many detailed, complex, and controversial thermal models have been proposed for the larger planets Venus and Earth. Beyond noting the almost certain extensive internal convection continuing to the present, the details of these models are beyond the scope of this text.

In all these models, an important point must be kept in mind. By their nature, computer models use numbers and so they often can appear to be quantitative and precise. They are not. Our ideas of the real values of the conductivity, heat capacity, extent of convection, and many other important parameters inside any moon or planet are never better than rough guesses. The number of phase changes and chemical reactions that can go on inside a body, all of which add or subtract heat to the energy budget, provide another uncertainty that is difficult to model. And, of course, the more complex the computer model is, the more open it is to logic or programming errors.

Thermal models can show us general trends: For example, they tell us that internal heat sources cannot account for the activity seen on some small moons, so other heat sources must be investigated. Thermal models can make predictions about the general structure of a moon or planet's interior, such as the existence or general size of a lithosphere (big or small, early or late in its history). But precise numerical statements ("the lithosphere must have been 11.3 km thick 3.2 billion years ago") are to be taken lightly. Although computer models use numbers, they are not quantitative but merely qualitative indicators.

## SUMMARY

Moons and planets can be heated up by accretion, decaying radionuclides, and in some cases the dissipation of tidal stresses.

Once the heat is generated, it flows from the warm interior to the cold surface of a body, where it can radiate into space. If the temperature inside a body is warm enough, the material there may melt and erupt onto the surface, forming new surface rocks or ice flows. Even before it melts, it is possible that the interior of a body may start to convect. This convection stirs up the chemical elements inside a body, fostering chemical reactions. It cools off the deep interior, bringing this heat up to warm the parts of the body closer to the surface. And it can break up pieces of the surface of a body, moving them about to form cracks and grooves, mountains and valleys.

## STUDY QUESTIONS

1. Several moons may have been heated by the dissipation of tidal stresses. Name two.

2. Name one advantage, and one disadvantage, to steady-state thermal models.

3. Consider two icy moons of similar size. If one has been heated up, name one thing you'd expect to see on its surface that would not be present in the unheated moon.

4. What chemical might be present in an icy moon that would lower the melting point of ice and encourage flows of liquid on its surface?

5. Name one moon that shows evidence of considerable thermal evolution.

## FURTHER INFORMATION...

# Solid-State Convection

A parcel of molten lava can move itself, and the heat it carries, faster than the heat can flow out of the lava. Thus as a lava moves upwards it carries its heat with it, removing that heat from the interior. Recall that this **convection** is how planetary atmospheres and regions inside the Sun transport heat.

Warm rock can start convecting even if it isn't molten yet. This happens when the strength of the energy flow is greater than the strength of the rock. This solid-state convection occurs if the temperature gradient is very steep, if the viscosity of the material is low, or if the system is very large with the temperature changing over a very long distance.

One way to predict whether or not convection takes place is to look at the Rayleigh number, defined in Equation 10.32. If heat is produced within the convecting region at a rate $Q$, where $K$ is the thermal conductivity and the other terms are as defined, the Rayleigh number becomes

$$Ra_Q = \frac{\rho^2 g \alpha Q d^5}{\eta \kappa K} \qquad (12.10)$$

If the Rayleigh number is much greater than a few hundred, then one can expect convection to take place.

An important point to notice in both definitions is the strong dependence on $d$, the thickness of the convecting region. A thermal gradient across a region like the mantle of a planet is likely to be unstable against convection, even if the viscosity $\eta$ of the warm rock is on the order of $10^{20}$ or higher, simply because the planets are so big. On the other hand, smaller moons and asteroids may be much more stable against convection.

Our models for convection in an atmosphere (Chapter 10) treated the atmosphere as if it were an ideal gas. Obviously, warm soft rock behaves rather differently. Furthermore, atmospheres can be treated (locally, at least) like simple slabs of air; convection inside a body involves motion in three dimensions across regions of changing gravitational acceleration $g$. It is quite challenging to model such flow. As a result, the temperature in a convecting region of a body can be very difficult to predict.

An easy way to model convective heat flow is to assume that a convecting region of a body transports heat just as if it were conducting heat, but with a conductivity some factor greater than the nonconvecting regions. This factor, the ratio of heat flow due to convection versus heat flow by conduction, is called the **Nusselt number**. For many simple thermal models, merely assuming a Nusselt number of 10 or 100 can give reasonable results; a better approximation sets

$$Nu \approx \sqrt[3]{Ra} \qquad (12.11)$$

Recall, however, that even our value of $Ra$ is only an approximation.

In many cases solid-state convection is the most efficient way to carry heat up from the interior of a body. It also mixes components inside a body, it carries the products of chemical reactions from the interior of a body to its surface and back again, and it can even push bits of crust around the surface from place to place. Convection is the way that the energy of heat in a body physically transforms the interior and surface of that body.

## 12.4 PROBLEMS

1. There's a reason why each planet's moons bear the names they are given. Most of the names are of characters from mythology. Try to identify at least one character from each planet's moons. (An encyclopedia or a copy of *Bulfinch's Mythology* might help.) What seems to be the logic behind each set of names?

2. Which moons do you think may have significant amounts of ammonia in them, judging from their surface features? Which moons may have organic compounds?

3. Tides are symmetric. The bulge on the far side of a planet is almost exactly the same size as the bulge underneath the tide raiser. Prove this statement mathematically. (Hint: Solve for the difference in force from the center to the far side of the planet, and show that the result is similar to the difference in forces, derived in the chapter, between the center and the near side of the planet.)

4. What is the acceleration of the Sun's gravity at the center of Venus? At the near side of Venus? At the far side of Venus? Is the difference in these forces (center minus near and far minus center) nearly the same?

What effect do these tides have on the atmosphere of Venus?

5. If all Phobos's kinetic energy is added to Mars when it crashes into the surface, how much will this change the length of the day on Mars? (Hint: What is the orbital kinetic energy of Phobos? Compare it to the kinetic energy of Mars's spin. The kinetic energy of a uniform spinning sphere is $\frac{1}{5}Mv^2$, where $M$ is the mass of the sphere and $v$ the speed at its equator.)

6. A popular best seller in the 1970s called *The Jupiter Effect* predicted dire consequences when Jupiter, Venus, and Earth were in conjunction and the effects of their tides on the Sun were combined. Which of these three planets raises the greatest tide on the Sun? How big is the effect of the other two tides, by comparison? Are you surprised that, contrary to the predictions of this book, the world is not wracked with earthquakes when the planets approach such an alignment?

7. The Io–Europa–Ganymede resonance is stable. The effect of the mutual perturbations of these three moons keeps the three of them in their resonances. But how could such an odd coincidence of orbits occur in the first place? (Hint: Assuming that all three moons formed out of resonance to start with, describe how tides raised on Jupiter by these moons could have changed their periods until the resonance was formed. Which moon would have changed its period most quickly? Which resonant pair would have been formed first by tides, Io–Europa or Europa–Ganymede? Why?)

8. In a science fiction novel by Robert Heinlein, *Farmer in the Sky* (about a boy who goes to work on a farm on Ganymede), a giant "Ganymede-quake" occurs when the 4 Galilean moons of Jupiter are lined up on the same side of Jupiter. Comment on the scientific accuracy of this story.

9. The *Voyager* spacecraft observed that Io was giving off energy at a rate of $2 \, \text{J/s} \cdot m^2$. By contrast, the Apollo astronauts measured the heat flow out of the Moon to be $0.02 \, \text{J/s} \cdot m^2$.
   (a) How much heat is radiating altogether from each moon?
   (b) If these moons have been radiating heat at this rate for $4.6 \times 10^9$ y, how many joules of heat has each moon radiated?
   (c) How do these energies compare with the kinetic energy in the orbit of each of these moons? With the rotational energy of the planets that they orbit?
   (d) If the surface of the Moon is at 300 K and that of Io is 100 K, what is the temperature at a depth of 1 km inside each moon? Assume the conductivity of rock on both moons is $5 \, \text{J/m} \cdot s \cdot K$. (Hint: For a distance of merely 1 km, you can assume that the heat flow and thermal gradient are constant; just

set "heat out" to the observed heat output derived in part *a*.)

10. Of the seven planet-sized rocky bodies in the solar system (Mercury, Venus, Earth, Moon, Mars, Io, and Europa), which ones do you expect have had vigorous convection in their interiors? Do you think any are convecting today? (Hint: Consider the surface features seen on these bodies.)

11. *Computer Spreadsheet Problems.* Using the data given in the appendix for the five largest moons of each planet and assuming spherical moons where necessary:
    (a) Find the ratio of masses for each moon to the mass of its parent planet. Is there any range of mass ratios that seems to be favored?
    (b) Compare the distances of the moons from their parent planets in kilometers and also in terms of the parent planet (for example, the Moon is 30 Earth radii from Earth; Io is 5 Jovian radii from Jupiter). By each standard, which moons are closest and which farthest from their parent planets? Is there any range of distances that seems to be favored?
    (c) What moon has a surface area closest to the area of the continental United States? Of the state of Texas? Of Manhattan?
    (d) The speed that a satellite must attain to reach a circular orbit about its main body equals $\sqrt{GM/r}$, where $G$ is the gravitational constant ($6.67 \times 10^{-11}$ in SI units), $M$ is the mass of the body being orbited (in kg), and $r$ is the distance from the center of the body being orbited (in meters, not km!). A major league pitcher can hurl a fast ball at 40 m/s (90 mph). From which moons could such a pitcher hurl a baseball into orbit?

## 12.5 FOR FURTHER READING

The University of Arizona series has produced three books on satellites. The first, now somewhat dated (1977) but still containing valuable background papers, is *Planetary Satellites,* edited by Joseph Burns. In 1982 a special volume on Jupiter's moons, *Satellites of Jupiter* was edited by David Morrison. In 1986 the volume *Satellites* appeared, edited by Joseph Burns and Mildred Shapley Matthews. All books are published by the University of Arizona Press in Tucson.

An advanced discussion of tides from the point of view of their geophysics can be found in advanced geophysics textbooks such as Frank D. Stacey, *Physics of the Earth* (New York: Wiley, 1977). Chapters on tides and resonances also appear in the University of Arizona Press books.

No good single reference for modelling the thermal histories of moons or planets exists, to our knowledge, and even technical papers published on the subject tend to emphasize results, not techniques. The granddaddy of all icy satellite models, however, is John S. Lewis's paper "Satellites of the outer planets: their physical and chemical nature," which appeared in *Icarus*, volume 15, pp. 174–85 (1971). Written years before the *Voyager* encounters, it anticipated nearly every major development in icy satellite studies. Our approach to steady-state thermal models is derived from this paper.

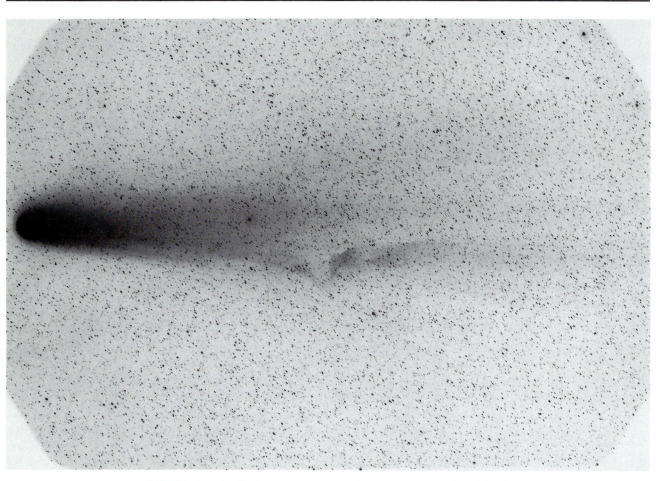

**FIGURE 13.1** Halley's comet, as seen from Easter Island on March 21, 1986.

# CHAPTER 13

# Comets and the Solar Wind

## 13.1 COMETS

In previous chapters we have explored the Sun, the planets, and their moons. We now turn our attention to the space between the planets.

What lies between us and the planets? Medieval philosophers thought that a pure vacuum was an impossibility, and nineteenth-century physicists agreed, arguing that for light waves to travel from the stars to us, they must pass through a medium (just as water waves need water or sound waves need air). Experiments in the latter part of that century (most notably the experiment of Michelson and Morley of 1887) failed to find any evidence of this medium, however. Relativity and quantum physics were responses to the need to explain how light could travel through empty space.

But "outer space" is not really empty. It is filled with tenuous plasmas, tiny dust particles, and electromagnetic fields. How do we know this? How can we study it? And how does it affect the planets?

Our first clue to the nature of space between the planets came from observing comets. Comets' orbits, like those of interplanetary space probes, are subject to gravitational forces from both the Sun and nearby planets. Comet tails provided the first evidence for the plasma from the Sun, called the **solar wind**, that fills interplanetary space. And

the interaction between comets and the solar wind, which produces those tails, is but one example of the interactions possible between the solar wind and planets.

### The Nature of Comets

In a world where daily life is fraught with uncertainty, where a farmer's crop can depend on the whims of the weather or a city's future can be changed by the random act of a far-off king, the predictability of the stars can be a great solace. A homesick traveller can look up at night and see the same stars she grew up with; a mystic can find a perspective on the greater meaning of life by contemplating the subtle, unchanging points of light in the night sky.

Thus the appearance among these tiny, pure, predictable sparkling stars of a garish and ominous cloud of light, 20 times the size of the Moon (or even larger!), arriving without warning, moving without reason through any part of the sky, changing its appearance from night to night, disappearing after a week or so into the sunlight and then reappearing again inevitably had a startling and awe-inspiring effect on anyone familiar with the predictability of the stars. A person might see at most one or two such apparitions in a lifetime. They came, like bad news, completely by surprise, unsettling and unexpected.

These balls of light, marked by long wispy clouds of luminescence trailing off in a direction always pointed away from the Sun, looked like "hairy stars" to the ancients and thus they were named, from the Latin word for "hair," **comets**.

Classical astronomers were torn between two possible theories about the nature of comets. One school (including Aristotle) held that comets were atmospheric phenomena, clouds of fire in the upper air. Others, including the Roman poet Seneca, believed that they were objects travelling among the planets themselves. In 1577 Tycho Brahe and the German astronomer Michael Mästlin observed a comet against background stars from two different sites in Europe; that this comet showed no visible parallax (recall Chapter 8) indicated that it was located at a distance much farther away than the Moon, suggesting that it was a traveller out in space.

Nonetheless, the idea that comets were connected with Earth's atmosphere continued to be hotly debated. In Galileo's book *The Assayer*, where he promoted his philosophy of experimentation and rational proof as the foundation of science, he also included a long diatribe of sarcastic attack against his astronomical rivals who were foolish enough to believe that comets moved among the planets!

Isaac Newton's physics and Edmund Halley's calculations were the final proofs that comets indeed travelled in orbits subject to the same forces and obeying the same laws as the orbits of planets. Although most comets appeared to have orbits indistinguishable from parabolas (recall Chapter 9), Halley showed that several comets recorded in history were in fact recurring visits of the same comet whose orbit he calculated after it appeared in 1682. He predicted its return visit in 1758. In a triumph of astronomy, the comet indeed appeared on Christmas 1757 and reached its closest approach to the Sun only a month off from the date that Halley calculated. This comet now bears his name (see Figures 13.1 and 13.3).

With the improvement of mathematics (especially due to the German mathematician Carl Friedrich Gauss), astronomers in the early nineteenth century finally began to plot comet orbits with considerable accuracy. Johann Encke, a student of Gauss, noted that faint comets seen in 1786, 1795, and 1805 all had similar orbital elements. He determined that it was the same comet, one whose period was in fact a mere 3.3 years! Encke's Comet is still the comet with the shortest known period.

Comets such as those named after Encke and Halley are **short-period comets**. With the invention of the telescope, fainter comets could be seen; nowadays some five to ten new comets a year are discovered, mostly by amateur astronomers. The majority of them are **long-period comets**, whose orbits carry them past the Sun once in a million years. Another difference between short-period and long-period comets is the inclination of their orbits. Most short-period comets lie close to the ecliptic plane, the plane of the planets, but long-period comets can occur anywhere in the sky and are as likely to be in retrograde orbits as in prograde orbits.

Unpredictable as they are, long-period comets are also more likely to be bright and active. But only rarely, about once a decade or so, will astronomers in some part of the world be treated to the sight of a bright naked-eye comet.

## Comet Structure

Comets are balls of dust and volatile ices that expand into a large cloud of gas when they come within about 3 AU of the Sun. This gas becomes ionized by solar ultraviolet (UV) radiation; the ions and the dust are swept out into a tail by the **solar wind**, a gas of protons and electrons blowing out from the Sun.

By studying comets with telescopes, radar, and most recently spacecraft, we've been able to get a fairly good picture of how a comet is put together. By far the best view of a comet came in 1986, when several nations launched space probes to Halley's Comet.

For most of its quiescent existence out beyond the planets, a comet is simply a small piece of ice and dust roughly 1 to 10 km across. This iceball becomes the **nucleus** of the comet as the comet approaches the Sun and the warming ice starts to evaporate very rapidly. The gaseous water, and other chemicals such as cyanide ($HCN$) and methyl cyanide ($CH_3CN$), form a cloud of gas called a **coma**, extending some 10,000 km about the nucleus. At the outer edge of this coma, the solar UV light starts chemical reactions leading to the formation of molecules such as CN, CO, NH, OH, and other very reactive species. Still farther out, beyond 1 million km from the nucleus of the comet, these substances become ionized. $H_3O^+$, $CO^+$, $N_2^+$, $CH^+$, and $OH^+$ have all been observed.

Because these ions have a net electric charge, they can be moved about by the Sun's magnetic field. This is the brilliant **ion tail** of gas seen from Earth, which extends tens of millions of kilometers away from the comet, always pointing in a direction away from the Sun (Figure 13.2).

In the nucleus of the comet, where the ice is evaporating, some amount of rocky dust is mixed into the ice. As the ice boils off the surface, the dust is free to move away from the comet as well; in some comets, this second distinct **dust tail** is also visible. It too is pointed away from the Sun. Because the dust particles are much larger and heavier than individual molecules or ions, it is not surprising that they would respond much more sluggishly to any forces acting on them. Indeed, the two tails tend to separate and point in slightly different directions. This dust is not ionized, as is the ion tail, so it is not affected by the Sun's magnetic field. What causes it to be pushed away from the Sun?

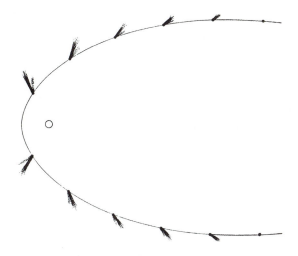

**FIGURE 13.2** The tail of a comet is pushed away from the Sun by the pressure of the solar wind. Ionized gas is carried by the Sun's magnetic field and is more easily moved than dust, so as the comet approaches the Sun the gas tail and the dust tail are pushed in slightly different directions and separate.

**FIGURE 13.3** Halley's comet, as photographed by the *Giotto* spacecraft at a distance of 18,000 km. The nucleus is the dark object to the right of the photograph; the darkest spot may be just a spot on the nucleus. The bright regions are jets of gas reflecting sunlight.

One possibility is that sunlight itself can exert a force on the grains of dust. The difficulty is that most of the dust grains are so tiny that they are actually smaller than the wavelength of visible light (about 0.5 $\mu$m). In such a case the force of sunlight would not be very efficient at moving the dust in any direction, simply because most of the energy in the sunlight bypasses the dust altogether. There must be something else coming from the Sun that is moving this dust tail around. These dust tails thus were our first evidence for the existence of a **solar wind**.

## The Comet Nucleus

With the return of Halley's comet, astronomers finally got a close-up view of the nucleus of the comet itself, the chunk of material that is the source of the gas and dust visible from Earth. The nature of the nucleus turned out in many ways to be quite different from earlier ideas.

Before Halley was visited, most astronomers referred to comets as "dirty snowballs", because both ice and dust were seen to be emanating from them. However, actual photographs of the nucleus by the *Giotto* spacecraft, launched by the European Space Agency, showed that the nucleus of Halley was remarkably dark, reflecting only 2.7% of the light it received from the Sun. It has been referred to as a "charcoal peanut" (Figure 13.3). As well as being darker, it was also larger than expected. Instead of the expected radius of 1 km, Halley was closer to 10 km across, similar to the size of Mars's moons. How could this dark, rocky appearance be consistent with the large quantities of volatile materials seen in the coma and tail?

The answer to this question seems to be found in the *structure* of the material that forms a comet, not just its chemical composition. We noted above that dust tails are sometimes seen coming from comets; indeed, the results from *Giotto* indicated that very tiny dust particles, less than $10^{-8}$ m in diameter, were even more evident than had been expected. In a comet nucleus these tiny particles must be glued together by ices; apparently this produces a very loose, fluffy network of grains and gaps. Thus the darkness of Halley's nucleus is at least partly explained because the light that hits the nucleus does not hit a smooth surface, but hits one that is full of holes that trap light, much as a velvet cloth traps light that a flat black sheet of cloth might reflect. Indeed, solid material may make up less than 20% of the volume of the comet; the rest might just be empty space.

Furthermore, dark dust stays behind as the ice evaporates, coating the surface of the comet with a rocky "armor" that prevents the ice deeper down from escaping. When this happens, the production of a tail from that part of the comet stops. Like most objects in space, the nucleus of a comet spins; swirling pinwheel patterns can be seen in the coma near the head of the comet when one part of the evaporating ice breaks through this armor and streams out of a spinning

## FURTHER INFORMATION...

# Light Pressure and Solar Wind Pressure

Just as the wind on Earth moves dust around by pushing against its surface, in space **surface forces** arising from either plasma or sunlight striking the dust can accelerate the dust. Like all forces, the net acceleration that results depends on the mass of the dust particle; from Newton's second law, $a = F/m$. The bigger the particle, the more surface area it has, but the more mass it has, too. If we assume that particles are spherical, then we know that $F$ may increase as the surface area increases; $F \propto \pi b^2$ where $b$ is the radius of the dust particle. But the mass increases as the volume increases; $m \propto \frac{4}{3}\pi b^3$. Thus the amount a dust particle is accelerated should vary as $1/b$; the smaller the dust, the more it gets moved. In general, when we refer to dust we'll be thinking of particles whose radii are 10 $\mu$m.

Light from the Sun can itself push dust particles. The force of gravity, varying as $1/r^2$, attracts dust towards the Sun; the intensity of light pushing dust away from the Sun also varies as $1/r^2$. The solar wind, an outflowing of plasma from the Sun, also has a net outward pressure that drops as $1/r^2$. Thus the effect of **radiation pressure** and **solar wind pressure** is to counter the effect of gravity slightly. Dust particles in effect act as if they were orbiting a star slightly less massive than the Sun. For a given distance from the Sun these dust particles will orbit at a slightly slower than Keplerian velocity.

To derive the force available from this effect, consider first the solar wind. It flows with a velocity $v_{sw}$ and a density $nm_p$, where $n$ is the number of particles per cubic meter in the wind and $m_p$ is the mass of a proton, the primary component of the solar wind's mass. Its energy is kinetic energy; thus $\frac{1}{2}nm_p v_{sw}^2$ is the energy density of the wind. However, as we saw in Chapter 10, an energy density is the same as a pressure, a force per unit area. Multiply this by the cross-sectional area of the dust grain $A$ (equal to $\pi b^2$ for spherical grains), divide by the mass of the grain ($\frac{4}{3}\pi\rho b^3$ for spherical dust grains of density $\rho$), and you find the acceleration $a$ due to solar wind pressure on a stationary particle. Once the particle starts moving with the solar wind, of course, the wind pressure it sees will drop (falling to zero if the dust grain is moving exactly at the solar wind

speed). Thus we replace $v_{sw}$ with $v_{sw} - v_r$ ($v_r$ is the radial velocity of the particle) and arrive at

$$a_{swp} = \frac{3}{8}\frac{nm_p(v_{sw} - v_r)^2}{\rho b} \qquad (13.1)$$

For the pressure of sunlight, we make three adjustments. First, we replace the energy density term. Recall from Chapter 10 that the solar constant $F$ describes how much energy falls on a square meter in a second; this much energy must be contained in the light filling a volume 1 m$^2$ at its base, with a length equal to the distance light travels in 1 s. Hence the energy density of sunlight—the pressure it can exert—must be the solar constant divided by the speed of light.

Second, we must take note of an added complication with light. Light can only affect grains that absorb or scatter it. Dust grains smaller than the wavelength of light tend to forward-scatter light, rather than back-scatter. The efficiency, $Q$, with which small particles scatter light depends strongly on the optical properties of the material in question. As long as the dust grains are larger than a few $\mu$m, $Q$ is 1; for most materials, $Q$ falls to 0.5 by the time the grains have a radius of 0.5 $\mu$m (the wavelength of green light), but metallic grains can be sensitive to light at much smaller sizes. To account for this size variation, we multiply our acceleration by $Q$.

Finally, to account for the velocity of the dust grain itself, we multiply by $1 - v_r/c$, thus arriving at an equation for radiation pressure:

$$a_{rp} = \frac{3}{4}\frac{QF[1 - (v_r/c)]}{c\rho b} \qquad (13.2)$$

For particles much larger than a light wave, the radiation pressure turns out to be some four orders of magnitude larger than the solar wind pressure. However, for particles much smaller than a light wave, the radiation pressure drops to zero as $Q$ approaches zero, and so the impact of solar wind ions becomes important.

comet nucleus. However, because the comet is not made of well-packed ice and rock but is more like a sponge full of void spaces, this armor may have to be several meters thick before it can effectively trap incoming sunlight. Thus the outer surface of the nucleus may be free of ice but very porous.

## Interplanetary Dust

Cometary dust eventually spreads out through interplanetary space. Tiny craters can be seen in Moon rocks where this dust has hit the lunar surface. Specially equipped aircraft flying in the stratosphere of Earth can collect this dust as it falls into our atmosphere. By knowing how to look, it can be found in ocean sediments as well. Each dust grain tends to be a few microns in diameter (a millionth of a meter) and has a composition very similar to carbonaceous chondrites.

Interplanetary dust tends to be concentrated more or less along the ecliptic, the plane in which Earth and the other planets orbit the Sun. On very dark nights, it is possible to see with the naked eye a slight glow in the sky along the ecliptic, running through the constellations of the zodiac. This glow, caused by the reflection of sunlight off the interplanetary dust, is brightest in two areas. One, exactly opposite the direction of the Sun, is called the **gegenschein** (which might be translated, from the German, as "back-shine"), and the other, near the Sun, so that it is seen only near sunset and sunrise, is called the **zodiacal light**.

The *Pioneer* and *Voyager* spacecraft, which passed through the asteroid belt to the outer solar system, carried experiments designed to detect this dust. Hundreds of dust encounters were recorded, enough to show that dust exists, but not so much as to endanger the safety of future spacecraft.

Between the time when the dust forms a tail of a comet and when the other small forces acting on this dust spread it out into interplanetary space, the dust tends to be found in streams that follow the orbits of the parent comets. Earth crosses the paths of several of these streams of cometary particles every year. When this happens, we observe a meteor shower as these dust particles hit Earth's atmosphere and burn up. It is possible to associate certain annual meteor showers with particular comets; this is done in Table 13.1. These showers occur annually, around the dates given.

Notice that Halley and Encke appear in this listing twice: Earth passes through both the incoming and outgoing portions of their orbits.

Although the meteor rate of the Taurids is not as great as for other meteor showers, nonetheless a considerable amount of larger material appears to be associated with comet Encke. At least three known asteroids, Hephaistos, 1982 TA, and 1984 KB, have been identified with

**TABLE 13.1**

Meteor Showers and Comets

| Shower | Comet | Occurrence | Meteors/h |
|---|---|---|---|
| Quadrantids | 1491 I | Jan. 1–4 | 80 |
| Lyrids | Thatcher | Apr. 19–24 | 35 |
| Aquarids | Halley | May 1–8 | 40 |
| β Taurids | Encke | June 24–July 6 | — |
| Perseids | Swift-Tuttle | Aug. 9–17 | 40 |
| Orionids | Halley | Oct. 15–25 | 45 |
| Taurids | Encke | Oct.–Nov. | 5 |
| Leonids | Temple-1 | Nov. 15–20 | 40 |
| Geminids | Phaethon | Dec. 7–15 | 60 |
| Ursids | Tuttle | Dec. 17–24 | 5 |

Note: Showers are named for the constellation from which they appear to radiate. A comet is a candidate source of the meteor shower if its orbital elements are close to the meteors' orbital elements.

this meteor stream. Hephaistos is the largest of the Apollo (Earth-crossing) asteroids. Indeed, it has been estimated that 50 or more Apollo asteroids may be associated with this stream. In addition, in 1975 lunar seismograms recorded an unusual number of meteors hitting the surface of the Moon at a time when it was passing through this stream. The famous Tunguska impact on Earth, thought to be due to a small body that levelled thousands of square kilometers of forest when it hit Siberia in 1908, occurred during one of the Encke meteor showers; the impactor might also have been a fragment broken off from comet Encke.

## SUMMARY

Comets are collections of ices and dust. They have much more eccentric orbits than asteroids. When these objects approach the Sun, the ices evaporate into a gas cloud, which in turn becomes ionized by solar radiation and swept away from the Sun. Dust from these comets can also form tails, separate from the ionized gas tail. Meteor showers result when Earth passes through the path of a comet and collects this dust in its atmosphere.

## STUDY QUESTIONS

1. What causes the tail of a comet to be always pointed away from the Sun?

2. Why do some comets have two tails?

3. Give one reason why we think that meteors—so-called "shooting stars"—are caused by bits of dust from comets.

**FURTHER INFORMATION...**

# The Poynting–Robertson Effect

Besides moving towards or away from the Sun, the dust particles are also in orbit around the Sun. This motion produces what turns out to be the most important effect of all for perturbing interplanetary dust: the **Poynting–Robertson Effect**. (The British physicist John Henry Poynting calculated the momentum of light waves in 1884; the American astronomer Howard Robertson applied it to interplanetary dust in 1937.)

Consider an automobile driving through a rainstorm. Even if the rain falls straight down, the car's forward motion means that the front of the car is hit by more raindrops than the rear of the car. In the frame of reference of the car, it seems to stand still while the rain appears to be falling from the front to the back of the car, giving it a slight push backwards. The same effect applies to dust grains orbiting in sunlight. The light hits the dust with a component of momentum opposite to the direction in which the dust moves.

(An alternative but equally valid way of looking at the situation is to recognize that when the light hits the dust grain, it is first absorbed then reemitted in all directions. But the light emitted in the direction of motion is blue shifted, and the light in the trailing direction is red shifted, relative to its original wavelength. The recoil from the more energetic blue-shifted photon is stronger than that from the red-shifted photon.)

The net effect of the Poynting–Robertson effect, therefore, is to rob energy from orbiting dust particles. As a result, they will tend to circularize their orbits, and slowly spin into the Sun.

The equation describing the net acceleration on the dust grain is similar to that for radiation pressure. The fraction of the pressure that acts in the direction opposite to its motion is simply the ratio of its velocity to the velocity of the light photons. Hence

$$a_{PR} = -\frac{3}{4} \frac{QF}{\rho b} \frac{v}{c} \qquad (13.3)$$

Notice the negative sign, needed because the direction of the force is opposite to the direction of the dust grain velocity.

Dust of 10 $\mu$m at Earth's orbit will spiral into the Sun on a time scale of 1 million years. This is considerably less than the age of the solar system; it follows that any dust in the solar system today must be of relatively recent origin. Smaller dust moves even more quickly. However, the effect of $Q$ again limits the power of the Poynting–Robertson effect on grains much smaller than 1 $\mu$m.

An analogous effect occurs for solar wind impacts. **Solar wind drag** is much less effective for larger particles but may become important for smaller dust grains or around stars with a much stronger solar wind. (**T Tauri** stars, mentioned in Chapter 14, have a solar wind eight orders of magnitude stronger than our Sun's; the Sun itself may have gone through a T Tauri phase early in its history.) The equation for solar wind drag acceleration is

$$a_{swd} = \frac{3}{8} \frac{nm_p v_{sw}^2}{\rho b} \frac{v}{v_{sw}} = \frac{3}{8} \frac{nm_p v_{sw} v}{\rho b} \qquad (13.4)$$

---

4. How big is the nucleus of comet Halley?
5. What is the albedo of the nucleus of comet Halley?

## 13.2 THE SOLAR WIND

The most spectacular thing about a comet is its tail. But the comet cannot make a tail all by itself; rather, the tail of a comet is the result of the interaction of the comet with the Sun. Indeed, it was the presence of comet tails that gave the first clues, even before satellites were sent into space, that there was something filling the interplanetary medium, the space between the planets. That something is called the **solar wind**. But what is the solar wind?

## Acceleration of the Corona

Recall from Chapter 2 that the Sun is a ball of hot gases fueled by the reactions fusing hydrogen into helium. The intense energy produced by the Sun warms the planets as sunlight. Besides the energy of sunlight, however, the heat of the Sun has another outlet.

Above the surface of the Sun is a region called the **corona**. The corona can be seen from Earth with the naked eye whenever an eclipse occurs and the bright disk of the Sun itself is obscured (recall the photograph of an eclipsed Sun, Figure 1.1). Its brightness is less than the brightness of the blue sky, so it is difficult to observe from Earth except during a solar eclipse. Measurements of the light emitted by the corona indicate that the particles there are heated to nearly 1 million degrees, far hotter than any molecules can withstand. Thus the corona must be a thin atmosphere of free electrons and protons. Astronomers are still not satisfied that they understand exactly how this corona becomes so hot. The energy to heat it up may be carried from the interior of the Sun by the Sun's magnetic fields or by violent shock waves through the gas.

The protons and electrons within the corona are attracted to the Sun by the Sun's gravity. Like any gas, the particles move at a variety of speeds, with an average speed proportional to their temperature. Some particular particles move faster than their average thermal speed, some slower. But because they are so hot, the average speed at which they travel is so fast that some of them are moving faster than the local escape velocity of the Sun. These particles can break away from the Sun's gravity and begin to move out into interplanetary space.

But the force of gravity is an inverse-square force. The farther these particles move away from the Sun, the weaker the force of the Sun's gravity becomes. In the frame of reference of the gas, this drop in acceleration towards the Sun looks like a gain in acceleration away from the Sun. Acting as if they were accelerated away, they move still farther away, feel even less attracted by the Sun, and move yet farther out. Eventually, at a distance from the Sun equal to about 5 solar radii, the motion out away from the Sun dominates any random thermal motions in the other directions. Beyond this point, the primary motion of the gas is simply to travel out away from the Sun.

Thus every second tons of the Sun's atmosphere are literally boiled off into space and stream out away from the Sun at a very high velocity, filling the regions between the planets with a very hot, but extremely tenuous, plasma. This ionized gas, blowing away from the Sun, is what we call the **solar wind**.

## Structure of the Solar Wind

The very idea that there was an ionized "wind" blowing away from the Sun originated in the 1950s, before any spacecraft had ever left Earth to sample interplanetary space. The concept arose as a way of explaining the shape of comet tails. By observing its spectrum, it was known that a comet's tail consists of hot ionized gases and fine dust particles. Because this tail was observed to be pointed away from the Sun at all times, astronomers realized that it

was not simply made up of particles left behind as a comet passes, but rather it looked like material blown off a comet by some sort of wind moving away from the Sun.

How would such a wind behave? Just by making a few simple assumptions it was possible see that this wind would have to have certain predictable properties.

For example, consider a handful of solar wind particles flowing through an imaginary volume, in the shape of the box in Figure 13.4. How many particles flow into the small end of the volume in 1 s? That depends on how fast they are going, how big an area they're going through, and how densely packed the particles are in that part of the wind. If you double any one of those factors, you'll double the number of particles entering the volume.

But the same number of particles must leave through the bigger side as enter through the smaller side; otherwise, we'd have particles piling up inside the box. Thus we can say that the product of these three factors—the speed, the area, and the density of the gas—must be the same at both ends of the box.

Notice that the bigger side of the volume is the side away from the Sun. These curved sides are actually parts of imaginary spheres enclosing the Sun. Mathematically, we can say that the area of each of these sides is proportional to $4\pi r^2$, the area of a sphere, assuming that everything is spherically symmetric. Call the speed of the particles $v$ and the number of protons or electrons per unit of volume $n$; we conclude that this product, $n \cdot 4\pi r^2 \cdot v$, must always be a constant. Because $4\pi$ is already a constant, we can just write $nr^2v = $ constant.

What happens a long way away from the Sun? Common sense tells us that light years away from the Sun (when $r^2$ is very large) the amount of gas in the wind has to eventually drop to zero. Otherwise we would have the entire universe filled with the gas from just our own Sun. The

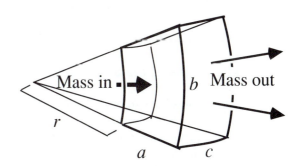

**FIGURE 13.4** Conservation of mass. In a section of a spherical volume of sides *a*, *b*, and *c*, the amount of mass entering the inner side must equal the mass leaving the outer side. Because the area of the side increases, the density of the fluid must decrease.

number of particles, $n$, must drop towards zero as $r$ approaches infinity.

What about the velocity? It can't keep increasing indefinitely. Either it levels off to some constant value or else it drops back down to zero. But recall that density times speed times area ($nr^2v$) is constant. When $r$ becomes very large, we saw already that $n$ becomes very small; so it is possible that these two changes could cancel each other nicely. The velocity does not have to change at all. Of course, it is possible that both $n$ and $v$ drop to zero when $r$ goes to infinity, but this would mean that neither could drop very quickly, and so there would still have to be a lot of gas—$n$ would be large—a great distance away from the Sun.

In fact, a more detailed mathematical analysis of this problem using the equations of fluid mechanics shows that either the speed of the wind approaches a constant or else, if it falls with distance from the Sun, it must vary at least as $1/r^2$. This latter situation would mean that the density of the wind, $n$, would have to be constant, which we argued before was unlikely. So the only reasonable conclusion seems to be that the velocity of the solar wind, $v$, has some constant value.

Thus, even before space probes were ever launched, comet tails acted as probes of interplanetary space. They indicated that there must be a plasma wind blowing away from the Sun with a constant velocity, with a density that obeys an inverse square law. (This is just like the flux of light from the Sun, as discussed in Chapter 10. Recall that light photons are also conserved, and they too travel at a constant speed: $c$, the speed of light.) In fact, now that we have space probes to measure the solar wind between the planets, this prediction is confirmed.

For "quiet Sun" conditions (see Chapter 2) the average speed of solar wind particles at Earth and beyond turns out to be about 400 km/s (nearly a million mph). This speed is roughly constant, and the plasma density $n$ is observed to be proportional to $1/r^2$. At Earth, the density of the solar wind is roughly 5 protons per cubic centimeter, or $5 \times 10^6 \, \mathrm{m}^{-3}$. That is 19 orders of magnitude less dense than the air on Earth, virtually a vacuum.

## The Magnetic Field and the Solar Wind

The density of the solar wind is so small that normally one might think that it could be neglected altogether. However, the solar wind is a plasma, a gas made of ionized particles. Because these particles are charged, they affect, and are affected by, any local magnetic field. Either the field drags the plasma around or else the plasma drags the field around. It depends on which one has more energy. In the case of the solar wind, the plasma wins.

Like Earth, the gas giant planets, and most stars, there is a magnetic field generated inside the Sun. Recall that it is the Sun's magnetic field, 10,000 times as strong as Earth's, at its surface that accounts for the presence of sunspots (discussed in Chapter 2).

One might expect that the plasma in the Sun's corona would serve to contain this field and prevent it from being felt outside the Sun itself. But pictures of the Sun taken by astronauts in space using cameras sensitive to high-energy light waves such as x-rays (which don't penetrate Earth's atmosphere and so can't be seen by telescopes on Earth) have shown that the corona of the Sun has holes in it. It isn't a smooth sphere of ionized gas, but more like an irregular veil with torn places, usually near the Sun's poles. Through these holes, the magnetic field lines tend to be bundled together, something like they were bundled together at the sunspots, and pour out into space.

If the Sun's field were a simple dipole, then at least in the ecliptic plane, where the planets are, it would be directed from pole to pole, in the $\theta$ direction. Furthermore, like a dipole its strength would fall quickly, as $1/r^3$. But if the field passes through an energetic plasma, the plasma can drag magnetic field lines along with it as it moves, changing both its direction and its strength. The Sun's magnetic field is dragged away from the Sun exactly in this way by the solar wind. One way to picture this effect is to think of the magnetic field lines as if they were rubber bands trapped in molasses. As the molasses moves across the rubber bands (like the solar wind plasma moves across the $\theta$-direction field lines), the rubber bands become stretched into a direction parallel to the motion of the molasses. In the same way, the motion of the solar wind pulls the $\theta$-direction field lines out away from the Sun, creating a radial $r$-directed component to the magnetic field.

Furthermore, the Sun spins with a period of about 27 Earth days. (In fact, the spin rate is slightly faster at the Sun's poles than it is at its equator. Remember that the Sun is a big ball of gas and does not necessarily spin at the same rate at all points.) Material from the Sun flowing out with the solar wind flows straight away from the Sun; but the magnetic field lines, tied to material poured out from the Sun at different times, tend to become twisted around the Sun as the Sun spins. The field lines look just like the stream of water that comes out of a spinning water sprinkler: Each bit of water goes straight away from the sprinkler, but as it spins, the streams of water appear to be curved (Figure 13.5). This creates a component of the magnetic field in the $\phi$-direction, the angular direction around the Sun's equator. The result is that the magnetic field around the Sun has a distinctive spiral shape called the **Parker spiral**, after Eugene Parker, an American physicist whose paper in 1958 first worked out the structure of the solar wind. The field lines start out near the Sun radiating away from it, but like the spiral of a pinwheel they tend to appear more and more curved as one moves farther away from the Sun.

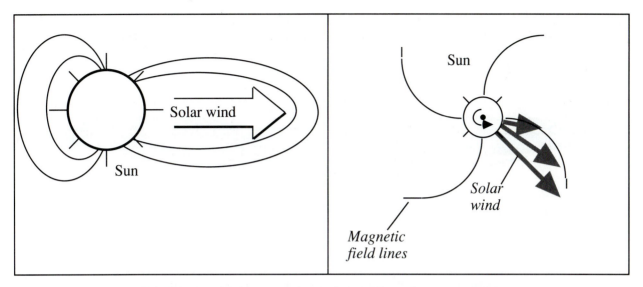

**FIGURE 13.5** The Sun's magnetic field is "frozen" into the outflowing solar wind, which distorts it away from a simple dipole shape. Left: A view from the plane of the Sun's equator illustrates how outflowing plasma stretches field lines. Right: A view from over the Sun's north pole shows how the field lines get further twisted into a spiral shape as the Sun spins.

Recall that the magnetic field is carried by the solar wind plasma, and consider the parcel of plasma illustrated in Figure 13.4; but now add the lines of magnetic field. The tighter together field lines are packed, the stronger the magnetic field, and the field strength in any direction at the location of the parcel is proportional to the density of field lines passing through the appropriate side of the parcel.

Recall that the parcel has dimensions $a$, $b$, and $c$. Because the parcel moves with a constant velocity, $a$ stays constant. However, $b$ and $c$ are each parts of the circumference of a circle that grows as $2\pi r$. Thus the side bounded by $b$ and $c$, which changes its area as $b \times c$, grows as $r^2$. The magnetic field lines that pass through this side are the lines of the $r$-component of the total $B$ field, which we'll call $B_r$. Because the area of the $b \cdot c$ side is growing as $r^2$, the strength of $B_r$ must be dropping as $1/r^2$. On the other hand, the magnetic field through the $a \cdot b$ side of the parcel, $B_\phi$, only varies as $1/r$ because this side grows only as $b$ grows. In both cases we see that the strength of the field falls much more slowly than the field from a simple dipole. The magnetic field carried by the solar wind can be effective at a much greater distance from the Sun than a simple dipole field.

In fact, the strength of the $B_\phi$ component of the field depends on how much the original field has been moved and twisted by the solar wind. Recall our rubber band and molasses picture again. The amount that the rubber band gets stretched by the molasses depends on several factors: how strong the rubber is (analogous to the strength of the original magnetic field at the Sun), how gooey the molasses is (analogous to the electrical resistance of the solar wind plasma, which controls how strongly it holds the magnetic field), and how fast the molasses is moving (analogous here to the speed of the solar wind.)

For the solar wind case, we don't really know the initial strength of the field or the conductivity of the plasma near the Sun very well. But these factors are the same for both $B_r$ and $B_\phi$. Only the relevant velocities are different for the two different magnetic field directions. $B_r$ is controlled by the speed of the solar wind, $v$, while $B_\phi$ is determined by the apparent speed of the Sun's spin, $v_\phi$. So if we just want to compare the relative strengths of the $B_r$- and $B_\phi$-components of the field, all we need to notice is that the ratio of $B_r$ to $B_\phi$ is the same as the ratio of $v$ to $v_\phi$.

At the equator of the Sun, the $\phi$ velocity at a distance $r$ from the Sun is the angular spin rate of the Sun, $\Omega$, times $r$. In three dimensions, as we travel above or below the plane of the Sun's equator, we must replace $r$ with $r \sin \theta$, where $\theta$ is the angle from the Sun's north pole. The Sun spins in one direction; relative to the Sun, the plasma seems to be moving in the opposite direction. Because it is opposite to the motion of the Sun, we use a negative sign: $v_\phi = -\Omega r \sin \theta$.

Thus we find the strength of the field in the $\phi$-direction, relative to its strength in the $r$-direction:

$$B_\phi = -B_r \frac{\Omega r \sin \theta}{v} \qquad (13.5)$$

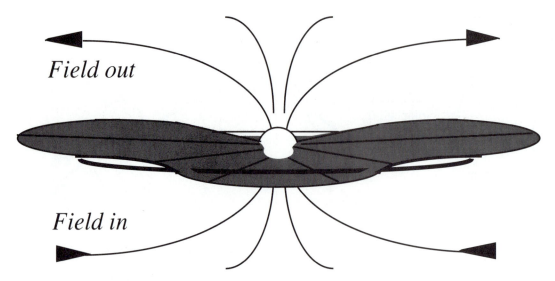

**FIGURE 13.6** The neutral sheet looks like a warped record. The heavy line represents the orbit of Earth. As the neutral sheet spins with the Sun, we on Earth see, alternately, the top and bottom of the sheet. As we pass from one side to the other, we appear to pass from one sector of the solar wind to another.

Recall that $B_r$ varies as $1/r^2$; from this equation it follows that $B_\phi$ varies as $1/r$. This agrees with what we found above.

Notice that as $r$ increases, $B_\phi$ eventually becomes bigger than $B_r$. As the density of the plasma drops, the strength of the field must also thin out; but, as these equations show, the bits of the field that are pointed in the direction of the spiral don't thin out as quickly as those pointed out away from the Sun. As a result, by the time the solar wind reaches Earth, the magnetic field that it carries points roughly as much in the direction going around the Sun as in the direction radiating away from the Sun. Beyond the orbit of Earth, the magnetic field carried by the solar wind is primarily in the $\phi$-direction.

## Sector Structure

The field generated inside the Sun starts out looking something like the field from a bar magnet, a dipole field. The field lines from the coronal holes to the north of the Sun's equator have one polarity, while those coming from the south have the opposite polarity.

Once they are trapped by the solar wind, the field lines become twisted into a spiral. Above and below the Sun's equator, the spiral shape is the same; but the polarity is different. At any give moment, the field lines above the Sun's equator might be spiralling out away from the Sun, while those lines below the equator are directed in towards the Sun. There is a thin boundary between the two hemispheres, called the **neutral sheet**, where the two magnetic field polarities meet and connect, closing the loop of

the original dipole field lines. But, as can be seen in Figure 13.6, this boundary is not a perfectly flat plane along the Sun's equator. Instead, because of irregularities in the shape of the coronal holes, the neutral sheet tends to be warped and tilted by up to roughly 15° from the equator of the Sun.

At the same time, the orbits of the planets do not exactly lie in the equatorial plane of the Sun. The result is that several times over a 27-day period, one rotation of the Sun, every planet may cross through this warped neutral sheet from one region of magnetic polarity to another. From the point of view of the planet it appears to be going into different spiral-shaped sectors of the Sun's magnetic field. Over a 27-day period, it may experience two, four, six, or more sectors, depending on the warp of the sheet. Four is most common.

## Prominences, Flares, and Solar Storms

The solar wind as we have described it up to this point represents a simple ideal, and in fact this is not a bad description during the periods of "quiet Sun," when the number of sunspots is at a minimum. However, at the peaks of the 11-year sunspot cycle, the region above the Sun's surface becomes quite active.

Recall that sunspots are regions of strong magnetic fields and that these spots tend to come in pairs. At one spot, the field lines exit the Sun, and at the other spot they reenter. The magnetic field lines near the Sun look something like a loosely wound ball of string. Magnetic field lines erupting from the Sun's surface are places where plasma tends

**FOR THE RECORD...**

# Plasma

In everyday life, we are familiar with material existing in three states: solids, liquids, or gases. The state of a substance depends on its temperature, which is a measure of how much energy each molecule has. If the thermal energy of a substance is stronger than the energy holding molecules together, bonds in a solid can bend or break and the material will melt or vaporize. But consider a temperature so great that not only are the bonds between molecules overwhelmed, but the very bonds uniting electrons to the nuclei of their atoms are also broken. Instead of a gas consisting of neutral balls of atoms, we have a **plasma** of charged particles.

In the Sun, made primarily of hydrogen, the resulting plasma consists mostly of protons and electrons. Other ions, such as the nuclei of helium, oxygen, carbon, and the other elements, are present in their relative cosmic abundances. These nuclei are positively charged, having lost some of or all of their electrons, and act much like protons do. In effect, this plasma looks like two gases, one of protons and one of electrons, mingled together.

However, there are certain important differences between a plasma and a regular gas. First, where neutral gas molecules merely bounce off one another, plasma particles can repel or attract one another by the electric charge that each particle carries.

Second, the charged particles can be controlled by whatever electric or magnetic fields are present. The plasma arranges itself to cancel out any external electric fields, and it is tightly bound to magnetic field lines. Thus if you try to move either kind of field through a plasma, you wind up dragging the whole plasma along with the field.

Last, as ordinary sound waves can propagate through neutral gases, plasmas with a magnetic field present can also carry a new variety of wave called an **Alfvén wave**. The speed and efficiency with which these waves can propagate depends on whether the wave travels along the magnetic field lines or across field lines. All these waves can carry energy from one end of a plasma to another; thus they are important when trying to understand how a plasma gets hot or cools off.

All these factors make plasmas quirky and unpredictable substances. However, they are potentially very important substances. Fusion power plants will become feasible once we understand how to contain and control large amounts of extremely hot plasma. On a more mundane level, plasmas already play a role in our lives as the glowing gases inside fluorescent bulbs and neon lights.

---

to concentrate. If we observe the Sun in a particular wavelength (the spectrum line $H\alpha$) that is emitted by hot atomic hydrogen but not by the plasma, the strings of plasma along the magnetic field lines connecting the sunspots appear as dark lines, while areas near the sunspots themselves appear brighter. The dark lines are called **filaments**, and the bright regions are **plages**. When the sunspot is sitting on the limb of the Sun, the dark lines reveal themselves to be long arcing ribbons of gas called **prominences**. (There is a good example of a prominence in Figure 2.1.) In among the bright plages may be a cluster of thin bright lines, jets of hot gas, called **spicules**, typically 1000 km wide and as much as 10,000 km high, which can flicker on and off in 15 minutes.

About 10% of the time, sunspots can be very active as the magnetic fields inside them become complex and twisted. When two sets of strong magnetic fields running in opposite directions get pushed together by the turbulent motion of the material in the Sun, the field lines annihilate

each other in a process called **reconnection** and turn their energy into a sudden burst of heat. This produces bright, energetic solar flares. These flares dump a large amount of energy (x-ray and gamma-ray radiation, and hot plasma) out into the corona and from there into the solar wind.

Finally, along the neutral sheet separating the sectors of the solar wind there may be energy released as the opposing magnetic field lines reconnect and annihilate each other. The energy that used to be contained in the magnetic field lines now goes into heating the solar wind plasma.

This results in unusually high numbers of very energetic solar wind particles. Especially during periods of solar maximum, when the sunspot cycle is at its peak, the greatest numbers of high-energy particle streams occur along the boundary between the sectors. During this period the solar wind becomes quite turbulent and irregular.

Even during the quietest of times, energetic particles from the Sun are poured onto the planets. Planets with magnetic fields, like Earth, trap these particles in their fields and

**FURTHER INFORMATION...**

# The Mathematics of the Solar Wind

The behavior of comet tails gave the first indication that a solar wind existed. In the mid 1950s an application of basic mathematical physics was able to outline not only that this wind existed, but how it must behave.

The starting point in any physical derivation is the assumed starting conditions. For a first model, it seemed reasonable to demand that such a wind must be steady, not varying with time, and that the only external force acting on the wind worth worrying about was the force of gravity. Finally, it seemed reasonable to assume that the plasma that made up the wind could be treated as an ideal plasma (an ionized ideal gas); indeed, the near-vacuum conditions of space are as close to an ideal-gas situation as one is likely ever to find in nature. Thus the equations of fluid motion could be greatly simplified and analyzed.

As we saw in the text, conservation of mass gave us the equation $nr^2v = $ constant, where $n$ is the number density of the plasma, $v$ is the velocity of the wind, and $r$ is the distance from the Sun. We can assume that we know the conditions of the wind at some position $r_0$, where the number density is $n_0$ and the velocity is $v_0$, so that this can be written as

$$nr^2v = n_0r_0^2v_0 \tag{13.6}$$

The ideal gas law for a plasma can be written as

$$P = 2nkT \tag{13.7}$$

where $P$ is the pressure of the plasma, $k$ is the Boltzmann constant, and $T$ is the temperature. The factor of 2 is necessary because a plasma consists of both electrons and protons, whereas the number density $n$ counts only protons (which make up the bulk of the mass).

The basic equation of fluid mechanics, the **Euler equation**, is a variation of Newton's second law $F = ma$. It states that in a steady-state flow, the forces (per unit volume) that can make a fluid change its velocity from one place to another must equal $\rho v(\partial v/\partial r)$ (note that the units do equal a mass times acceleration per unit volume). The forces acting on our fluid are gravity and any gradient in the pressure of the fluid itself; thus our equation becomes

$$\rho v \frac{dv}{dr} = -\frac{GM\rho}{r^2} - \frac{dP}{dr} \tag{13.8}$$

Here, $G$ is the gravitational constant and $M$ the mass of the Sun; $\rho$ is the mass density of the plasma. Notice that the system has spherical symmetry; the only spatial variable is $r$ and thus we need not worry about partial derivatives or vector notation.

This gives us three equations, but we have four variables: $v$, $n$, $P$, and $T$ that we need to find as a function of $r$. For our fourth equation, we simply note that plasmas are notoriously good conductors of heat (as well as of electricity), and thus the temperature $T$ can be assumed to be a constant.

We first solve the conservation of mass equation (13.6) for $n$ and insert this into the ideal gas equation (13.7):

$$P = 2kT \left( \frac{r_0^2 v_0 n_0}{r^2 v} \right) \tag{13.9}$$

Next, noting that the mass density of the plasma $\rho = m_p n$, where $m_p$ is the mass of a proton, we divide the Euler equation (13.8) by $\rho$ and substitute in Equation 13.9 for $P$:

$$v \frac{dv}{dr} + \frac{GM}{r^2} + \frac{1}{nm_p} \frac{d}{dr} \left[ \frac{2kTr_0^2 v_0 n_0}{r^2 v} \right] = 0 \tag{13.10}$$

Next we define two new quantities. The first, $v_p$, is the average thermal speed of the plasma, the speed at which an average proton travels. From the laws of thermodynamics we know that the square of this speed is equal to

$$v_p^2 = \frac{3kT}{m_p} \tag{13.11}$$

Next we define a "critical radius":

$$r_c \equiv \frac{GMm_p}{4kT} \tag{13.12}$$

Notice that this does indeed have units of distance. When the temperature is $10^6$ K, a typical temperature for thin, hot plasma, then the critical radius is about 5.5 solar radii. However, if the temperature were a factor of six higher, then the critical radius would be inside the Sun, and the average speed of the atmosphere would exceed the escape velocity of the Sun.

With these definitions we can substitute into Equation 13.10 and, after taking the appropriate derivatives, arrive at the following differential equation for the velocity of the solar wind as a function of position:

$$\left(v^2 - \frac{2}{3}v_p^2\right)\frac{1}{v}\frac{dv}{dr} = \frac{4}{3}\frac{1}{r}v_p^2\left(1 - \frac{r_c}{r}\right) \quad (13.13)$$

This equation cannot be easily solved, but we don't have to solve it. Simply examining it can reveal a wealth of information about how the speed of the solar wind must behave as a function of position.

For instance, notice that the term $dv/dr$ appears on the left-hand side. We know from basic calculus that when this term equals zero, the function of $v(r)$ must go through an extremum; in other words, when $r$ is equal to the critical radius, $r_c$, then the velocity must either be at a maximum or a minimum value. There's a third possibility, of course; at the critical radius, the velocity $v$ could be exactly equal to $\sqrt{\frac{2}{3}}v_p$, so that both sides of the equation equal zero without $dv/dr$ being zero. Indeed, two different curves could have this property, one rising toward a constant velocity and one decreasing asymptotically to zero, so that one can imagine four families of solutions to this equation. They are illustrated in Figure 13.7.

The first set of curves, family 1, describes the case where the solar wind velocity is a minimum at the critical radius. However, all the curves in this family have the velocity approaching infinity at the surface of the Sun, which is not physically reasonable. For the same reason, we can rule out curve 2, the one curve that goes through the critical point with an ever-decreasing velocity.

Thus only two possible solutions remain. Curve 3, the "solar wind" solution, postulates that the wind velocity starts near zero at the surface of the Sun, increases to exactly $\sqrt{\frac{2}{3}}v_p$ at the critical radius, and then flattens out, approaching a constant value. Family 4 consists of any curves that start at zero, reach their maximum value at the critical radius, and then drop to zero again. It's called the "solar breeze" solution.

How can we choose between these solutions? Recall from the text that if the velocity of the solar wind drops too quickly with distance, then conservation of mass demands that the density of the wind cannot drop as fast; indeed, if the velocity falls as $1/r^2$, then the density would be constant everywhere, all the universe would have to be filled with solar wind particles, and this is unreasonable. So let us look at an extreme case of family 4, where the solar wind maximum velocity is much, much smaller than the plasma speed $v_p$. Presumably such a case would have a very gentle slope at large values of $r$. Is this so?

Look again at Equation 13.10. Divide both sides by $v^2$ and look at the case just described. With $v_p \gg v$ and $r \gg r_c$ we can keep only the largest terms on each side and arrive at

$$-\frac{2}{3}\left(\frac{v_p}{v}\right)^2\frac{1}{v}\frac{dv}{dr} = \frac{4}{3}\frac{1}{r}\left(\frac{v_p}{v}\right)^2$$

which becomes

$$-\frac{1}{v}dv = 2\frac{1}{r}dr \quad (13.14)$$

Upon integrating this, we see that $v = 1/r^2$, which we had just argued was unreasonable. Thus the only physical solution is the "solar wind" solution. This means that we not only know the behavior of the wind velocity far from the Sun—it approaches a constant—but we can even estimate

**FURTHER INFORMATION...**

# The Mathematics of the Solar Wind (continued)

that speed, given a good guess for the temperature of the plasma. The best guess in the 1950s was a speed of 800 km/s, which is only a factor of two larger than the actual measured speed of the solar wind!

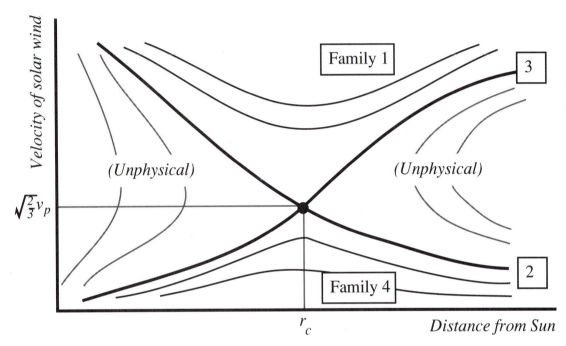

**FIGURE 13.7** Four families of solutions exist to describe the velocity of the solar wind, *v*, as a function of position *r*. They are described in the text.

the shape of the planet's field becomes warped in response to the pressure of the solar wind. Planets without fields, like Venus or the Moon, are hit directly by these energetic protons and electrons, which can sometimes result in changes in the chemistry of their surfaces or atmospheres.

During a solar storm, planets receive an unusual amount of solar wind particles. Earth's magnetic field and ionosphere can be suddenly warped into a new shape by the extra force of this unusual wind, disrupting radio communications. The extra particles are carried by Earth's magnetic field into the north and south poles of Earth, where they hit the atmosphere and ionize gas there. As the atoms cool down and deionize, they emit light, called the **aurora borealis** ("Northern Lights") in the northern hemisphere and the **aurora australis** ("Southern Lights") in the southern hemisphere. Similar auroras have been seen in the polar regions of Jupiter and the other gas giant planets. In some

way still not understood, these solar cycles also affect the weather on Earth and perhaps other planets.

These solar storms can have serious effects in telephone and electric lines on Earth. A **geomagnetic storm** is a change in Earth's magnetic field caused by an unusually dense or energetic solar wind. Sudden electric surges of over 1000 volts (V) have been recorded in trans-Atlantic telephone cables during geomagnetic storms. On March 13, 1989, a storm created a power surge in northern Canada that knocked out electricity throughout Quebec for more than 8 hours; 6 million people were without electricity for much of that day. Solar storms can create voltages exceeding 10,000 V on satellites orbiting in Earth's magnetosphere, severely challenging the electronic equipment aboard. Bursts of high-energy protons from the Sun in October 1989 led to measurable levels of radiation among passengers in airliners flying near the poles. If an astronaut had been on the

Moon during that time, she would have died from radiation poisoning.

# SUMMARY

The solar wind is a stream of protons and electrons blowing from the Sun's corona into interplanetary space. The wind has an average velocity of about 400 km/s. Its density (during quiet Sun periods) is roughly $5 \times 10^6$ particles/m$^3$ by the time it gets to Earth.

The Sun's magnetic field is pulled away from the Sun by this outflowing plasma, and the field lines become twisted into a spiral shape. Because the northern and southern hemispheres of the Sun have fields of opposite polarity, there is a neutral sheet in the solar wind where the polarity of the magnetic field carried by the solar wind changes. As the Sun spins, this neutral sheet passes through Earth and the other planets; thus, from the point of view of the planets, the solar wind appears to have sectors of alternating polarity.

When the Sun is most active, the solar wind and its magnetic field become dense and turbulent. This causes auroras, radio interference, and possibly changes in the weather on planets.

# STUDY QUESTIONS

1. In what direction does the solar wind flow: out, away from the Sun, or around the Sun, in the direction the Sun spins?
2. What is the windspeed of the solar wind?
   (a) 6 m/s
   (b) 800 m/s
   (c) 400,000 m/s
3. What do you call a gas that is so hot that the electrons are separated from the nuclei of the atoms?
4. Name one effect of solar storms that we might notice on Earth.
5. What appears to change direction when Earth passes from one sector of the solar wind to another sector?

# 13.3  IONOSPHERES AND MAGNETOSPHERES

The solar wind blows a constant stream of energetic protons and electrons out from the Sun with a density of 5 particles/cm$^3$ at Earth (the density drops inversely with the square of the distance from the Sun) and a velocity of 400 km/s. What happens when this wind hits a planet?

There are three distinct cases.

1. The wind may hit the surface of a bare rock, like an asteroid or the Moon.
2. The wind may hit a planet with an atmosphere. In this case, the ions of the solar wind interact with the ions of the rarefied topmost layer of the planet's atmosphere called the **ionosphere**.
3. The wind may encounter a planet with both a magnetic field and an atmosphere. These interactions are the most complex of all.

## Bare Planets

When the solar wind travels out from the Sun, it has trapped within it the magnetic field of the Sun. When it moves past a bare planet, it carries this magnetic field past the planet. As we saw in Chapter 10, a magnetic field moving past a conductor sets up a current in that conductor.

Thus, if the planet were made of some good conductor such as a metal, we would expect a substantial current to start flowing through the planet. As shown in Figure 13.3, the direction of the current would be from north to south, given the orientation of the solar wind as illustrated. By convention, the flow of positive charged particles defines the direction of the current, so this means that electrons from the solar wind would be attracted to the south pole of the planet. They would push electrons through to the north pole, where they would rejoin the solar wind. Because the solar wind magnetic field has a "sector structure," it changes sign every 7 days or so; and this current sloshes back and forth through the planet as it responds to the changing sectors of the solar wind that it sees.

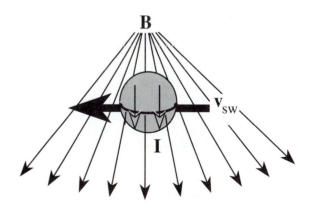

**FIGURE 13.8** An electrically conductive planet moving through the solar wind would have a current flowing from pole to pole.

Of course, with the possible exception of certain asteroids, metal planets are not all that common. However, all materials have a certain finite electrical conductivity. All bare planets therefore experience some sort of current. Because no material is a perfect conductor, some of the energy of the current is always dissipated as heat. As a result, the planets can be heated up by this solar wind. Only reasonably high conductivities do the trick, however; if the conductivity is too low, too little current flows. Given the weak solar magnetic field today and the low conductivity of rocky materials, this is not a significant heat source at the present time.

But it has been proposed that magnetic fields were much stronger in the early solar system. Carbonaceous chondrites, for instance, have remanent magnetic fields of $10^{-4}$ tesla, some five orders of magnitude stronger than the present-day interplanetary field. Cold rocky material is too poor a conductor of electricity for even this field to have created much current; however, if any asteroids were warmed slightly by other effects (collisions or short-lived radioactive nuclides, for instance), then currents induced by the solar wind magnetic field may have been an important heat source in those asteroids, perhaps producing the basaltic material seen in meteorites.

Another effect of this current is to induce a small magnetic field in the planet. By measuring the small field set up in the Moon by the solar wind, it has been possible to estimate the electrical conductivity of the Moon, which is indicative of its composition and internal temperature. Based on this, it seems possible that the Moon has a small (200–400 km radius) metallic core. In this way an induced magnetic field provides another tool for determining the bulk composition and structure of planets.

Finally, if there's nothing to stop the solar wind ions, they impact directly into the surface of the body. As we mentioned before, such impacts can be a way to erode the ice off the surface of comets and icy moons. But recall that the protons in the solar wind are also the nuclei of hydrogen atoms; this constant addition of hydrogen onto a rocky planet's surface may also be responsible for subtly altering its chemical composition. Over time, it may also destroy crystals in the surface minerals, so that they do not reflect light as well and thus seem darker than freshly exposed rocks. The whole surface of the Moon may be constantly darkened by solar wind impacts; freshly disturbed material spreading out from younger craters on the Moon overturns the dark surface rock, leaving the bright rays seen around craters such as Tycho and Copernicus.

## Ionosphere Interactions

The second case involves the solar wind hitting a planetary atmosphere. To see what happens here, we must first examine what the upper reaches of an atmosphere look like.

The uppermost uncharged layer of an atmosphere is called the **thermosphere**. This is the layer directly heated by solar UV radiation. The more powerful ultraviolet photons not only heat and break apart molecules, but they also ionize the atoms they hit, breaking off electrons and producing a plasma layer above the thermosphere, called the **ionosphere**, a highly ionized and conducting gas at the top of the atmosphere.

We've known about the ionosphere of Earth since the beginning of the twentieth century when it was found that this conducting layer of plasma allows world wide radio transmission by reflecting radio waves around the curvature of Earth. An outline of the various layers of plasma that have been seen is shown in Figure 13.9.

When the solar wind hits a planet with an atmosphere, the impact of its ions tends to continue the ionization started by the solar UV radiation. The point of contact is called the **ionopause**. The ions in the ionosphere are hit by solar wind ions travelling at 400 km/s, faster than the speed of sound in the plasma. The result is reminiscent of the hypervelocity impacts of meteorites that create craters on rocky planets: A shock wave is formed. In the ionosphere, this shock wave is a standing wave separating the denser ionospheric plasma from the tenuous but fast-moving solar wind. As the wind

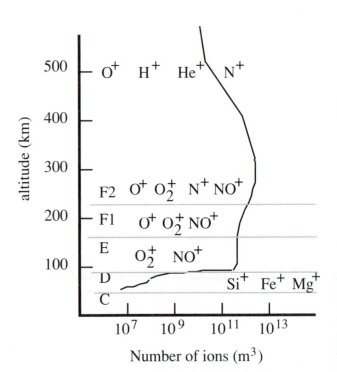

**FIGURE 13.9** Ion density in Earth's atmosphere as a function of altitude. The various layers, as named, have distinctive plasma densities and species of elements present. The C layer consists of more complex ions, such as $H_3O^+$, $NO_3(H_2O)_n^-$, and $HCO_3(H_2O)_n^-$.

sweeps around the planet, it carries away ions into a long tail that can extend for tens of millions of kilometers.

This tail should be very familiar to us; what we've described happening on a planet like Venus or Mars, with an atmosphere but no magnetic field, is also exactly what happens to comets. As the cometary ice evaporates, it produces an atmosphere around the comet that is ionized and carried away into a tail by the solar wind. Likewise, Mars and Venus have "tails"; however, unlike comets, the upper atmospheres of these planets are much less dense than the gas from ices evaporating round comets, and so planetary tails can only be spotted by sensitive detectors flying near the tails.

## Magnetosphere Interactions

What happens if the planet has its own magnetic field? This is the case on Jupiter and the other outer planets and on Earth.

A parcel of solar wind moves with a velocity $v$ and has a mass density of $n \cdot m_p$, where $n$ is the number of protons in a cubic meter and $m_p$ is the mass of a proton. (Electrons are ignored because their mass is 2000 times smaller than a proton's mass.) Recall that kinetic energy is $\frac{1}{2}mv^2$ and that pressure is the same thing as an energy density, energy per unit volume. The force that this wind exerts on a surface, its **ram pressure**, is thus

$$P_{\text{ram}} = \tfrac{1}{2}nm_p v^2 \qquad (13.15)$$

When it hits a planet's magnetic field, the solar wind has to push against the field lines. The field becomes compressed until the energy density of the field is equal to the ram pressure of the solar wind. The energy density of a magnetic field, $u$, in SI units is derived in many elementary physics texts:

$$u = \frac{1}{2\mu_0}B^2 \qquad (13.16)$$

where $B$ is measured in teslas and $\mu_0$ is a constant designed to keep the units straight, equal to $4\pi \times 10^{-7}$ kg $\cdot$ m $\cdot$ coulomb$^{-2}$. (Magnetic fields are one realm of physics where it is very important to keep in mind which system of units one uses. The equation for magnetic energy density looks a bit different in cgs units. However, the derivation that follows can be repeated in any set of units you wish to choose.)

Consider Earth's magnetic field. Recall that it looks like the field of a dipole whose strength varies as $1/r^3$, where $r$ is the distance from the center of Earth.

At some point the force of the solar wind must equal the energy density of the compressed dipole field. We can

estimate this position by setting the ram pressure equal to *twice* the original magnetic energy density (as given in Equations 13.15 and 13.16). Why twice? The solar wind distorts the original dipole field, pushing the field lines in closer to the planet; thus the energy density of the field holding off the ram pressure is the energy density of the original field plus the energy density of the field lines pushed into that original field. From this formulation we arrive at

$$\frac{B^2}{\mu_0} = \frac{1}{2}nm_p v^2 \qquad (13.17)$$

Next we substitute for $B$ by writing an equation for Earth's field strength:

$$B = B_0 \left(\frac{R}{r}\right)^3 \qquad (13.18)$$

where $B_0$ is the strength of the field at the surface of Earth and $R$ is the radius of Earth. With $B_0 = 3 \times 10^{-5}$ Tesla and $n = 5 \times 10^6$ protons/m$^3$, we can solve for the position of the standoff point, $r_{so}$:

$$\frac{B_0^2}{\mu_0}\left(\frac{R}{r_{so}}\right)^6 = \frac{1}{2}nm_p v^2$$

$$r_{so} = R \left(\frac{B_0}{v}\right)^{\frac{1}{3}} \left(\frac{2}{\mu_0 nm_p}\right)^{\frac{1}{6}} \qquad (13.19)$$

We find that the standoff point marking the point of balance between the solar wind and Earth's magnetic field occurs at about 10 Earth radii.

On the far side of Earth, the solar wind has the opposite effect. The flowing plasma tends to pull out the magnetic field lines into a long, complicated tail. A similar effect occurs at Jupiter; indeed all the outer planets visited by *Voyager* have distinct regions where the plasma realm of the solar wind ceases, and the plasma realm of the planet's magnetic field begins. The boundary between the solar wind region and a planet's magnetic field region is called the **magnetopause**. However, actually deriving the standoff point for Jupiter and the location of its magnetopause is more complex because of the way its rapid spin distorts its magnetic field.

## Particle Motions in Radiation Fields

The region between the ionosphere and the magnetopause, which we call the **magnetosphere**, is filled with both a distorted dipole magnetic field and a hot plasma. The plasma is much thinner in density than the ionosphere plasma but the individual particles are much more energetic. They make up the **Van Allen belts**.

(In 1957, after two successful Soviet Sputniks had been put into orbit, the U.S. Army rushed together a rocket to match the Russians. James Van Allen proposed putting an off-the-shelf Geiger counter into the payload of the rocket. The first successful American satellite, *Explorer 1*, was launched in January 1958 with Van Allen's experiment on board. When the device unexpectedly recorded a flurry of charged particles at the highest part of the orbit, one of his students suggested that "space must be radioactive!" Van Allen correctly interpreted the results as evidence for a belt of charged particles trapped in Earth's magnetic field.)

How do particles in a radiation field behave? Recall from Chapter 10 that a charged particle moving past a magnetic field line feels a force that is perpendicular to both the direction the particle is moving and the direction of the magnetic field line. The strength of this force depends on both how fast the particle is moving and how strong the magnetic field is. Around a planet with a dipole field, a charged particle undergoes three basic types of motion.

**Cyclotron motion** occurs as the particle moves in a direction perpendicular to the direction of the magnetic field. The force that results from this motion is perpendicular to both the field line and the original motion. Just as we saw with bodies in circular orbits, this perturbing force does not change the energy of the plasma particle or its speed. It merely changes its direction. But no matter what new direction the plasma particle moves, it still is forced sideways to change direction again. The result is that the particle goes in circles, orbiting around the magnetic field lines (Figure 13.10).

We know from Chapter 10 that the strength of the magnetic force is equal in magnitude to $qvB$, where $q$ is the charge and $v$ the velocity of the particle moving perpendicular to a magnetic field $B$. From Chapter 8 we learned that a body undergoes uniform circular motion of radius $r$

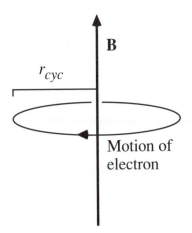

**FIGURE 13.10** Cyclotron motion. The magnetic field creates a force perpendicular to the direction of the electron's motion, bending its path into a circle around a field line.

when it is subjected to a force of $mv^2/r$. Equate these forces and one can solve for the radius of the circular motion due to the magnetic field, called the **cyclotron radius**:

$$r_{cyc} = \frac{mv}{qB} \qquad (13.20)$$

The stronger the field, the smaller the radius of the circle in which the particle travels. But the more momentum ($mv$) the particle has—the bigger the particle or the faster it's moving—the larger the circle will be.

If this particle travels in a circle of circumference $2\pi r_{cyc}$ at a velocity $v$, then we can find the **cyclotron frequency** $f$ of the particle:

$$f_{cyc} = \frac{v}{2\pi r_{cyc}} = \frac{qB}{2\pi m} \qquad (13.21)$$

Notice that because the cyclotron radius also depends on velocity, the velocity terms divide out. The cyclotron frequency of a particle does not depend on its speed.

**Bounce motion** is when a particle in cyclotron motion about a magnetic field line is, in effect, tied to that field line. It cannot move away from the line; it can only move around it in a circle. This circle can "slide along" the field line itself, however. Thus particles are free to move along the line, following it as it twists and bends, going through regions of stronger and weaker field. The motion of such a particle resembles a sort of corkscrew path, as it winds around the field line.

The angle between the direction it moves and the direction of the field line, the angle of the corkscrew, is called the **pitch angle**, $\alpha$ (Figure 13.11). Because only the component of the particle's velocity perpendicular to the $B$ field counts in calculating the cyclotron radius, we must replace $v$ in the cyclotron radius equation (Equation 13.20) with the quantity $v \sin\alpha$. Our final equation for the cyclotron radius $r_{cyc}$ becomes

$$r_{cyc} = \frac{mv\sin\alpha}{qB} \qquad (13.22)$$

As the particle moves around the field line in circular motion, it also moves up or down the field line with the velocity $v\cos\alpha$.

Around a planet like Earth with a dipole field, the field line that is out in space near the equator eventually bends into Earth at the magnetic north and south poles. As it gets closer to the poles, the strength of the field increases. As noted above, a particle can follow a field line as it curves in its dipole shape down into the magnetic pole of the planet. What happens to the motion of the particle as the strength of the magnetic field changes?

## FURTHER INFORMATION...

# The Adiabatic Invariant

We stated in the text that the product of the magnetic field times the area enclosed by the path of a charged particle in cyclotron motion, $B\pi r^2$, must be constant. Where does this result come from?

The force that a moving charged particle feels from a magnetic field is always perpendicular to its motion; therefore, by its nature the magnetic force cannot act in the direction of motion. It cannot change the angular momentum of the particle. Recall (from Chapter 4) that the angular momentum of a body in circular motion is given by the formula $L = mv_\perp r$. If $L$ and $m$ are constant, as $v_\perp$ is increased $r$ must be decreased.

But $r = mv_\perp/qB$, so $v_\perp = qBr/m$. Thus if $mv_\perp r$ is constant, then $qBr^2$ is constant, or $(q/\pi)B\pi r^2$ is a constant. The terms in parentheses are constants already, so $B\pi r^2$ must be constant, too. The product of the magnetic field and the area enclosed by the cyclotron motion is constant. That's what we wanted to prove.

Strictly speaking, this derivation works only for a slow and reversible change in the magnetic field. Even though the force of a magnetic field is always perpendicular to the motion of a charged particle, the force arising from the *change* in a magnetic field acts in the direction of the motion. This force must be small compared with the static magnetic force for the derivation to be valid. Therefore, the change must occur very slowly compared with the cyclotron motion of the particle, and "reversibly," that is, without making a significant change in the energy of the particle.

The idea of a slow and reversible change is one we have used before, when we described how pressure, temperature, and density changed in a convecting atmosphere. The term used for such changes is **adiabatic**. In magnetic fields, $B\pi r^2$ is known as the **adiabatic invariant**.

---

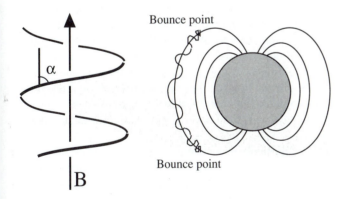

**FIGURE 13.11**   The pitch angle $\alpha$ and bounce motion of a particle in a planetary magnetic field.

The physics of magnetic fields includes a concept called the **adiabatic invariant**: The strength of the magnetic field going through the circle made by the particle in cyclotron motion, times the area of that circle, must always remain constant. Thus

$$\pi r_{cyc}^2 B = \text{constant} \qquad (13.23)$$

From Equation 13.22 it follows that the term $\sin^2\alpha/B$ must be constant. Thus

$$\sin^2\alpha \propto B \qquad (13.24)$$

What happens if $B$ changes? As it becomes bigger, the angle $\alpha$ must also increase. But $\alpha$ can only get so big before the term $\sin\alpha$ equals 1, its maximum value. At this point, the pitch angle $\alpha$ equals 90°. If the field line goes into a region where the $B$ field is so great that the angle $\alpha$ becomes 90°, then the particle stops and won't move down the field line any farther. (In this way, hot plasmas in fusion reactors are confined by cleverly designed magnetic fields, sometimes called "magnetic bottles.")

A particle with a shallow pitch angle thus is stopped by the "magnetic bottle" effect as it approaches the pole. Once stopped, it can turn around and travel down the field line back through space, until it is stopped at the other pole of the field. This motion is called the **bounce motion** of the particle. The points where particles are stopped and turned around are called **mirror points**. Typical particles in Earth's magnetosphere take about a second or so to bounce from pole to pole.

However, particles that start with a steep enough pitch angle plow into the atmosphere before they reach their mirror points. Such particles are thus lost from the radiation belts. When these energetic particles encounter molecules in the atmosphere, like nitrogen or oxygen, the molecules become

**FURTHER INFORMATION...**

# Electromagnetic Forces on Dust

The cometary dust particles mentioned at the beginning of this chapter represent another case of interactions with the solar wind, much as ring dust particles are affected by planetary magnetic fields (as mentioned in Chapter 11).

An electrostatically charged dust particle moving in the presence of a magnetic field (either a planet's field or the Sun's field carried by the solar wind) experiences significant Lorentz forces. Furthermore, these charged dust particles experience an electrostatic attraction or repulsion (depending on the relative sign of their charge) with the local plasma field. Because the dust is usually travelling at a speed different from the plasma, significant drag can result.

The strength of the magnetic Lorentz force depends on the charge $q$ on the dust grains. In a planet's plasma, all grains should be at close to the same (negative) electrostatic potential, determined by the temperature of the plasma. In interplanetary space, UV photons from the Sun will cause **photoemission** of electrons from the grains; again, a constant electric potential is achieved (this time positive). In either case, because the electric potential of a sphere of radius $b$ is $V$, proportional to $q/b$, we find that the force $F \propto q \propto b$. Thus the acceleration in this case would be $a \propto F/m \propto b/b^3 \propto 1/b^2$. On the other hand, the acceleration from collisions between plasma and uncharged dust should vary as $1/b$, as we found when we calculated the effect of sunlight and solar wind pressures on dust. Thus the collision accelerations are greater than the Lorentz accelerations on larger particles (a few microns in radius or so), but for smaller dust grains the Lorentz acceleration dominates.

The direction of the Sun's magnetic field, as seen near the plane of the planets, appears to alternate its direction as a result of the sector structure of the solar wind. This makes

a complete discussion of its effects on interplanetary dust very complicated and beyond the scope of this text. The bottom line, however, is that this **Lorentz scattering** of dust tends to "stir up" clouds of micron-sized dust over a time scale of 100,000 years, without preferentially moving the dust in any particular direction. However, because the strength of the Sun's magnetic field falls off only as $1/r$, this force can be effective at a considerable distance from the Sun.

The strong dependence of this force on particle radius means that very tiny particles, with radii less than a hundredth of a micron, can be picked up and carried along with any magnetic fields present. These tiny dust particles are carried out of the solar system altogether by the outflowing solar wind; they are said to experience **Lorentz blowoff**. On the other hand, dust particles larger than 10 $\mu$m feel a negligible amount of force from the solar magnetic field.

Dust particles between these two size ranges experience wildly complex, at times chaotic, motion as their Keplerian orbits are strongly perturbed by the magnetic forces. A wide variety of motions with exotic names such as **gyrophase drift** can result. In addition, such particles are strongly susceptible to **Coulomb drag**. Attempting to quantify the strength of this drag is complicated because charged dust particles can interact with each other as well as the plasma and strongly alter the temperature and density of the plasma in which they sit. For instance, the lifetime of Jupiter's dust ring under Coulomb drag has been estimated to be anywhere from several thousand years to as little as 10 years.

The study of the behavior of dust particles subject to both gravity and electromagnetic forces is a science barely 10 years old. It is called **gravitoelectrodynamics**.

---

ionized. After the encounter, the ionized gas molecules recombine, giving up their ionization energy as light. This glowing light produces the characteristic colors of the **auroras**.

**Drift motion** is a third, longer-scale motion that occurs in planetary magnetospheres. Drift motion arises because the particles move slightly closer to, then farther away from, the planet as they spin about magnetic field lines in their cyclotron motion. As illustrated in Figure 13.12, the

magnetic field that they see closer to the planet is slightly stronger than the field at the outer part of their circles. As a result, they turn with a slightly tighter radius on the inside part of their cycle than on the outside. This change in cyclotron radius produces a net drift.

Given the current orientation of Earth's field, electrons in the Van Allen belts tend to drift eastward and protons to drift westward. It takes a few hours for particles to drift completely around Earth in this manner.

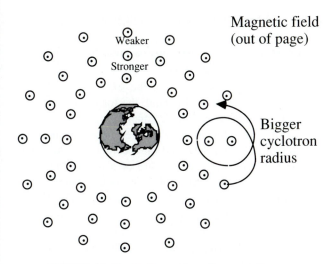

Magnetic field (out of page)

Bigger cyclotron radius

**FIGURE 13.12** Drift motion. Because the cyclotron radius varies inversely with the strength of the magnetic field, the particle travels farther on the outer part of its cyclotron motion and as a result tends to drift around the planet.

### Corotating Plasma

Recall that the ionized upper region of a planet's atmosphere, the ionosphere, consists of tiny electrons and somewhat more massive protons travelling at very high velocities. Because the protons are more massive than the electrons they tend to settle slightly deeper than the electrons into the planet's gravity field. As the protons and electrons are separated, a small electric field is produced. This electric field extends out into the region of the magnetosphere and weakly attracts the particles in the radiation belts.

The ionosphere is just an extension of the atmosphere of the planet. The atmosphere spins at the same rate as the planet spins, and the ionosphere spins at the same rate as the atmosphere. The electric field set up in the ionosphere ties the radiation belts to this rotation. Thus they too spin as the planet spins.

For a planet like Earth, this effect is not too significant. But consider that the farther one is from the planet, the faster the plasma must move to keep up with the planet's spin. On Jupiter, which spins very quickly and has a dense plasma and an extensive magnetosphere, this effect becomes enormous. The plasma on Jupiter is especially dense because of the large number of sulfur and sodium ions spewed off the surface of its volcanic moon, Io. These ions, 30 or 40 times more massive than hydrogen atoms, tend to get thrown out away from Jupiter by the strong centrifugal force arising from Jupiter's rapid spin rate. As the plasma moves, it pulls out the magnetic field with it. The result, seen in Figure 10.10, is that Jupiter has a very distorted magnetosphere. Far from the planet its magnetic field no

longer resembles a dipole field; rather, it is more reminiscent of the Sun's magnetic field as it is pulled out by the flow of the solar wind.

### SUMMARY

Planets interact with the interplanetary plasma and magnetic field of the solar wind in different ways, depending on whether or not they have atmospheres or magnetic fields. Bare rocks might have small currents set up inside them as the magnetic field moves past the planet. Atmospheres have an ionosphere, a plasma that can absorb the impact of the solar wind; ions from this ionosphere get picked up and carried away by the solar wind, forming a tail like a comet tail. Magnetospheres hold off the solar wind several planetary radii away from the planet's surface.

Within the magnetospheres, energetic particles make up radiation belts tied to the planet's magnetic fields. These particles spin around the field lines in cyclotron motion, bounce between the poles, and drift about the planet. The plasma about the planet rotates as the planet rotates; a planet's rapid rotation can change the shape of its magnetic field.

### STUDY QUESTIONS

1.  The plasma of the solar wind hits a small, airless, rocky asteroid, causing an electrical current to flow in this body. Name one effect that this could have on the asteroid.

2.  A planet with an atmosphere always has a highly ionized plasma at the top of this atmosphere. What is this layer called?

3.  The standoff point is the place where the pressure of a planet's magnetic field is just strong enough to balance what other pressure?

4.  The magnetosphere of Jupiter looks like a large, warped pancake (especially compared with Earth's dipole field). What is it about Jupiter that causes its field to be so oddly shaped?

5.  Name one way in which the plasmas in space can be noticed by ordinary people on Earth.

### 13.4 PROBLEMS

1.  *The Oort Cloud.* With a little bit of logic and arithmetic, one can determine a number of properties of comets. Here, in increasing order of difficulty, are four puzzles to try:

(a) *How many comets are there?* Say that roughly 10 long-period comets enter the inner solar system every year and spend 1 year of their million-year orbits here. This doesn't change from year to year. How many comets, total, are there out in space?

(b) *Where are all these comets?* Assuming that these long-period comets have periods of a million years, find *a*, their average distance from the Sun in AU. Given that *a* is the average distance and assuming that these comets come to about 1 AU of the Sun at their closest approach, what is their maximum distance from the Sun?

(c) *Is the space between the stars swarming with comets?* By the rules of planetary orbits, comets spend most of their time at their maximum distance from the Sun; we can assume that virtually all comets are at this maximum distance. If the comets are uniformly distributed at this distance in all directions from the Sun, roughly how far apart are any two comets likely to be? (The theory that most comets are to be found in a sphere of material many thousands of AU from the Sun was first proposed by the Dutch astronomer Jan Oort in 1950. This hypothetical cloud of comets is called the **Oort cloud**.)

(d) *What is the source of short-period comets?* About 100 short-period comets, those with periods of less than 200 years, have been discovered. They make up a distinctly different population of comets. Most of them have low-inclination prograde orbits, with periods of around 15 years. The source of these comets is called the **Kuiper belt**. Where is the Kuiper belt located? How many short-period comets might be visible from Earth in any given year?

2. A comet may make hundreds of passes about the Sun before it uses up all its ices. Eventually, however, it should exhaust its ice. Speculate on what such a "dead" comet might look like.

3. What is the average density of the solar wind at Jupiter?

4. "Every second, tons of the Sun's atmosphere are literally boiled off into space." Is this true?

(a) How much mass, on average, does the Sun lose every second to the solar wind?

(b) Assuming that this wind is as old as the Sun, how much mass has it lost over the past 4.5 billion years?

(c) How much mass does the Sun lose, every second, due to fusion? (Hint: Take the solar luminosity from the appendix and use the Einstein formula $E = mc^2$.)

(d) Assuming that the fusion rate has been the same over the history of the Sun, how much mass has

it lost over the past 4.5 billion years due to this process?

(e) How do these mass losses compare to the total mass of the Sun today?

5. What is the mass density of the solar wind at Earth? The ram pressure of the wind, the force with which it pushes against a surface, is equal to one half the mass density times the velocity squared. Compare the ram pressure of the solar wind on a comet at 1 AU from the Sun with the ram pressure of a 5 m/s breeze on Earth (where the density of air is about 1 kg/m$^3$). You can neglect the motion of the comet itself. (Why?)

6. Speculate on the eventual fate of material blown off the Sun in the solar wind. Do you think other stars have winds? What might happen to the material in their winds?

7. The edge of the solar system can be defined as the place where the solar wind stops and interstellar gas begins. This takes place where the pressure of the interstellar gas is equal to the ram pressure of the solar wind.

(a) Assume that interstellar gas has a density of 1 proton/cm$^3$ and a temperature of 10,000 K. Use the ideal gas law (Equation 10.6) to find the pressure of this gas.

(b) Assume quiet Sun values: Set the density of the solar wind at Earth to 5 protons/cm$^3$ and its velocity at 400 km/s. Find the ram pressure as a function of *r*, the distance from the Sun.

(c) Equate your results from parts *a* and *b* and solve for *r*.

(This boundary, where a quickly moving low-density fluid meets a static, higher-density fluid, can be illustrated in any sink. Let a stream of water flow from a faucet onto a large flat surface, and notice how the circular disk of outflowing water stops abruptly at a thicker level of standing water.)

8. *Decimetric radiation from Jupiter.*

(a) What is the strength of Jupiter's magnetic field at 2 Jupiter radii? (The field can be approximated as a dipole field this close to Jupiter.) What is the cyclotron frequency for an electron at this position? For a proton?

(b) If an electron moves at one half the speed of light in Jupiter's magnetic field, 2 Jupiter radii from the center of the planet, what is the size of its cyclotron radius? (Ignore relativistic effects for the moment.)

(c) We saw in Chapter 2 that an electron moving back and forth creates electromagnetic waves. Electrons in cyclotron motion are, in effect, moving back and forth. What frequency of electromagnetic waves do you expect to see from Jupiter? (Hint: See part *a*.)

**(d)** If the electron or proton is travelling at a very high speed $v$, the cyclotron frequency as defined in this chapter must be multiplied by a factor of $\gamma^2$, where $\gamma = 1/\sqrt{1 - v^2/c^2}$ and $c$ is the speed of light. This produces **synchrotron radiation**. In the 1950s, radio waves with a frequency of $10^9$ Hz (cycles per second) were discovered being emitted from Jupiter. Two rival theories to explain these radio waves were (1) they were due to cyclotron radiation from particles in Jupiter's magnetic field, or (2) they were an example of synchrotron radiation. Given what we know now about the strength of Jupiter's magnetic field and your answers to parts $b$ and $c$, can you make any judgment about either theory? Can you estimate the speed of the electrons?

**(e)** Why is this radiation called "decimetric"? (Hint: The wavelength is equal to the speed of light divided by the frequency.)

## 13.5  FOR FURTHER READING

The University of Arizona Press has produced a book of review papers, *Comets*, edited by Laurel Wilkening. Though pre-Halley (1982), it contains a wealth of background information. The post-Halley view is brought together in a two-volume collection of papers following an International Astronomical Union colloquium (IAU Coll. 116), in *Comets in the Post-Halley Era*, edited by R. L. Newburn, Jr., M. Neugebauer, and J. Rahe (Dordrecht: Kluwer Academic Publishers, 1989).

John C. Bradt, *An Introduction to the Solar Wind* (San Francisco: W. H. Freeman, 1970), is still perhaps the most readable place to begin further solar wind studies. *The Sun in Time*, edited by C. P. Sonett, M. S. Giampapa, and Mildred Shapley Matthews (1991), is another University of Arizona anthology of review papers covering many topics involving solar wind studies.

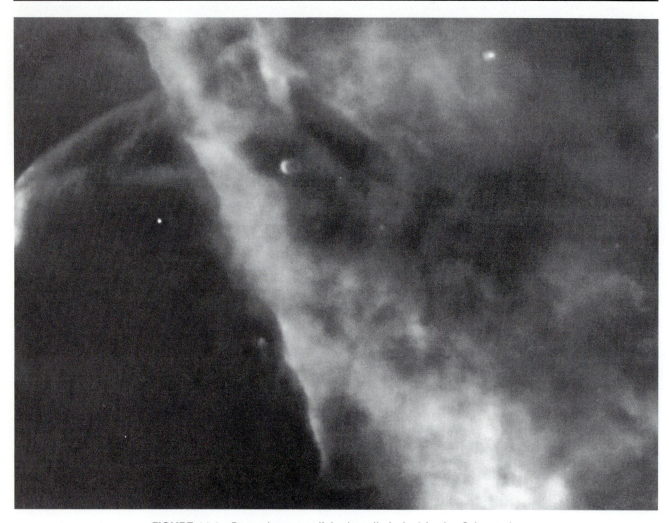

**FIGURE 14.1** Protoplanetary disks (small circles) in the Orion nebula imaged by the *Hubble Space Telescope*. This image is 0.5 light-years across.

# CHAPTER 14

# The Origin of the Solar System

## 14.1 THE FUNDAMENTAL PHYSICAL QUESTIONS

The origin of the solar system remains the single largest unsolved problem in planetary science, and as long as it is considered as one problem it will probably remain unsolved. However, it is not one problem, but a multitude of problems.

For most of human history, the question was insolvable because no one knew nearly enough about the solar system to begin to guess what processes must have taken place when it was formed. That state of affairs does not hold today. Instead, we have been flooded with new information about the planets. The difficulty now is to decide which facts about the chemistry and physics of the planets are generally significant constraints that must be met before any theory can be considered satisfactory, and which facts merely reflect chance happenings in a small part of a much larger system. It is possible to take any given oddity in the solar system and build an elaborate theory that can explain that oddity. The challenge is to explain all the oddities with the same theory.

We should start by listing the general trends seen throughout the entire the solar system and then ask what sort of theory is a good starting point to explain them. Once we have a theory that can explain the general trends, we can ask what needs to be done to fit it to specific, peculiar questions.

### Constraints on Origin Theories

One of the reasons a question such as, "How did the solar system form?" is so difficult to answer is that it is such a broad question. Instead, let us first look only into the physical structure of the solar system. Several specific questions come to mind, questions that any origin theory must successfully answer:

1. Why do all the planets orbit in the same direction?
2. Why do most, but not all, of them spin the same way they orbit, the same way the Sun spins?
3. Why do all the planets orbit in nearly (but not exactly) the same plane?
4. Why do all the planets orbit in nearly (but not exactly) circular orbits?

The planets all orbit the Sun in the same direction, and that's the same direction that the Sun itself spins. It's also the direction that the planets and most of their moons spin. The only exceptions among the planets are Venus, which barely spins at all (but does, in fact, spin "backwards"), and Uranus and Pluto, which are tipped over onto their sides. What's more, all the orbits (except Pluto's) lie in nearly the same plane, near the plane of the Sun's equator, and all the orbits (except Pluto's and Mercury's) are nearly circular.

For the Sun and its nine planets to move together in such similar paths is probably no coincidence. It seems reasonable to believe instead that some common process formed the whole system, all together, at more or less the same time. If this is true, then it means that the formation of the planets is intimately connected with the formation of the Sun itself. That means that we can use the theories of how stars are formed and observe places where young stars are being formed today to give us an idea of the kinds of conditions we could expect to be prevalent when our star, the Sun, and its planets were being formed.

## Observations of Young Stars

We saw in our study of nucleosynthesis (Chapter 2) that old stars spew dust and gas into space as they die. What happens to this gas and dust? Throughout the galaxy, especially along the spiral arms, we see clouds of glowing gas, dozens of light-years in diameter, called **diffuse nebulas**. One example is the Orion nebula, a fuzzy dot visible to the naked eye in the "sword" of the constellation Orion. Even a modest pair of binoculars shows that this spot is an irregular cloud of light in which a handful of stars are embedded.

Telescopes reveal many such nebulas throughout the galaxy. In most of these diffuse nebulas, clusters of stars are also seen. Furthermore, there seems to be a trend: The more stars visible, the thinner the nebula appears to be. Finally, we observe **open clusters** of stars (for example, the Pleiades), stars completely formed and shining, with mere wisps of gas still remaining around them.

From these observations, one can infer that the diffuse nebulas are regions in space where clumps of new stars are being formed. A nebula disappears when its stars are fully grown, when the stars have gathered all its gas and dust into themselves. Several arguments support this idea.

First, we note that it is unlikely to find a loose collection of hundreds of stars together in one place, unless they were all formed together. As each star of the cluster moves in its own orbit around the galactic center, it drifts away from the others. Given the positions of these stars relative to the center of the galaxy, one can use the laws of celestial mechanics to calculate that the whole collection of stars should be spread apart by this action in much less than a billion years. Thus, because these stars are all still together implies that they were all formed fairly recently out of the same large gas cloud.

Some of the stars in these clusters are quite massive (stars of types O, B and A; see the box on star types in Chapter 2), and the most massive of these stars usually have already evolved off the main sequence and into the red giant phase. That's not surprising, because we can calculate from theory that very large stars ought to shine for less than a hundred million years or so before they turn into red giants. But many of these large stars are *not* red giants yet. To calculate the age of the cluster, we look for the brightest star that is still on the main sequence. The age of this star then gives us the age of the cluster. We conclude that most of these clusters are less than a few hundred million years old.

When we look more carefully at stars embedded in nebulas of gas and dust and stars in open clusters, we see the sort of activity predicted by theorists for very young stars. For example, some of these stars are bright infrared sources, indicating that the energy of these stars is heating up surrounding clouds of dust. (Remember from Chapter 2 that hot stars radiate chiefly visible and ultraviolet light, while cooler solid material around the stars radiates infrared light.)

Stars of a type called **T Tauri** stars, named after star "T" in the constellation Taurus (the prototype for the class), seem to have very strong stellar winds similar to the solar wind described in Chapter 13. Large amounts of electrons and protons, and large amounts of dust, are being blown away from these stars and out into space.

Indeed, some of these stars emit large flares of x-rays. Such energetic activity is to be expected when a star starts to settle down into the main sequence, and the final energy of accretion that heated the collapsed cloud of gas and dust (just as accretion heats a planet, as we saw in Chapter 5) is replaced by the energy of atomic fusion.

These observations strongly suggest that the Sun and the planets were formed, together with a collection of other stars, out of a large diffuse cloud of gas and dust. The particular fragment of the gas and dust cloud that formed the Sun and planets is commonly called the **solar nebula**.

We can draw further conclusions. If the Sun was formed as part of an open cluster, then the early solar system was surrounded by many other stars, much closer to it than stars typically are today. We can infer that the early Sun experienced a period of intense solar wind, an outflow of gas and dust a billion times more intense than the current solar wind. We might expect that the early Sun was very active, emitting large flares of x-rays and energetic particles. This suggests certain things we might look for:

- Nearby stars that have already entered a red giant phase might inject isotopically different grains of stardust into the solar nebula; do we see evidence for such grains in the oldest meteorites? Possibly; some isotope anomalies might come from such a source, and tiny grains called **fremdlinge—**

German for "stranger"—have been found in certain meteorites.

- If more stars once existed closer to the solar system than is the case today, is there any evidence that their gravity perturbed the orbits in the early solar system? Yes; Earth's Moon and the moons of the outer solar system all seemed to have suffered from a great deal of impact cratering early in their histories. The gravity of other nearby stars might have perturbed the orbits of small planetesimals enough to send an increased number of such bodies onto collision courses with the evolving planetary systems.

- If the Sun went through a T Tauri solar wind phase or emitted large amounts of x-rays and flares, do we see evidence for that in meteorites today? Yes and no. Some meteorites have remanent magnetism consistent with their being formed in the presence of a strong solar wind, but there is no direct evidence that they experienced any unusual flux of solar wind particles. However, some isotope anomalies in the meteorites may be the result of hyperactive solar flares in the early Sun.

## Jeans Collapse

It seems reasonable to expect that the solar system, like any other star system, was formed out of a collapsing cloud of gas and dust. But what makes this cloud collapse? Why do stars form?

The British astronomer James Jeans (who also calculated Jeans escape for atmospheres, described in Chapter 7) deduced that gravity and gas temperature can interact to cause a gas cloud to collapse if that cloud is sufficiently large. But just how large is "sufficiently large"?

Consider a spherical cloud of gas of density $\rho$ and temperature $T$. The internal random motions of the molecules in the cloud—what we measure as the temperature of the gas—work to make the cloud expand. How fast the individual molecules move, and thus how rapidly the cloud expands, depends on the temperature of the gas: For a given temperature, the internal motions are the same for any size cloud. But on the other hand, the mass of the cloud creates a force of gravity that can lead to the cloud's collapse. The bigger the cloud, the more mass it has; and the greater the mass, the greater the force of gravity acting to pull the cloud together.

So there must be some critical size, $\lambda$, such that a cloud with a radius larger than $\lambda$ will have enough mass to collapse, in spite of its internal energy. Jeans showed that for uniform, spherical clouds this **critical Jeans length** is

$$\lambda = \sqrt{\frac{\pi v_s^2}{G\rho}} \qquad (14.1)$$

Here $G$ is the gravitational constant and $\rho$ is the density of the gas. The velocity $v_s$ is the speed of sound in the gas, a typical velocity for individual molecules in the gas. (Recall our discussion of Jeans escape in Chapter 7.) It is a function of the temperature of the gas, $T$, and the mass of the molecule, $m$; $k$ is Boltzmann's constant. Thus

$$v = \sqrt{\frac{3kT}{m}} \qquad (14.2)$$

In our galaxy, the typical density of gas is only $10^{-21}$ kg/m$^3$ and so, even though the temperature may be only a few degrees above absolute zero, the critical Jeans length is about 500 light-years. This would make a cloud so large that, even with such a low density of gas, it would have enough material for 100,000 stars. In fact, we do find clumps of 100,000 stars, all of similar age, gathered together into tight spherical groupings. They're called **globular clusters**.

Why did such a cluster break into 100,000 parts when it formed rather than make one huge star? The answer may be to note that, while it is collapsing, the gas cloud is also increasing its density. As a result, the critical Jeans length within the cloud shrinks faster than the cloud itself shrinks. As the Jeans length becomes a fraction of the length of the cloud, the cloud fragments into several smaller clouds. This process ought to continue for as long as the clouds keep collapsing, until eventually hundreds of thousands of stars are formed from one cloud.

Why aren't the Sun and all the other stars in the galaxy part of such globular clusters? Other motions of the cloud ought to scatter the collapsing pieces while they are still forming; for example, as the cloud orbits the center of the galaxy, following Kepler's orbital laws, the outer parts of the cloud move more slowly than the inner parts, so that the cloud will eventually be spread apart.

This is the mechanism by which open clusters like the Pleiades are gradually spread apart today. Thus the question is not, Why aren't all stars part of globular clusters? but rather, Why do any globular clusters stay together long enough to form in the first place?

Two factors may be operating. First, globular clusters are in fact very old, formed before the rest of the galaxy took shape, when there would not have been other concentrations of mass around to pull them apart while they were forming. Indeed, although the galaxy today is in the shape of a disk, globular clusters seem to be scattered around the center of the galaxy in a spherically symmetric distribution. The second factor is that globular clusters are much larger than open clusters. The typical open cluster has a few hundred stars, not hundreds of thousands. Thus it has less gravity holding it together.

Because open clusters are so much smaller than the mass you find inside a Jeans length for typical interstellar conditions, some additional force must be responsible for starting the collapse that eventually formed our own solar nebula. One suggestion has been that the shock wave from a nearby supernova could compress a small gas and dust cloud, increasing its density enough to start a collapse. (Recall from Equation 14.1 that as the density goes up, the size of the Jeans length shrinks.) This nearby supernova could account for some (but not all) of the isotope anomolies seen in meteorites. Another possibility is that a nebula gets compressed when it passes through the spiral arm of a galaxy. This passage occurs about once every 200 million years, and some isotopes (but not all) in meteorites seem to indicate a 200-million-year lag between nucleosynthesis and the formation of the meteorites.

## The Solar Nebula and the Fundamental Questions

From looking at how stars form we have concluded that the collapse of a cloud of gas and dust must eventually have evolved into hundreds of disks, each with roughly a star's mass of material. Our solar system emerged from one such disk, a nebula with a mass at least as great as the Sun's plus a little more for the planets.

Why do we say that this gas cloud will form a disk? If the cloud spins (which seems inevitable, as we discuss below), the centrifugal force within the disk should tend to support gas around the equator of our original collapsing sphere. Gas at the poles, however, can continue to collapse in towards the center without being held out by centrifugal force. Thus as it collapses a gas sphere becomes a flattened disk.

This spinning disk of gas goes a long way to explaining most of our basic questions about the structure of the solar system. If all the planets and the Sun formed from the same spinning disk, then it is only to be expected that they all orbit and spin in the same direction: the direction the original gas cloud was spinning. Furthermore, it is no longer a mystery why all the planets orbit in the same plane. We can conclude that all the planets formed together in the plane of the solar nebula disk.

However, we still have to ask why the orbits are so circular. Isn't it likely that bits of the disk could be moving towards or away from the center, as this gas cloud swirls around the Sun? Furthermore, we have to ask what caused our general trends to be less than perfect. If we'd expect all the planets to spin the same way, then what happened to Uranus, Pluto, and Venus? If we expect all the planets to be formed in the same disk, then why are the orbits slightly

askew? Finally, how do we go from a disk of gas and dust to nine very distinct, and very different, planets? Clearly, the idea of a solar nebula is an obvious starting point. But we need to know more about how the nebula evolved.

## Evolution of the Solar Nebula

A disk with one or two times the mass of the Sun and dimensions roughly the same as the present-day solar system has a much higher density than any interstellar cloud. The center of such a disk clearly is the area with the densest concentration of gas; the Sun forms here. What about the rest of the disk? Is it possible for it to continue to fragment into smaller, planet-sized lumps?

It seems both possible and likely. The majority of stars seen today are in binary or multiple star systems. Our solar system is somewhat unusual for having only one sun. Thus it has been suggested that Jupiter, Saturn, and the other gas giant planets are small fragments of the original solar nebula. It has also been suggested that the terrestrial planets may represent the rocky cores of protoplanets nearer the Sun, protoplanets whose gaseous outer layers eventually were pulled by tides from the Sun into giant tidal lobes, to be stripped off the rocky cores and into the growing Sun or blown off into interstellar space.

Appealing as this picture might appear at first, this idea has difficulty explaining the chemical differences between the planets. For example, why should Neptune be so much more dense than Jupiter if it's just a smaller version of a lump of solar nebula? Why should the density of Mercury be so different from that of Mars, and why should it have such a different chemical composition if they were both cores of similar solar nebula fragments?

On the other hand, physicists studying the stability of rotating gas disks insist that such a disk may well be unstable against breakup into these protoplanetary lumps. Perhaps giant protoplanet fragments of the nebula were formed and even processed interstellar grains into the chemical peculiarities we see in different meteorite classes today, but then were totally dissipated by the growing Sun without leaving any planets behind. In other words, such protoplanets could have been created and destroyed without having any direct connection with any of the planets we see in the solar system today.

How could the Sun destroy a protoplanet? Recall Figure 11.22 of the restricted three-body problem. Between the larger mass (here, the growing proto-Sun) and the somewhat smaller protoplanet is the *L1* point. The Sun, being at the center of the system, grows more quickly than the protoplanet, and as its mass increases, the position of the *L1* point shifts closer to the protoplanet. As the *L1* point touches the edge of the protoplanet's lump of gas, material is stripped off the protoplanet, passes through the *L1* point, and is added to the Sun. As the Sun grows, the *L1* point moves closer to the protoplanet, and more material is stripped. Eventually, most of the protoplanet could be disrupted, and consumed, in this way.

In this case, the disk may have eventually evolved into a young Sun surrounded by a large, broad ring of dust left over from the disrupted protoplanets. The force of gravity would pull bits of the dust ring together into rocks, and eventually planetesimals, snowballing into asteroid-sized bodies that, over time, would collide and stick together—**accrete**—into the planets we see today. The craters we see on the terrestrial planets are evidence that some sort of accretion process must have gone on at least during the later stages of planetary formation.

Meanwhile, what happened to the leftover gas? Some would have been pulled in to make the Sun; it might have been captured by protoplanets in the outer part of the solar system, far enough from the Sun to survive, at least partly, the "stripping" described above, thus creating the gas giant planets. Some gas might have been blown away from the solar system altogether in a massive solar wind, similar to the solar winds observed blowing from stars like T Tauri.

Again, this hypothesis suggests things we might look for in the solar system today:

- Do the planets appear to have been accreted from smaller bodies? Yes; there's hardly a solid surface in the solar system without craters, evidence of an intense bombardment early in its history.
- Is there evidence that gas of a solar nebula composition was ever trapped in meteorite grains? Yes, especially noble gases like neon and argon, which have isotope abundances in some meteorites quite different from those seen in planetary atmospheres but similar to what is observed in the Sun and solar wind today.
- Is there any evidence that the gas giant planets once had gas clouds much bigger than the planets seen today? Yes; something like an extended atmosphere must have been around Jupiter when it slowed down and captured the asteroidlike bodies that make up its irregular satellites; the same is true of Saturn. Furthermore, as we'll see below, the Galilean satellites themselves look like they were formed in a "mini-solar nebula."
- How well mixed are the chemical elements from planet to planet? This is a key question. We will address it at length below.

## The Angular Momentum Problem

One consequence of this theory is harder to deal with, however. If a gas cloud light-years across collapses into a collection of stars, and if a solar nebula as big as Pluto's orbit collapses into a Sun, the laws of physics predict that both systems should behave in one way quite different from what is actually observed.

It is a law of physics that the **angular momentum** of a spinning body must be conserved. Mathematically, the definition of angular momentum (recall from Chapter 4) is

$$L = mrv \qquad (14.3)$$

where $m$ is the mass of the spinning body, $r$ is the distance of this mass from the axis that it is spinning around, and $v$ is the speed at which the body is moving around this axis. If angular momentum and mass are conserved and $L$ and $m$ are constant, then making $r$ smaller must make $v$ larger. In the classic example, an ice skater spins faster when she pulls in her arms. What works for ice skaters also works for gas clouds. If the cloud had any spin at all when it started to collapse, then it should spin much faster as the collapse proceeds.

Let's start with a spherical cloud that has as much mass as our Sun, and whose density is just large enough for it to be 1 Jeans length across. (This makes $\rho = 10^{-17}$ kg/m$^3$.) Such clouds, called **cold molecular clouds**, have actually been observed in space. With this density it must have a volume of $2 \times 10^{47}$m$^3$ to have one solar mass of material; so its radius is $3.6 \times 10^{15}$ m, or about a third of a light-year.

Now put this cloud in orbit about the galaxy, about halfway out in the galactic disk (where our Sun is located), that is, at a distance of $a = 3 \times 10^{20}$ m from the center of the galaxy. For a body in a circular orbit moving through the outer parts of the galaxy, the angular speed of the orbiting stars is roughly constant regardless of position; however, this means that in a reference frame rotating with the center of the galaxy (recall Chapter 11), the cloud appears to be spinning once every time it orbits the galaxy (Figure 14.2). For a cloud a third of a light-year in radius to turn around once every 200 million years (the observed orbital period of our Sun about the galaxy), the outer edges of the cloud must be travelling at a speed of 3.6 m/s.

The angular momentum of 1 kg of material at the edge of this cloud, relative to its center, thus is

$$\begin{aligned} L_{kg} &= (1\,\text{kg}) \cdot (3.6 \times 10^{15}\,\text{m}) \cdot (3.6\,\text{m/s}) \\ &= 1.3 \times 10^{16}\,\text{kg} \cdot \text{m}^2 \cdot \text{s}^{-1} \end{aligned} \qquad (14.4)$$

A rough estimate for total angular momentum of the whole cloud can be found by multiplying Equation 14.4 times the mass of the Sun, $2 \times 10^{30}$ kg. This is only a rough approximation because clearly the angular momentum of the interior of the rotating cloud, where $r$ is less than the total radius of the cloud, starts with a smaller amount of

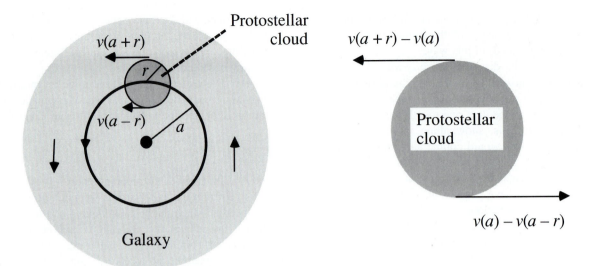

**FIGURE 14.2** Even if the galaxy were rotating everywhere at a constant rate, from the point of view of the gas cloud the inner parts would seem to be moving more slowly than the outer parts, and so the cloud would appear to be spinning.

angular momentum. Thus, looking merely to the nearest order of magnitude, rounding low, we find

$$L_{\text{cloud}} \approx 10^{46} \text{ kg} \cdot \text{m}^2 \cdot \text{s}^{-1} \qquad (14.5)$$

If $L$ stays constant when $r$ shrinks to 1 star radius, then $v$ must increase accordingly; when $r$ is set to the present-day radius of the Sun, $7 \times 10^8$ m, we can solve for

$$v = \frac{L}{mr} = 2 \times 10^6 \text{ m/s} \qquad (14.6)$$

This is about 1% of the speed of light!

Clearly, this does not happen. In fact, the Sun's spin today gives it an angular momentum of only $10^{41}$ kg·m²/s, roughly 1/100,000 of our estimated original angular momentum. Most of the original angular momentum must be lost. Where is it?

Some of it is in the planets. Jupiter, the most massive of the planets, orbits the Sun at a distance of about $8 \times 10^{11}$ m. The angular momentum of its orbit is $L = 10^{43}$ kg·m²/s, 100 times that of the Sun. Virtually all the angular momentum of the solar system can be found in the orbits of Jupiter and the other planets. Yet even this is still 1/1000 of the angular momentum of the collapsing cloud.

## Losing Angular Momentum

So we are faced with two questions. Where did 99.9% of the angular momentum of the collapsing cloud go? And how did the planets get most of what's left in the solar system? These are questions with no universally accepted answers. There are, however, general concepts that may be important clues to finding those answers.

The first point to note is that the spin of the Sun and the orbit of Jupiter are not lined up with the spin axis of the galaxy. In fact, there seems to be no correlation in general between the direction stars spin and the way they orbit the galaxy. This is quite unlike the case of the planets and the Sun, where the spin axis of each planet but Uranus, and virtually all their major satellites, is within 25° of the spin axis of the Sun. Thus in the galaxy there must be some large-scale process that throws the spin of the collapsing cloud fragments into random orientations.

The best examples of such random motions in a fluid, in our own experience on Earth, are turbulent eddies. **Turbulence,** the chaotic swirling motion of quickly moving fluids, grows naturally in rapidly flowing liquids and gases; it seems quite reasonable to expect extensive turbulence in a collapsing gas cloud, with eddies that add angular momentum from one section of the cloud and remove it from another. It may be that only those sections of the collapsing cloud that have lost their angular momentum through turbulence can succeed in becoming stars.

Another point to consider is that we have assumed a constant mass in all these calculations. In fact, it is possible (indeed, likely) that as the gas cloud collapsed it expelled large amounts of gas and dust. Such fragments, travelling at ever greater distances from the solar system, could carry away most of its angular momentum as long as they were "tied" to the spinning disk of the solar nebula. (Once they are completely detached, they only carry away their own angular momentum and no longer affect the rest of the disk.) Such angular momentum could be transmitted outwards through the disk by the viscosity of the gas, by turbulent eddies within the disk, or by a magnetic field that could be tied to a thin, partially ionized gas. Any or all of these processes, and more, may have been operating on the nebula as it collapsed.

What would happen to this gas, spun off from the solar nebula? It may well be the source of the Oort cloud of comets (see Problem 1 in Chapter 13).

## SUMMARY

Young stars are observed today to be forming in clouds of gas and dust. The planets all orbit the Sun in the same direction, the direction the Sun is spinning, implying that they were all formed from the same spinning gas cloud that formed the Sun. This cloud has been termed the solar nebula. The Sun would have formed in the center of this cloud; other regions of gas and dust, or just dust, may have collapsed due to their own gravitational attraction into planetesimals and planets. Planets may have been formed directly from such collapsing fragments, or they may have grown from the repeated collisions of smaller fragments. The intense cratering history on the oldest surfaces of the Moon and Mercury is evidence that these collisions did take place during the last stages of planetary accretion. These steps are sketched out in Figure 14.3.

## STUDY QUESTIONS

1. To understand the origins of planets, why do we observe regions where stars are forming?

2. For typical interstellar conditions, how many stars might be formed from a Jeans-collapse cloud?

3. What do we call the cloud of gas and dust from which the Sun and planets formed?

4. Young stars are formed from clouds of gas. After the gas is dissipated, the young stars are still all close to one another. What do we call these gatherings of stars?

5. What planet has the majority of the solar system's angular momentum?

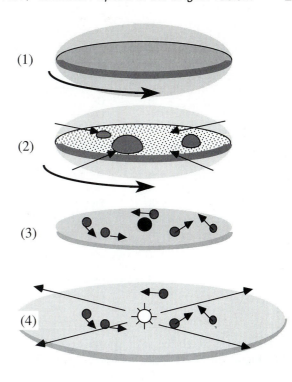

(1)

(2)

(3)

(4)

(5)

**FIGURE 14.3** Five stages in the evolution of the solar nebula. (1) Starting as a disk-shaped cloud of gas and dust. . . (2) the cloud collapsed into fragments. . . (3) that began to orbit about the largest fragment, the proto-Sun. (4) As the Sun ignited, the energy from its solar wind and the pressure of its light combined to blow away the rest of the gas. . . (5) leaving behind rocky planetesimals, which would eventually collide and form into planets.

## 14.2  CHEMICAL ASPECTS OF THE ORIGIN PROBLEM

The Sun and the planets seem to have been formed out of the same sort of material. The Sun, Jupiter, Saturn, and most stars seem to share the same general trend of cosmic abundances in their relative proportions of each chemical element. If we assume that the rocky planets and meteorites merely represent the nongaseous elements, then we can ask if they too reflect these cosmic abundances.

Each isotope of each element, with few exceptions, seems to be in a constant proportion with other isotopes of

**TABLE 14.1**

The 20 Most Abundant Elements and the Compounds They Form

| Element | Compound | Type | Element | Compound | Type |
|---------|----------|------|---------|----------|------|
| H | $H_2$ | A | Ar | Inert gas | A |
| He | Inert gas | A | Al | $Al_2O_3$ | L |
| O | $CO, H_2O$ | A | Ca | CaO | L |
| C | $CO, CH_4$ | A | Na | $Na_2O$ | L |
| Ne | Inert gas | A | Ni | Ni | S |
| N | $N_2, NH_3$ | A | Cr | $Cr_2O_3$ | L |
| Mg | MgO | L | Mn | MnO | L |
| Si | $SiO_2$ | L | P | $PH_4, P_2O_5$ | A, C |
| Fe | $Fe, FeS, FeO$ | S, C, L | Cl | HCl | A |
| S | $H_2S, FeS$ | A, C | K | $K_2O$ | L |

Note: Compounds are given in the order of decreasing formation temperature within the solar nebula. The types of compounds include: A, atmophiles; L, lithophiles; C, chalcophiles; and S, siderophiles.

the same element in all the solar system materials we've tested—Earth and Moon rocks, meteorites, even the solar wind—to one part in 1000. This implies that all the material in the solar system came out of the same well-mixed pot. However, at the part-per-10,000 level, significant differences do show up in isotopes of certain elements from certain meteorites, as discussed in Chapter 9. Many theories have been developed to try to explain this observation, but we still do not entirely understand everything this is telling us about the formation of the solar system.

According to the solar nebula theory, the Sun was formed at the center of a disk of gas and dust, in the hottest and densest region. Convection, carrying the heat of the Sun's accretion outwards, should lead to gradually falling temperatures farther out in the disk, and the motions of the convecting parcels of gas and dust should serve to keep the disk well stirred.

As the disk was collapsing it would warm up, just as gravitational potential energy heats up accreting planets. However, after accretion of the nebula was complete and the nebular gas had begun to disperse into space, the materials in the solar system would cool off. Thus, between these two times, there was a time in the solar nebula when the temperatures reached a maximum point. Many detailed theoretical studies since the 1960s have attempted to estimate exactly how hot the solar nebula may have become and what these maximum temperatures might have been at various distances from the center of the disk.

Once given the temperature and pressure at some point in the solar nebula, it should be possible to determine the chemical species that should exist at that point. The abundances of these chemical species in turn should be reflected in the compositions of the meteorites and the planets that were formed from a given region of the solar nebula.

## The "Loves" of an Element

Take a piece of the Sun and cool it down. What chemical species might you expect to form? Table 14.1 lists the 20 most abundant elements and the compounds they are most likely to form with each other. These compounds can be divided into four classes, depending on where they are most likely to be found in a planet.

**Atmophiles** (literally, "lovers of the air") are gases, compounds that stay unfrozen until the coldest outer reaches of the solar system are reached. Any frozen atmophile (not just water) we call an **ice**. Some ices do exist even as close to the Sun as Earth; but for the most part, atmophiles are found in the atmospheres of planets.

**Lithophiles** are "rock-loving" compounds. The most common lithophile elements combine with silicon and oxygen to make silicate minerals; rocks are simply a mixture of such minerals (see Chapter 3). There are many rock-forming minerals without silicon, too. These include many very oxidized minerals such as calcium carbonate ($CaCO_3$), magnesium sulfate ($MgSO_4$), and magnetite ($Fe_3O_4$).

It is also conceivable that oxygen could have been completely absent when solid minerals were being formed. In such a case, sulfide or phosphide minerals such as oldhamite (CaS) and schreibersite ($Fe_3P$) could form. Even in the presence of oxygen, troilite (FeS) forms, and it is a very common mineral in many meteorites. Elements that tend to be found in these sulfide minerals are called **chalcophiles** ("copper-lovers"). These minerals are usually found with metal rather than with rocks.

Pure metals, such as iron and nickel, belong to the class of elements called **siderophiles**, or "iron-lovers." Rare metals such as iridium, platinum, or gold may be found along with iron and nickel.

As can be seen in Table 14.1, some elements may exist in several different compounds, including compounds in several different classes. But which compound is present at any given time depends on the chemical conditions. For instance, if the temperature, pressure, and other chemicals present all favor sulfur reacting to form troilite, then it is unlikely that there would be any sulfur present in magnesium sulfate.

## Equilibrium Condensation Theory

It should be possible to tell which of these compounds are present and which are absent by simultaneously solving (on a computer) all the different chemical equilibrium equations for the formation of each possible compound for any temperature and pressure in the solar nebula. Figure 14.4 shows the highlights of these calculations.

The easiest (and traditional) way of understanding this graph is to pretend that we had a piece of solar nebula that was very hot—over 2000 K—and slowly let it cool. It's not likely that any part of the solar nebula ever got this hot, except near the Sun or deep inside giant gaseous protoplanets (assuming they ever existed). But one principle of chemical equilibrium is that the chemical species are completely determined by the local temperature and pressure. As long as nothing is physically removed from the system, it doesn't matter whether we got to that local temperature by cooling from a higher temperature or heating from a cooler one. But, it is easier to picture what's happening if we start out hot and cool down.

At very low pressure, everything is vaporized at 2000 K; everything is a gas. As we start to cool the gas, at roughly 1800 K (the precise temperature depends on the pressure of the nebula gas) the first, most **refractory** materials (those with very high melting and vaporization temperatures) condense out of the gas and become solids. They include metals such as tungsten (the stuff of light bulb filaments) and, at slightly cooler temperatures, refractory oxides such as corundum ($Al_2O_3$) and perovskite ($CaTiO_3$). Recall that materials of this composition were found in the white inclusions of the Allende meteorite. Many trace elements, including the rare-earth elements and uranium and thorium, also condense into solids at this time. The last two elements are important because they are mildly radioactive; as they decay, they heat up any solid material around them.

At about 1300 K, metallic iron and nickel condense, as well as the first major silicate mineral, enstatite ($MgSiO_3$). (Recall the unusual composition of the enstatite meteorites.) Also near this temperature, the previously condensed calcium and aluminum minerals react with silicon and oxygen in the gas to form the calcium-rich plagioclase mineral, anorthite ($CaAl_2Si_2O_8$).

Just below 1000 K, the other feldspars of sodium and potassium, albite ($NaAlSi_3O_8$) and orthoclase ($KAlSi_3O_8$),

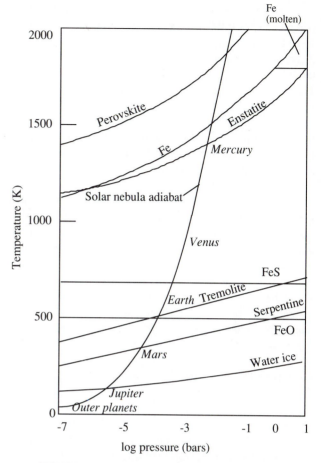

**FIGURE 14.4** The equilibrium condensation sequence in a solar nebula. The lines mark the P–T conditions where various representative minerals begin to condense or are formed by a reaction with a preexisting condensate and the gas. An adiabat for the solar nebula is shown, with suggested condensation ranges of each planet, Mercury through Jupiter and the outer planets.

condense out of the gas. At 700 K, the metallic iron reacts with hydrogen sulfide in the gas to form the first major chalcophile, troilite (FeS). At this point, reactions of the iron metal with water and enstatite also start to become important; this gradually introduces iron oxide into the silicate minerals. By 500 K, all the metallic iron has been consumed, and the iron is now present either as FeS or FeO in rocks. Material found in ordinary chondrites may have been formed under these conditions.

Some trace elements, such as thallium and lead, also condense at these lower temperatures. These elements are called **volatile** elements, from the Latin word *volare*, "to fly"; even a moderate amount of heating induces them to "fly" out of the solid back into the nebular gas. Atmophiles are, obviously, the most volatile of the elements.

At about 500 K, the first hydrous mineral, tremolite ($Ca_2Mg_5Si_8O_{22}(OH)_2$), becomes stable. This is a mineral resembling a calcium-rich pyroxene that has OH molecules incorporated into its crystal structure; thus any rock containing this mineral has the elements of water present in it. The amount of water that can be contained in a rock this way is limited by the amount of calcium available in the solar nebula. However, at around 400 K, the mineral serpentine ($Mg_3[Si_2O_5](OH)_4$), familiar to us as asbestos, can form. Because olivine (formed by the reaction of enstatite and FeO) is more abundant than calcium, much more water can be incorporated into rocks.

Still, given the large cosmic abundance of oxygen compared with the rock-forming elements, there's lots of water left in the solar nebula even after these hydrous minerals form. At roughly 200 K, this water vapor condenses into water ice. At 125 K, ammonia in the gas can react with this water to form ammonium monohydrate. At 100 K, methane can be incorporated into voids in the water-ice crystal structure to form methane **clathrate**. By 50 K, other clathrates incorporating Ar or $N_2$ may form, and methane ice itself may freeze.

## Compositions of the Planets

How do these calculations relate to the compositions of the planets? The curve noted in Figure 14.4 shows an **adiabat**, a line that shows the temperatures and pressures that might exist within the solar nebula, given certain starting conditions. Closer to the Sun, higher temperatures and pressures should occur; farther away, the gas becomes colder and thinner.

We would expect that a planet formed at any given spot in the nebula would consist only of those materials that were solids in that part of the nebula. Thus for Mercury the theory of equilibrium condensation in the nebula predicts that refractory metals and oxides, metallic iron, enstatite, and anorthite should be present. Mercury would be poor in potassium and sodium, iron sulfide, or iron oxide and should not have any water. Venus should contain feldspars with sodium and potassium, but still have no iron sulfide and no water. Earth would have tremolite and therefore some water; it should also have iron sulfide and some iron oxide in the rocky minerals as well as metallic iron. Mars should have serpentine and thus much more water; it should also be rich in iron oxide and completely devoid of iron metal. At Jupiter and beyond, ice-rich planets should be formed. How do these predictions match our understanding of the compositions of the planets?

We know that Mercury is unusually dense, probably because it has a larger iron-to-silicate ratio than cosmic abundances predict. Because the position of Mercury is close to the condensation curve of both enstatite and iron, it is conceivable that the condensation of iron could be slightly favored over that of enstatite. Alternatively, it has been proposed that the nebula gas (which was denser here than around any of the other planets) may have swept away the lighter dust grains of rocky material more than the denser grains of iron, thus increasing the relative iron-to-rock concentration in the region where Mercury formed.

Mars, at the other extreme, is the least dense of the four terrestrial planets. This is consistent with the presence of water and with iron present as an oxide or a sulfide, instead of the much denser metal phase.

The absence or presence of iron sulfide is also important because it dramatically lowers the melting point of iron, allowing a liquid core of an iron–iron sulfide eutectic to form. The convection of such a core is thought to produce the magnetic field of Earth. This theory is consistent with the absence of fields on Venus (no sulfur) and Mars (no metallic iron) and the presence of a field on Earth; the field of Mercury, however, does not fit the theory.

There is little or no iron oxide in Mercury's rocks, as far as we can tell. We don't know yet about potassium and sodium, because these are difficult to measure with just reflectance spectra. On Venus, sodium and potassium have been detected by the Soviet space probes that landed on its surface, and the observed water content of the atmosphere is very low, nearly 100,000 times less abundant than on Earth. Even more water than Earth has been inferred for Mars, at least in its past; this water is possibly frozen below its surface today. The red color of Mars is due to oxides of iron in the Martian dust. The observed densities of Venus, Earth, and Mars are all consistent with the predictions of this theory.

## Chemistry of the Outer Solar System

The outer solar system is chemically quite different from the realm of the terrestrial planets. The gas giant planets obviously have a different composition; in some way they represent pieces of the solar nebula that were captured into planet form. The planets themselves are much bigger and much farther apart from each other than the inner planets. The distance from Mercury to Mars is less than 2 AU; the gap between Jupiter and Saturn alone is more than twice that distance.

The presence of ice can result in some significant differences in the way accreting materials behave. For one thing, as soon as ices become stable, we triple the amount of solid material available to form a planet, because there's twice as much water as rocky material in the solar nebula. The more material that's available, the sooner it will find other solid material to run into and accrete. In addition, ice near its melting point is "sticky"; if grains of ice collide in the outer solar system, they're more likely to stick together (whereas rocky materials might bounce or shatter) and snowball into a bigger body, which can then accrete yet more material.

As a result, we expect that planets should accrete first in the region where there's the most solid, sticky material available, at the distance from the Sun where ice is first stable. That's where Jupiter is today. If a large enough planet accreted before the Sun's early solar wind swept away the solar nebula gas, then this planet's gravity could capture a substantial amount of that gas. This may be the way in which the gas giant planets were formed.

There's another complication that may occur in the outer solar nebula. Chemicals react very slowly when the temperature is cold, and if the reactions are too slow, chemical equilibrium will not be achieved. It is possible that chemical species that ought to be favored, according to the laws of equilibrium chemistry, will not form at all.

As an example, the nitrogen in the solar nebula ought to exist as $N_2$ gas in the warmer parts of the nebula and as $NH_3$ gas, ammonia, where the nebula is cooler. The ammonia eventually should be dissolved into water, and there is reason to believe that such a solution could be an important constituent in the icy moons of the outer planets (see Chapter 12). Likewise, carbon should be present as carbon monoxide gas, CO, in the warmer parts of the nebula and as methane, $CH_4$, where the nebula is cooler.

However, the rate at which nitrogen gas reacts to form ammonia, and the rate at which carbon monoxide reacts to form methane, is very slow when the pressure of the gas is very low. And most nebula models predict that the pressure of the solar nebula beyond Jupiter is not much more than a billionth of an atmosphere, close to a good vacuum by laboratory standards. At this low pressure, calculations suggest that the reactions to form ammonia and methane take longer than a million years to complete. By that time, the gas of the solar nebula would be completely dissipated by the T Tauri wind of the early Sun.

In the higher pressure regions around the planets, however, the reaction rates are quick enough to allow the formation of substantial amounts of these compounds. Thus we conclude that the moons should be rich in methane and ammonia, while comets (formed in the solar nebula itself, not near a gas giant planet) should be rich in carbon monoxide ice. We used this idea in Chapter 11 to argue that Pluto was formed near a gas giant, not as a planet by itself.

## Problems with the Theory

With the exception of the Mercurian magnetic field, there is no gross inconsistency between this theory and our knowledge of the planets. However, there are many chemical trends among the planets that this theory does not explain.

The origin of the inner planets' atmospheres is still not completely understood. We know that the atmospheres of the terrestrial planets are not just some leftover bit of solar nebula because they are poor in isotopes such as $^{36}Ar$ or $^{22}Ne$ that are plentiful in the Sun (and so presumably in the solar nebula), while being rich in gas isotopes such as $^{40}Ar$ that are produced by the radioactive decay of lithophilic isotopes such as $^{40}K$. It must be that some gases were trapped in the accreting rocks and then were outgassed to the surfaces of the rocky planets after they were formed. Oxygen may have come from hydrous minerals, and carbonaceous material in the planet could react with this water to produce carbon dioxide. Nitrogen is rather easily dissolved in iron meteorites. But all the details of how the chemical compositions of these atmospheres actually evolved into what we see today remain to be worked out.

In addition to the higher density of Mercury, the unusually low density of the Moon must be accounted for. Perhaps more embarrassingly, the mere fact that the Moon's composition is different from Earth's, even though they are both the same distance from the Sun today, shows that simply correlating composition with position is probably too simple-minded to reflect all the forces that influenced the chemical composition of the planets.

(As mentioned in Chapter 3, a recent theory that may explain the Moon's composition is that it was formed when a large planetesimal struck an already-formed Earth, erupting material from Earth's mantle into a ring of rock that eventually formed the Moon. The nineteenth-century idea that the Moon was erupted from the Pacific Ocean basin has been ruled out on a number of grounds. Moon rocks are far older and of a completely different composition than the rocks of Earth's oceanic crust. Likewise, forming the Moon elsewhere and then capturing it by Earth can't be done without doing violence to the laws of celestial mechanics or without stressing and cracking both the Moon and Earth in ways that clearly never happened.)

The basaltic meteorites also appear to have come from planets that contain calcium feldspars but little sodium or potassium, implying a condensation temperature between 1000 K and 1200 K. However, both the Moon and the basaltic meteorite parent body also appear to have been rich in iron oxide; FeO makes up a significant part of both kinds of rocks. But according to our theory, iron oxide should not occur in minerals unless the equilibrium temperature was below 500 K.

The distribution of mass in the solar system appears to follow a general trend (see Problem 4), with planets getting progressively larger out towards Jupiter then smaller as one moves farther from the Sun. But the small size of Mars and the lack of a planet in the asteroid belt are obvious problems with this trend. Something more than simple accretion must have been going on in this region of the solar system. Mars is the planet that was formed closest to Jupiter, whose gravity could have perturbed material into or out of the zone where Mars-forming planetesimals were accreting. But any material—more than the mass of Earth, if the general trend of planetary masses is to be believed—that could be perturbed into a Mars-like orbit, could be sent to all the other inner planets. If there is too much mixing of accretion material from the outer solar system into the region where the terrestrial planets were formed, then the very chemical differences from planet to planet that are the key point of the whole theory will be eliminated.

The increasing density of gas giant planets as one moves away from the Sun is also surprising. If Jupiter, Saturn, Uranus, and Neptune are all lumps of the solar nebula, then why are Uranus and Neptune so much denser than Jupiter and Saturn? This difference argues against all four planets being the remnants of giant gaseous protoplanets; instead, they all started as large balls of rock and ice formed early enough to capture some leftover solar nebula gas before it was all dissipated from the Sun. Was there just less gas around when the outer two planets were formed? Did they form after Jupiter and Saturn? Or is there something more fundamentally different about their origin?

So the equilibrium condensation theory has its problems. Does this mean that the theory is completely wrong? Perhaps it is. On the other hand, it may be that each planet had some unique process that acted to change slightly its chemical content away from the general trend predicted by this theory.

Mercury is unusual for being so close to the Sun, where the nebula was most dense and so gas drag could separate out denser iron grains from less-dense rock grains. Earth and Venus may have received part of their atmospheres as a thin veneer of volatile rich material from the region where Mars and the asteroids were forming. The Moon may have been formed elsewhere in the solar system, perhaps closer to the orbit of Mars, and was perturbed into a violent collision with Earth, leaving its iron metal in Earth and losing its volatile atmophiles such as water in the process. The accretion of Mars and the asteroids could have been disrupted by the gravitational perturbations of an early-forming Jupiter. The equilibrium condensation theory may be the starting point for understanding all the planets, but it fully explains none of them.

## SUMMARY

According to the present theories for the origin of the solar system, the planets formed out of a gas of solar composition called the solar nebula. Knowing the temperature and pressure of this gas at any distance from the center where the Sun formed, one can use the laws of chemical equilibrium to determine which compounds should be solid and which ones gaseous near the regions where each planet formed. We can assume that the terrestrial planets were formed out of the solid condensates. Thus this theory can predict the chemical composition of each of these planets.

These predictions seem to be in general agreement with many of the observed characteristics of these planets, but in many details the theory seems to be inadequate (Table 14.2). Most particularly, the composition of the

***TABLE 14.2***

The Compositions of the Planets per the Equilibrium Condensation Theory

| Planet | $T_{cond}$ | Minerals | Problems |
|--------|--------|----------|----------|
| Mercury | 1700 | Metallic iron, enstatite | Iron core too big |
| Venus | 900 | ...plus feldspars, olivine | Sulfur in atmosphere |
| Earth | 600 | ...plus FeS, some FeO, tremolite | Sulfur hidden in core? |
| Moon | 600 | ...same as Earth | No water, low alkalis |
| Mars | 400 | ...plus, serpentine, all FeO | Lost water since formed? |
| Asteroids | 300 | Hydrated minerals | M class unoxidized |
| Jupiter | 200 | ...plus water ice | Captured solar nebula |
| Saturn | 125 | ...plus ammonia ice | Captured solar nebula |
| Uranus | 100 | ...plus methane clathrate | Too much water today? |
| Neptune | 75 | ...more methane? | Too much water today? |
| Pluto | 50 | ...methane? or carbon dioxide? | Big comet or lost moon? |
| Comets | 25 | Water, carbon dioxide, nitrogen ices | Kinetically inhibited |

Notes: The condensation temperature, $T_{cond}$, is an estimate of the highest temperature experienced by solid material at that planet's distance from the Sun. The rightmost column notes some of the ways in which the planet differs from the prediction, that is, the iron core seen in Mercury is larger than simple equilibrium chemistry alone predicts.

Moon, the density and magnetic field of Mercury, and the atmospheres of the inner planets are different in several respects from that predicted by the theory. The relatively small sizes of Mars and the asteroids also goes beyond the predictions of this theory. Nor does the theory explain why Uranus and Neptune are more dense than Jupiter and Saturn. Differences between predicted and observed compositions may represent a basic weakness with the theory, or they may reflect aspects of each planet's unique formation history.

## STUDY QUESTIONS

1. Refractory materials are those that are present as solids in the solar nebula, even at high temperatures. Name one.

2. Name a planet that we expect to be rich in volatile materials.

3. Name one of the basic assumptions used in our predictions for the chemical compositions of the planets.

4. Venus is poor in water, while Earth is rich. How does the equilibrium condensation theory explain this?

5. Name one case where the equilibrium condensation theory fails.

## 14.3  IO: A CASE STUDY IN PLANETARY EVOLUTION

What is in the solar system, and how did it get to be the way it is today? These have been the questions motivating our study throughout this book. Given all that we have learned, let us now apply it to the case of one specific body: Io, a complex and fascinating moon of Jupiter. The model developed in this section is sketchy and in some parts still controversial. We present it not as the final truth about Io, but rather as an example of how the study of a moon or planet can begin.

The *Voyager* flyby of Jupiter in 1979 gave us our first close-up view of Io, a moon with fountaining volcanoes and hot spots, strangely colored plains, and large surface features that appeared to be young and constantly changing (Figure 14.5). Even over the time of 1 year, between *Voyager 1* in 1979 and *Voyager 2* in 1980, distinct changes in the appearance of the surface and the number of active volcanoes were seen.

This colorful and active planet is quite different from Earth's Moon, even though both Io and the Moon have similar radii (1815 km for Io, 1738 km for the Moon) and

**FIGURE 14.5** Io as photographed by *Voyager*. The spots on the surface are active volcanoes.

similar densities (3.6 g/cm$^3$ for Io, versus 3.4 g/cm$^3$ for the Moon). Two factors may have led to these differences.

First, the two bodies must have started with significantly different chemical compositions. Their similarity in density is almost certainly just a coincidence, a point worth remembering when using density to guess at the composition of a planet. Whereas the Moon is depleted in both dense iron and light volatiles, Io is likely to be relatively rich in both.

Second, both now and probably during a substantial part of its history, Io must have been subjected to much more heating. Tidal heating arising from the stresses of tides that Jupiter raises on Io is believed to be the source of the heat that produces the very active volcanoes seen there today. The measured flow of heat out of Io is 100 times that measured by the Apollo astronauts out of the Moon.

What are the possible combinations of elements that could have made up Io initially? How would all this energy have caused these chemicals to react? What does the interior of Io look like today? These all are questions that chemical and physical models of this planet try to answer. By observing how one can go about answering these questions, we can see how the evolution of any planet can be studied, and we can see how one interesting planet in particular, Io, may have changed radically over the age of the solar system.

### Starting Conditions

The chemical composition of Io was probably determined by the chemistry of the gas around Jupiter early in its history. We know that Jupiter has a composition very similar to that of the Sun. It is believed that the major moons of Jupiter also formed out of such a gas. Because they are all solid bodies, not gas planets like Jupiter, their chemical composition would reflect the composition of those materials that existed as solids in chemical equilibrium with the gas.

The four Galilean moons have densities that decrease regularly as one goes farther from Jupiter. This is consistent with the idea that each satellite accreted successively greater amounts of low-density volatile material such as water ice. Jupiter must have been quite warm as it was forming, and its heat would have kept water ice from forming near Io. Europa, which shows an icy surface but has a density like that of rocky material, may have been formed from minerals like tremolite or serpentine, which are rich in water. As the satellite evolved, this bound water would have been driven out of the rocks in its interior and frozen on the surface. Judging from their low densities, Ganymede and Callisto appear to have accreted water ice directly from the proto-Jovian nebula.

In the denser nebula gas around a planet, condensation of some minerals can occur at a higher temperature than in the solar nebula. Tremolite now condenses at a temperature of about 750 K, compared with the 500 K condensation point in the solar nebula (Figure 14.6). The condensation temperature of the major sulfur-rich mineral, iron

sulfide, is still 700 K, however. This means that, unlike in the solar nebula, water-bearing and sulfur-bearing minerals condense at about the same time, and any reactions inside Io involving sulfur should also involve water. Serpentine is condensed at temperatures below 550 K. Once formed, serpentine has a much greater abundance than tremolite because there is much more iron and magnesium in the gas than there is calcium. Iron oxide is also condensed at about this temperature. Water ice does not freeze out of the gas until the temperature drops below 300 K.

Observations of the surface of Io can put some more specific constraints on the temperature of the gas where it was formed. Sulfur and sulfur dioxide have been seen on its surface, so one can infer that the gas must have been cool enough for a sulfur-bearing compound such as iron sulfide, FeS, to be solid and present in the satellite. Water is completely absent, on the other hand. Whatever water was carried into Io by minerals such as tremolite or serpentine must have been driven off the surface into space, perhaps by high-energy particles from Jupiter's magnetosphere. Because Europa did not lose all its ice, it is likely that it started with more water than Io. Perhaps it was formed where serpentine was condensed, while the water in Io came only from tremolite.

## Heat Sources in Io

Io's spectacular volcanoes (Figure 14.7) are evidence of significant heating inside the moon. Io's interior must have been warmed by the decay of radioactive isotopes such as uranium, thorium, and potassium in its rocks, as discussed

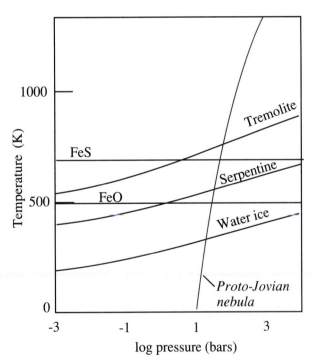

**FIGURE 14.6** The condensation sequence near Jupiter. Here the pressure is higher than at the solar nebula, and so iron sulfide and water-bearing rocks like talc and tremolite may be stable at similar temperatures.

**FIGURE 14.7** Io close up, looking down on a volcano. The dark lines radiating away from the spot in the bottom of the picture are lava flows, presumably rich in sulfur.

in Chapter 5. In addition, however, Io is subjected to tidal stresses that today appear to be producing 100 times as much heat as the radioactive elements would produce.

Recall our discussion of tidal heating in Chapter 12. The tidal heating of Io arises from its interactions with both Jupiter and Europa. The orbit of Europa is in a resonance with that of Io such that Europa's gravity tugs on Io in exactly the same place in Io's orbit, orbit after orbit. At the same time, tides from Jupiter are pulling at Io, trying to reduce its orbit's eccentricity. The tidal forces of Jupiter stress and heat Io as long as Io's orbit is eccentric, and this eccentricity is enforced by Io's resonance with Europa. As far as we can tell, Jupiter, Io, and Europa have been in this same configuration for much of the age of the solar system, so it is possible that Io has experienced this strong tidal heating for most of its history.

The result of all this heat is that Io's interior has probably been molten or nearly molten for most of its existence. This heat inside the planet is sufficient to drive any light, gaseous components of the planet (such as water) to its surface very early in its history. The heat also should cause the rocks to convect; they should overturn slowly, with warm rocks from deep inside the planet welling up to the surface while cooler surface rocks are being dragged down into the planet. This convection would tend to stir up the interior. Gases flowing through the rocks to the surface, convection mixing the rocks themselves, and the high interior temperature can all contribute to chemical reactions inside Io.

## Chemical Reactions in Io

We see sulfur and sulfur dioxide ice on the surface of Io today. Where did these chemicals come from?

The problem is this: If Io were formed at a temperature in the 550–700 K range, then strict equilibrium between the Jovian nebula and the condensed material would allow both iron metal and iron sulfide (Fe and FeS) to be present originally. But iron sulfide is more chemically reduced than the sulfur and sulfur dioxide seen on Io today. Thus if our understanding of Io's initial composition is correct, then chemical reactions inside Io must have liberated enough oxygen to react with iron sulfide and produce free sulfur and sulfur dioxide.

Metallic iron is also chemically reduced; it reacts with any free oxygen to make iron oxide. Thus it competes with iron sulfide for oxygen. Furthermore, direct heating of iron and iron sulfide inside Io, without accounting for the other chemical reactions, leads to the melting of an Fe–FeS eutectic composition. A dense molten iron–iron sulfide mixture would then drain to the center of the planet to form a metal rich core with no sulfur appearing on its surface at all. Thus our chemical reactions must have occurred before all the sulfur was drained away into a core and lost from the upper regions of the satellite.

What sort of reactions might have occurred?

The oxygen we need could come from water. Recall that our condensation calculations for a dense Jovian nebula imply that some water will already be present, in the mineral tremolite, at the temperature where iron sulfide condenses. If Io were formed from material originally rich in sulfur and water, then tidal heating would drive water out of hydrous minerals such as tremolite, or epsomite and serpentine if they are also present. As the water was released some small fraction of it would dissociate into hydrogen and oxygen gases, $H_2$ and $O_2$, with the amount of gas produced governed by the rules of chemical equilibrium, dependent on the temperature reached inside Io.

Unfortunately, the pressure of oxygen produced by this reaction at the relatively low temperature inside Io is quite small, because in equilibrium the hydrogen produced recombines with the oxygen to make water as quickly as the oxygen is produced. However, if the hydrogen is removed by some other process, more water dissociates to replace the lost hydrogen, and this produces more oxygen.

Thus to make sulfur we need free oxygen, and to free up enough oxygen to make the sulfur we need to get rid of hydrogen. The easiest way to lose the hydrogen is for hydrogen-bearing gas to **outgas**, to expel itself from the interior of the planet to the surface and escape into space.

At low temperature, the pressure of hydrogen itself is not great enough for it to percolate quickly through Io and escape to space. But carbonaceous meteorites contain carbon; the carbon makes up about 1–5% of the mass of CM and CI chondrites. Any carbon present in Io could react with the released hydrogen gas to form methane ($CH_4$) if the temperature is below 1000 K. However, it reacts with oxygen to make carbon dioxide ($CO_2$) if the temperature exceeds 1000 K; so we don't want the interior to be too hot. Equilibrium calculations predict that this methane should have a pressure of about $10^{10}$ Pa (about 100,000 atmospheres). Because the lithostatic pressure at the center of Io is only $5 \times 10^9$ Pa, this gas can easily overcome the pressure of the rocks above it and escape to the surface.

So the carbon escapes with the hydrogen, which frees up the oxygen, which reacts with the iron sulfide, to make iron oxide plus free sulfur. Further reaction of oxygen with the sulfur would produce sulfur dioxide gas ($SO_2$). Because each carbon atom takes four hydrogens with it, each carbon atom frees two oxygens for reaction with metallic iron.

Furthermore, each atom of iron weighs about five times as much as a carbon atom. It follows, therefore, that each gram of carbon can free enough oxygen to oxidize 10 times its mass in iron. If the original composition were 1% carbon, then iron representing 10% of the mass of Io could be oxidized.

After all the methane has been outgassed, it takes higher temperatures to produce more oxygen to react with any metal that's left. Even at higher temperatures, equilibrium chemistry calculations indicate that the hydrogen gas formed along with the oxygen has a pressure only on the order of $10^8$ Pa, which may not be enough to escape from the deep interior of Io. Thus it is possible that only the upper 300 km or so of Io may be completely oxidized.

If enough free oxygen exists after the iron metal is oxidized, rising temperatures eventually produce carbon dioxide (if any carbon is left) or else react with the iron sulfide to produce liquid sulfur and gaseous sulfur dioxide in the warm interior of Io. This appears to be the chemical state just below the surface of Io today. All methane, carbon dioxide, hydrogen sulfide, and water must have been completely outgassed and have escaped to space by now; none of these chemicals is seen on Io.

## Io's Core

Let's assume that Io started out with the composition of a typical CM meteorite: by mass, 1% carbon, 5% sulfur, 10% water, anywhere from 10% to 50% iron metal, and the rest iron and magnesium silicate minerals.

The eutectic composition (the mixture of sulfur and iron that has the lowest melting point) is roughly 25% sulfur, 75% iron (by mass) at the higher pressures deep inside Io, and richer in sulfur near the surface of Io where the pressure is lower. If all the sulfur were carried away into a molten iron–iron sulfide core, then there would be none left to form the volcanoes we see on the surface of Io. On the other hand, if all the sulfur in Io were present as elemental sulfur, then it would make a crust of sulfur 50 km thick. Given the heat produced inside Io, such a crust would be mostly molten and the surface would probably flow like warm plastic. That's not what we see today, however; there are distinct mountains and cliffs visible on its surface. The crust of Io appears to be made mostly of rock, with only a thin coating of sulfur and sulfur dioxide ice.

To explain why we don't have a thick sulfur crust on Io, we are faced with four alternatives:

1. Io started with less sulfur than we've postulated,
2. Io has most of its sulfur still distributed evenly throughout the planet,
3. Io has lost most of its original sulfur to space, or
4. Io has most of its sulfur in a sulfur rich core, but some of it is still in the crust.

Let's look at each possibility in turn.

*1. A sulfur-poor Io.* Sulfur starts out condensed in the chemical form of iron sulfide, FeS. Suppose that Io started out without much iron sulfide, but still rich in metallic iron (needed to account for its density). If there were any eutectic melting, the composition of the melt would be more sulfur-rich than the body as a whole, and so it would drain all the sulfur into a core, leaving only metal behind. Any iron sulfide that did not melt must still be oxidized from iron sulfide into iron oxide and free sulfur.

That reaction requires a lot of free oxygen, because the metallic iron competes with FeS for whatever free oxygen is around. To get much free oxygen we need water. But sulfur-bearing and water-bearing minerals condense at roughly the same temperature; all water-rich meteorites are also rich in sulfur. So we expect a body rich in water to be rich in sulfur as well. But that implies a lot of FeS, contradicting our first assumption. That argues against the first possibility.

*2. Evenly distributed sulfur.* Could the sulfur seen on the surface of Io be typical of the sulfur abundance throughout the moon? This seems unlikely, given all the heating that Io experiences and that sulfur is a very chemically active element.

*3. Io lost its sulfur.* There is no known mechanism that could strip Io of all its sulfur (see Problem 15). Thus the third possibility seems unlikely. From this we conclude that

*4. Io must have a core involving most, but not all, of its sulfur.*

## Putting It All Together

Given these constraints on the chemistry of Io, we can summarize its likely history. The following tale may not be completely correct, nor is it necessarily the only possible history that matches the facts; instead, think of it as one possible way in which Io may have evolved.

Io started with a composition something like a CM carbonaceous chondrite, with significant amounts of water, iron sulfide, metallic iron, and a small amount of carbon. Early heating to moderate temperatures, below 1000 K, resulted in vigorous outgassing of methane, leading to the oxidation of much of the metallic iron inside Io, until the carbon was all used up. This may have happened even before Io was locked into a tidal resonance with Jupiter; the heat of accretion and radioactive nuclides inside Io could have provided enough heat to begin its evolution.

As the temperature inside continued to increase, hydrous minerals in the upper 100 km or so continued to outgas, leading to further oxidation of crustal material. Deeper inside, however, the pressure of the rock succeeded in confining hydrogen gas, preventing the further oxidation of metal. At a temperature of about 1275 K, a eutectic melt of iron and iron sulfide was formed, draining sulfur into the deep interior from all but the top 100 km of this moon.

Later, at still higher temperatures, solid-state convection began to bring sulfur-free rocky material from the interior to within the upper 100 km, where any remaining water in the rocks could react. The hydrogen escaped, while the oxygen reacted with the sulfur in the upper 100 km and

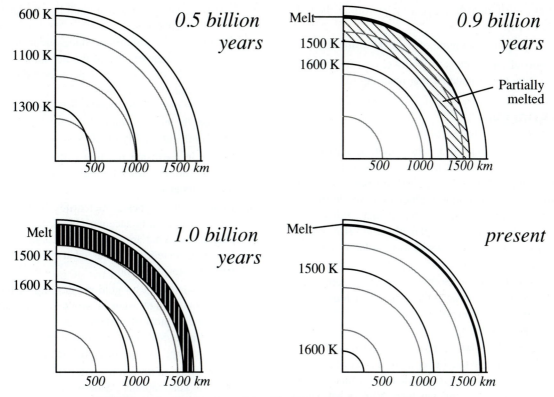

**FIGURE 14.8** Thermal models of Io. The model assumes that Io was heated by both tides and long-lived radioactive elements. Tidal heating varies as one goes from pole to equator. The first three illustrations show temperature profiles (solid lines) at various radii (shaded lines) inside Io over the first billion years. The last illustration shows the eventual outcome of the heating, assuming a sulfur-rich crust with low conductivity.

any remaining iron sulfide or metallic iron, to form sulfur dioxide with sulfur and iron oxide also present. (This convection would also serve to cool off the interior, as seen in Figure 14.8.)

At present, we have a crust that erupts and recycles sulfur from the original upper 100 km of the moon; a mantle of silicates, poor in sulfur but rich in iron oxide; and an iron sulfide core, of roughly 1000 km radius, containing 20% of the mass of Io, including more than 90% of its sulfur (Figure 14.9). Chemical reactions at the core–mantle interface may be slowly oxidizing this core, bringing fresh sulfur to the surface.

## STUDY QUESTIONS

1. There are many similarities between Io and our Moon. Name one.

2. There are many differences between Io and our Moon. Name one.

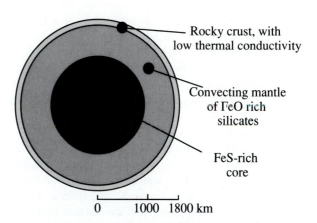

**FIGURE 14.9** Cross section of Io today, according to the model described in this chapter. A rocky crust about 100 km thick overlies a convecting mantle and a 1000-km FeS-rich core.

3. What chemical element plays a major role in the volcanoes of Io?

4. What is the primary source of the heat inside Io?

5. What sort of meteorite is probably closest in chemical composition to the material that made up Io?

## 14.4 PROBLEMS

1. You've just discovered a new planet, as massive as Earth, orbiting at twice the distance from the Sun as Pluto. Unlike all the other planets, this planet's orbit is retrograde; that is, it moves around the Sun in the opposite direction from all the other planets. What is the angular momentum of this planet's orbit? How would such a discovery affect our ideas of solar system formation?

2. If you could observe another planetary system about another star, what basic traits would you look for to help in our understanding of the origins of planetary systems in general?

3. Assume that the solar nebula was 100 AU in radius, 10 AU thick, and contained two solar masses of gas and dust. What is the average gas density in this nebula? If the temperature of this gas were 50 K, what would the Jeans length be? How many lumps ("giant gaseous protoplanets"?) would this nebula fragment into?

4. *A Minimum Mass Solar Nebula (A computer spreadsheet problem).*
   (a) In a spreadsheet, take the mass of each planet and add enough hydrogen and helium to bring that mass back up to solar composition; this is the mass of solar nebula needed to make that planet. (Do not waste time trying to calculate this in detail; instead, if the planet is not already at solar composition, guess, to an order of magnitude, what fraction of the solar nebula is missing and just multiply the observed planetary mass by the inverse of that fraction. Or, in other words: Rocky material makes up what fraction of the mass of the Sun?) Don't forget the asteroid belt! Treat it as if it were a single planet twice the size of Ceres.
   (b) Spread this mass into a disk. Calculate the surface density of this disk. (Assume that the planets were formed at their present locations. Assume that all the material in each planet came from a "feeding zone" extending from half the distance to the next inner planet to half the distance to the next outer planet. Find the area of this annular disk. Divide the mass from part (a) by this area.)
   (c) Find the gas density for the nebula at each planet by assuming that the vertical thickness of the nebula at any point is one tenth the distance from the Sun to that point.
   (d) From the equilibrium condensation theory, guess a likely temperature of the nebula at each planet.
   (e) Find the critical Jeans length for the nebula at the location of each planet.
   (f) Make plots of the nebula surface density and Jeans length versus distance from the Sun.
   (g) Comment on any trends you see.

5. Where in Earth are the four different kinds of compounds (atmophiles, lithophiles, chalcophiles, siderophiles) located?

6. Consider three types of meteorites discussed in Chapter 9: irons, carbonaceous chondrites, and enstatite chondrites. What are likely condensation temperatures for these meteorites? Near what planets would these temperatures be found?

7. If you wished to go gold-prospecting in the asteroid belt, what type of asteroid would you choose to mine: a C-type, typical of carbonaceous chondrite material, an S-type, which may be typical of stony meteorites, or an M-type, typical of metal-rich meteorites? Why?

8. Over what temperature range are lithophiles condensed? Chalcophiles? Siderophiles? Atmophiles?

9. The atmosphere of Jupiter appears to be gas from the solar nebula that was captured by the gravity field of the growing planet. Venus, on the other hand, has an atmosphere very different from the solar nebula. What does this suggest about the relative times at which each of these planets formed in relation to when the solar nebula gas was dissipated?

10. It has been suggested that, as soon as a solid particle condensed from the solar nebula, it was accreted into a planet and removed from chemical contact with the gas. If this occurred, the condensation sequence outlined in the chapter would no longer apply. For example, no iron sulfide would be formed because all the iron would have been accreted into a planet and thus could not react with hydrogen sulfide gas in the nebula. Find three other minerals, found in the equilibrium condensation sequence, that would not be formed in this "disequilibrium" case.

11. Recall the discussion in Chapter 9 concerning the origin of chondrules. Given the solar nebula scenario for the origin of the solar system, devise a hypothesis for when and where chondrules may have been formed. (Many possibilities exist!) Is there any key experiment or observation that would conclusively rule out your hypothesis?

12. A chemical reaction produces a gas deep inside a planet. Name three different possible fates for this gas. Under what circumstances might each occur?

13. Name two differences between the Moon and Io that resulted in different evolutions for each planet. In terms of these differences, is Earth more like Io or the Moon? Do you suppose Earth's evolution was more like Io's or the Moon's?

14. Name three similarities between Io and Mars. Name three differences. Describe how any one of the similarities or differences would result in similar or different trends in the evolution of these two bodies.

15. If Io is 5% sulfur by mass, how many kilograms of sulfur are there in Io? How many kilograms of sulfur would you have to lose every second, on average, if you were to get rid of this sulfur over the age of the solar system (4.5 billion years)? The observed rate of sulfur loss due to sputtering is about 1000 kg/s; at this rate, how much sulfur has been lost over the age of the solar system?

16. What are some of the unresolved questions in planetary science?

## 14.5  FOR FURTHER READING

In the January 1990 issue of *Reviews of Modern Physics* (volume 62, pp. 43–112), Stephen G. Brush, a historian of science, examines "Theories of the origin of the solar system 1956–1985." It's an intriguing and comprehensive critique of how planetary scientists approach these issues, as viewed by an educated outsider.

The University of Arizona Press in Tucson has run several conferences on the origin of the solar system, with papers gathered into the volumes *Protostars and Planets*, edited by Tom Gehrels (1978); *Protostars and Planets II*, edited by David C. Black and Mildred Shapley Matthews (1985); and *Protostars and Planets III*, edited by Eugene Levy, Jonathan Lunine, and Mildred Shapley Matthews (1993).

For a discussion of the equilibrium condensation theory by one of its chief proponents, consult John Lewis and Ronald Prinn, *Planets and Their Atmospheres* (New York: Academic Press, 1984).

# Planets

## A.1 PHYSICAL PROPERTIES

Data on the physical parameters of the planets are taken from the 1993 edition of *Astronomical Almanac*, except for Neptune data, from Tyler and others, *Science*, volume 246, p. 1466 (1989), Pluto data, from Beletic and others, *Icarus*, volume 79, p. 38 (1989), and Bond albedo data.

The Bond albedo is the ratio of total light reflected against total light received by a planet's surface over all wavelengths and directions. It is a difficult number to derive, requiring many observations at different wavelengths and viewing geometries. Data on the Bond albedoes of the planets are from Veverka and others, in *Mercury* (Tucson: University of Arizona Press, 1988), p. 37; Colin and others, in *Venus* (Tucson: University of Arizona Press, 1983), p. 20; Kieffer and others, in *Mars* (Tucson: University of Arizona Press, 1992), p. 1; and Pearl and Conrath, *Journal of Geophysical Research*, volume 96, p. 18921 (1991). For Pluto and the satellites we report geometric albedoes measured in visible light near opposition.

| Planet | Radius (km) | Mass (kg) | Density (g/cm$^3$) | Spin Period (Earth days) | Pole Tilt (degrees) | Bond Albedo |
|---|---|---|---|---|---|---|
| Mercury | 2,439 | $3.3022 \times 10^{23}$ | 5.43 | 58.65 | 0 | 0.12 |
| Venus | 6,052 | $4.8690 \times 10^{24}$ | 5.24 | $-243$ | 177.3 | 0.75 |
| Earth | 6,378 | $5.9742 \times 10^{24}$ | 5.515 | 1.0 | 23.45 | 0.36 |
| Mars | 3,397 | $6.4191 \times 10^{23}$ | 3.94 | 1.026 | 25.19 | 0.25 |
| Jupiter | 71,492 | $1.8988 \times 10^{27}$ | 1.33 | 0.414 | 3.12 | 0.34 |
| Saturn | 60,268 | $5.6850 \times 10^{26}$ | 0.69 | 0.444 | 26.73 | 0.34 |
| Uranus | 25,559 | $8.6625 \times 10^{25}$ | 1.27 | $-0.718$ | 97.86 | 0.30 |
| Neptune | 24,765 | $1.0278 \times 10^{26}$ | 1.64 | 0.671 | 29.56 | 0.29 |
| Pluto | 1,152 | $1.28 \times 10^{22}$ | 2 | $-6.387$ | 118 | 0.3 |

## A.2 ORBITAL PROPERTIES

Data on the parameters of the planets' orbits are taken from the 1993 edition of *Astronomical Almanac*. The angles $\Omega$ and $\omega$ and the position of the perihelion are measured relative to the First Point in Aries, the line made by the Sun and Earth at the moment of vernal equinox.

| Planet | Orbital Period | Semimajor Axis of Orbit (AU) | Eccentricity | Inclination (degrees) | $\Omega$ (degrees) | $\omega$ (degrees) | Perihelion (degrees) |
|---|---|---|---|---|---|---|---|
| Mercury | 87.97 days | 0.387 | 0.2056 | 7.005 | 48.34 | 29.1 | 77.44 |
| Venus | 224.7 days | 0.723 | 0.0068 | 3.395 | 76.70 | 54.8 | 131.5 |
| Earth | 365.25 days | $\equiv 1.0$ | 0.1670 | $\equiv 0.0$ | $\equiv 0.0$ | 102.8 | 102.8 |
| Mars | 686.98 days | 1.524 | 0.0934 | 1.850 | 49.58 | 286.4 | 336.0 |
| Jupiter | 11.86 yr | 5.203 | 0.0483 | 1.305 | 100.47 | 275.2 | 15.68 |
| Saturn | 29.46 yr | 9.53 | 0.054 | 2.487 | 113.67 | 339.3 | 93.0 |
| Uranus | 84.01 yr | 19.2 | 0.047 | 0.772 | 74.0 | 99.0 | 173.0 |
| Neptune | 164.1 yr | 30.1 | 0.006 | 1.772 | 131.7 | 271.3 | 43.0 |
| Pluto | 248.5 yr | 39.8 | 0.254 | 17.14 | 110.3 | 113.7 | 224.0 |

# APPENDIX B

# Planetary Satellites

## B.1  PHYSICAL PROPERTIES

Data on the physical parameters of the satellites are taken from the 1993 edition of *Astronomical Almanac*, except for Charon data, which are derived from Tholen and Buie,

*Astronomical Journal*, volume 96, p. 1977 (1988) and from Beletic and others, *Icarus,* volume 79, p. 38 (1989), and for the masses and densities of Janus and Epimetheus, which are taken from Yoder and others, *Astronomical Journal*, volume 98, p. 1875 (1989).

| | Satellite | Radius (km) | Mass (kg) | Density ($g/cm^3$) | Geometric Albedo |
|---|---|---|---|---|---|
| **Earth** | | | | | |
| | Moon | 1738 | $7.35 \times 10^{22}$ | 3.36 | 0.12 |
| **Mars** | | | | | |
| MI | Phobos | $13.5 \times 10.8 \times 9.4$ | $9.63 \times 10^{15}$ | 1.6 | 0.06 |
| MII | Deimos | $7.5 \times 6.1 \times 5.5$ | $1.93 \times 10^{15}$ | 1.8 | 0.07 |

| | Satellite | Radius (km) | Mass (kg) | Density (g/cm³) | Geometric Albedo |
|---|---|---|---|---|---|
| **Jupiter** | | | | | |
| JXVI | Metis | 20 | $9.49 \times 10^{16}$ | 2.8 | 0.05 |
| JXV | Adrastea | $12.5 \times 10 \times 7.5$ | $1.90 \times 10^{16}$ | $\approx 4$ | 0.05 |
| JV | Amalthea | $135 \times 83 \times 75$ | $7.22 \times 10^{18}$ | 2.1 | 0.05 |
| JXIV | Thebe | $55 \times 45$ | $7.60 \times 10^{17}$ | 1.6 | 0.05 |
| JI | Io | 1815 | $8.89 \times 10^{22}$ | 3.55 | 0.61 |
| JII | Europa | 1569 | $4.75 \times 10^{22}$ | 2.94 | 0.64 |
| JIII | Ganymede | 2631 | $1.48 \times 10^{23}$ | 1.94 | 0.42 |
| JIV | Callisto | 2400 | $1.07 \times 10^{23}$ | 1.85 | 0.20 |
| JXIII | Leda | 8 | $5.70 \times 10^{15}$ | 2.66 | — |
| JVI | Himalia | 93 | $9.49 \times 10^{18}$ | 2.82 | 0.03 |
| JX | Lysithea | 18 | $7.60 \times 10^{16}$ | 3.11 | — |
| JVII | Elara | 38 | $7.60 \times 10^{17}$ | 3.31 | 0.03 |
| JXII | Ananke | 15 | $3.80 \times 10^{16}$ | 2.69 | — |
| JXI | Carme | 20 | $9.49 \times 10^{16}$ | 2.83 | — |
| JVIII | Pasiphae | 25 | $1.90 \times 10^{17}$ | 2.90 | — |
| JIX | Sinope | 18 | $7.60 \times 10^{16}$ | 3.11 | — |
| **Saturn** | | | | | |
| SXV | Atlas | $20 \times 10$ | — | — | 0.9 |
| SXVI | Prometheus | $70 \times 50 \times 50$ | — | — | 0.6 |
| SXVII | Pandora | $55 \times 45 \times 35$ | — | — | 0.9 |
| SXI | Epimetheus | $70 \times 60 \times 50$ | $5.6 \times 10^{20}$ | 0.67 | 0.8 |
| SX | Janus | $110 \times 100 \times 80$ | $2.0 \times 10^{21}$ | 0.64 | 0.8 |
| SI | Mimas | 196 | $4.55 \times 10^{19}$ | 1.44 | 0.5 |
| SII | Enceladus | 250 | $7.39 \times 10^{19}$ | 1.13 | 1.0 |
| SIII | Tethys | 530 | $7.39 \times 10^{20}$ | 1.19 | 0.9 |
| SXIV | Calypso | $17 \times 11 \times 11$ | — | — | 0.6 |
| SXIII | Telesto | $17 \times 14 \times 13$ | — | — | 0.5 |
| SIV | Dione | 560 | $1.05 \times 10^{21}$ | 1.43 | 0.7 |
| SXII | Helene | $18 \times 16 \times 15$ | — | — | 0.7 |
| SV | Rhea | 765 | $2.50 \times 10^{21}$ | 1.33 | 0.7 |
| SVI | Titan | 2575 | $1.35 \times 10^{23}$ | 1.89 | 0.21 |
| SVII | Hyperion | $205 \times 130 \times 110$ | $1.71 \times 10^{19}$ | 1.4 | 0.3 |
| SVIII | Iapetus | 730 | $1.88 \times 10^{21}$ | 1.15 | 0.05–0.5 |
| SIX | Phoebe | 110 | — | — | 0.06 |

| | Satellite | Radius (km) | Mass (kg) | Density (g/cm³) | Geometric Albedo |
|---|---|---|---|---|---|
| **Uranus** | | | | | |
| UVI | Cordelia | 13 | — | — | — |
| UVII | Ophelia | 15 | — | — | — |
| UVIII | Bianca | 21 | — | — | — |
| UIX | Cressida | 31 | — | — | — |
| UX | Desdemona | 27 | — | — | — |
| UXI | Juliet | 42 | — | — | — |
| UXII | Portia | 54 | — | — | — |
| UXIII | Rosalind | 27 | — | — | — |
| UXIV | Belinda | 33 | — | — | — |
| UXV | Puck | 77 | — | — | — |
| UV | Miranda | 240 | $0.75 \times 10^{20}$ | 1.30 | 0.27 |
| UI | Ariel | 579 | $1.35 \times 10^{21}$ | 1.59 | 0.34 |
| UII | Umbriel | 586 | $1.28 \times 10^{21}$ | 1.44 | 0.18 |
| UIII | Titania | 790 | $3.48 \times 10^{21}$ | 1.66 | 0.27 |
| UIV | Oberon | 762 | $2.93 \times 10^{21}$ | 1.51 | 0.24 |
| **Neptune** | | | | | |
| NVIII | Naiad | (29) | — | — | — |
| NVII | Thalassa | (40) | — | — | — |
| NV | Despina | 75 | — | — | 0.059 |
| NVI | Galatea | 80 | — | — | 0.063 |
| NIV | Larissa | $104 \times 89$ | — | — | 0.056 |
| NIII | Proteus | $218 \times 208 \times 201$ | — | — | 0.064 |
| NI | Triton | 1350 | $2.15 \times 10^{22}$ | 2.1 | 0.7 |
| NII | Nereid | 170 | $2 \times 10^{19}$ | 1 | 0.155 |
| Pluto | | | | | |
| | Charon | 615 | $1.95 \times 10^{21}$ | 2.0 | 0.36 |

## B.2 ORBITAL PROPERTIES

Data on the orbital parameters of the satellites are taken from the 1993 edition of *Astronomical Almanac*. Corrections to the orbits of the newly discovered moons of Uranus come from William Owen and S. Synott, *Astronomical Journal*, volume 93, p. 1268 (1987), and corrections to the orbits of the moons of Neptune come from Owen and others, *Astronomical Journal*, volume 96, p. 1514 (1990). Data for Charon are derived from Beletic and others, published in *Icarus*, volume 79, p. 38 (1989).

| | *Satellite* | *Semimajor Axis of Orbit (km)* | *Period (Earth days)* | *Eccentricity* | *Inclination (degrees)* | *Precession (degrees/yr)* |
|---|---|---|---|---|---|---|
| **Earth** | | | | | | |
| | Moon | 384,400 | 27.321 | 0.0549 | 18.28–28.58 | 19.34 |
| **Mars** | | | | | | |
| MI | Phobos | 9,378 | 0.319 | 0.015 | 1.0 | 158.8 |
| MII | Deimos | 23,459 | 1.262 | 0.0005 | 0.9–2.7 | 6.614 |
| **Jupiter** | | | | | | |
| JXVI | Metis | 128,000 | 0.295 | — | — | — |
| JXV | Adrastea | 129,000 | 0.298 | — | — | — |
| JV | Amalthea | 181,000 | 0.498 | 0.003 | 0.4 | 914.6 |
| JXIV | Thebe | 222,000 | 0.674 | 0.015 | 0.8 | — |
| JI | Io | 422,000 | 1.769 | 0.004 | 0.04 | 48.6 |
| JII | Europa | 671,000 | 3.551 | 0.009 | 0.47 | 12.0 |
| JIII | Ganymede | 1,070,000 | 7.154 | 0.002 | 0.21 | 2.63 |
| JIV | Callisto | 1,883,000 | 16.69 | 0.007 | 0.51 | 0.643 |
| JXIII | Leda | 11,094,000 | 238.7 | 0.148 | 26.07 | — |
| JVI | Himalia | 11,480,000 | 250.6 | 0.158 | 27.63 | — |
| JX | Lysithea | 11,720,000 | 259.2 | 0.107 | 29.02 | — |
| JVII | Elara | 11,737,000 | 259.7 | 0.207 | 24.77 | — |
| JXII | Ananke | 21,200,000 | 631 | 0.169 | 147 | — |
| JXI | Carme | 22,600,000 | 692 | 0.207 | 164 | — |
| JVIII | Pasiphae | 23,500,000 | 735 | 0.378 | 145 | — |
| JIX | Sinope | 23,700,000 | 758 | 0.275 | 153 | — |
| **Saturn** | | | | | | |
| SXV | Atlas | 137,670 | 0.602 | 0.000 | 0.3 | — |
| SXVI | Prometheus | 139,353 | 0.613 | 0.003 | 0.0 | — |
| SXVII | Pandora | 141,700 | 0.629 | 0.004 | 0.0 | — |
| SXI | Epimetheus | 151,422 | 0.694 | 0.009 | 0.34 | — |
| SX | Janus | 151,472 | 0.695 | 0.007 | 0.14 | — |
| SI | Mimas | 185,520 | 0.942 | 0.020 | 1.53 | 365.0 |
| SII | Enceladus | 238,020 | 1.370 | 0.0045 | 0.00 | 156.2 |
| SIII | Tethys | 294,660 | 1.888 | 0.000 | 1.86 | 72.25 |
| SXIV | Calypso | 294,660 | 1.888 | — | — | — |
| SXIII | Telesto | 294,660 | 1.888 | — | — | — |
| SIV | Dione | 377,400 | 2.737 | 0.002 | 0.02 | 30.85 |
| SXII | Helene | 377,400 | 2.737 | 0.005 | 0.0 | — |
| SV | Rhea | 527,040 | 4.518 | 0.001 | 0.35 | 10.16 |
| SVI | Titan | 1,221,830 | 15.945 | 0.029 | 0.33 | 0.5213 |
| SVII | Hyperion | 1,481,100 | 21.277 | 0.104 | 0.43 | — |
| SVIII | Iapetus | 3,561,300 | 79.33 | 0.028 | 14.72 | — |
| SIX | Phoebe | 12,952,000 | 550.48 | 0.163 | 177 | — |

| | Satellite | Semimajor Axis of Orbit (km) | Period (Earth days) | Eccentricity | Inclination (degrees) | Precession (degrees/yr) |
|---|---|---|---|---|---|---|
| **Uranus** | | | | | | |
| UVI | Cordelia | 49,750 | 0.335 | 0 | 0.1 | 550 |
| UVII | Ophelia | 53,760 | 0.376 | 0.01 | 0 | 419 |
| UVIII | Bianca | 59,170 | 0.435 | 0.001 | 0 | 299 |
| UIX | Cressida | 61,770 | 0.464 | 0 | 0 | 257 |
| UX | Desdemona | 62,660 | 0.474 | 0 | 0.2 | 245 |
| UXI | Juliet | 64,360 | 0.493 | 0.0005 | 0 | 223 |
| UXII | Portia | 66,100 | 0.513 | 0 | 0 | 203 |
| UXIII | Rosalind | 69,930 | 0.5585 | 0 | 0.3 | 167 |
| UXIV | Belinda | 75,255 | 0.624 | 0 | 0 | 129 |
| UXV | Puck | 86,004 | 0.762 | 0 | 0.3 | 81 |
| UV | Miranda | 129,390 | 1.413 | 0.0027 | 4.2 | 19.8 |
| UI | Ariel | 191,020 | 2.520 | 0.0034 | 0.3 | 6.8 |
| UII | Umbriel | 266,300 | 4.144 | 0.0050 | 0.36 | 3.6 |
| UIII | Titania | 435,910 | 8.706 | 0.0022 | 0.14 | 2.0 |
| UIV | Oberon | 583,520 | 13.463 | 0.0008 | 0.10 | 1.4 |
| **Neptune** | | | | | | |
| NVIII | Naiad | 48,230 | 0.294 | 0.00033 | 4.74 | 626 |
| NVII | Thalassa | 50,070 | 0.3115 | 0.00016 | 0.21 | 551 |
| NV | Despina | 52,530 | 0.335 | 0.00014 | 0.066 | 466 |
| NVI | Galatea | 61,950 | 0.429 | 0.00012 | 0.054 | 261 |
| NIV | Larissa | 73,550 | 0.555 | 0.00138 | 0.20 | 143 |
| NIII | Proteus | 117,650 | 1.122 | 0.00044 | 0.04 | 28.8 |
| NI | Triton | 354,760 | 5.877 | 0.00002 | 157.35 | 0.5232 |
| NII | Nereid | 5,511,000 | 360.1 | 0.751 | 27.6 | 0.039 |
| **Pluto** | | | | | | |
| | Charon | 19,640 | 6.387 | 0.0001 | 98.3 | — |

# Other Useful Data

Source: Cesare Emiliani, *Dictionary of the Physical Sciences* (New York: Oxford University Press) 1987.

| *Constants and Conversion Factors* | |
|---|---|
| Astronomical Unit (AU) | $1.495978707 \times 10^{11}$ m |
| Sun, radius ($R$) | $6.9599 \times 10^{8}$ m |
| Sun, mass ($M$) | $1.9891 \times 10^{30}$ kg |
| Sun, luminosity | $3.826 \times 10^{26}$ J/s |
| Solar constant at 1 AU | $1.360 \times 10^{3}$ J/s·m$^2$ |
| Speed of Light ($c$) | $2.99792458 \times 10^{8}$ m/s |
| Electron mass | $9.10939 \times 10^{-31}$ kg |
| Proton mass | $1.672623 \times 10^{-27}$ kg |
| Neutron mass | $1.67929 \times 10^{-27}$ kg |
| Atomic Mass Unit | $1.66054 \times 10^{-27}$ kg |
| Alpha particle mass | $6.64466 \times 10^{-27}$ kg |
| Electronic charge ($e$) | $\pm 1.602177 \times 10^{-19}$ coulombs |
| Gravitational constant ($G$) | $6.67206 \times 10^{-11}$ m$^3$/kg·s$^2$ |
| Boltzmann constant ($k$) | $1.38066 \times 10^{-23}$ J/K |
| Stefan–Boltzmann constant ($\sigma$) | $5.6705 \times 10^{-8}$ J/m$^2$· s · K$^4$ |
| Planck constant ($h$) | $6.626075 \times 10^{-34}$ J · s |
| Gas constant ($R$) | 8.3145 J/K· mole |
| Avogadro's number ($N$) | $6.022136 \times 10^{23}$ molecules/g · mole |
| 1 bar | $= 10^{5}$ Pa |
| 1 g/cm$^3$ | $= 10^{3}$ kg/m$^3$ |
| 1 kiloton | $= 4.1868 \times 10^{12}$ J |
| 1 sidereal year | $= 3.1558150 \times 10^{7}$ s |

Formulas:

Area of a sphere (radius $r$):

$$A = 4\pi r^2$$

Area of an oblate spheroid ($a \times a \times b$) where $e = (a - b)/a$:

$$A = 2\pi a^2 + \pi \frac{b^2}{e} \ln \frac{1 + e}{1 - e}$$

Area of a prolate spheroid ($a \times b \times b$) where $e = (a - b)/a$:

$$A = 2\pi b^2 + 2\pi \frac{ab}{e} \arcsin e$$

Volume of a sphere:

$$V = \tfrac{4}{3}\pi r^3$$

Volume of an ellipsoid ($a \times b \times c$):

$$V = \tfrac{4}{3}\pi abc$$

Volume of a cone (radius $r$ and height $h$):

$$V = \tfrac{1}{3}\pi r^2 h$$

Moment of inertia, general form:

$$I = \int r^2 dm$$

Moment of inertia, uniform sphere with mass $M$ and radius $r$:

$$I = \tfrac{2}{5}Mr^2$$

Moment of inertia, uniform shell with inner radius $r_i$, outer radius $r_o$:

$$I = \frac{2}{5}M\frac{(r_o^5 - r_i^5)}{(r_o^3 - r_i^3)}$$

# Figure Credits

Grateful acknowledgments are given to the facilities, and especially the people, of the National Space Science Data Center, without whom this book would be largely without photos. All photos taken from American space probes, as well as the Soviet Venera probes, are courtesy of NASA.

The principal investigators of these space probes are as follows:

*Apollo 8* **Experiment Principal Investigator** Dr. Ronald J. Allenby

*Apollo 15* **Experiment Principal Investigator** Dr. Frederick J. Doyle

*Galileo* **Experiment Principal Investigator** Dr. Michael J. S. Belton

*Lunar Orbiter* **Experiment Principal Investigator** Mr. Leon J. Kosofsky

*Mariner 10* **Experiment Principal Investigator** Dr. Bruce Murray

*Magellan* **Experiment Principal Investigator** Dr. Gordon H. Pettengill

*Skylab* **Experiment Principal Investigator** Dr. D. M. Packer

*Viking* **Lander Experiment Team Leader** Dr. Raymond Arvidson

*Viking* **Orbiter Experiment Team Leader** Dr. Michael H. Carr

*Voyager* **Experiment Team Leader** Dr. Bradford A. Smith

## Individual photo credits:

**Figure 1.1**  D. M. Davis

**Figure 1.3**  Karen Sepp

**Figure 2.1**  *Skylab*, 74-HC-260

**Figure 2.4**  D. M. Davis

**Figure 3.1**  Lick Observatory photograph #L3

**Figure 3.2**  *Apollo 15*, 71-HC-1140

**Figure 3.4**  *Galileo*, P37299

**Figure 3.5**  *Apollo 16* sample 6001800, photo 72-H-681

**Figure 3.6**  Lunar Orbiter, IV-187M

**Figure 3.8**  W. T. Schaller (1436), courtesy U.S. Geological Survey

**Figure 3.11, top**  *Apollo 17* sample 7621500, photo 73-5986

**Figure 3.11, bottom**  Calkins (704), courtesy U.S. Geological Survey

**Figure 3.18**   *Apollo 15*, metric 1580

**Figure 3.20**   M. W. Schaefer

**Figure 3.21**   Wayne C. Williams

**Figure 3.23**   Lunar Orbiter, V-125M

**Figure 4.1**   *Mariner 10*, P14580

**Figure 4.3**   *Mariner 10*, P15043

**Figure 4.4**   *Mariner 10*, FDS 27463

**Figure 4.7**   Karen Sepp

**Figure 5.1**   *Magellan*, P37948

**Figure 5.2**   *Magellan* mosaic, P39225

**Figure 5.3**   data set courtesy R. Steven Nerem, NASA/GSFC

**Figure 5.4**   *Venera 14*, YG06848

**Figure 5.5**   *Magellan*, P38340

**Figure 5.6**   *Magellan*, P39916

**Figure 5.7**   *Magellan*, P36711

**Figure 5.9**   *Mariner 10*, P14400

**Figure 5.11**   *Magellan*, P38088

**Figure 6.1**   *GOES* satellite photo, courtesy Fritz Hassler, NASA/GSFC

**Figure 6.4**   courtesy Hong Kie Thio, California Institute of Technology Seismological Laboratory

**Figure 6.7**   courtesy Paul D. Lowman, Jr., NASA/GSFC

**Figure 6.10**   U.S. Geological Survey, no. 321

**Figure 6.11**   J. B. Garvin

**Figure 7.1**   *Viking* photomosaic, MC-9 NE

**Figure 7.2**   *Viking 2* lander, P17689

**Figure 7.3**   *Viking 2* lander, P21841

**Figure 7.4**   data set courtesy U.S. Geological Survey

**Figure 7.5**   *Viking* photomosaic, U.S. Geological Survey, Flagstaff

**Figure 7.6**   *Viking* orbiter, 575B60

**Figure 7.7**   *Viking* orbiter, 646A28

**Figure 7.8**   M. W. Schaefer

**Figure 7.9**   *Viking* photomosaic, 211-5730

**Figure 7.10**   *Viking* orbiter, P18116

**Figure 7.11**   *Viking* orbiter, P19131

**Figure 7.12**   *Viking* orbiter photomosaic, 211-4987

**Figure 7.13**   *Viking* orbiter, 97A62

**Figure 7.18**   *Viking* photomosaic, MC-18 NW

**Figure 7.19**   *Viking* orbiter, 608A45

**Figure 7.20**   *Viking* orbiter, 032A18, and *Viking* orbiter photomosaic, 211-5066

**Figure 7.22**   *Viking* orbiter, P16952

**Figure 7.24**   Karen Sepp

**Figure 8.1**   *Apollo 8*, AS8142383

**Figure 9.1**   *Galileo*, P40449, courtesy Clark Chapman

**Figure 9.2**   courtesy Edward Olson, Field Museum of Natural History, Chicago

**Figure 9.3**   courtesy Larry Grossman, University of Chicago, reproduced, with permission, from the *Annual Review of Earth and Planetary Sciences*, volume 8, ©1980 by Annual Reviews, Inc.

**Figure 10.1**   *Voyager 2*, P21082

**Figure 10.5**   *Voyager 1* and *Voyager 2*, P21599

**Figure 10.6**   *Voyager 2*, P21737

**Figure 11.1**   *Voyager 1*, P23254

**Figure 11.2**   *Voyager 2*, P34622

**Figure 11.5**   *Voyager 1*, P23352

**Figure 11.6**   *Voyager 2*, P29466

**Figure 11.7**   *Voyager 2*, P34712

**Figure 11.8**   *Voyager 2*, P21774

**Figure 11.11**   *Voyager 2*, P23925

**Figure 11.12**   Karen Sepp

**Figure 11.20**   courtesy Clark R. Chapman

**Figure 12.1**   *Voyager 2*, P29541

**Figure 12.2**   *Viking* orbiter, P18612

**Figure 12.3**   *Voyager 2*, P21758

**Figure 12.4**   *Voyager 2*, P21751

**Figure 12.5**   *Voyager 2*, P21745

**Figure 12.6**   *Voyager 2*, P34714

**Figure 12.10**   M. W. Schaefer

**Figure 12.11**   *Voyager 1*, P23210, and *Voyager 2*, P23956

**Figure 12.14**   *Voyager 2*, P29520 and P29522

**Figure 13.1**   William Liller, observing on behalf of the International Halley Watch/Large-Scale Phenomena Discipline "Island Network," photo courtesy Malcolm Niedner

**Figure 13.2**   Wayne C. Williams

**Figure 13.3**   copyright (1986) Max-Planck-Institüte für Aeronomie, Lindau/Harz, Germany; courtesy Dr. H. U. Keller; image taken by the Halley Multicolour Camera on board ESA's *Giotto* spacecraft.

**Figure 14.1**   *Hubble Space Telescope*, STScI-PRC92-29B, courtesy C. Robert O'Dell and NASA

**Figure 14.5**   *Voyager 1*, P21487

**Figure 14.7**   *Voyager 1*, P21286

# Index

## A

Aberration, stellar, 150
Absorption, 30–31
Abundances, cosmic, 28, 31–32, 47, 60
Acceleration, 25, 155
    instantaneous, 224–26
    normal, 224–26
    radial, 225–26, 228
    transverse, 225–27
Accrete, 287
Accretion, energy of, 99
Accuracy, 147
Achondrites, 169–70, 174, 186
    basaltic, 169–70, 174, 186
    enstatite, 169–70
Acid, sulfuric, 83, 89
Adiabat, 192, 197, 200, 215, 277, 291–92
Adiabatic invariant, 277
Adolphus, Gustavus, 149
Adrastea, 239–40
Adsorbed gas, 90
Aerosol, 87
Agassiz, Louis, 112
Age of Enlightenment, 167
Airy, George B., 75
Alba Patera, 124
Albedo, 175, 194
Albert of Saxony, 15
Albert the Great, 15
Albite, 48

Aldrovandi, Ulysses, 15
Alexander, James, 12
Alexander of Aphrodisias, 15–16
Alexandria, 8–9, 14
Alfvén wave, 269
Algebra, 156
Allende, 171, 174
Alpha particle, 22
Alpha Regio, 85
Alvarez, Luis, 112
Amalthea, 239–40
Ammavaru, 98
Analysis, Fourier, 113, 146
Ananke, 239–40
Andre de Luc, Jean, 16
Anemones, 87
Angular momentum ($L$), 73, 287–88
Angular momentum per unit mass ($h$), 162
Anomalies, gravity, 75–76, 80
Anorthite, 45, 48, 51
Anorthosite, 45, 52
Anorthositic gabbro, 47 (*See also* Anorthosite; Gabbro)
Anticline, 128–29
Aphelion, 69, 183
Aphrodite Terra, 85–86
*Apollo,* 42, 110
*Apollo 8,* 46, 142
*Apollo 10,* 46
*Apollo 11,* 46, 171
*Apollo 12,* 46

*Apollo 13,* 46
*Apollo 14,* 46
*Apollo 15,* 42, 46
*Apollo 16,* 46
*Apollo 17,* 46
Apollo asteroids, 178, 183, 263
Aquinas, Thomas, 146
Arachnoid, 82, 86–87
Arc-minute, 14
Area, cross-sectional, 194
Argument of pericenter ($\omega$), 154, 158, 224
Ariel, 241, 243, 252–54
Aries, First Point in (*See* First Point in Aries)
Aristarchus of Samos, 6–9, 11, 14, 143
Aristotle, 8, 15, 144, 155, 260
Artemis Chasma, 87
Aryabhata, 145
Ascending node, longitude of the [*See* Longitude of the ascending node ($\Omega$)]
Asteroids, 166, 174–78
    composition classes of, 176
    C-type 176
    location of, 176
    sizes of, 175–76
    S-type, 176
    Trojan, 234 (*See also* Trojan asteroids)
Asthenosphere, 94, 96, 107, 127
Astronomical Unit (AU), 5–7, 11, 13, 64